International Conference on Physics and Technology of Advanced Materials (ICPTAM) in conjunction with 8th Nanoscience and Nanotechnology Symposium (NNS) 2024

Journal of Physics: Conference Series Volume 2980

Bali, Indonesia
8-10 October 2024

ISBN: 979-8-3313-2025-6
ISSN: 1742-6588

Printed from e-media with permission by:

Curran Associates, Inc.
57 Morehouse Lane
Red Hook, NY 12571

Some format issues inherent in the e-media version may also appear in this print version.

Institute of Physics
Dirac House, Temple Back
Bristol BS1 6BE UK

Phone: 44 1 17 929 7481
Fax: 44 1 17 920 0979

techtracking@iop.org

Additional copies of this publication are available from:

Curran Associates, Inc.
57 Morehouse Lane
Red Hook, NY 12571 USA
Phone: 845-758-0400
Fax: 845-758-2633
Email: curran@proceedings.com
Web: www.proceedings.com

TABLE OF CONTENTS

Journal of Physics: Conference Series **2980** (2025) 011001 doi:10.1088/1742-6596/2980/1/011001

Preface to the Proceedings of the International Conference on Physics and Technology of Advanced Materials (ICPTAM) 2024 in conjunction with 8th Nanoscience and Nanotechnology Symposium (NNS)

The **International Conference on Physics and Technology of Advanced Materials (ICPTAM) 2024**, held in conjunction with **the 8th Nanoscience and Nanotechnology Symposium (The 8th NNS)**, took place from 7-10 October 2024 in Bali, Indonesia. With the overarching theme of "Advanced materials for future technology", this event brought together leading scientists, engineers, and professionals to discuss cutting-edge advancements in materials science and nanotechnology in celebration of international year of quantum science and technology. The event features 2 keynote speakers, 6 plenary speakers, 14 Invited speakers, and 90 contributed oral presenters, which come from several countries: Japan, Malaysia, Indonesia, South Korea, United Kingdom, Thailand, Taiwan, China, and India.

This year's conference provided a unique platform for participants to exchange knowledge, foster collaboration, and showcase innovative research on topics ranging from functional advanced materials, nanomaterials and nanotechnology, computational materials and modelling, nanoelectronics and nanodevices, energy materials, environmental and green materials, biomaterials and biodevices, quantum computing. This event also highlighted the crucial intersection of physics and materials science in shaping the future of technology and society.

We would like to express our sincere appreciation to Prof. Ir. Wahyu Srigutomo, Dean of the Faculty of Mathematics and Natural Sciences (FMIPA), Bandung Institute of Technology (ITB), for his invaluable support and guidance in preparing this scientific event. We also extend our heartfelt gratitude to the Physical Society of Indonesia, MRS-id, and the Department of Electrical Engineering, Udayana University, for their generous support and collaboration, which played a pivotal role in ensuring the success of this conference and creating an engaging and stimulating environment.

We would like to extend our deepest gratitude to our Keynote Speakers, Plenary Speakers, and Invited Speakers including Prof. Osamu Sugino (The University of Tokyo), Prof. Hadi Nur (Universitas Negeri Malang), Prof. Chang-Ho Yang (Korea Advanced Institute of Science and Technology (KAIST)), Prof. Pei-Wen Li (National Yang Ming Chiao Tung University), Prof. Halim Kusumaatmaja (The University of Edinburgh), Prof. Santi Maensiri (Suranaree University of Technology), Prof. Jun Xu (Nanjing University/ Nantong University), Prof. Kosuke Sugawa (Nihon University), Prof. Yutaka Matsuo (Nagoya University), Prof. Hirofumi Tanaka (Kyusu Institute of Technology), Prof Andriyan B. Suksmono (Bandung Institute of Technology), Prof. Kuwat Triyana (Gadjah Mada University), Assoc Prof. Khoirul Anwar (Telkom University), Dr. Lisman Suryanegara (National Research and Innovation Agency), Dr. Indri Badria Adilina (National Research and Innovation Agency), Prof. Kouichi Takase (Nihon University), Prof. Isnaeni (National Research and Innovation Agency), Prof. Trio Adiono (Bandung Institute of Technology), and Arie Wibowo, Ph.D (Bandung Institute of Technology), whose presentations offered profound insights into the latest developments in their respective fields. We also acknowledge the valuable contributions of all authors whose work forms the basis of these proceedings.

Special thanks go to the members of the advisory board, i.e., Prof. Ariando (National University of Singapore), Prof. Kouichi Takase (Nihon University, Japan), Prof. Seda Aksoy Esinoglu (Istanbul Technical University), Prof. Dr. Ahmad Taufiq (Universitas Negeri Malang), Dr.Ing. Hutomo Suryo Wasisto (Infineon), Assoc. Prof. Satria Zulkarnaen Bisri (Tokyo University of Agriculture and Technology), Prof. Atsushi Kohno (Fukuoka University), Prof. Dr. Eng. Ferry Iskandar (Bandung Institute of Technology), Prof. Dr. Eng. Khairurrijal (Bandung Institute of Technology), Prof. Dr. Eng. Mikrajuddin Abdullah (Bandung Institute of Technology), Dr. Neni Surtiyeni (Bandung Institute of Technology), Prof. Dr. Pepen Arifin (Bandung Institute of Technology), Prof. Dr. Toto Winata (Bandung Institute of Technology), Ronny Sutarto, Ph.D (Canadian Light Source), all organizing committee, and all reviewers for their tireless efforts in ensuring the success of this conference. Their dedication to maintaining the highest standards of scientific rigor and organizational excellence was instrumental in making ICPTAM 2024 a memorable and impactful event.

We are grateful to our sponsors, including Horiba, Nexus Analytics, PT. LAPI ITB, PT. Antam, PT. Agronesia, PGS, Garuda Indonesia, and other sponsors and partners for their generous support and collaboration, which enabled us to host this conference in such a vibrant and inspiring setting.

Finally, we thank all participants for their enthusiastic engagement and valuable contributions. It is our hope that the discussions and collaborations initiated here will continue to thrive and lead to new breakthroughs in advanced materials and related fields.

We look forward to welcoming you to future editions of the ICPTAM.

Bali, January 2025

Chairman of ICPTAM 2024
Prof. Dr. Eng. Yudi Darma
Head of the Physics and Technology of Advanced Materials Research Group
Department of Physics, Faculty of Mathematics and Natural Sciences
Bandung Institute of Technology
yudi@itb.ac.id

List of Advisory Board, and Technical Committee of the the International Conference on Physics and Technology of Advanced Materials (ICPTAM) 2024 in conjunction with 8th Nanoscience and Nanotechnology Symposium (NNS), are available in this Pdf.

Journal of Physics: Conference Series **2980** (2025) 011001 doi:10.1088/1742-6596/2980/1/011001

THE INTERNATIONAL CONFERENCE ON PHYSICS AND TECHNOLOGY OF ADVANCED MATERIALS (ICPTAM) 2024 IN CONJUNCTION WITH 8th NANOSCIENCE AND NANOTECHNOLOGY SYMPOSIUM (NNS)

ADVISORY BOARD

Prof. Ariando (National University of Singapore, Singapore)
Prof. Kouichi Takase (Nihon University, Japan)
Prof. Seda Aksoy Esinoglu (Istanbul Technical University, Turkey)
Prof. Dr. Ahmad Taufiq (Uiversitas Negeri Malang, Indonesia)
Dr.Ing. Hutomo Suryo Wasisto (Infineon, Germany)
Assoc. Prof. Satria Zulkarnaen Bisri (Tokyo University of Agriculture and Technology, Japan)
Prof. Atsushi Kohno (Fukuoka University, Japan)
Prof. Dr. Eng. Ferry Iskandar (Bandung Institute of Technology)
Prof. Dr. Eng. Khairurrijal (Bandung Institute of Technology)
Prof. Dr. Eng. Mikrajuddin Abdullah (Bandung Institute of Technology)
Dr. Neni Surtiyeni (Bandung Institute of Technology)
Prof. Dr. Pepen Arifin (Bandung Institute of Technology)
Prof. Dr. Toto Winata (Bandung Institute of Technology)
Ronny Sutarto, Ph.D (Canadian Light Source)

ORGANIZING COMMITTEE

Chairman : Prof. Dr. Yudi Darma (Bandung Institute of Technology, Indonesia)
Vice chair : Prof. Dr. Fatimah Arofiati Noor (Bandung Institute of Technology, Indonesia)
Secretariat : Dr. rer. nat. Akfiny Hasdi Aimon (Bandung Institute of Technology, Indonesia)
Treasurer : Pipit Fitriani, Ph.D. (Bandung Institute of Technology, Indonesia)
Publication : Dr. Fitri Aulia Permatasari (Bandung Institute of Technology, Indonesia)
Logistics : Dr. Dhewa Edikresnha (Bandung Institute of Technology, Indonesia)
Event : Dr. M. Arief Mustajab E. Maryono (Bandung Institute of Technology, Indonesia)
IT and Website: Dr. Dian Ahmad Hapidin (Bandung Institute of Technology, Indonesia)

CO-ORGANIZING COMMITTEE

Prof. Ida Ayu Dwi Giriantari (Udayana University, Indonesia)
Dr. Wayan Gede Ariastina (Udayana University, Indonesia)
Prof. Dewi Wirastuti (Udayana University, Indonesia)

Journal of Physics: Conference Series **2980** (2025) 011002 doi:10.1088/1742-6596/2980/1/011002

Peer Review Statement

All papers published in this volume have been reviewed through processes administered by the Editors. Reviews were conducted by expert referees to the professional and scientific standards expected of a proceedings journal published by IOP Publishing.

- **Type of peer review:** Double Anonymous
- **Conference submission management system:** Morressier
- **Number of submissions received:** 44
- **Number of submissions sent for review:** 44
- **Number of submissions accepted:** 38
- **Acceptance Rate (Submissions Accepted / Submissions Received × 100):** 86
- **Average number of reviews per paper:** 2
- **Total number of reviewers involved:** 8
- **Contact person for queries:**
  Name: Fitri Aulia Permatasari
  Email: fauliap@itb.ac.id
  Affiliation: Institut Teknologi Bandung

Journal of Physics: Conference Series **2980** (2025) 012001
doi:10.1088/1742-6596/2980/1/012001

Green synthesizes of Silver Nanoparticles (AgNP) from Henna Leaf Extract (Lawsonia Inermis) and Its Chemical and Physical Properties as Photo-sensitizers in The DSSC

Ida Usman[1], Aisyah Dani Saputri[1], Waode Sukmawati Arsyad[1*], La Ode Ahmad Nur Ramadhan[2], La Aba[1], Rahmat Hidayat[3] and Toto Winata[3]

[1] Department of Physics, Halu Oleo University, Kendari, Indonesia
[2] Department of Chemistry, Halu Oleo University, Kendari, Indonesia
[2] Department of Physics, Bandung Institute of Technology, Bandung, Indonesia

*E-mail: wdsukmawati@uho.ac.id

Abstract. To substitute an expensive inorganic dye, we green synthesized silver nanoparticles (AgNP) from Henna leaf extract as photosensitizers in DSSC. Henna dye was macerated for 24 hours using ethanol and filtered. To synthesize AgNPs, 10 ml of Henna dye was added to 90 ml of $AgNO_3$ solution (0.01 M, 0.05 M, and 0.3 M) and stirred. The FTIR analysis of the Henna dye showed three main Lawsone functional groups (O-H phenol, C=O, and C=C). As for AgNP, the new peak with high intensity emerged at a frequency of 1381 cm^{-1}, indicating the existence of Ag. The FTIR results upheld by the diffractogram of the AgNP that showed the crystalline peak at the 2θ position of (29.5, 32.6, 38.0, and 44.1)° associated with Miller indices of [210],[122],[111], and [200], respectively. The evaporation process increases the crystallinity of the AgNP. The nanoparticle size increased from 124.8 nm to 189.1 nm with the $AgNO_3$ concentration from 0.01M to 0.1M characterized by PSA. From the Henna dye absorbance spectrum, the peak at (238 and 370) nm originated from Lawsone molecules. AgNP showed the absorbance peak at 420 nm and 430 nm for unevaporated and evaporated Henna, respectively. These peaks' intensity increased with the irradiation time of up to 72 hours, associated with the shift of the plasmon surface resonance to the lower energy. The energy gap ranges from (2.08-2.09) eV and (2.58-3.69) eV for Henna dye and AgNP, respectively. It is in the range for light harvester material in DSSC.

1. Introduction

Nanoparticles are particles sized around one to 100 nm [1] and have been used in cosmetics, medicine, mechanics, energy, optics, and many more. Nanomaterials can divided into several categories depending on their size, chemical properties, agglomerations, and particle morphology [2]. Based on the materials, nanoparticles are divided into carbon basics, inorganic, organic, and composites [3]. One of the most developed nanomaterials is metal nanoparticles.

Metal nanoparticles (metal NP) gain attention due to the surface plasmon effect, an excitation of electromagnetic waves generated by the fluctuation of charge density at the interface of plasma and dielectric materials [4]. The surface Plasmon effect enhances the optical, electrical, and catalytic properties of metal NP, and thus increases its ability to convert light into current [5]. One of the metal NP that has a high surface Plasmon effect is silver nanoparticles (AgNP) [6].

There are two methods to synthesize AgNP, including bottom-up and top-down methods. The bottom-up approach orchestrates nanoparticles from their smaller constituent atoms or molecules involving chemical and biological processes. The top-down method directly synthesizes nanoparticles by reducing the size of bigger materials (bulk) using a physical process [7]. The physical and chemical process hurts the environment, and health, and needs larger energy, therefore, AgNP synthesis is done with the biological process utilizing natural ingredients such as microorganisms and plants that can act as both reduction and capping agent [8, 9].

The biological method for synthesizing AgNP is generally known as a bio-synthesis method, which is the method to synthesize nanoparticles (NP) from biological materials specifically, plants, and microorganisms (fungi, bacteria, and algae). The method that specifically uses plants is called the green synthesis method since this method is more environmentally friendly, abundant, economical, and toxic-free. Parts of plants have secondary metabolite compounds such as alkaloids, lawsone, flavonoids, and terpenoid [10]. Molecules from metabolite compounds can bond and reduce the metal ions, for example, Ag^{+} to Ag^{0}, thus serving as reduction and capping agents. Henna leaves (Lawsonia inermis) possess a secondary metabolite compound that contains Lawsone.

Henna leaf extract is commonly used as nail and hair dye, and in healthcare used as an analgesic and anti-inflammation [11], anti-fungal [12], and anti-bacterial [13]. Henna leaves are rich in Lawsone [14] and phenolic compounds [15] that are effective for silver reduction in the NP synthesis process [16, 17]. AgNP synthesized from Henna leaf extract increases photoabsorption due to localized surface Plasmon resonance (LSPR) [18], hence it can be used as the photosensitizer in dye-sensitized solar cells (DSSC). Furthermore, Lawsone has carboxyl and hydroxyl functional groups [19] that can be bonded with the TiO_2 surface [20, 21].

2. Experimental

2.1 Henna's dye extraction

Henna leaves were blended in a blender. The refined Henna leaves (75 gr) were added into 100 ml of ethanol (Merck) while stirred until homogeneous and macerated in a closed and wrapped jar for 24 hours. The solution was then filtered (Whatman paper) and stored in a dark and closed bottle in a refrigerator.

2.2 AgNPs synthesized from Henna dye

The AgNP synthesis process follows the procedure in [22]. 10 ml of Henna dye extract was added to 90 ml of 0.01M of $AgNO_3$ in 100 ml ethanol, and incubated for 15 minutes in a closed and wrapped jar. The same step was done for 0.05M and 0.1M of $AgNO_3$. The AgNP solution was evaporated in a rotary evaporator at 70°C until it became more concentrated, stored in a dark bottle, and kept in a refrigerator before use.

2.3 Henna dye and AgNPs characterization

The molecular structure, crystalline structure, and optical properties of the Henna dye and the AgNPs (unevaporated and evaporated) were characterized using FTIR, XRD, and UV-Vis spectrometer. The AgNPs size was characterized using Particle Size Analyzer (PSA).

Journal of Physics: Conference Series **2980** (2025) 012001
doi:10.1088/1742-6596/2980/1/012001

Table 1. The list of the sample

Sample Name	Sample Code	Sample Name	Sample Code
Unevaporated Henna dye	HNE	0.1M AgNP-Unevaporated	HNE 0.1
Evaporated Henna dye	HE	0.01M AgNP-evaporated	HE 0.01
0.01M AgNP-Unevaporated	HNE 0.01	0.05M AgNP-Evaporated	HE 0.05
0.05M AgNP-Unevaporated	HNE 0.05	0.1M AgNP-Evaporated	HE 0.1

3. Results and Discussions

3.1 Dye extraction from Henna leaves and the AgNPs synthesized from the extract with $AgNO_3$ variation.

For simplicity, the names of the samples were tabulated in Table 1. The dye extracted from Henna leaves (unevaporated and evaporated) and the AgNP produced from it are shown in Figure 1.

Figure 1. Unevaporated (a) and evaporated (b) henna dye and the respecting AgNP synthesized from it with $AgNO_3$ variation concentration.

Unevaporated and evaporated Henna dye shows yellow-brown and no color change in the evaporation process. The extract was then utilized to synthesize AgNP with different $AgNO_3$ concentrations, specifically (0.01, 0.05, and 0.1) M. The color of the solution was changed from yellow-brown to greenish brown with different denseness (Figure 1a). This is an indication of a chemical reaction during the AgNP formation. The AgNP that was evaporated resulted in a darker color (Figure 1b).

3.2 The FTIR analysis of AgNPs from Henna extract

The FTIR spectrum unevaporated and evaporated Henna extract and AgNP was shown in Figure 2 (a and b). We can identify O-H (phenol) group at 3427 cm^{-1} [23], two carbonyls (C=O) at (1722 and 1624) cm^{-1} [24], and the (C=C) group in aromatic benzene rings at 1436 cm^{-1} [25] which are the main functional group for Lawsone [26], chemically known as 2-hydroxy-1,4 Napthoquinone. The broad O-H and C=O absorption peaks at 1722 cm^{-1} indicate the presence of a carboxylic acid functional group. The presence of aromatic rings in the chemical structure of henna leaf extract is further supported by a sharp absorption peak at 769 cm^{-1}, which corresponds to the out-of-plane bending vibrations of C-H in aromatics [27].

The ability of Lawsone to reduce silver ions comes from the formation of an intermediate complex, where the electron density obtains shifted towards the silver ions because of its hydroxyl groups (–OH) [28]. The Henna leaf extract also showed the presence of heterocyclic compounds (alkaloids and glycosides) that can act as capping agents for nanoparticles [28, 29]. The evaporated Henna dye showed an increase in the absorbance intensity which is an indication of the multiply in the compounds in the extract due to solvent evaporation [30].

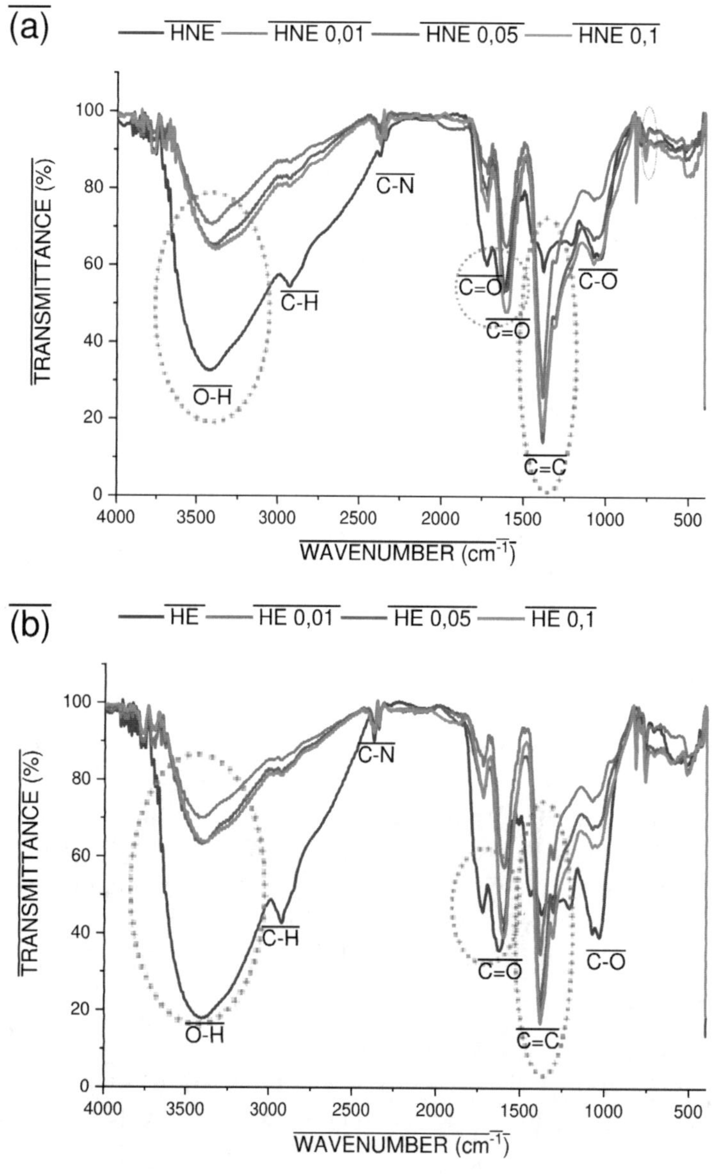

Figure 2. The FTIR spectra of unevaporated and evaporated Henna dye and AgNP with different $AgNO_3$ concentrations. The circled peak marks the point where the bond characteristic changes.

There were some noticeable differences in the FTIR spectra of Henna dye and AgNP which confirmed the formation of the AgNP. The absorbance intensity at a frequency of 1381 cm^{-1} might be originated from the C-N bond (amine), N-H bond from type I and II amides, C=C vibration in the aromatic benzene ring, bending vibration of N=O from the nitro group, or stretching vibration of C=O carbonyl group [22] was increased. The increase in intensity at the frequency of 1381 cm^{-1} shows the reaction between C or O atoms with N atoms in the form of C-N or N=O bonds, which indicates that N atoms might be responsible for the stabilization of AgNP since the electron donation from N atoms to Ag will obstruct the resonance of the amides group [22]. Furthermore, we observed the change in the intensity of the absorption band at a frequency of 825 cm^{-1} which shows the interaction between the extract and silver ions. The intensity of O-H from carboxylic acid functional groups decreased since the O-H group reduced the silver ions in $AgNO_3$.

The intensity of the absorption peak of AgNP escalates with the $AgNO_3$ concentration, this is because the increase of the $AgNO_3$ concentration will enlarge the AgNPs formed.

Figure 3 shows the estimated chemical structure formed between Lawsone molecules and $AgNO_3$ to produce AgNPs. The arrow indicated complex compounds with the coordinated covalent bond.

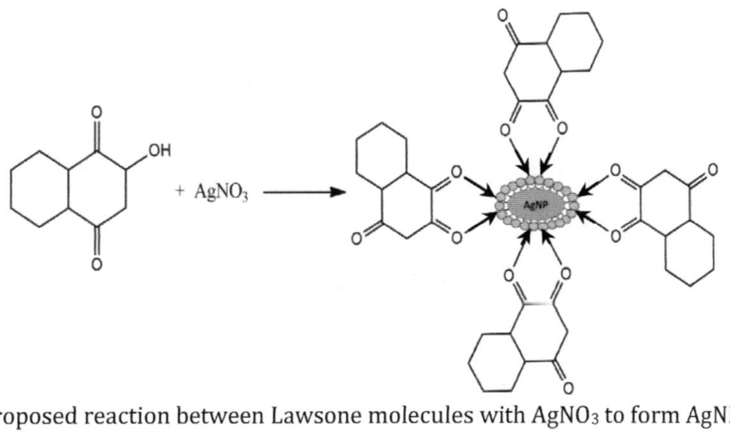

Figure 3. The proposed reaction between Lawsone molecules with $AgNO_3$ to form AgNP

3.3 Crystalline structure and size analysis of AgNPs from Henna extract

The XRD spectrum of unevaporated and evaporated Henna extract and AgNP was shown in Figure 4 (a and b). We confirmed the formation of AgNP with Ag crystalline peak at 2θ position of $(29,5, 32,6, 38,0, \text{dan } 44,1)^0$ that correspond with crystal plane with miller indices of [210], [122], [111], and [200] based on fcc cubic crystal structure with JCPDS file number 04-0783 for silver [25, 26]. The peak positions slightly shift indicating the presence of strain in the crystal structure, which is a characteristic feature of nanocrystallites [25]. From the calculated crystallinity index [27], we found that the evaporation process increased the crystallinity of the AgNPs from 12.68% to 45.06% for HNE 0.1 and HE 0.1, respectively.

The AgNPs' size was measured using PSA, and the result is shown in Figure 5 (a and b). The average size of AgNPs increased with the $AgNO_3$ concentration, specifically from 124.8 nm to 189.1 nm for (0.01 and 0.1) M HNE, respectively. This result was supported by Htwe et al., [28], where the precursor concentration affected the NPs size. The size of the AgNPs that we obtained was slightly above the size of standard NPs which is typically 1-100 nm, however, several cases also reported the NPs with a size above 100 nm [29, 30, 31], thus one can say that AgNPs have

Journal of Physics: Conference Series **2980** (2025) 012001 doi:10.1088/1742-6596/2980/1/012001

been formed from henna leaves extract. 0.1 M is the maximum concentration at which $AgNO_3$ dissolves completely in the henna leaf extract solution. Higher concentrations, such as 0.3 M and 0.5 M, do not allow the $AgNO_3$ to dissolve completely. Different types of extracts yield nanoparticles of varying sizes. In our previous study [22], where we synthesized AgNPs using purple cabbage extract at a concentration of 0.1 M, we obtained an average nanoparticle size of 50 nm. In contrast, in this study, using henna leaf extract at the same concentration, we obtained an AgNP size of 189.1 nm. To achieve an optimal photo-sensitizing effect, the resulting nanoparticles should ideally be smaller than the average TiO_2 particle size, which is typically in the range of 20-30 nm [34]

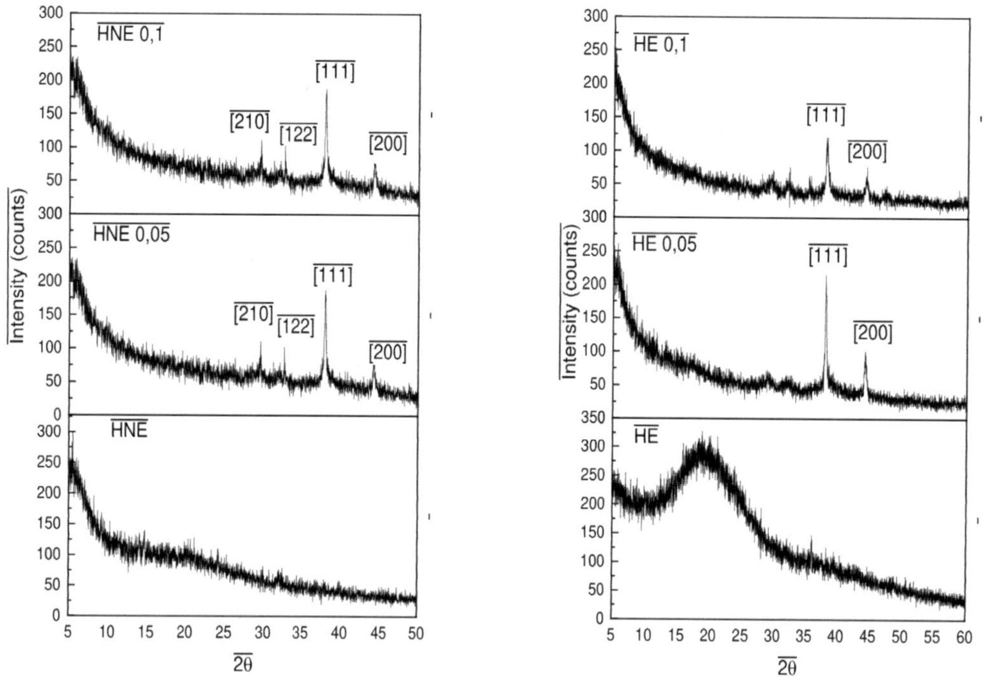

(a) (b)

Figure 4. The XRD of unevaporated and evaporated Henna dye and AgNP with different $AgNO_3$ concentrations.

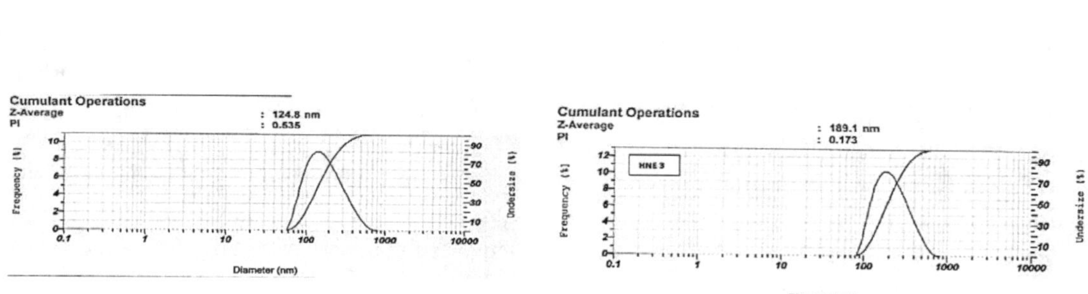

(a) (b)

Figure 5. PSA spectrum for AgNP for 0.01 M (a), and 0.1 M (b) $AgNO_3$.

3.4 Optical properties analysis of AgNPs

The absorbance spectrum of unevaporated and evaporated AgNP formulated from it are shown in Figure 6.

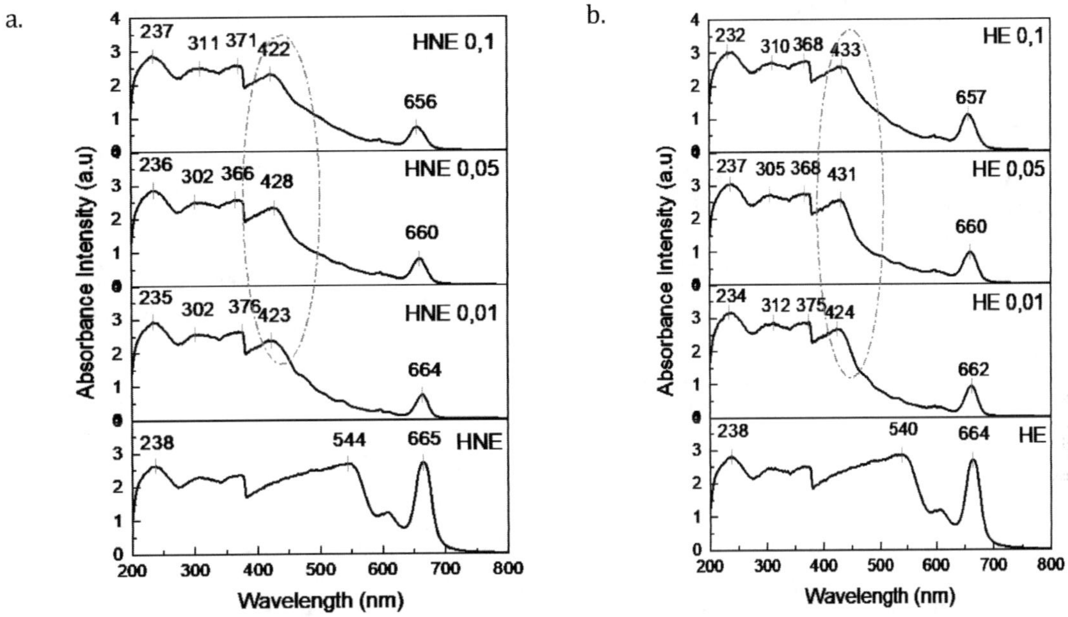

Figure 6. Absorbance spectra of unevaporated and evaporated Henna dye and AgNPs with different $AgNO_3$ concentrations. The circled peak marks the point where the absorption peak changes, which shows the absorbance peak of AgNPs.

There are differences between the absorption of unevaporated and evaporated Henna dye and the AgNPs. The absorption peak at 540 nm disappeared and a new peak emerged at 420 nm and 439 nm, which were the absorption peak for AgNPs [39, 40]. The increase of $AgNO_3$ concentration decreased the absorption peak at 660 nm. There is no significant difference between the absorbance spectrum of unevaporated and evaporated AgNPs.

We estimated the unevaporated and evaporated Henna dye and the respecting AgNPs energy gap using the Tauc method in our developed program [41], the results were tabulated in Table 2.

Table 2. The estimated energy gap of unevaporated and evaporated Henna dye and AgNps for different $AgNO_3$ concentrations.

Sample Name	Energy Gap (eV)	Sample Name	Energy Gap (eV)
HNE	2,08	HE	2,09
HNE 0,01	3,67	HE 0,01	2,58
HNE 0,05	3,68	HE 0,05	2,58
HNE 0,1	3,69	HE 0.1	3,63

Journal of Physics: Conference Series **2980** (2025) 012001 doi:10.1088/1742-6596/2980/1/012001

Quite large differences in energy gap (E_g) were observed between unevaporated and evaporated AgNPs, for which the E_g were 29.7%, 29.9%, and 1.36% smaller for all concentration variations. The decrease in E_g due to the evaporation process might be related to more $AgNO_3$ that was being reduced [42], as well as the surface plasmon effect that was marked with the emergence of the absorption peak at 430 nm. [43]. The estimated E_g in this research is much bigger than the E_g of the standard Ruthenium dye in DSSC which is 1.99 eV.

3.4.1 Irradiation effect analysis to the localized surface plasmon.

To investigate whether there is a shift in the surface plasmon effect, we measured the absorbance of the evaporated AgNPs solution for time variation of (0, 24, 48, and 72) hours, and the results are shown in Figure 7 (a, b, c).

For all concentrations (Figure 7a, 7b, and 7c), irradiation time caused plasmon surface resonance to shift to a lower energy level (longer wavelength). This correlated with the aggregation due to irradiation that caused the delocalization of conduction electrons near the particle surface and shared between neighbouring particles [44].

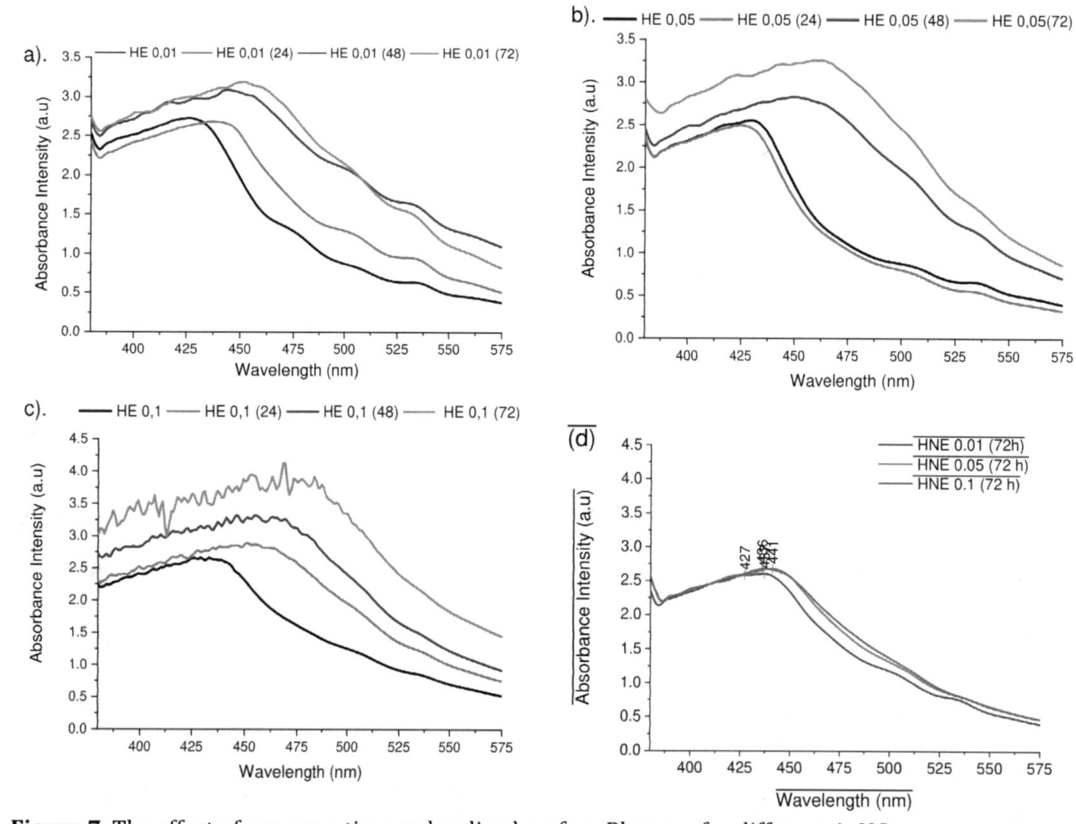

Figure 7. The effect of exposure time on localized surface Plasmon for different $AgNO_3$ concentration variations, 0.01 M (a), 0.05 M (b), and 0.1 M (c).

Figure 7d shows the absorbance spectrum of AgNPs synthesized from unevaporated henna with varying concentrations of $AgNO_3$ (0.01 M, 0.05 M, and 0.1 M), which were exposed to light for 72 hours. The data indicate that increasing the concentration of $AgNO_3$ slightly enhances the

Journal of Physics: Conference Series **2980** (2025) 012001 doi:10.1088/1742-6596/2980/1/012001

absorbance intensity of the AgNPs. An increase in absorbance intensity indicates the formation of more AgNPs [45] as well as an increase in the size of the AgNPs [46]. This was in good agreement with the PSA results in Figure 5. This confirmed that precursor concentration ($AgNO_3$) could be used to control the size of the AgNPs [46].

Conclusions

We have extracted Lawsone from Henna leaves and green synthesized AgNPs from it, which was confirmed through its FTIR, XRD, and absorbance spectrum. The evaporation process of the dye and AgNP solution increases index crystallinity and lowers the energy gap value. The higher the $AgNO_3$ concentration, the bigger the AgNPs' size and the energy gap. Based on these results, it cannot be concluded which one has better, higher, or lower gap energy for use as a sensitizer in DSSC because the sample has not been tested directly on a DSSC. However, one requirement that must be met is that no matter how large or small the gap energy is, the excited state level of the gap energy of the sensitizer or coloring agent must be higher than the conduction band edge of the oxide semiconductor used as the photo-anode (TiO_2 or ZnO) to ensure efficient electron injection process. To be able to use AgNPs as a sensitizer in DSSC, it must be avoided using iodine-based electrolytes since it can react with Ag to form soluble AgI so that AgNPs cannot bond to the TiO_2 surface. Apart from that, one solution is to coat the sensitized TIO_2 layer with a conductive polymer layer on top to prevent the reaction between Ag and iodine from the electrolyte.

Acknowledgment

This research was funded by DRTPM DIKTI through Fundamental Research Scheme No. 049/E5/PG.02.00.PL/2024 and No. 24/UN29.20/PG/2024.

References

[1] M. Naito, M. Hosokawa and T. Yokoyama, 2018, Nanoparticle Technology Handbook, Elsevier.

[2] M. Willems, 2005, "Nanoparticles: Roadmap Report," VITO NV.

[3] J. Jeevanandam, A. Barhoum, S. Chan Y, D. A and K. Danquah M, 2018, "Review On Nanoparticles and Nanostructured Materials: History, Sources, Toxicity and Regulations," *Beilstein Journal of Nanotechnology,* 9(1), 1050–1074.

[4] S. Husain, S. Husain, K. Megasari, E. Suharyadi and K. Abraha, 2012, "Deteksi Biomolekul pada Fenomena Surface Plasmon Resonance (SPR) Terlapisi Nanopartikel Magnetik Fe3O4," in *Scientific Meeting of HFI.*

[5] S. Saravanan, R. Kato, M. Balamurugan, S. Kaushik and T. Soga, 2017, "Efficiency Improvement in Dye Sensitized Solar Cells by the Plasmonic Effect of Green Synthesized Silver Nanoparticles," *JS: AMD,* 2(4), 429-437.

[6] G. Bryant and M. Pelton, 2013, Introduction to Metal-Nanoparticle Plasmonics, USA: Wiley-Blackwell.

[7] G. Cao, 2005, Nanostructures and nanomaterials: Synthesis, properties & applications, London: Imperial College Press.

[8] F. Ahmed S, M. Mofijur, N. Rafa, T. Chowdhury A, S. Chowdhury, M. Nahrin, S. Islam A and C. Ong H, 2022, "Green approaches in synthesising nanomaterials for environmental nanobioremediation: Technological advancements, applications, benefits and challenges," *Environ. Res.,* 204, 112255.

[9] S. Iravani, H. Korbekandi, V. Mirmohammadi S and B. Zolfaghari, 2014, "Synthesis of silver nanoparticles: chemical, physical and biological methods," *Research in Pharmaceutical Sciences,* 9(6), 85–406.

[10] S. Julianto T, 2019, "Fitokimia Tinjauan Metabolit Sekunder dan Skrining Fitokimia," *In Journal of Chemical Information and Modeling,* 53(9).

[11] T. A. Talab, N. Alfuraiji and A. E. Al-Snafi, 2022, "THE ANALGESIC AND ANTI-INFLAMMATORY EFFECT OF LAWSONE ISOLATED FROM LAWSONIA INERMIS," *ScienceRise, Pharm. sci.,* 1(35), 77-84.

[12] S. Meena, B. R. Meena and K. Sharma, 2019, "Lawsonia Inermis (Henna) Extract: A Potent Natural Antifungal against Plant Pathogenic Fungi," *IJRASET,* 7(V), 2015-2018.

[13] M. F. Bafghi, S. Salary, F. Mirzaei, H. Mahmoodian, H. Meftahizade and R. Zareshahi, 2022, "Antibacterial and anti-trichomunas characteristics of local landraces of Lawsonia inermis L.," *BMC complement. med. ther,* 22(203).

[14] J. Sankar, M. Kini and P. Sangle, 2023, "Extraction of Lawsone Active from the Henna Leaves (Lawsonia," *Med & Analy Chem Int J,* 7(2), 1-7.

[15] S. El-Massoudi, A. Zinedine, J. M. Rocha, M. Benidir, I. Najjari, L. El-Ghadraoui, M. Benjelloun and F. Errachidi, 2023, "Phenolic Composition and Wound Healing Potential Assessment of Moroccan Henna (Lawsonia inermis) Aqueous Extracts," *Cosmetics,* 10(3), 92.

[16] L. Natrayan, M. Jayakrishna, K. Shanker, G. Muthu, S. Kallappan and G. Velmurugan, 2024, "Green synthesis of silver nanoparticles using lawsonia inermis for enhanced degradation of organic pollutants in wastewater treatment," *Global NEST Journal,* 26(3), 05463.

[17] S. Senthilkumaran, 2024, "Phytosynthesis of Eco-Friendly Silver Nanoparticles Using Lawsonia Innermis (L) and Their Biomedical Applications," *J. pharm. bioallied sci,*16(2), S1274-80.

[18] S.-C. G. Kiruba-Daniel, N. Mahalakshmi, J. Sandhiya, K. Nehru and M. Sivakumar, 2013, "Rapid synthesis of Ag nanoparticles using Henna extract for the fabrication of Photoabsorption Enhanced Dye Sensitized Solar Cell (PE-DSSC)," *Adv. Mat. Res.,* 678(2013), 349-60.

[19] A. Pallipurath, J. M. Skelton, A. Delori, C. Duffy, A. Erxleben and W. Jones, 2015, "Crystalline adducts of the Lawsone molecule (2-hydroxy-1,4-naphthaquinone): optical properties and computational modelling," *CrystEngComm,* 17, 7684-7692.

[20] E. Tilahun and Y. Dessie, 2022, "Recent Progress, Advancements, and Efficiency Improvement Techniques of Natural Plant Pigment-Based Photosensitizers for Dye-Sensitized Solar Cells," *J. Nanomater.,* 2022, 1024100.

[21] M. A.-M. Al-Alwani, N. A. Ludin, A. B. Mohamad, A. A.-H. Kadhum, M. M. Ba-Abbad and K. Bin-Sopian, 2016, "Optimization of dye extraction from Cordyline fruticosa via response surface methodology to produce a natural sensitizer for dye-sensitized solar cells," *Results Phys.,* 6(C), 520–29.

[22] Arsyad Waode S, Cassandra F, Asharuddin M, Suryani, Ramadhan LOAN and Hidayat R, 2022, "Green synthesis of silver nanoparticles from anthocyanin extracts of purple cabbage (brassica oleracea var capitata) and its characteristics for dye-sensitized solar cells (DSSC) application," *J. Phys.: Conf. Ser.,* 2274, 012001.

[23] S. Saadaoui, A. Mohamed, B. Moufida, G. Rached, S. Emanueles, M. Salvo, A. Alessandra and A. Rosaria, 2017, "Performance of natural dye sensitized solar cells by ZnO nanorod and nanowall enhaced photoelectrodes.," *Belstein J. Nanotechnol,* 8, 287-295.

[24] W. Nik, F. Zulkifli, O. Sulaiman, K. Samo and R. Rosliza, 2012, "Study of Henna (Lawsonia inermis) as Natural Corrosion Inhibitor for Aluminum Alloy in Seawater," *IOP Conference Series: Materials Science and Engineering,* 1(2), 723-728.

[25] A. Hamdy and N. El-Gendy, 2013, "Thermodynamic, adsorption and electrochemical studies for corrosion inhibition of carbon steel by henna extract in acid medium," *Egypt. J. Pet,* 22, 17-25.

[26] H. Hajar, F. Zulkifli, M. Sabri and W. Nik, 2016, "Protection against Corrosion of Aluminum Alloy in Marine Environment by Lawsonia inermis," *International Journal of Corrosion,* 4891803.

[27] S. Wahyuningsih, L. Wulandari, M. Wartono, H. Munawaroh and A. Ramelan, 2017, "The Effect of pH and Color Stability of Anthocyanin on Food Colorant," *IOP Conference Series: Materials Science and Engineering,* 193, 012047.

[28] A. Gupta, S. R. Bonde, S. Gaikwad, A. Ingle, A. K. Gade and M. Rai, 2014, "Lawsonia inermis-mediated synthesis of silver nanoparticles: activity against human pathogenic fungi and bacteria with special reference to formulation of an antimicrobial nanogel," *IET Nanobiotechnology,* vol. 8(3), 172-178.

[29] S. Iravani, 2011, "Green synthesis of metal nanoparticles using plants," *Green Chem,* 13, 2638–2650.

[30] B. Stuart, 2004, Infrared Spectroscopy: Fundamentals and Applications, UK: John Wiley & Sons.

[31] R. Priyadharshini, G. Prasannaraj, N. Geetha and P. Venkatachalam, 2014, "Microwave-mediated extracellular synthesis of metallic silver and zinc oxide nanoparticles using macro-algae (Gracilaria edulis) extracts and its anticancer activity against human PC3 cell lines," *Appl. Biochem. Biotechnol,* 174, 2777–2790.

[32] K. Roy, C. Sarkar and C. Ghosh, 2014, "Green snthesis of silver nanoparticles using fruit extract of Malus domestica and study of its antimicrobial activity," *Dig. J. Nanomater. Biostruct,* 9, 1137–1147.

[33] M. Jaworska, K. Sakurai, P. Gaudon and E. Guibal, 2003, "Influence of chitosan characteristics on polymer properties. I: Crystallographic properties," *Polym. Int,* 52, 198-205.

[34] Y. Htwe, W. Chow, Y. Suda and M. Mariatti, 2019, "Effect of Silver Nitrate Concentration on the Production of Silver Nanoparticles by Green Method," *Materials Today: Proceedings,* 17(2019), 568–573.

[35] I. Latumakulita and Suparno, 2022, "Characterization of Silver Nanoparticle Electrolysis Method with UV-Vis Spectrometer, Atomic Absorption Spectrophotometer, and Particle SizeAnalyzer," *Kasuari: Physics Education Journal,* 5(1), 42-52.

[36] T. Lestari, F. Tahlib, J. K. Sukweenadhi and C. Avanti, 2019, "Physical Characteristic and Antibacterial Activity of Silver Nanoparticles from Green Synthesis Using Ethanol Extracts of Phaleria macrocarpa (Scheff.) Boerl Leaves," *Traditional Medicine Journal,* 24(1), 16-21.

[37] A. Abed, 2017, "Synthesis and study silver nanoparticles by chemical method," *Diyala Journal for Pure Sciences,* 13(1), 2518-9255.

[38] J. Salimian, S. Osfouri, R. Azin and T. Jalali, 2022, "Impacts of paste preparation methods on the porous TiO2 nanostructure properties and naturally dye-sensitized solar cells performance," *JMR&T,* 18, 4816-4833.

[39] Y. Meng, 2015, "A Sustainable Approach to Fabricating Ag Nanoparticles/PVA Hybrid Nanofiber and Its Catalytic Activity," *Nanomaterials,* 5, 1124-1135.

[40] D. Acharya, M. Singha, P. Pandey, B. Mohanta, J. Rajkumari and L. P. Singha, 2018, "Shape dependent physical mutilation and lethal effects of silver nanoparticles on bacteria," *Scientific Reports,* 8(201).

[41] V. Mursyalaat, V. I. Variani, W. O. S. Arsyad and M. Z. Firihu, 2023, "The development of program for calculating the band gap energy of semiconductor material based on UV-Vis spectrum using delphi 7.0," *J. Phys. Conf.,* 2498(1), 012042.

[42] L. Niu, Q. Li, F. Wei and H. wang, 2003, "Electroactivity of electrochemicaly synthesized poly(aniline boronic acid) as a function of Ph: Role of self-doping," *J Electroanal. Chem,* 544, 121-125.

[43] Zhu H, Zeng H, Subramanian V, Masarapu C, Hunga KH and Wei B, 2008, "Anthocyanin-sensitized solar cells using carbon nanotube films as counter electrodes," *Nanotechnology,* 19, 465204-465208.

[44] L. Mahmudin, D. Darwis, E. Suharyadi, A. Utomo and K. Abraha, 2021, "The effect of the concentration of the stabilizer in the formation of a silver nanoparticle on the phenomenon of Surface Plasmon Resonance (SPR) as an active material for biosensor," *Journal of Physics: Conference Series,* 1763, 012064.

[45] S. D. Mulyana, R. Sari and A. S. Rijal, 2024, "Green Synthesis of Silver Nanoparticles with Bioreductant from Lime Juice Powder (Citrus aurantifolia): Effect of Concentration and pH," *Indones. J. Chem,* 24(5), 1445 - 1455.

[46] M. Skiba, A. Pivovarov, V. Vorobyova, T. Derkach and I. Kurmakova, 2019, "PLASMA-CHEMICAL FORMATION OF SILVER NANOPARTICLES:THE SILVER IONS CONCENTRATION EFFECT ON THE PARTICLE SIZE AND THEIR ANTIMICROBIAL PROPERTIES," *J. Chem. Technol. Metall,* 54(2), 311-318.

The Influence of Annealing Temperature on the Grain Size of a Nickel Metal Catalyst Suitable for Carbon Nanotube Growth

Rustan Ruslan[1], Abd. Wahidin Nuayi[1,2], Ajeng Eliyana[1,3], Fatimah Arofiati Noor[1] and Toto Winata[1*]**

[1] Physics and Technology of Advanced Materials Research Division, Department of Physics, Faculty of Mathematics and Natural Sciences, Institut Teknologi Bandung, Bandung, Indonesia
[2] Physics of Materials Group, Department of Physics, Faculty of Mathematics and Natural Sciences, Universitas Negeri Gorontalo, Jln. Jend. Sudirman No. 06 Kota Gorontalo, Gorontalo, 96128 Indonesia
[3] Physics of Material and Instrumentation Research Group, Departement of Physics, faculty of science, Institut Teknologi Sumatera, Jl. Terusan Ryacudu, Jati Agung, Lampung Selatan, Indonesia

E-mail: twinata@itb.ac.id*, fatimah@itb.ac.id**

Abstract. The catalyst plays a key role in the growth of carbon nanotubes (CNTs) functioning as a guiding medium in the growth process. In this research, a nickel (Ni) catalyst was on a 7101 glass substrate using the thermal evaporation method. First, the effect of the thickness of the layer on its morphology was investigated, with thicknesses of 30, 40, 50, and 60 nm examined. An optimum density was obtained at 50 nm, at which the Ni particles attained the highest average diameter of 4.23 nm, and polydispersity index lower than the other samples, at 5.87%. Energy-dispersive X-ray spectroscopy results also showed a positive linear relationship between the thickness of the layer and both its mass and purity. Samples at this optimum thickness were annealed at 300 °C, 400 °C, or 500 °C for 4 hours to refine the crystal structure of the catalyst and produce small, uniform features. The samples annealed at 500 °C had the greatest average particle diameter of 5.92 nm and the lowest polydispersity of 5.76%. With a Ni catalyst that has a low polydispersity and thus a high level of uniformity, CNTs with a good level of uniformity and quality can also be obtained.

Keywords: Catalysts, Nickel, CNT, Evaporation and Annealing

1. Introduction

Since their discovery by Iijima in 1991 [1], carbon nanotubes (CNTs) have become a prominent research topic in nanotechnology and advanced materials research. These carbon-based tubular structures offer several unique characteristics, such as high mechanical strength, high thermal and electrical conductivity, and an ability to form intricate nanoscale structures, making them an attractive choice for various applications in electronic and optical devices, including in solar cells [2–5]. Carbon nanotubes (CNTs) have the potential to transform current technologies and pave the way for next-generation materials and devices that are more effective and efficient.

One important aspect of the growth of CNTs is the role of catalysts, which function as initiators in the growth process. The reduction of catalyst dimensions to the nanoscale will increase the surface area of the catalyst and consequently enhance the catalytic activity in certain reactions [6]. According to [7], the thicker the catalyst, the shorter the CNTs produced. This is related to the diffusion length of carbon within the catalyst particles: for thicker catalysts, the carbon diffusion length is greater and the growth rate is slower, resulting in shorter CNTs [8]. Therefore, the quality and characteristics of CNTs are strongly dependent on the properties of the catalyst used in their growth, making the selection and optimization of the catalyst an important area of CNT research [9,10]. This optimization procedure entails not only the catalyst material selection but also the parameters and deposition technique.

One of the methods commonly used in catalyst deposition is thermal evaporation. This method is unlikely to cause damage to the substrate, has a high deposition rate, and allows for precise thickness control [11]. Thermal evaporation also offers scalability and consistency, making it an ideal technique for producing large quantities of catalyst with uniform properties. Several recent studies indicate that variations in the thickness of the catalyst layer can produce CNTs with different morphologies and the structures [12,13]. These variations can be used to tailor CNTs to meet specific requirements for different applications.

In this study, the material used as a catalyst source was nickel (Ni), which was grown using the thermal evaporation method. Nickel (Ni) has a high carbon solubility and a high carbon diffusion value [8,14,15]. Additionally, Ni is likely to have a sufficiently wide temperature window for the growth of CNTs with carbon precursors due to its high melting point and low equilibrium vapor pressure. This makes it an effective and versatile option for a range of different CNT synthesis environments. Therefore, Ni is considered the most suitable catalyst for CNT growth [14]. The use of Ni as a catalyst in this study aims to leverage these properties to optimize the growth conditions and achieve high-quality CNTs with desirable attributes for advanced applications.

After deposition, to achieve a nanocatalyst structure, an annealing treatment is required. This involves heating the material to a specific temperature to alter the crystal structure and grain size. In [16], it was found that the annealing temperature has a significant impact on the formation of nano-sized nickel catalyst grains. Smaller grain sizes tend to provide more active sites for the growth of CNTs, leading to CNTs of higher quality [17,18]. Moreover, the annealing process can be finely tuned to control the distribution and size of these grains, influencing the overall growth dynamics and final properties of the CNTs. This makes annealing a critical step in the preparation of catalysts for CNT synthesis.

In this study, the thickness of the nickel catalyst layer was optimized, followed by an examination of the effect of annealing temperature on the grain size of the formed nickel catalyst. By understanding the effect of the thickness of the catalyst layer and the annealing temperature on the grain size of the catalyst, these parameters can be used to control the quality and characteristics of the CNTs that are intended to be grown. These findings should advance the field of nanomaterials in general and pave the way for the development of precisely engineered CNTs with unique properties suited to particular technological applications, particularly those involving high-performance materials such as energy storage, electronics, and nano-optics.

2. Experimental Methods

The catalyst growth was carried out in two stages: a Ni thin film was grown on a glass substrate using thermal deposition equipment from LJUHV Taiwan, and then it was annealed using a furnace. The work was carried out at the National Innovation Research Agency, Bandung, Indonesia. The first step was to prepare a suitable substrate. Corning 7101 glass was cleaned with an ultrasonic cleaner for 15 minutes in deionized water (DI), acetone, methanol, and finally deionized water again. The substrate was then dried by spraying the surface of the glass with a nitrogen gas gun until there were no remaining traces of water. Next, the growth of nickel catalyst was carried out with thicknesses of 30, 40, 50, and 60 nm using the thermal evaporation method; the growth parameters are shown in Table 1. The nickel catalyst was then characterized using scanning electron microscopy (SEM) and energy-dispersive X-ray spectroscopy (EDS) to observe the morphology and composition of the catalyst layer, and measurements of the diameter and particle size distribution were performed to obtain the optimum thickness.

Table 1. Nickel catalyst growth parameters.

Parameter	Unit	Value
Glass substrate dimensions (w × l)	mm	76.2 × 25.4
Ni wire source		
Diameter	mm	0.5
Purity	%	99.99
Mass	mg	160, 190, 220, 250
Thickness	nm	30, 40, 50, 60
Chamber pressure	Torr	6.4×10^{-7}

The next step was to perform an annealing treatment on the nickel catalyst with the optimum thickness of 50 nm. The sample was annealed at three different temperatures, 300 °C, 400 °C, and 500 °C, for 4 hours. Next, it was characterized using SEM/EDS to observe the morphological shape and composition, as well as to measure the diameter and particle size distribution in the samples with different annealing temperatures.

3. Results and Discussion

3.1 Optimization of Ni Layer Thickness

The growth of the Ni layer was carried out at thicknesses of 30, 40, 50, and 60 nm. This was intended to observe the influence of layer thickness on the formed nickel particles. The results of the SEM observations on the growth of nickel nanocatalysts at these thicknesses show changes in morphology (Figure 1). At 30, 40, and 50 nm, many grains are observed, but the appearance is somewhat different at a thickness of 60 nm. This is due to the early growth of nickel crystals that tend to form clusters or agglomerates on the substrate. At a thickness of 30 or 40 nm (Figures 1 (a) and (b), respectively), this growth is still in the early and intermediate stages, in which grains begin to form and move closer together. As the layer thickens to 50 nm (Figure 1 (c)), more material is available to form grains, resulting in more nucleation and crystal growth and an increase in the number of grains. However, the results are slightly different at a thickness of 60 nm (Figure 1 (d)), at which the number of grains decreases or they disappear entirely. This is made possible by several factors, including the coalescence of small particles into larger and flatter structures, as well as surface diffusion, which causes nickel atoms to fill the spaces between the particles.

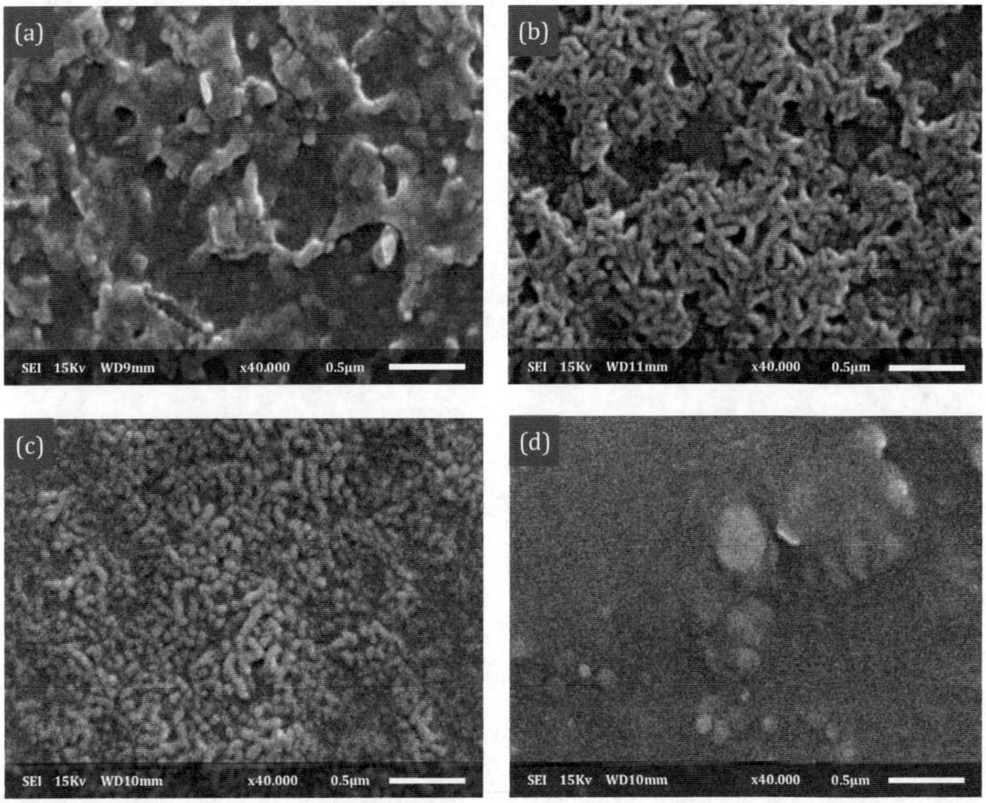

Figure 1. SEM morphological form of Ni nanocatalysts with thicknesses of (a) 30 nm, (b) 40 nm, (c) 50 nm, and (d) 60 nm.

Next, the nickel particles observed in Figure 1 were mapped, and size analysis was conducted on the four samples using the ImageJ software. To facilitate interpretation, the data from this application were processed by creating histograms with Gaussian curves using the Origin software. Figure 2 displays the results of the particle size analysis, detailing the average diameter and polydispersity index of the sample's particles. The polydispersity index is a parameter that describes how uniform the particle size of a sample. The smaller the percentage of the polydispersity index of a sample, the better the level of uniformity of the size of a sample. It can be seen that the Ni nanocatalyst sample has a diverse particle diameter distribution. In the sample with a layer thickness of 30 nm, a large number of particles in the diameter of 0–10 nm was observed, with a total particle count of 10821. The smallest diameter obtained was 0.7 nm, and the largest was around 139 nm. The average diameter was 4.18 nm with a polydispersity index of 6.34%. Furthermore, the sample with a thickness of 40 nm also had the largest number of particles in the diameter range of 0–10 nm, totaling 4879 particles. In this case, the smallest and largest diameters were 0.7 nm and 177.4 nm, with an average diameter of 3.40 nm and a polydispersity of 6.55%.

In the sample with a thickness of 50 nm, the highest frequency of particles was observed in the diameter range of 0–10 nm, at 3265. The smallest diameter was approximately 0.7 nm and the largest was 143.62 nm. The average diameter was 4.23 nm and the polydispersity was 5.87%. For the sample with a thickness of 60 nm, the largest number of particles was also in the range of 0–10 nm, totaling 9670. The smallest diameter was also about 0.7 nm and the largest was 89.21 nm. The average diameter was 2.5 nm and the polydispersity was 13.37%.

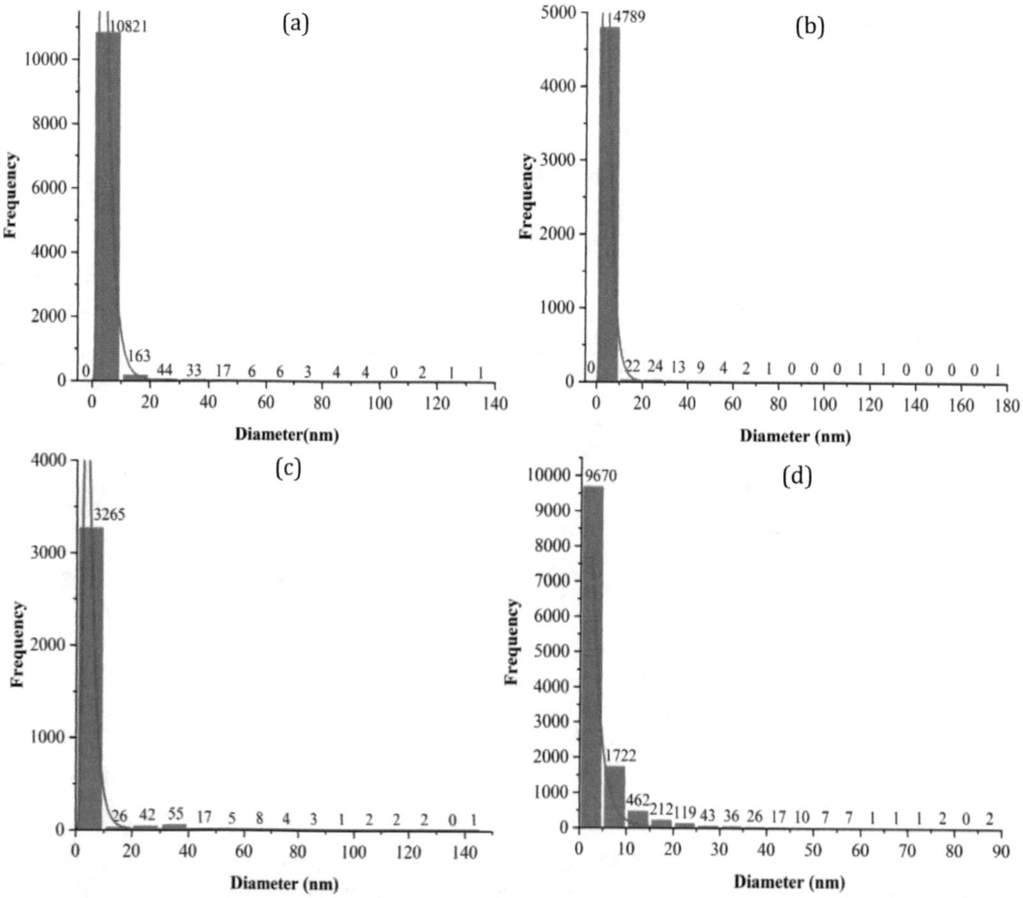

Figure 2. Histograms of the particle diameter distributions in the Ni nanocatalyst samples with thicknesses of (a) 30 nm, (b) 40 nm, (c) 50 nm, and (d) 60 nm.

Of the four thicknesses studied, the smallest average diameter was obtained from the 60 nm sample, but it had the highest polydispersity index. The polydispersity of the catalyst is an important parameter for CNT growth; it describes the uniformity of particle sizes within a component. The smaller the polydispersity, the more uniform the particle size distribution within a component. Even distribution of the catalyst thickness and homogenous grain sizes result in CNTs with consistent diameters and improve thermal stability throughout the annealing process [19]. It was found that the sample with a thickness of 50 nm had the smallest polydispersity at 5.87%, which means that this sample had a high level of homogeneity. Thus, based on the polydispersity index, the optimum sample thickness among the four thickness changes was 50 nm. In addition to the morphology and particle diameter distribution, the percentage composition of the elements in the Ni catalyst layer was also confirmed, both in mass and atomic percentages, as shown in Tables 2 and 3. It can be seen that, besides Ni, other elements such as Na, Si, O, Mg, and Ca, which are constituents of the glass substrate used, were also identified.

Table 2. Results of EDS characterization for the mass percentage composition of elements in the Ni catalyst layer.

	Mass composition (%)					
	Element					
Thickness (nm)	O	Na	Mg	Si	Ca	Ni
30	6.97	6.92	2.51	50.99	7.95	24.67
40	5.63	5.78	2.36	46.94	7.64	31.64
50	4.16	5.24	1.94	42.72	6.71	39.23
60	4.49	4.64	1.46	39.29	6.67	43.45

Table 3. Results of EDS characterization for the atomic percentage composition of elements in the Ni catalyst layer.

	Atomic composition (%)					
	Element					
Thickness (nm)	O	Na	Mg	Si	Ca	Ni
30	13.30	9.20	3.15	55.45	6.06	12.83
40	11.35	8.11	3.13	53.89	6.14	17.37
50	8.88	7.79	2.73	52.02	5.72	22.85
60	9.85	7.09	2.10	49.12	5.84	25.99

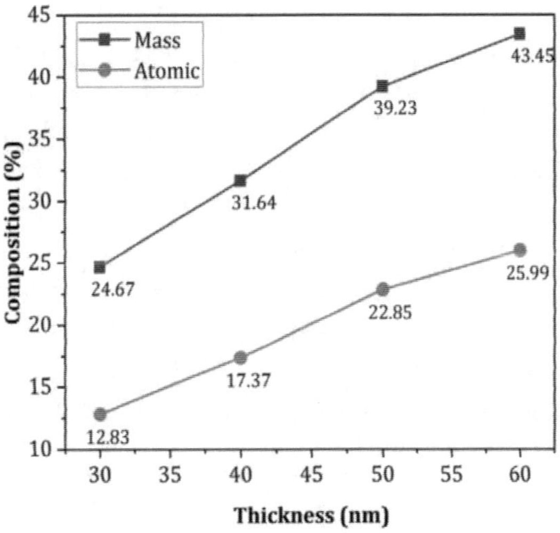

Figure 3. Relationship between layer thickness and mass or atomic composition of the Ni catalyst.

In Tables 2 and 3, it can be observed that as the percentage of nickel increases, the percentages of elements such as O, Na, and Si, which are components of the glass substrate, decrease. This is due to the increasing amount of nickel deposited on the glass substrate. It can be confirmed in Figure 3 that the mass percentage of nickel (marked with blue squares) increases significantly as the layer thickness increases. From 30 nm to 40 nm, the mass percentage rises from 24.67% to 31.64%; from 40 nm to 50 nm, the mass percentage reaches 39.23%, and from 50 nm to 60 nm, the mass percentage increases again to 43.45%. This indicates that an increased layer thickness leads to an increase in the absolute amount of nickel deposited in the sample, suggesting that adding nickel layers will increase the total weight of the sample, as seen in Figure 3. A similar trend is also observed in the atomic percentage (shown with red circles), which shows that the atomic percentage of nickel also increases with the layer thickness. It can be seen that from a thickness of 30 nm to 40 nm, the atomic percentage rises from 12.83% to 17.37%; from 40 nm to 50 nm, the atomic percentage grows to 22.85%, and from 50 nm to 60 nm, the atomic percentage rises again to 25.99%.

3.2 The Influence of Annealing Temperature

To optimize the annealing temperature, samples with the optimal thickness of 50 nm underwent annealing at a temperature of 300 °C, 400 °C, and 500 °C for a duration of 4 hours. The SEM images presented in Figures 1 and 4 show that the surface of the Ni catalyst film has an uneven structure that resembles mounds more than a smooth surface. This variation in topography is caused by the irregular scattering of particles from the SEM light source. Atoms of Ni are absorbed onto the substrate surface, then diffuse and interact, resulting in the formation of polyatomic clusters in the shape of islands [16]. In addition, the interactions between catalyst atoms are stronger than those between catalyst atoms and the constituent atoms of the substrate [20].

The SEM images in Figure 4 were taken after annealing Ni catalyst samples with a thickness of 50 nm (see Figure 1 (c)). Before annealing, the morphology includes large and elongated grains as well as numerous blank spots. The large and elongated grains on the sample are the result of the deposition or initial growth of nickel particles that are not yet evenly spread across the surface, along with the presence of an irregular structure. The blank spots may indicate contamination or another layer covering certain parts of the sample, resulting in an uneven distribution. For the sample annealed at a temperature of 300 °C, the morphology appears smoother, although some blank spot areas are still visible. This indicates a change in the surface structure, but it is not yet fully uniform. At a temperature of 300 °C, the sample undergoes a recrystallization process in which the irregular crystal structure or deformation in the nickel grains can begin to self-repair. This process can result in a smoother surface due to the rearrangement of nickel atoms into a more orderly and stable crystal structure. At this temperature, it is also possible that the diffusion rate of nickel atoms into the surface of the sample may be slower than at higher temperatures, reducing the formation of new grains or the agglomeration of nickel particles, which consequently results in a smoother surface [21].

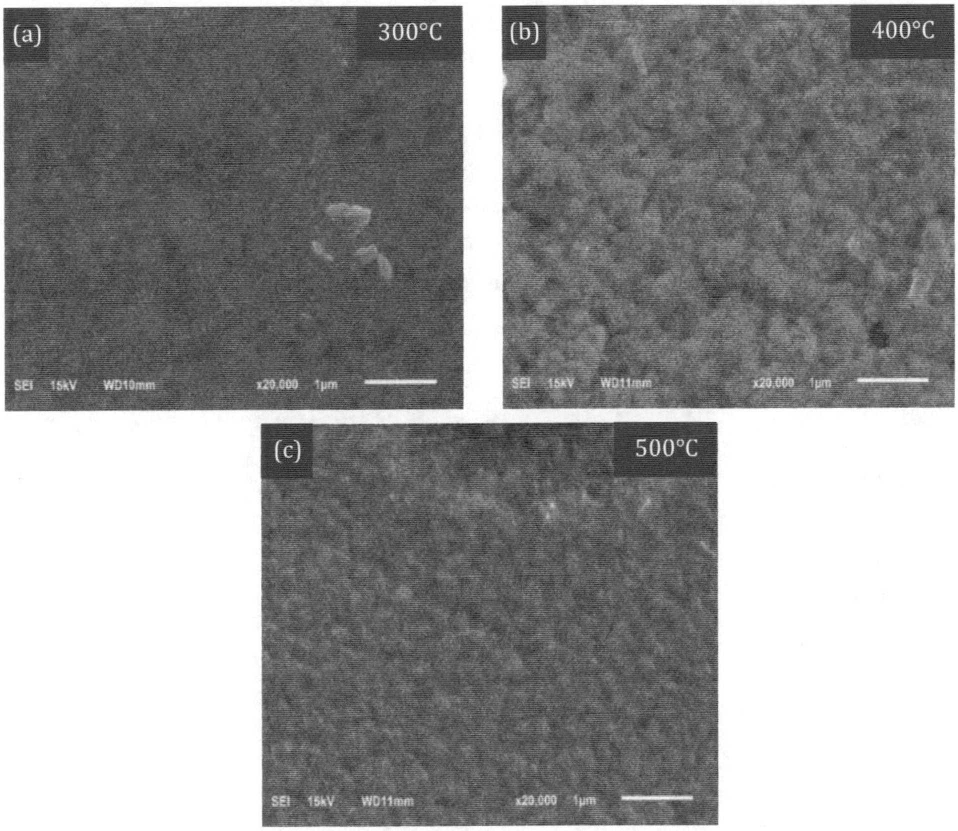

Figure 4. SEM morphological form of 50 nm thick Ni nanocatalyst after annealing treatment for 4 hours at temperatures (a) 300 °C, (b) 400 °C, and (c) 500 °C, at a magnification of 20,000×.

At a temperature of 400 °C, there is an increase in the number of large, irregular grains, and there are still empty areas indicating that the annealing process has not been uniform throughout the sample. At this temperature, there is higher thermal activity than before, which can cause nickel particles to move and interact more intensively, leading to the agglomeration of particles into larger grains [22]. Additionally, the distribution of the thin nickel layer before the annealing process is uneven across the sample, so during the annealing process, areas with higher nickel concentrations tend to experience more agglomeration, resulting in larger grains, while areas with lower concentrations may remain relatively fine. However, the sample that was annealed at a temperature of 500 °C showed smaller and more uniform grains, although there were still some blank spots. This indicates an improvement in the homogeneity of nickel distribution after the annealing process at a higher temperature. At this temperature, it seems that the diffusion rate of nickel atoms into the surface of the sample becomes faster, and further recrystallization of the nickel crystal structure can occur. Nickel atoms that were initially trapped or bound to larger grains can move and form smaller and more uniform grains across the surface of the sample [21].

Next, based on the SEM results, an analysis of the size and distribution of the Ni catalyst particles was conducted. Figure 5 shows that the distribution of particle diameters varies between the three samples, with the highest numbers of particles in the diameter range of 1–10 nm for samples annealed at 300 °C, 400 °C, and 500 °C, totaling, respectively, 8658, 7542, and 7686 particles. In the sample annealed at 300 °C (Figure 5 (a)), the smallest particle diameter was 1.42 nm and the largest was 89.52 nm. The average diameter was 4.88 nm and the polydispersity was 18.82%. For the sample annealed at 400 °C (Figure 5 (b)), the smallest and largest diameters were 1.41 nm and 192.12 nm, respectively. In this case, the average diameter was 5.11 nm and the polydispersity was 7.73%. The average diameter of this sample was greater than in the sample annealed at 300 °C, but it had a lower polydispersity. Furthermore, in the sample annealed at 500 °C (Figure 5 (c)), the smallest diameter was approximately 1.41 nm, and the largest was 132.05 nm. The average diameter was 5.92 nm and the polydispersity was 5.76%.

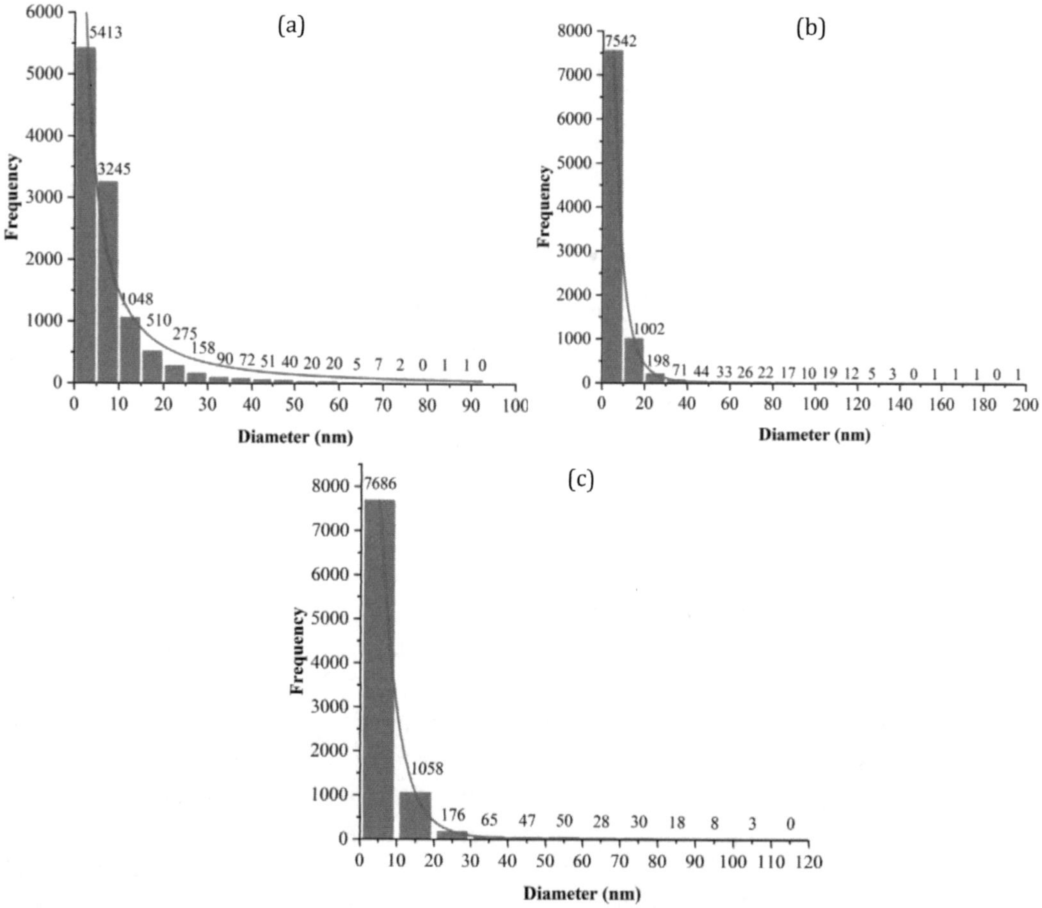

Figure 5. Histograms of the particle diameter distributions in the Ni nanocatalyst sample with a thickness of 50 nm, after annealing treatment for 4 hours at temperatures of (a) 300 °C, (b) 400 °C, and (c) 500 °C.

Although the sample annealed at a temperature of 500 °C had a larger average diameter than the other two samples, it had a lower polydispersity index of 5.76% compared with the samples annealed at temperatures of 300 °C and 400 °C, which had polydispersity indices of 18.82% and 7.73%, respectively. This indicates that the sample annealed at 500 °C had a high level of homogeneity. Thus, this result serves as one of the reference parameters for a good Ni catalyst sample to be used in the growth of CNTs. Catalysts with uniform grain size produce CNTs with consistent diameters, thus ensuring uniformity of the physical and electronic properties of CNTs [23,24].

Next, the elemental compositions of the Ni catalyst layers annealed at the three different temperatures are shown in Tables 4 and 5. Similar to what is shown in Tables 2 and 3, it can be observed that in addition to Ni, other elements such as Na, Si, O, Mg, and Ca, which are components of the glass substrate, were also identified. The observed composition varied with the annealing temperature.

Table 4. Results of EDS characterization for the mass percentage composition of elements in Ni catalyst layers with a thickness of 50 nm, after annealing at various temperatures for 4 hours.

	Mass (%)					
	Elements					
Annealing (°C)	O	Na	Mg	Si	Ca	Ni
300	6.38	5.58	1.84	44.20	7.59	34.41
400	6.76	5.07	1.89	38.85	6.56	40.87
500	6.34	5.76	2.05	41.19	7.37	37.30

Table 5. Results of EDS characterization for the atomic percentage composition of elements in Ni catalyst layers with a thickness of 50 nm, after annealing at various temperatures for 4 hours.

	Atomic (%)					
	Elements					
Annealing (°C)	O	Na	Mg	Si	Ca	Ni
300	13.01	7.92	2.45	51.33	6.18	19.12
400	14.26	7.43	2.62	46.68	5.52	23.49
500	13.14	8.30	2.79	48.61	6.10	21.06

The results of the EDS characterization for the nickel catalyst samples with a thickness of 50 nm that were annealed for 4 hours at temperatures of 300 °C, 400 °C, and 500 °C show variations in the elemental composition on the surface of the samples (Figure 6). For the sample annealed at 300 °C, the nickel mass percentage was recorded at 34.41%, and the atomic percentage was 19.12%. When the annealing temperature was increased to 400 °C, there was a significant increase in the nickel mass percentage to 40.87% and the atomic percentage to 23.49%. However, at 500 °C, the nickel mass percentage decreased to 37.3% and the atomic percentage also dropped to 21.06%.

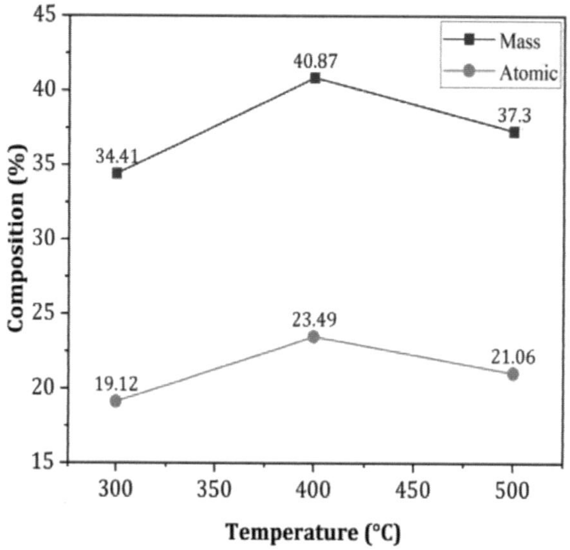

Figure 6. Relationship between annealing temperature and the mass or atomic composition of 50 nm layers of Ni catalyst.

The results show that increasing the annealing temperature to 400 °C enhances the concentration of nickel on the surface of the sample, which may be due to a reduction in contaminants or other elements as well as the diffusion of nickel from the interior of the material to the surface. However, at 500 °C, there is a decrease in nickel concentration, which could be attributed to several factors, including the oxidation of the sample under those conditions and the recrystallization process that leads to a more uniform distribution of nickel rather than a concentration at the surface. In other words, the sample that was annealed at a temperature of 500 °C had a high level of homogeneity, although its Ni content was slightly lower.

4. Conclusions

Experiments in which nickel catalysts were grown using the thermal evaporation method with four variations in thickness showed that the optimum thickness is 50 nm. The nickel catalyst produced under these conditions exhibits better morphology, smaller grains, and a higher level of uniformity, as shown by its lower polydispersity index. The mass and atomic percentage in the nickel sample also increased with the thickness of the layer on the substrate used during growth. The annealing treatment on the 50 nm thick sample achieved optimum conditions at an annealing temperature of 500 °C for 4 hours. The nickel catalyst obtained under these conditions also showed better morphology, with a small diameter of about 5.92 nm and improved uniformity, as evidenced by a low polydispersity index of about 5.76%. The smaller the polydispersity index of a sample, the more uniform its particles, indicating that the sample is of good quality for use in the growth of CNTs.

Acknowledgment

This research was supported by "Penelitian, Pengabdian kepada Masyarakat, dan Inovasi (PPMI)" in the fiscal year 2024. RR acknowledges the support from the Indonesia Endowment Funds for Education (LPDP), the Ministry of Finance of the Republic of Indonesia.

References

[1] Iijima S 1991 © 19 9 1 Nature Publishing Group 그라첼거 *Nature* **354** 56–8
[2] Sibiński M, Jakubowska M, Znajdek K, Słoma M and Guzowski B 2011 Carbon nanotube transparent conductive layers for solar cells applications *Opt. Appl.* **41** 375–81
[3] Ferguson A J, Blackburn J L and Kopidakis N 2013 Fullerenes and carbon nanotubes as acceptor materials in organic photovoltaics *Mater. Lett.* **90** 115–25
[4] Endo M, Strano M S and Ajayan P M 2008 Potential applications of carbon nanotubes *Top. Appl. Phys.* **111** 13–62
[5] Venkataraman A, Amadi E V, Chen Y and Papadopoulos C 2019 Carbon Nanotube Assembly and Integration for Applications *Nanoscale Res. Lett.* **14**
[6] Wei Y Y, Eres G, Merkulov V I and Lowndes D H 2001 Effect of catalyst film thickness on carbon nanotube growth by selective area chemical vapor deposition *Appl. Phys. Lett.* **78** 1394–6
[7] Han J hee, Kim H J, Yang M H, Yang C W, Yoo J B, Park C Y, Song Y H and Nam K S 2001 Effects of thickness of Ni layer deposited on glass substrate on the growth and emission properties of carbon nanotubes *Mater. Sci. Eng. C* **16** 65–8
[8] Lim Y D, Avramchuck A V, Grapov D, Tan C W, Tay B K, Aditya S and Labunov V 2017 Enhanced Carbon Nanotubes Growth Using Nickel/Ferrocene-Hybridized Catalyst *ACS Omega* **2** 6063–71
[9] Ding E X, Geng H Z, Mao L H, Wang W Y, Wang Y, Luo Z J, Wang J and Yang H J 2016 Recent research progress of carbon nanotube arrays prepared by plasma enhanced chemical vapor deposition method *Mater. Sci. Forum* **852** 308–14
[10] Veronese G P, Rizzoli R, Angelucci R, Cuffiani M, Malferrari L, Montanari A and Odorici F 2007 Effects of Ni catalyst-substrate interaction on carbon nanotubes growth by CVD *Phys. E Low-Dimensional Syst. Nanostructures* **37** 21–5
[11] Savale P A 2016 Physical Vapor Deposition (PVD) Methods for Synthesis of Thin Films: A Comparative Study *Sch. Res. Libr. Arch. Appl. Sci. Res.* **8** 1–8
[12] Chen X, Pang X and Fauteux-Lefebvre C 2023 The base versus tip growth mode of carbon nanotubes by catalytic hydrocarbon cracking: Review, challenges and opportunities *Carbon Trends* **12**
[13] Yesilbag Y O, Tuzluca Yesilbag F N, Huseyin A, Tuzluca M, Ismail I and Ertugrul M 2021 Synthesis and characterization of graphene/carbon nanotube hybrid: effects of Ni catalyst thickness and H2 flow rate on growth and morphological structure *J. Mater. Sci. Mater. Electron.* **32** 7943–55
[14] Huang Z P, Wang D Z, Wen J G, Sennett M, Gibson H and Ren Z F 2002 Effect of nickel, iron and cobalt on growth of aligned carbon nanotubes *Appl. Phys. A Mater. Sci. Process.* **74** 387–91
[15] Rao R, Sharma R, Abild-Pedersen F, Nørskov J K and Harutyunyan A R 2014 Insights into carbon nanotube nucleation: Cap formation governed by catalyst interfacial step flow *Sci. Rep.* **4** 1–6
[16] Nuayi A W, Noor F A, Sustini E, Irzaman and Winata T 2023 Effect of the annealing process on Ni catalyst film grown using the vacuum thermal evaporation method *Fullerenes Nanotub. Carbon Nanostructures* **0** 1–14
[17] Li H, Shi C, Du X, He C, Li J and Zhao N 2008 The influences of synthesis temperature and Ni catalyst on the growth of carbon nanotubes by chemical vapor deposition *Mater. Lett.* **62** 1472–5
[18] Li Y, Yang P, Xiao Y, Liu Y, Chen Y, Yang H, Feng H, Zhang J and Shu B 2024 Effect of carbon nanotube content and annealing temperature on corrosion performance of carbon nanotube/Ni composite layer *Mater. Res. Express* **11** 46503
[19] Pezone R, Vollebregt S, Sarro P M and Unnikrishnan S 2020 The influence of H2 and NH3 on catalyst nanoparticle formation and carbon nanotube growth *Carbon N. Y.* **170** 384–93
[20] Sengupta J and Jacob C 2010 The effect of Fe and Ni catalysts on the growth of multiwalled carbon nanotubes using chemical vapor deposition *J. Nanoparticle Res.* **12** 457–65
[21] Kostin V N, Perov V N, Mikhailov L V., Serbin E D and Vasilenko O N 2022 Magnetic Analysis of Recrystallization Processes during Annealing of Cold-Deformed Nickel *Russ. J. Nondestruct. Test.* **58** 992–9
[22] Yang C, Huang H, Thorogood G J, Jiang L, Ye X, Li Z and Zhou X 2016 The Effect of Grain Size and Dislocation

Journal of Physics: Conference Series **2980** (2025) 012002 doi:10.1088/1742-6596/2980/1/012002

Density on the Tensile Properties of Ni-SiCNP Composites During Annealing *J. Mater. Eng. Perform.* **25** 726–33

[23] Yang L, Zhao T, Jalil A, Yin Y, Shu Y and Jiang T 2024 Controlled growth of single-walled carbon nanotubes with narrow diameter distribution and high areal density on Mn-doped ZrO2-supported Fe catalyst *Ceram. Int.* **50** 43442–51

[24] Wang S G, Zhang Q, Zhong G F, Yoon S F, Ahn J and Yang D J 2003 The Effects of Catalyst Grain Sizes on the Diameter of MPECVD Grown Patterned Carbon Nanotubes *Mater. Res. Soc. Symp. - Proc.* **772** 69–74

Fabrication of Fe_3O_4/PVDF Magnetic Nanofiber Membranes for Magnetic Field Sensor

Nabila Putri Aulia[1], Lya Rizka Herawati[1], ST. Ulfawanti Intan Subadra[1], Arif Hidayat[1,*], Samian[2], Ahmad Taufiq[1]

[1]Department of Physics, Faculty of Mathematics and Natural Sciences, Universitas Negeri Malang, Indonesia
[2]Department of Physics, Faculty of Science and Technology, Universitas Airlangga, Indonesia

*E-mail: arif.hidayat.fmipa@um.ac.id

Abstract. This study investigates the fundamental performance of Fe_3O_4/Polyvinylidene Fluoride (PVDF) magnetic nanofiber membranes as a magnetic field sensor. The nanofiber membranes were fabricated using the electrospinning method. The success of nanofiber formation was confirmed by XRD and FTIR characterizations. The nanofiber had an average diameter of 544.7 nm. The results of the magnetic property analysis showed that the nanofiber was superparamagnetic with a saturation magnetization value of 6.1 emu/g. Furthermore, the sensor sensitivity was also evaluated with a value of 0.98 mV/mT. Interestingly, the sensor's sensitivity shows good and prospective performance, making it suitable for new magnetic sensor candidates in the future.

Keywords: Fe_3O_4/PVDF; membrane; nanofiber; sensor; magnetic field

Introduction

Magnetic field sensors have been widely used in many scientific and industrial applications, including biomedical detection [1], aviation industry [2], space research [3], and geophysics [2]. The methods that are widely developed as sensors today utilize the Hall effect [4], magneto-transistor [5], magneto diode [5], fluxgate [6], and magnetoresistive [7] to detect magnetic fields. However, these devices have some drawbacks related to power consumption, cost, miniaturization, low stability, low multiplexing capability, and long-distance detection. Another disadvantage is the sensitivity to several external magnetic field sources, which causes measurement interference and is susceptible to electromagnetic interference. Therefore, innovations related to the performance of magnetic field sensors need to be carried out. Considering the above problems, new materials are needed to overcome the weaknesses of nanoparticles. In this study, a nanofiber membrane was developed and applied as a magnetic field sensor device. The nanofiber structure allows the sensor to be small in size but has high sensitivity to changes in magnetic fields [8], thus enabling accurate detection. In addition, it also has resistance to electromagnetic interference [9], [10], high selectivity and sensor detection range [11], and has good mechanical properties compared to ferrite [12], making this technology suitable for use in various environments.

Nanofiber membranes were developed by combining Fe_3O_4 nanoparticles with polyvinylidene fluoride (PVDF) polymer. Fe_3O_4 nanoparticles have a magnetic saturation of 92 to 100 Am^3/kg, which is considered a permanent magnetic material [13]. Meanwhile, PVDF is a piezoelectric polymer with high elasticity and easy processing, making this material suitable for various technological applications, such as sensors [14]. The combination of Fe_3O_4/PVDF is possible to have high magnetization, high magnetic permeability, and good mechanical properties [15], so that this composite can provide a very high potential magnetoelectric (ME) effect for magnetic field sensor applications [16]. This multiferroic magnetoelectric (ME) material has received more attention because it is promising for multifunctional devices, one of which is a magnetic field sensor [17]. To create a magnetic field sensor with high sensitivity at a low cost, in this study, nanofiber membranes were synthesized by utilizing iron sand to produce Fe_3O_4 material. Fe_3O_4/PVDF magnetic nanofiber membrane synthesized using iron sand is expected to be a new sensor candidate in the future.

Method

In this study, natural sand was used as the main material for Fe_3O_4 synthesis. The supporting materials used were HCl solution (12 M, 37%) and NH_4OH solution (14 M, 25%). Furthermore, dimethylformamide (DMF), Polyvinylidene fluoride (PVDF), and Acetone were used as materials for membrane synthesis. Fabrication of Fe_3O_4/PVDF nanofiber membranes began by reacting iron sand with HCl and NH_4OH to form Fe_3O_4 nanoparticles. In detail, the steps for making Fe_3O_4 followed our previous study (18). Furthermore, the fabrication process of Fe_3O_4/PVDF membranes was carried out using electrospinning, where the PVDF solution obtained from mixing PVDF with DMF/acetone was added with Fe_3O_4 and followed by sonication using an ultrasonic bath for 10 minutes. Then, the mixed solution was put into a 10 mL syringe and shot into the collector to form a nanofiber membrane. Next, related to the characterization of membrane sensitivity, a light source is fired at the membrane subjected to a magnetic field. The light that penetrates the membrane will be received by an optical detector connected to a voltmeter. Furthermore, the recorded results will be converted into a measured voltage.

Results and Discussion

The diffraction pattern of Fe_3O_4/PVDF nanofiber is shown in Figure 1. The diffraction pattern is dominated by the PVDF peak at $2\theta = 20.2°$ which corresponds to the β (110) and β (200) planes, and at $2\theta = 36.1°$ which is assigned to the β(120) plane. This is in accordance with the results of the study by Chowdhury et al.; the diffraction pattern of Fe_3O_4 could not be directly identified, because the intensity of the Fe_3O_4 peak was not too high [18]. This result indicates that the nanofiber membrane mainly contains the β phase crystal structure of PVDF. This is because the higher concentration of PVDF results in the coexistence of the β phase due to the solute-solvent interaction during the electrospinning process [19]. Due to the high electric voltage applied during the electrospinning method, the random electric dipoles in the PVDF solution align, resulting in the formation of the β phase crystal structure which affects the increase in piezoelectric properties [18].

Journal of Physics: Conference Series **2980** (2025) 012003 doi:10.1088/1742-6596/2980/1/012003

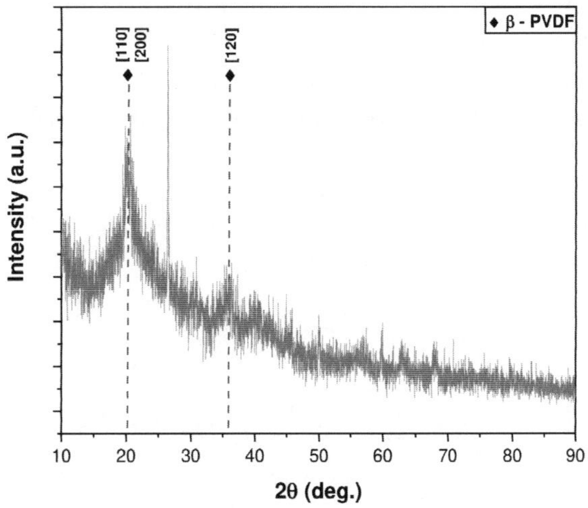

Figure 1. XRD Pattern of Fe_3O_4/PVDF Nanofiber Membrane

The FTIR spectrum of Fe_3O_4/PVDF nanofiber is shown in Figure 2. The formation of Fe_3O_4 nanoparticles can be indicated by the emergence of Fe-$O_{octahedral}$ and Fe-$O_{tetrahedral}$ functional groups at wave numbers of 428 cm^{-1} and 763 cm^{-1}, respectively. These results are in accordance with previous studies that reported that Fe-O bonds in octahedral sites appear at 451 cm^{-1} [20] and Fe-O in tetrahedral sites at 707 cm^{-1} [20]. In addition, the presence of PVDF was confirmed by C-H and C-F bonds. The bands at 612 cm^{-1} (CF_2) [18] and 979 cm^{-1} (CH) [18] were associated with the α phase of PVDF. The vibration bands at 842 cm^{-1} (CH_2) [18], [21], 887 cm^{-1} (CF_2), 1286 cm^{-1} (CF) [18], [22], 1417 cm^{-1} (CH_2) [18], [22] correspond to the β phase of PVDF. The symmetric and asymmetric vibrations of the CH_2 bond of PVDF correspond to the bands at 3020 cm^{-1} [18], [21] dan 2978 cm^{-1} [18], [21]. PVDF-DMF/Acetone is also a source of the stretching vibration of the amide group C=O at 1743 cm^{-1}[23].

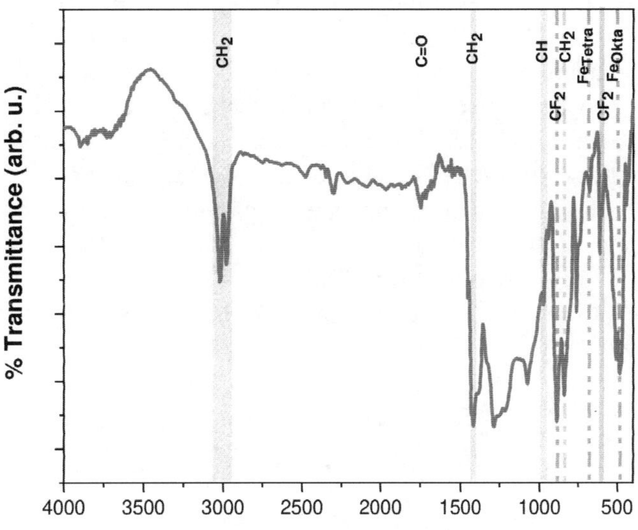

Figure 2. FTIR Spectrum of Fe_3O_4/PVDF Nanofiber Membrane

The magnetic properties of Fe_3O_4/PVDF nanofibers are presented with hysteresis curves in Figure 3. Qualitatively, all samples have an S-shaped hysteresis curve, indicating that Fe_3O_4/PVDF nanofibers exhibit superparamagnetic behaviour [24]. This is evidenced by the analysis results with a coercivity field value (Hc) approaching zero, which is 0.0187 ± 0.0003 T. Further quantitative analysis shows that the remanent magnetization value (Mr) and saturation magnetization value (Ms) are 0.0004 ± 0.0030 emu/g and 6.06 ± 0.01 emu/g. Because the dispersed nanoparticles may not be evenly distributed in the solution during the electrospinning process, resulting in a lower Ms value of the nanoparticles [25].

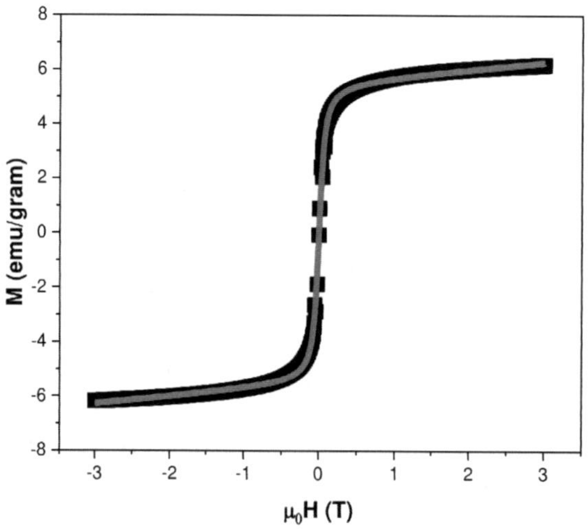

Figure 3. Hysteresis Curve of Fe_3O_4/PVDF Nanofiber Membrane

The surface morphology and average diameter distribution of Fe_3O_4/PVDF nanofiber membrane nanofiber are presented in Figures 4(a) and 4(b). Based on Figure 4(a), the surface of small fibers that form the membrane is filled with Fe_3O_4 spherical particles. The histogram in Figure 4(b) shows that the average diameter of the nanofiber is 544.7 ± 24.8 nm, with the presence of beads produced due to the low viscosity of the solution. This phenomenon is similar to the results of Zheng's study [25], where this condition causes imperfections in the nanofiber structure [26]. This condition occurs because the PVDF chain does not experience optimal stretching due to excess solvent collected in the collector, thus damaging the material formed [27]. Based on the EDX analysis data for the Fe_3O_4/PVDF nanofiber membrane presented in Table 1, it shows that the nanofiber contains carbon (C), fluorine (F), iron (Fe) and oxygen (O) elements, with the highest concentration of carbon (C) and fluorine (F), thus confirming that the membrane is dominated by PVDF [28], [29].

Journal of Physics: Conference Series **2980** (2025) 012003 doi:10.1088/1742-6596/2980/1/012003

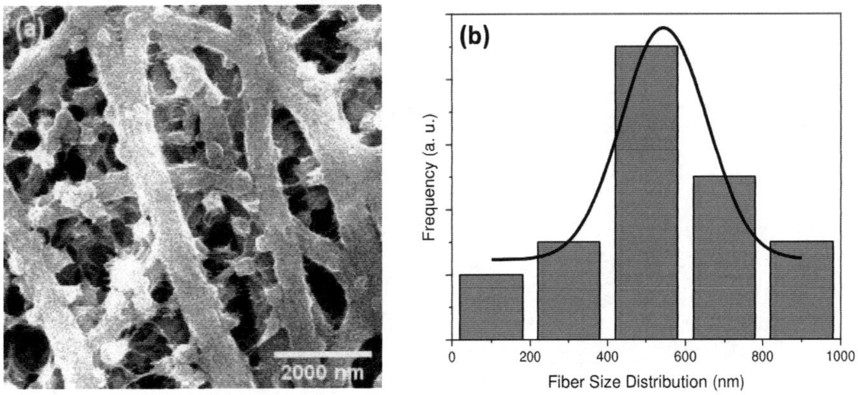

Figure 4. (a) Morphology and (b) Diameter Size of Fe_3O_4/PVDF Nanofiber Membrane

Table 1. Elemental Content of Fe_3O_4/PVDF Nanofiber Membranes

Element	Weight %
CK	51.8
OK	1.4
FK	45.4
FeK	1.4

The performance of the sensor was evaluated through its stability, sensitivity, and resolution. The sensor stability test was performed by measuring the output voltage for each sample every 30 seconds for 10 minutes by applying an external magnetic field at distances (4 mm, 14.5 mm and 27 mm) from the membrane, so that the sensor stability data was obtained as shown in Figure 5(a). Based on Figure 5(a), the stability measurements at distances of 4 mm, 14.5 mm, and 27 mm in quantitative analysis tended to be stable overall. This shows that the sensor has good accuracy and can provide consistent results over a long period of time and repeated use.

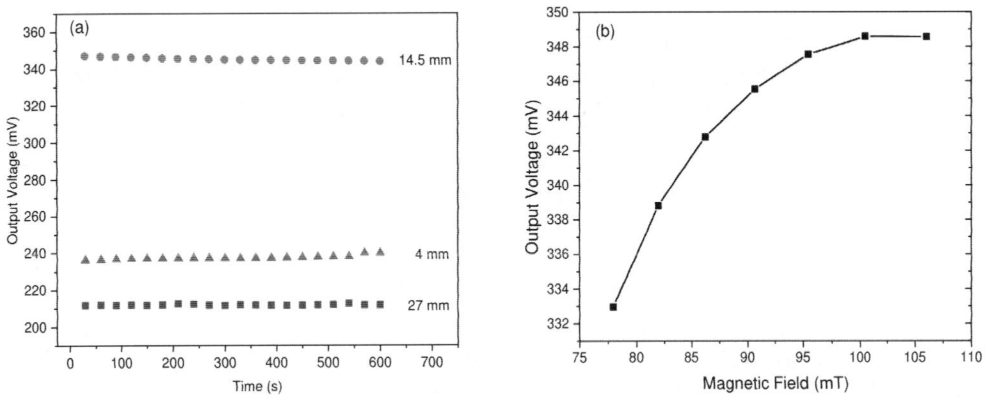

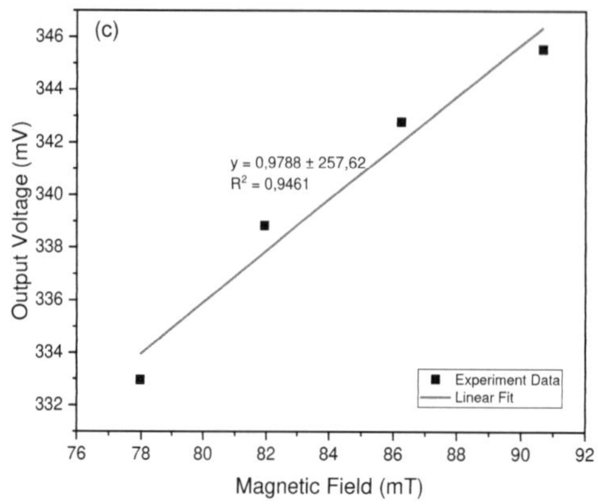

Figure 5. (a) Stability, (b) Range, (c) Sensitivity of Fe_3O_4/PVDF Nanofiber Membrane Sensor

Figure 5(b) shows the range of the sensor, which has a range of 78 mT-106 mT. Based on the fitting results between the the magnetic field and output voltage in Figure 5(b), the linear region of the sensor can be determined to determine the sensor sensitivity value. While for linearity follows Equation 1.

$$y = a + bx \qquad (1)$$

where b is representative of the sensor sensitivity value, x is the magnetic field. The sensor sensitivity is shown by the slope gradient of the graph in Figure 5(c). The sensor sensitivity in this study is 0.9788 mV/mT. The results of this sensor sensitivity are greater than the sensor sensitivity in previous studies, which is 0.28 mV/mT [30]. Next, another parameter analyzed is the sensor resolution. The sensor resolution is determined using Equation 2.

$$\text{Sensor Resolution} = \frac{\Delta V}{\Delta V/\Delta B} \qquad (2)$$

Where ΔV is $\frac{1}{2} NST\ Voltmeter$, $\Delta V/\Delta B$ is the sensor sensitivity. From the calculation, the sensor resolution was obtained as 0.051 mT; this result is smaller than the sensor resolution in previous research, which stated that the magnetic field resolution was 3.6 mT [30]. A good sensor is a sensor that has a small resolution value because a small resolution value indicates that the smaller the stimulus changes that can be detected by the sensor.

Conclusion

The synthesis of Fe_3O_4/PVDF nanofibers has been successfully carried out using the coprecipitation method. In terms of morphology, it shows that the fibers that make up the membrane are visible with an average nanofiber diameter of 544.7 ± 24.8 nm. In addition, the successful formation of nanofibers was confirmed by FTIR and XRD characterization. Although the diffraction pattern of Fe_3O_4/PVDF nanofibers is dominated by PVDF diffraction, the presence of Fe_3O_4 material is confirmed by Fe-Octahedral and Fe-Otetrahedral bonds in the FTIR characterization results. The sensor showed good performance with a sensitivity of 0.9788 mV/mT.

Acknowledgment

This research was funded by the Ministry of Education, Culture, Research and Technology through the 2024 PPS-PTM scheme with contract number 11.6.50/UN32.14.1/LT/2024.

References

[1] D. Murzin *et al.*, "Ultrasensitive Magnetic Field Sensors for Biomedical Applications," *Sensors*, vol. 20, no. 6, p. 1569, Mar. 2020, doi: 10.3390/s20061569.

[2] S. Tumanski, "Modern magnetic field sensors–a review," *Organ*, no. 10, pp. 1–12, 2013, [Online]. Available: http://www.red.pe.org.pl/articles/2013/10/1.pdf.

[3] J. Wu *et al.*, "Low temperature sensitive intensity-interrogated magnetic field sensor based on modal interference in thin-core fiber and magnetic fluid," *Appl. Phys. Lett.*, vol. 104, no. 25, Jun. 2014, doi: 10.1063/1.4884896.

[4] A. Roy, P. Sampathkumar, and P. S. Anil Kumar, "Development of a very high sensitivity magnetic field sensor based on planar Hall effect," *Meas. J. Int. Meas. Confed.*, vol. 156, p. 107590, 2020, doi: 10.1016/j.measurement.2020.107590.

[5] X. Yang, X. Zhao, Y. Bai, M. Lv, and D. Wen, "Two-Dimensional Magnetic Field Sensor Based on Silicon Magnetic Sensitive Transistors with Differential Structure," *Micromachines*, vol. 8, no. 4, p. 95, Mar. 2017, doi: 10.3390/mi8040095.

[6] M. Janosek, "Parallel Fluxgate Magnetometers," 2017, pp. 41–61.

[7] V. S. Luong *et al.*, "Magnetoresistive performances in exchange-biased spin valves and their roles in low-field magnetic sensing applications," *J. Sci. Adv. Mater. Devices*, vol. 3, no. 4, pp. 399–405, Dec. 2018, doi: 10.1016/j.jsamd.2018.09.004.

[8] L. Zhang, Y. Tang, and L. Tong, "Micro-/Nanofiber Optics: Merging Photonics and Material Science on Nanoscale for Advanced Sensing Technology," *iScience*, vol. 23, no. 1, p. 100810, Jan. 2020, doi: 10.1016/j.isci.2019.100810.

[9] T. Blachowicz, A. Hütten, and A. Ehrmann, "Electromagnetic Interference Shielding with Electrospun Nanofiber Mats—A Review of Production, Physical Properties and Performance," *Fibers*, vol. 10, no. 6, p. 47, May 2022, doi: 10.3390/fib10060047.

[10] D.-J. Kwon, I.-J. Kwon, J. Milam-Guerrero, S. B. Yang, J. H. Yeum, and H. H. Choi, "Aramid nanofiber-reinforced multilayer electromagnetic-interference (EMI) shielding composites with high interfacial durability," *Mater. Des.*, vol. 215, p. 110452, Mar. 2022, doi: 10.1016/j.matdes.2022.110452.

[11] J. Horne, L. McLoughlin, B. Bridgers, and E. K. Wujcik, "Recent developments in nanofiber-based sensors for disease detection, immunosensing, and monitoring," *Sensors and Actuators Reports*, vol. 2, no. 1, p. 100005, Nov. 2020, doi: 10.1016/j.snr.2020.100005.

[12] P. Saha, T. Debnath, S. Das, S. Chatterjee, and S. Sutradhar, "β-Phase improved Mn-Zn-Cu-ferrite-PVDF nanocomposite film: A metamaterial for enhanced microwave absorption," *Mater. Sci. Eng. B*, vol. 245, pp. 17–29, Jun. 2019, doi: 10.1016/j.mseb.2019.05.006.

[13] A. Bakhshi, M. Jalaly, and M. Vahedi, "The effect of GO–Fe_3O_4 hybrid coating on the magnetic field detection by a tapered optical fiber sensor," *Opt. Fiber Technol.*, vol. 74, p. 103134, 2022, doi: https://doi.org/10.1016/j.yofte.2022.103134.

[14] S. Guo, X. Duan, M. Xie, K. C. Aw, and Q. Xue, "Composites, fabrication and application of polyvinylidene fluoride for flexible electromechanical devices: A review," *Micromachines*, vol. 11, no. 12, pp. 1–29, 2020, doi: 10.3390/mi11121076.

[15] N. V. Lakshmi, P. Tambe, and B. Panda, "Surface modified iron oxide (Fe_3O_4) nanosheets reinforced PVDF nanocomposites: influence on morphology, thermal and magnetic properties," *Plast. Rubber Compos.*, vol. 51, no. 4, pp. 205–216, Apr. 2022, doi: 10.1080/14658011.2021.1966248.

[16] M. Bichurin *et al.*, "Magnetoelectric Magnetic Field Sensors: A Review," *Sensors*, vol. 21, no. 18, p. 6232, Sep. 2021, doi: 10.3390/s21186232.

[17] D. Viehland, M. Wuttig, J. McCord, and E. Quandt, "Magnetoelectric magnetic field sensors," *MRS Bull.*, vol. 43, no. 11, pp. 834–840, Nov. 2018, doi: 10.1557/mrs.2018.261.

[18] T. Chowdhury, N. D'Souza, and D. Berman, "Electrospun Fe_3O_4-PVDF Nanofiber Composite Mats for Cryogenic Magnetic Sensor Applications," *Textiles*, vol. 1, no. 2, pp. 227–238, Jul. 2021, doi: 10.3390/textiles1020011.

[19] I. Sriyanti *et al.*, "Case Studies in Chemical and Environmental Engineering Physicochemical and mechanical properties of polyvinylidene fluoride nanofiber membranes," *Case Stud. Chem. Environ. Eng.*, vol. 9, no. December 2023, p. 100588, 2024, doi: 10.1016/j.cscee.2023.100588.

[20] A. Taufiq *et al.*, "Synthesis of Fe_3O_4/Ag nanohybrid ferrofluids and their applications as antimicrobial and antifibrotic agents," *Heliyon*, vol. 6, no. 12, 2020, doi: 10.1016/j.heliyon.2020.e05813.

[21] M. A. Ramazanov, A. M. Maharramov, H. Shirinova, and L. Di Palma, "Structure and electrophysical properties of polyvinylidene fluoride (PVDF)/ magnetite nanocomposites," no. June 2019, 2018, doi: 10.1177/0892705718796542.

[22] Z. Zeng *et al.*, "Graphene Oxide Quantum Dots Covalently Functionalized PVDF Membrane with Significantly-Enhanced Bactericidal and Antibiofouling Performances," *Sci. Rep.*, vol. 6, no. February, pp. 1–11, 2016, doi: 10.1038/srep20142.

[23] Y. Liu *et al.*, "Free radical decay and micro-structure evolution of PVDF membranes in situ irradiated by synchrotron radiation X-ray," *Polym. Test.*, vol. 108, p. 107507, 2022, doi: 10.1016/j.polymertesting.2022.107507.

[24] G. Kirat, A. Erdoğan, and M. A. Aksan, "GMI-based biosensor for the detection and quantification of doxorubicin anticancer drugs labeled to Fe_3O_4 superparamagnetic nanoparticles," *Sensors Actuators A Phys.*, vol. 373, p. 115400, 2024, doi: https://doi.org/10.1016/j.sna.2024.115400.

[25] Z.-Q. Huang, F. Zheng, Z. Zhang, H.-T. Xu, and K.-M. Zhou, "The Performance of the PVDF-Fe_3O_4 Ultrafiltration Membrane and the Effect of a Parallel Magnetic Field Used during the Membrane Formation," *Desalination*, vol. 292, pp. 64–72, Apr. 2012, doi: 10.1016/j.desal.2012.02.010.

[26] Z. He, F. Rault, M. Lewandowski, E. Mohsenzadeh, and F. Salaün, "Electrospun PVDF nanofibers for piezoelectric applications: A review of the influence of electrospinning parameters on the β phase and crystallinity enhancement," *Polymers (Basel).*, vol. 13, no. 2, pp. 1–23, 2021, doi: 10.3390/polym13020174.

[27] G. C. Dias, T. S. P. Cellet, M. C. Santos, A. O. Sanches, and L. F. Malmonge, "PVDF nanofibers obtained by solution blow spinning with use of a commercial airbrush," *J. Polym. Res.*, vol. 26, no. 4, 2019, doi: 10.1007/s10965-019-1731-7.

[28] T. Chowdhury, N. D'souza, Y. H. Ho, N. Dahotre, and I. Mahbub, "Embedded corrosion sensing with zno-pvdf sensor textiles," *Sensors (Switzerland)*, vol. 20, no. 11, pp. 1–17, 2020, doi: 10.3390/s20113053.

[29] T. Chowdhury, N. D'Souza, and D. Berman, "Electrospun Fe_3O_4-PVDF Nanofiber Composite Mats for Cryogenic Magnetic Sensor Applications," *Textiles*, vol. 1, no. 2, pp. 227–238, 2021, doi: 10.3390/textiles1020011.

[30] A. S. Arifianto, "Magnetic field detection using fiber bundle and magnetic fluid as sensors," no. December 2019, pp. 1–5, 2020, doi: 10.1002/mop.32700.

Journal of Physics: Conference Series **2980** (2025) 012004 doi:10.1088/1742-6596/2980/1/012004

Optimization of Electrospinning Parameters Using Box-Behnken-based Design of Experiments (DoE) for the Fabrication of Ninjaflex/$CaCl_2$ Nanofibers as Potential Application in Flexible Sensors

Riris Sukowati[1,2], William Xaveriano Waresindo[1,2], and Dian Ahmad Hapidin[1*]

[1] Department of Physics, Faculty of Mathematics and Natural Sciences, Institut Teknologi Bandung, Bandung, Indonesia
[2] Doctoral Program of Physics, Faculty of Mathematics and Natural Sciences, Institut Teknologi Bandung, Bandung, Indonesia

*E-mail: dian.ahmad@itb.ac.id

Abstract. Motion monitoring sensors have recently garnered considerable research interest in wearable electronic devices and healthcare applications. Nanofiber-based electronic sensors show significant potential for such applications due to their flexibility. However, it is essential to optimize nanofiber formation parameters so that the resulting nanofiber has a minimum diameter that can increase the specific area to increase sensor sensitivity. This study employed the Box-Behnken-based Design of Experiments (DoE) to optimize the electrospinning process parameters for producing Ninjaflex/$CaCl_2$ nanofiber composites with controlled morphology. In this study, we carried out initial optimization with a 2^K factorial design on five parameters that influence fiber formation in the electrospinning process, such as concentration ratio, temperature, high voltage, needle-to-collector distance, and flow rate. Based on the initial optimization, the three most influential parameters are meticulously selected for further analysis using the Box-Behnken Design (BBD). The optimization analysis identified temperature as the most significant factor affecting the nanofiber diameter with p-value of 0.017. The optimal conditions for the minimum nanofiber diameter were a temperature of 50 °C, a Ninjaflex/$CaCl_2$ concentration ratio of 9:1, and a high voltage of 16,5 kV. The fabrication of Ninjaflex/$CaCl_2$ nanofiber composites was successfully achieved, demonstrating excellent elasticity and strength, highlighting their potential as flexible sensors. These results have significant implications for precisely controlling the morphological properties of Ninjaflex/$CaCl_2$ nanofiber composites, which are crucial for their application in flexible sensors.

Journal of Physics: Conference Series **2980** (2025) 012004 doi:10.1088/1742-6596/2980/1/012004

1. Introduction

The increasing commercialization prospects of various AI-based electronic products have spurred increasing interest in wearable electronic devices in health monitoring, sports training digitalization, and intelligent recognition [1]. This interest has led to various studies on the development of functional materials with the potential to become flexible sensors. These sensors play a crucial role in bridging the gap between artificial skin and human skin in terms of mechanical properties and detection functions, enabling the creation of electronic devices that can mimic the stretchability, flexibility, and other characteristics of human skin [2]. Materials with the potential to function as flexible sensors have been developed in various forms, including hydrogels [3], thin films [4], and nanofibers [5]. Nanofibers offer unique advantages such as a large specific surface area and high porosity which can significantly enhance sensor sensitivity [6]. Electrospun nanofiber membranes, with their high surface area-to-mass ratio, significant porosity, small pore sizes, flexibility, and outstanding mechanical properties, are poised to revolutionize numerous applications in vital technological fields. The most notable feature of the electrospinning process is its ability to produce nanofibers with diameters as small as tens of nanometers. These unique characteristics, combined with the reassuring simplicity and versatility of the technique, have instilled a sense of confidence and reassurance in the scientific community, leading to a surge in interest in electrospinning in recent years [7].

Nanofibers can be fabricated using various methods such as centrifugal jet spinning [8], needleless electrospinning [9], drawing [10], phase separation [11], self-assembly [12], and electrospinning [13,14]. These methods have different principles with their advantages and disadvantages. Electrospinning has received attention because of its various advantages, such as a simple synthesis process, providing controllable morphology, producing thin and homogeneous nanofibers, offering a continuous process, and being cost-effective compared to other methods [15]. By using the electrospinning method, homogeneous nanofibers with small diameters and controlled morphology can be formed.

Flexible sensors demand materials that can endure repeated deformation while retaining their functional and structural integrity. Ninjaflex, a thermoplastic polyurethane (TPU), stands out due to its superior elasticity and mechanical resilience [16]. Unlike commonly used materials such as polyvinylidene fluoride (PVDF) or polycaprolactone (PCL), which are valued for their piezoelectric and biocompatible properties [17], Ninjaflex (TPU) possesses outstanding mechanical properties, high elastotic, and flexibility, making it a promising candidate for flexible sensors when fabricated into nanofibers [18]. These properties enable it to endure bending, stretching, and compression, making it an ideal material for flexible sensors. As we strive to develop nanofibers with the potential to be flexible sensors, it is imperative to optimize their fabrication. This optimization is crucial in creating nanofibers with excellent mechanical properties and bead-free nanofiber morphology. However, for it to function as a flexible sensor, a material with good conductivity is essential. Therefore, a binding material, such as $CaCl_2$, is needed to enhance its conductivity. These unique properties present challenges for electrospinning and necessitate a customized optimization approach to achieve uniform

Journal of Physics: Conference Series **2980** (2025) 012004 doi:10.1088/1742-6596/2980/1/012004

nanofibers. This underscores the urgent need for parameter optimization in the fabrication of the Ninjaflex/$CaCl_2$ nanofiber composite [19].

The electrospinning process allows the optimization of various parameters such as concentration, temperature, flow rate, high voltage, and distance between the needle and the collector. Optimization of these parameters is necessary to produce high-quality nanofibers with a controllable minimum diameter. Research surface methodology design using Box-Behnken is usually used to optimize electrospinning parameters based on statistical analysis of several test variables. Unlike traditional one-variable-at-a-time methods or conventional optimization designs, RSM with Box-Behnken accommodates the complex analysis. Response Surface Methodology (RSM) integrates mathematical and statistical approaches, offering an effective tool for modeling and optimizing the influence of multiple independent variables on a given response [20]. The primary advantage of RSM lies in its ability to minimize the number of experimental runs while still providing statistically reliable results. Additionally, it can predict the mathematical relationship between influencing factors and the response value, identify the optimal experimental conditions, and enhance overall experimental efficiency [21,22].

Based on the various studies mentioned, no prior research has yet to focus on optimizing electrospinning parameters for fabricating Ninjaflex/$CaCl_2$ nanofibers with minimal diameters. Compared to previous studies on materials like PVDF and PCL, which often rely on empirical or less systematic optimization methods, our approach ensures a more efficient identification of optimal conditions to produce the minimum diameter of fibers. By addressing the specific interplay of parameters unique to Ninjaflex/$CaCl_2$ nanofiber, this study demonstrates the importance of material-specific optimization, contributing to advancements in developing electrospun nanofibers for flexible sensor applications.

In this research, we optimize five parameters affecting nanofiber formation using a 2^K Factorial design. The 2^K Factorial design is a highly efficient preliminary screening model that evaluates the impact of factors at two levels (upper and lower limits) to determine whether these factors significantly influence the response variable. By considering all combinations of factors at two levels, the 2^K factorial model effectively estimates the effect of each factor and their interactions. This method is statistically superior to a one-factor approach, which is often inefficient and fails to detect factor interactions [23,24].

Following the initial optimization, the three most influential parameters are meticulously selected for further analysis using the Box-Behnken Design (BBD). The Box-Behnken Design is employed to evaluate the coefficients in a quadratic mathematical model. Response Surface Methodology (RSM) is also applied to predict the optimal conditions for fabricating Ninjaflex/$CaCl_2$ nanofibers. Fifteen experiments are conducted with three levels for each variable, resulting in three variations of parameters. ANOVA is used to analyze the experimental data obtained from the Box-Behnken Design statistically. As a result, Ninjaflex/$CaCl_2$ nanofibers with the minimum possible diameters are produced, providing a high level of confidence in the results.

Journal of Physics: Conference Series **2980** (2025) 012004 doi:10.1088/1742-6596/2980/1/012004

2. Materials and Methods

2.1 Materials

Ninjaflex Filament 85A was purchased from Fenner Inc. (Stiegel St. USA). $CaCl_2$ analysis was purchased from Sigma Aldrich, Singapore. To dissolve the Ninjaflex/$CaCl_2$ Composite, the analytical solvent Dimethyl sulfoxide (DMSO) was used, which was purchased from Sigma Aldrich, Singapore.

2.2 Synthesis of Ninjaflex/$CaCl_2$ Composite Solution

The first step in nanofiber fabrication using the electrospinning technique involves preparing a Ninjaflex/$CaCl_2$ composite solution. Ninjaflex filament and $CaCl_2$ are initially ground into small pieces. The mixture of Ninjaflex and $CaCl_2$ is then placed in a sealed beaker at weight ratios of 9:1, 10:1, and 11:1. The next critical step is dissolving the Ninjaflex/$CaCl_2$ composite in Dimethyl sulfoxide (DMSO) for 18 hours, using a magnetic stirrer (ILMI-MS101, Indonesia) at 25°C. The magnetic stirrer ensures that a homogeneous Ninjaflex/$CaCl_2$ composite solution is achieved.

2.3 Ninjaflex/$CaCl_2$ Composite Nanofiber Fabrication

The homogenized Ninjaflex/$CaCl_2$ composite solution was then processed into nanofiber using the electrospinning method with the ILMI-N101 Electrospinning device (ITB, Indonesia), as shown in Figure 1. Initial parameter optimization was carried out using a 2^K factorial design for the five main parameters in the electrospinning process: concentration ratio, temperature, high voltage, flow rate, and distance between the needle and collector. Details of the experiments are shown in Table 1. After initial optimization using a 2^K factorial design, the three most influential factors were selected for further optimization using a Box-Behnken design to determine the most significant parameters. Details of the optimized parameters are shown in Table 2. The resulting Ninjaflex/$CaCl_2$ composite nanofibers were collected on a substrate, a crucial element placed on top of the collector. The electrospinning process was carried out at a humidity of 45% RH. The accumulated nanofibers were dried due to the electrospinning process. The preparations were then observed and analyzed using a digital microscope to determine the morphology and fiber diameter.

2.4 Parameter optimization using 2^K factorial design

Initial screening was carried out by optimizing the 2^K factorial design. Parameters optimized include Ninjaflex/$CaCl_2$ concentration ratio, high voltage, temperature, distance from needle to collector, and flow rate. The 2^K factorial design is a simple initial screening model for factor variables that only relies on two levels (upper and lower limits) with the aim of finding out whether the factor has a significant effect on the response variable. A detailed table of optimization parameters is shown in Table 1. To determine the three parameters that had the most influence on nanofiber formation, 8 experiments were carried out. The results of these experiments were then analyzed to identify the parameters that had the most significant impact on fiber morphology.

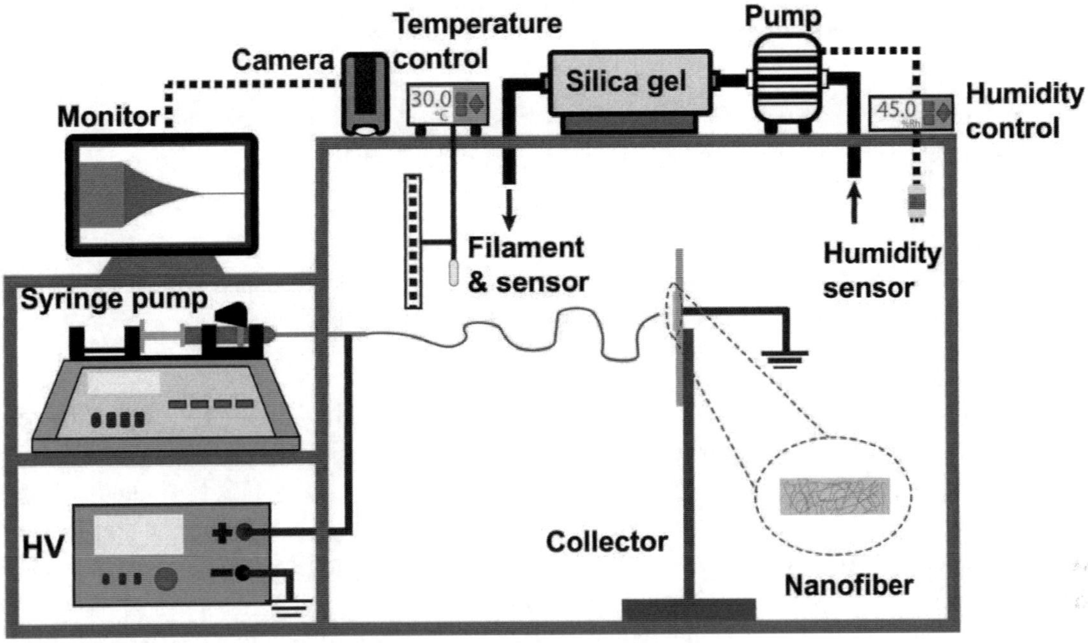

Figure 1. Schematic diagram of electrospinning apparatus.

Tabel 1. 2^K factorial design experiment.

Sample	High voltage (kV)	Flowrate (ml/h)	Distance (cm)	Concentration ratio (w/w)	Temperature (°C)
S1	16.5	0.75	15	11:1	50
S2	16.5	1.25	15	9:1	50
S3	16.5	1.25	19	9:1	30
S4	18.5	1.25	15	11:1	30
S4	16.5	0.75	19	11:1	30
S6	18.5	0.75	19	9:1	50
S7	18.5	1.25	19	11:1	50
S8	18.5	0.75	15	9:1	30

2.5 Parameter optimization Box Behnken design

After determining the three most influential parameters through optimization using a 2^K factorial design, the next step is to use a Box-Behnken design to determine the most significant factors in fiber formation. The optimized parameters were the Ninjaflex/$CaCl_2$ concentration ratio, high voltage, and chamber temperature. Optimization was carried out with 15 experiments, where each variable had three value levels, as presented in Table 2.

Tabel 2. Independent variable and levels used in the Box-Behnken Design.

Variable	Variable code	Level		
		-1	0	1
Concentration ratio (w/w)	X_1	9:1	10:1	11:1
Voltage (kV)	X_2	16.5	17.5	18.5
Temperature (°C)	X_3	30	40	50

2.6 Characterization

The morphology and diameter of the nanofibers were examined using an optical microscope (AmScope MU1000, USA). The average diameters of the fibers and beads were determined through image analysis using ImageJ software (version 64-bit Java 1.8.0_172, National Institutes of Health, USA). To ensure representative sampling, one hundred measurement points were randomly selected across the fibers. The experimental data were then statistically analyzed using Analysis of Variance (ANOVA) via Minitab 20.4 software, which provided p-values to assess the statistical significance of the independent variables on the dependent variables during characterization.

3. Result and Discussion

3.1 Optimization using 2^K factorial design

Optimization of the 2^K factorial design was carried out through eight experiments varying the parameters of concentration ratio, high voltage, flow rate, needle-to-collector distance, and temperature. Figure 2 shows the results of nanofiber images from a digital microscope for all eight samples. The resulting nanofibers did not show any beads in all samples and have continuous fibers. The results of the nanofiber diameter distribution analysis can be seen in Table 3.

Table 3. 2^K factorial design table and output response.

Runing order	High voltage (kV)	Flowrate (ml/h)	Distance (cm)	Concentration ratio (w/w)	Temperature (°C)	Nanofiber diameter (nm)
S1	16.5	0.75	15	11:1	50	470.50
S2	16.5	1.25	15	9:1	50	401.50
S3	16.5	1.25	19	9:1	30	565.00
S4	18.5	1.25	15	11:1	30	592.00
S4	16.5	0.75	19	11:1	30	595.51
S6	18.5	0.75	19	9:1	50	336.50
S7	18.5	1.25	19	11:1	50	401.00
S8	18.5	0.75	15	9:1	30	520.50

Journal of Physics: Conference Series **2980** (2025) 012004
doi:10.1088/1742-6596/2980/1/012004

Based on the response values measured in Table 3, statistical analysis is then carried out using analyze factorial design. The results of this process will identify three of the five process parameter variables that have the most significant influence on the response variable with a confidence level of 95% ($p < 0.05$). The results of statistical testing of the 2^K factorial design are presented in Figure 3 and Table 4.

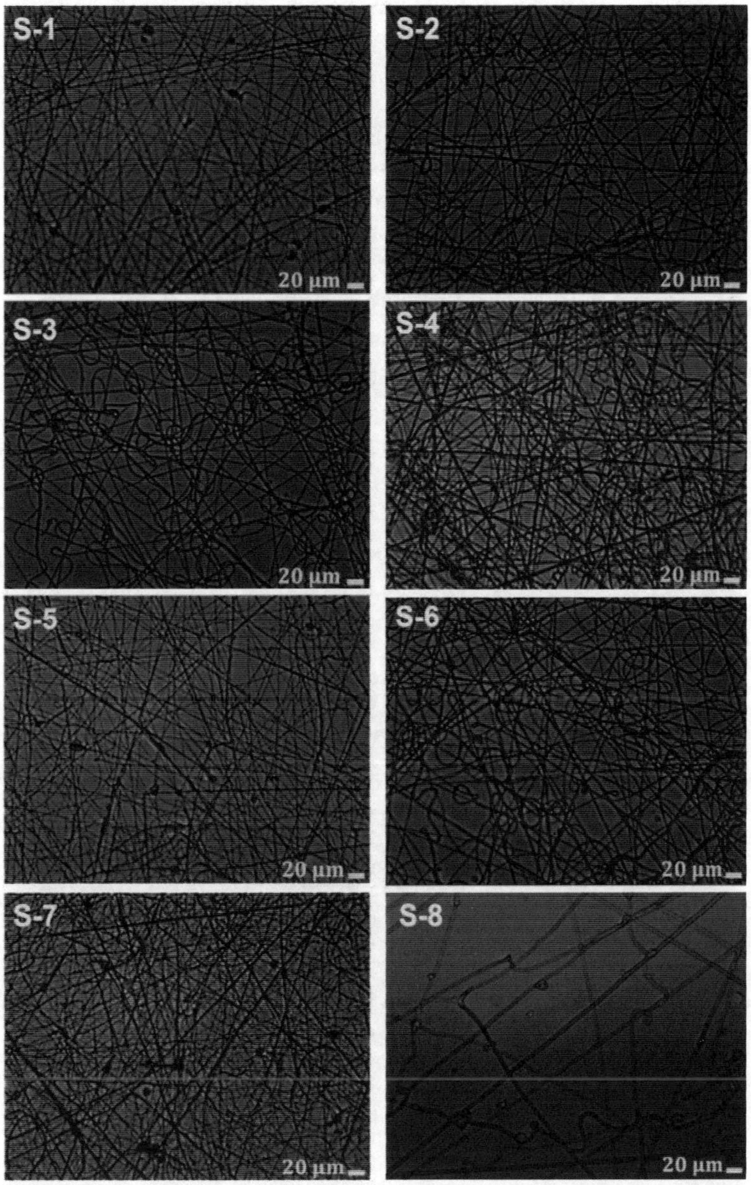

Figure 2. Microscope digital image of Ninjaflex/$CaCl_2$ nanofibers for sampel S1-S8 in initial optimization usin 2^K factorial design.

Table 4. Codded coefficient result.

Term	Effect	Coefficient	T-Value	P-Value	VIF
Constant		485.31	99.24	0.000	99.24
High voltage	-45.6	-22.81	-4.67	0.043	-4.67
Flowrate	9.12	4.56	0.93	0.449	-0.93
Distance	-21.62	-10.81	-2.21	0.158	-2.21
Concentration ratio	58.88	29.44	6.02	0.027	6.02
Temperature	-165.88	-82.94	-16.96	0.003	-16.96

Table 4 shows the results of regression analysis, which identifies the influence of factors on fiber diameter using the Response Surface Methodology (RSM) model. Based on the P-Value values, the three main factors that significantly influence fiber diameter are high tension (P-Value = 0.043), concentration ratio (P-Value = 0.027), and temperature (P-Value = 0.003), with all P-values being less than 0.05. The coefficient value shows that high voltage and temperature have a negative influence on fiber diameter, while concentration ratio has a positive impact. High voltage has an effect of -45.63, while temperature has the most significant impact of -165.88, confirming that increasing temperature significantly reduces fiber diameter. On the other hand, flow rate and distance did not show a significant effect on fiber diameter because the P-value values were each greater than 0.05, namely 0.449 and 0.158. All variables have a VIF of 1.00, indicating no multicollinearity problems in the model.

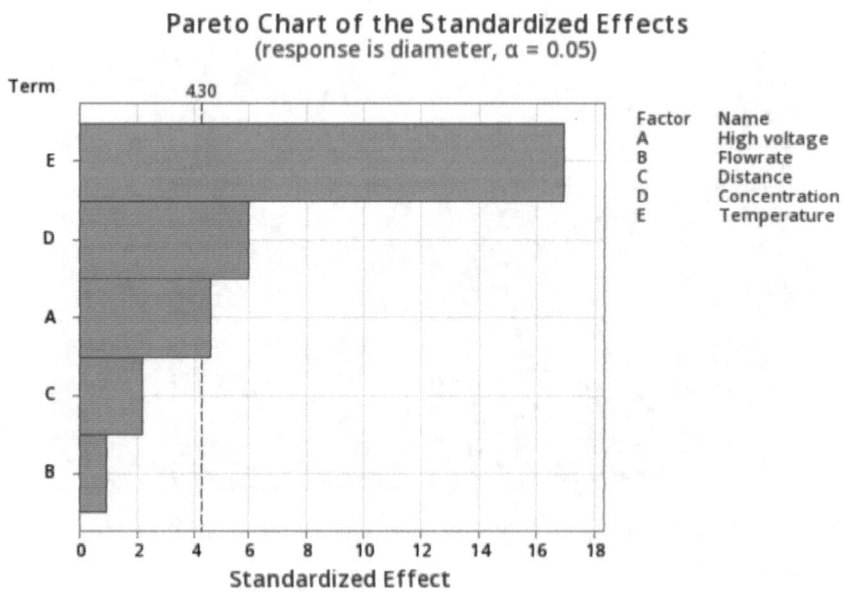

Figure 3. Illustration of the Pareto chart of the standardized effects for response to nanofiber diameter.

Figure 3 shows a Pareto graph of the effect. It shows that temperature (Temperature, E) has the biggest effect on fiber diameter, with a standardized effect that is much bigger than the critical significance line of 4.30 (α = 0.05). Other factors also significantly influence concentration ratio (concentration, D) and high voltage (high voltage, A), each exceeding the significance threshold value. Factors such as distance (Distance, C) and flow rate (Flowrate, B) show more minor effects, which are not statistically significant at the 95% confidence level because they are below the cutoff line. This analysis indicates that the third variables temperature, concentration, and high voltage dominate the diameter response, while distance and flow rate do not significantly contribute. These results are consistent with previous T-value and P-value findings, which confirm the influence of temperature and concentration ratio as the main factors in the model.

3.2 Optimization using Box-behnken design

Previous optimization results show that concentration ratio, high voltage, and temperature are the main factors influencing the formation of nanofibers via the electrospinning method. Figure 4 displays the morphology of Ninjaflex/$CaCl_2$ nanofibers from samples R1 to R15. The fibers have a uniform structure. Table 5 displays comprehensive information about the parameters employed and the diameter sizes of the fibers. The results of the optimization study were analyzed using Analysis of Variance (ANOVA). The ANOVA results showing the relationship between nanofiber diameter and electrospinning parameters are presented in Table 6. This research indicates that temperature has the most significant influence on nanofiber diameter, with a p-value of 0.017. Statistically, a p-value of less than 0.05 indicates a very significant effect (α=0.05). On the other hand, the concentration ratio has a p-value of 0.408, and the voltage shows a p-value of 0.909. This indicates that after further optimization, the ratio of $CaCl_2$ concentration and high voltage does not significantly influence fiber diameter because the p-value is greater than 0.05. Further analysis regarding the relationship between parameters and nanofiber diameter was carried out using the contour plot shown in Figure 5. This plot visually depicts the interaction between the parameters tested and their influence on fiber diameter.

Figure 5 (a) shows a contour plot illustrating the relationship between temperature and the concentration ratio on the diameter of the nanofibers. The concentration ratio is expressed as whole numbers for input into the Minitab software (e.g., a concentration value of 10 corresponds to a ratio of 9:1, 11 corresponds to 10:1, and 12 corresponds to 11:1). Where the diameter range is visualized through color gradation and the high voltage kept constant at 17,5 kV. Based on the contour pattern, it can be seen that the diameter tends to decrease with increasing temperature. Increasing temperature consistently shows a decrease in diameter. On the other hand, increasing concentration weakens with increasing diameter, as reflected in the darker color gradient at higher concentrations. This contour pattern also shows that temperature has a dominant effect on diameter compared to concentration, which is in line with the statistical analysis results, which show that the temperature is the only significant variable in the model. Overall, this plot strengthens the finding that temperature control is significant in optimizing the diameter of $CaCl_2$ nanofibers. Temperature plays a critical role in reducing solution viscosity by enhancing the polymer chain mobility, which facilitates jet stretching and results in smaller fiber diameters [25].

Journal of Physics: Conference Series **2980** (2025) 012004 doi:10.1088/1742-6596/2980/1/012004

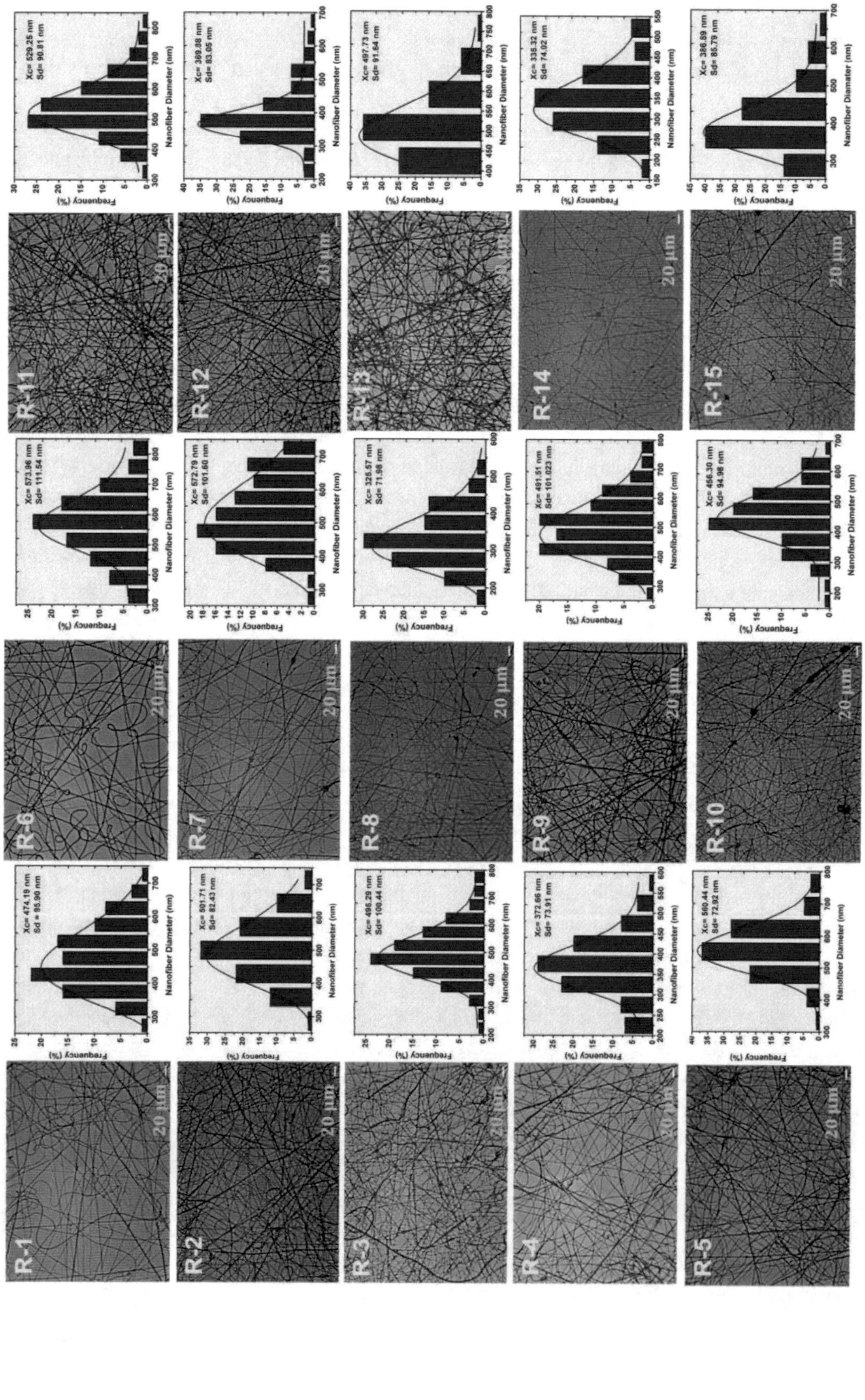

Figure 4. Microscope digital image of Ninjaflex/$CaCl_2$ nanofibers with diameter distribution graph.

Journal of Physics: Conference Series **2980** (2025) 012004 doi:10.1088/1742-6596/2980/1/012004

Table 5. Box-Behnken design table and output response.

Sample	Concentration ratio (wt.%)	Voltage (kV)	temperature (°C)	Nanofiber Diameter (nm)
R1	10:1	17.5	40	474.195
R2	10:1	17.5	40	501.713
R3	11:1	16.5	40	495.298
R4	10:1	17.5	40	372.66
R5	11:1	17.5	30	560.44
R6	10:1	18.5	30	573.967
R7	11:1	18.5	40	572.794
R8	10:1	18.5	50	325.572
R9	9:1	16.5	40	491.511
R10	9:1	17.5	50	456.3
R11	9:1	17.5	30	529.256
R12	9:1	18.5	40	369.881
R13	10:1	16.5	30	497.737
R14	10:1	16.5	50	335.324
R15	11:1	17.5	50	386.392

Table 6. ANOVA analysis of nanofiber diameter.

Source	DF	F-Value	P-Value
Model	9	2.03	0.225
Linear	3	4.44	0.071
X_1	1	12.49	0.017
X_2	1	0.81	0.408
X_3	1	0.01	0.909
Square	3	0.56	0.666
X_1^2	1	0.05	0.828
X_2^2	1	1.46	0.281
X_3^2	1	0.06	0.813
2-Way Interaction	3	1.10	0.430
X_1X_2	1	0.59	0.477
X_1X_3	1	0.43	0.542
X_2X_3	1	2.29	0.191

Journal of Physics: Conference Series **2980** (2025) 012004 doi:10.1088/1742-6596/2980/1/012004

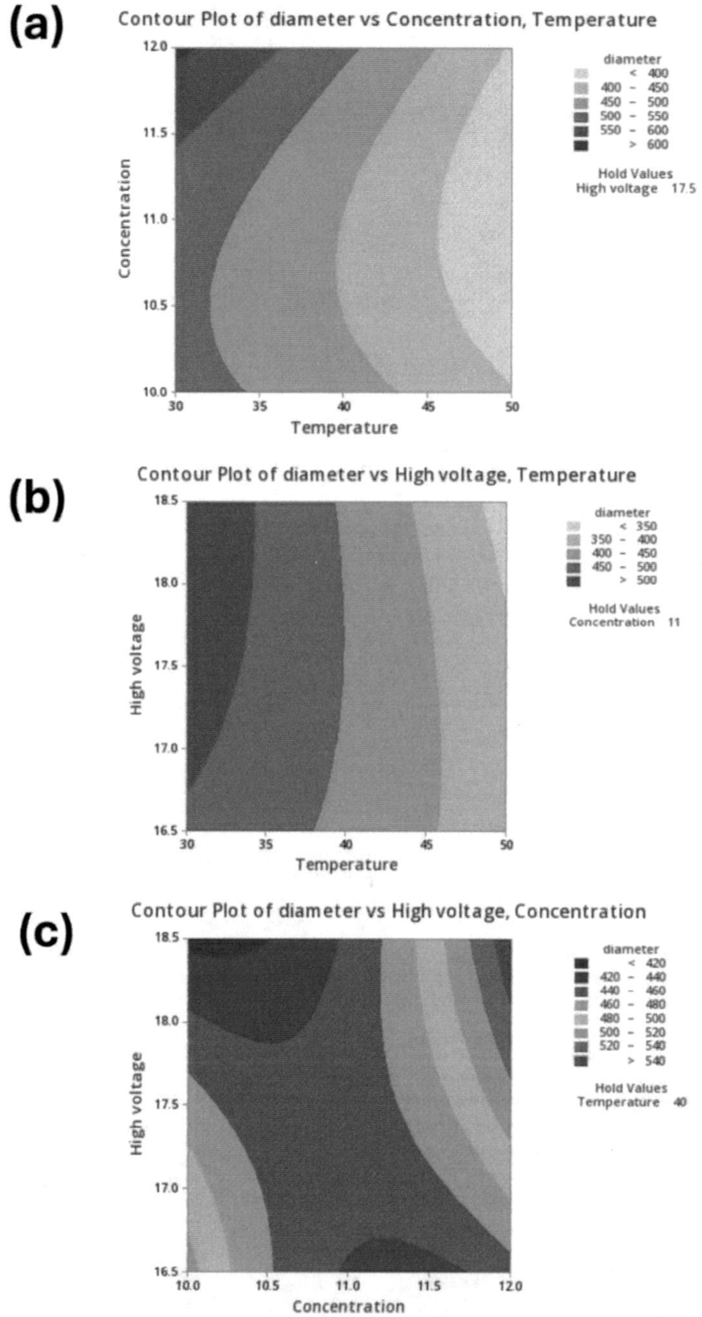

Figure 5. Contour plot of (a) Nanofiber diameter vs concentration ratio of Ninjaflex/$CaCl_2$ and temperature (b) Nanofiber diameter vs high voltage and temperature (c) nanofiber diameter vs high voltage and concentration ratio of Ninjaflex/$CaCl_2$.

Furthermore, Figure 5 (b) shows a contour plot depicting the relationship between diameter nanofiber, high voltage, and temperature, with the concentration ratio kept constant at 10:1. In this graph, the lower the temperature and high voltage, the larger the diameter of the Ninjaflex/$CaCl_2$ nanofiber formed, as seen from the darker color (>500 nm). Conversely, when the temperature increases to 50 °C and the voltage reaches about 18.5 kV, the Ninjaflex/$CaCl_2$ nanofiber diameter decreases to below 350 nm, marked by a lighter color. This pattern indicates that temperature significantly affects diameter formation, where increasing temperature is associated with a decrease in Ninjaflex/$CaCl_2$ nanofiber diameter.

In addition, Figure 5 (c) shows the contour plot of the relationship between nanofiber diameter with high voltage and the concentration ratio at a fixed temperature value of 40 °C. The colors on the plot indicate the distribution of diameter values with classified intervals from less than 420 nm to more than 540 nm. The dark blue area represents smaller diameters (<420 nm), while the dark green area represents larger diameters (>540 nm). The diameter tends to increase with increasing concentration ratio, while the effect of high voltage appears to be more complex with non-linear diameter variations at various concentration levels. In the electrospinning process, increasing the high voltage strengthens the electric field, enhancing the stretching and thinning of the polymer jet, which generally reduces the fiber diameter. However, this effect is highly dependent on the concentration of the polymer solution.

At lower concentrations, the polymer jet is more responsive to changes in voltage, leading to significant variations in fiber diameter. Conversely, at higher concentrations, the increased viscosity of the solution resists the stretching force, resulting in larger fiber diameters regardless of the applied voltage. At high concentrations, increased polymer chain entanglement can limit the stretching of the jet, reducing the impact of high voltage on diameter reduction [26,27]. The polymer solution concentration is a critical factor influencing nanofiber diameter during the electrospinning process. At low concentrations, the solution viscosity is insufficient to provide the molecular cohesion required for stable polymer jet formation. This often results in bead formation instead of uniform nanofibers. As the concentration increases, the viscosity also rises, enabling more stable jet formation and the production of smoother, uniform fibers. However, excessively high concentrations lead to very high viscosity, which resists the stretching forces of the electric field and results in larger fiber diameters [28,29]. This analysis suggests a relationship between concentration and high voltage on diameter changes.

Based on statistical analysis using Box-Behnken design, it was found that the results of the response optimization of the diameter of Ninjaflex $CaCl_2$ nanofibers aimed to minimize its value. The lower limit of the diameter was set at 325.572 nm, and the upper limit was 573.967 nm. Optimal conditions can be achieved at a combination of temperature of 50°C, the concentration ratio of 9:1 and high voltage of 16.5, with a predicted diameter of 356.853 nm. The multiple response prediction showed an estimated diameter of 356.9 nm with a 95% confidence interval (CI) for the diameter ranging from 265.2 nm to 448.5 nm, while the 95% prediction interval (PI) ranged from 191.3 to 522.5. These findings indicate that the optimized parameter configuration can produce a diameter close to the minimum target with a high level of accuracy and confidence in accordance with the limits set in the analysis.

4. Conclusion

The fabrication of Ninjaflex/$CaCl_2$ nanofiber composites was successfully carried out with homogeneous, bead-free, and continuous nanofibers. This study identified the optimal parameters for producing the smallest diameter Ninjaflex/$CaCl_2$ nanofibers at a temperature of 50 °C with a Ninjaflex/$CaCl_2$ concentration ratio of 9:1 and a high voltage of 16.5 kV. The small-diameter Ninjaflex/$CaCl_2$ nanofibers provided a large specific surface area, which offers significant advantages in their application as flexible sensors. Statistical analysis using ANOVA revealed that temperature had the most significant effect on the nanofiber diameter size. Although the experimental focus of this study is on optimizing electrospinning parameters, the inherent flexibility of Ninjaflex strongly supports its potential as a material for flexible sensors. Previous studies have validated the mechanical strength and elasticity of Ninjaflex and similar TPU-based materials, which can endure significant strain without structural degradation. These findings have important implications for producing Ninjaflex/$CaCl_2$ nanofiber composites with precise control over their morphological properties, which is crucial for flexible sensor applications.

Acknowledgements

Riris Sukowati & William Xaveriano Waresindo gratefully acknowledges the Indonesian Endowment Fund for Education (LPDP) for the scholarship.

References

[1] Cui T, Qiao Y, Li D, Huang X, Yang L, Yan A, Chen Z, Xu J, Tan X, Jian J, Li Z, Ji S, Liu H, Yang Y, Zhang X and Ren T-L 2023 Multifunctional, breathable MXene-PU mesh electronic skin for wearable intelligent 12-lead ECG monitoring system *Chemical Engineering Journal* **455** 140690

[2] Wu G, Wu L, Zhang H, Wang X, Xiang M, Teng Y, Xu Z, Lv F, Huang Z, Lin Y, Zhao Y, Zhao H, Xue C and Gao L 2024 Research progress of screen-printed flexible pressure sensor *Sens Actuators A Phys* **374** 115512

[3] Yang C, Liu J, Liu P, Wang W, Chen H, Bai L, Yang H, Yang L and Wei D 2024 Phytic acid extracted cellulose nanocrystals for designing self-healing and anti-freezing hydrogels' flexible sensor *Chemical Engineering Journal* 152276

[4] Li B, Sui N, Li M, Gu W, Yang W, Xu W and Zhao J 2024 High-sensitivity and energy-efficient chloride ion sensors based on flexible printed carbon nanotube thin-film transistors for wearable electronics *Talanta* **276** 126285

[5] Lin Q, Chen S, Huang Y, Wang H, Wu Y, Jin Y, Liu R, Shi Q and Liang Y 2024 High performance multifunctional piezoelectric PAN/UiO-66-NO2/MXene composite nanofibers for flexible touch sensor *Polymer (Guildf)* **304** 127162

[6] Sukowati R, Rohman Y M, Agung B H, Hapidin D A, Damayanti H and Khairurrijal K 2023 An investigation of the influence of nanofibers morphology on the performance of QCM-based ethanol vapor sensor utilizing polyvinylpyrrolidone nanofibers active layer *Sens Actuators B Chem* **386** 133708

[7] Lim C T 2017 Nanofiber technology: current status and emerging developments *Prog Polym Sci* **70** 1–17

[8] Stojanovska E, Kurtulus M, Abdelgawad A, Candan Z and Kilic A 2018 Developing lignin-based bio-nanofibers by centrifugal spinning technique *Int J Biol Macromol* **113** 98–105

[9] Kouhi M, Mobasheri M and Valipouri A 2023 Chapter 7 - Needleless electrospinning *Electrospun and Nanofibrous Membranes* ed A Kargari, T Matsuura and M M A Shirazi (Elsevier) pp 145–71

[10] Islam M S, Ang B C, Andriyana A and Afifi A M 2019 A review on fabrication of nanofibers via electrospinning and their applications *SN Appl Sci* **1** 1248

[11] Ji X, Li R, Liu G, Jia W, Sun M, Liu Y, Luo Y and Cheng Z 2021 Phase separation-based electrospun Janus nanofibers loaded with Rana chensinensis skin peptides/silver nanoparticles for wound healing *Mater Des* **207** 109864

[12] Rolandi M and Rolandi R 2014 Self-assembled chitin nanofibers and applications *Adv Colloid Interface Sci* **207** 216–22

[13] Rohman Y M, Sukowati R, Priyanto A, Hapidin D A, Edikresnha D and Khairurrijal K 2023 Quartz Crystal Microbalance Coated with Polyacrylonitrile/Nickel Nanofibers for High-Performance Methanol Gas Detection *ACS Omega* **8** 13342–51

[14] Sukowati R, Rohman Y M, Agung B H, Hapidin D A, Damayanti H and Khairurrijal K 2023 An investigation of the influence of nanofibers morphology on the performance of QCM-based ethanol vapor sensor utilizing polyvinylpyrrolidone nanofibers active layer *Sens Actuators B Chem* **386** 133708

[15] Zhang X, Ru Z, Sun Y, Zhang M, Wang J, Ge M, Liu H, Wu S, Cao C, Ren X, Mi J and Feng Y 2022 Recent advances in applications for air pollutants purification and perspectives of electrospun nanofibers *J Clean Prod* **378** 134567

[16] Alhazov D, Gradys A, Sajkiewicz P, Arinstein A and Zussman E 2013 Thermo-mechanical behavior of electrospun thermoplastic polyurethane nanofibers *Eur Polym J* **49** 3851–6

[17] Shehata N, Nair R, Boualayan R, Kandas I, Masrani A, Elnabawy E, Omran N, Gamal M and Hassanin A H 2022 Stretchable nanofibers of polyvinylidenefluoride (PVDF)/thermoplastic polyurethane (TPU) nanocomposite to support piezoelectric response via mechanical elasticity *Sci Rep* **12** 8335

[18] Wang K, Engelbrecht K and Bahl C R H 2023 Additive manufactured thermoplastic elastomers for low-stress driven elastocaloric cooling *Appl Mater Today* **30** 101711

[19] Wang K, Wu J Y, Wang R Z and Wang L W 2006 Effective thermal conductivity of expanded graphite-CaCl2 composite adsorbent for chemical adsorption chillers *Energy Convers Manag* **47** 1902–12

[20] Park J-Y, Shim W-G and Lee I-H 2011 Modeling and optimization of electrospun polyvinylacetate (PVAc) nanofibers by response surface methodology (RSM) *J Nanosci Nanotechnol* **11** 1359–63

[21] Wang C, Zuo Q, Wang L, Long B, Salleh K M, Anuar N I S and Zakaria S 2021 Diameter optimization of polyvinyl alcohol/sodium alginate fiber membranes using response surface methodology *Mater Chem Phys* **271** 124969

[22] Montgomery D C 2017 *Design and analysis of experiments* (John wiley & sons)

[23] Myers R H, Montgomery D C and Anderson-Cook C M 2016 *Response surface methodology: process and product optimization using designed experiments* (John Wiley & Sons)

[24] Smucker B, Krzywinski M and Altman N 2019 Two-level factorial experiments *Nat. Methods* **16** 211–2

[25] Huang Z-M, Zhang Y-Z, Kotaki M and Ramakrishna S 2003 A review on polymer nanofibers by electrospinning and their applications in nanocomposites *Compos Sci Technol* **63** 2223–53

[26] Bhardwaj N and Kundu S C 2010 Electrospinning: A fascinating fiber fabrication technique *Biotechnol Adv* **28** 325–47

[27] Pham Q P, Sharma U and Mikos A G 2006 Electrospinning of polymeric nanofibers for tissue engineering applications: a review *Tissue Eng* **12** 1197–211

[28] Li D and Xia Y 2004 Electrospinning of Nanofibers: Reinventing the Wheel? *Advanced Materials* **16** 1151–70

[29] Bhardwaj N and Kundu S C 2010 Electrospinning: A fascinating fiber fabrication technique *Biotechnol Adv* **28** 325–47

The Role of Collagen Addition on the Mechanical Properties of PVA/Chitosan Composite Nanofiber Membranes as Wound Dressing Materials

Ni Nyoman Rupiasih[1*], Dewa Ayu Sukma Pranastia[1], Made Sumadiyasa[1], I Made Sukadana[2], I Gede Eka Wiratnaya[3], Anung Syampurwadi[4] and Maykel Manawan[5]

[1] Department of Physics, Faculty of Mathematics and Natural Sciences, Udayana University, Bali 80361, Indonesia

[2] Department of Chemistry, Faculty of Mathematics and Natural Sciences, Udayana University, Bali 80361, Indonesia

[3] Faculty of Medicine, Udayana University, Bali 80361, Indonesia

[4] Research Center for Advanced Materials, National Research & Innovation Agency, Bandung Indonesia

[5] Faculty of Defense Technology, Indonesia, Defense University, Bogor 16810, Indonesia

*E-mail: rupiasih@unud.ac.id

Abstract. The increasing need for renewable natural resources in various applications such as in the medical field, a study has been conducted on using chitosan and collagen to prepare composite nanofiber membranes as wound dressing materials. The materials used were polyvinyl alcohol (PVA) solution of 10%, chitosan solution of 2%, and collagen solution added at concentrations of 2%, 4%, and 8%. The composite nanofiber membranes are manufactured by using an electrospinning technique with various PVA-chitosan-collagen ratios of 8:2:0, 8:2:2, 8:2:4, and 8:2:8. The names of the samples are P/Ch/C0, P/Ch/C2, P/Ch/C4, and P/Ch/C8, respectively. The membranes were characterized by FTIR, SEM, DMA (tensile strength), and swelling tests. The results indicated that collagen addition decreases the maximum tensile strength (UTS) and softens the membranes. The best membrane obtained was P/Ch/C2, with the average diameter and length being 145.55±8.64 nm and 24.19±1.03 μm respectively. Their mechanical properties are UTS of 1.52 MPa, Young's modulus (E) of 39.67 MPa, a percentage of elongation at break (BE) of 19.16%, and a percentage of swelling of 362%. It fulfills the standard of medical materials for a wound dressing application.

Keywords: PVA/chitosan, collagen solution, composite nanofiber membranes, electrospinning, wound dressing.

1. Introduction

Nanotechnology is the science and technology of engineering, designing, and manufacturing materials at the nanoscale. These materials are nanomaterials. Nanomaterials have excellent mechanical properties due to the surface area, volume, and quantum effects of nanoparticles [1,

2]. Based on their diameter, nanofibers can be considered nanomaterials because they geometrically fall into the group of one-dimensional nanoscale elements. Therefore, nanofibers have properties such as small pore size, high porosities and surface area, and surface morphologies that make them appropriate for biomedical uses [3].

Electrospinning is a technique for manufacturing nanofibers from almost any polymer solution under the effect of a high electrostatic field. The nanofibers have high surface areas and porosity, adjustable pore structure, and superior mechanical properties. These nanofibers can be made into materials for various functions and structural designs. Renewable natural polymers have advantages in their biodegradability and biocompatibility so it is widely used in the biomedical field [4]. The chitosan-based fibrous membranes made by the electrospinning technique have distinct characteristics such as high specific surface area and high porosity with multi-scale pore size distribution. Also, they have a three-dimensional (3D) porous structure that looks like the natural extracellular matrix, so they can be applied in wound healing [5].

Chitosan is a natural basic polysaccharide composed of N-acetyl glucosamine and it is derived from chitin [6]. Chitosan is a natural polymer that has good biocompatibility and biodegradability and antibacterial properties, so it is widely used to produce wound dressing materials [7]. However, natural polymers have disadvantages such as low mechanical strength and cannot be made into fibers. This can be overcome by adding synthetic polymers [7]. PVA is one of the polymers that can be added to improve the quality of chitosan nanofibers [8]. PVA is a synthetic polymer that is harmless, biodegradable, and good biocompatible properties [9].

Natural polymers such as collagen and chitosan have been widely useful as wound dressing materials [10]. Collagen is one of the proteins with a fiber-shaped structure [11]. Collagen has good functional characteristics in cell and tissue growth, so it is widely applied as a biomaterial for biotechnology and medicine [4]. Chitosan-collagen has great potential in the field of biomaterials, especially in pharmaceutical and medical applications. Currently, chitosan-collagen has been developed for antibacterial applications, wound dressings, and regenerative drugs [12].

Studies on chitosan blended with collagen have been conducted by many researchers. Thongchai et. al. [13] have studied the addition of collagen into chitosan blend film composites in physicochemical aspects of pharmaceutical materials. It was found that adding collagen caused the morphology of the composite to become rough, while the tensile strength decreased, and the elongation at break was slightly increased.

As mentioned above, chitosan modified with collagen has the advantage of biological properties as wound dressings. However, it is also important to study the impact of the variation in the composition on the mechanical properties and the swelling properties of the fibers. Therefore, this study aims to determine the impact of adding collagen on the mechanical properties of the PVA/chitosan nanofibers membranes and determine the best ratio of PVA/chitosan/collagen. It is expected to know the effect or the role of adding collagen with various compositions or certain ratios on the mechanical characteristics of PVA/chitosan/collagen nanofibers membranes, so it can be further developed for use in the biomedical field, especially as a material for wound dressings.

2. Materials and methods

2.1 Materials

The materials used were chitosan powder extracted from tiger prawn shells, PVA, acetic acid, and collagen powder extracted from fish. All materials were used directly without purification. Demineralized water (aqua DM) is used to prepare the solutions.

2.2 Preparation and electrospinning of PVA/chitosan/collagen composite nanofiber membranes
PVA solution (10% in aqua DM (w/v)), chitosan solution (2% in 2% acetic acid (w/v)), and collagen solution (2%, 4%, and 8% in 0.5 M acetic acid (w/v)) were prepared. PVA/chitosan/collagen composite nanofiber membranes with different ratios of volume PVA, chitosan, and collagen were denoted as P/Ch/C0 (8:2:0), P/Ch/C2 (8:2:2), P/Ch/C4 (8:2:4), and P/Ch/C8 (8:2:8).

The electrospinning process was performed on approximately 7 mL of sample solution for 3 h at room temperature and the fibers were collected on the collector wrapped with aluminum foil. The process was conducted with parameters of the distance between the spinneret tip and the collector was ± 10 cm, the flow rate of the solution was 1.00 μl/h, the relative humidity (RH) was 63.4%, and the voltage was 10 kV.

2.3 FTIR measurement
The nanofiber membranes were analyzed using FTIR (Shimadzu IRPrestige-21) spectrophotometer. The measurements were taken in transmittance mode and the background was taken before each measurement. For each sample, 100 scans were recorded within the range of wavenumber from 400 and 4000 cm^{-1}, at room temperature and a resolution of 4 cm^{-1}.

2.4 SEM measurement
Scanning electron microscopy (SEM, Inspect-S50, FEI) was used to investigate the surface morphology of nanofiber membrane samples. The size of the sample was ± 1 cm^2. The average lengths and diameters of fibers were obtained by measuring the lengths and diameters of 100 fibers that were randomly selected from SEM images (the SEM images obtained from different parts of samples) by using ImageJ and Origin software.

2.5 Mechanical measurement
The mechanical test of nanofiber was carried out using dynamic mechanical analysis (DMA) SDTA861e Mattler Toledo following the ASTM D 5023 standard method. The test was carried out at a constant speed of 0.5 N/s and at room temperature.

2.6 Cross-link process and swelling measurement
The nanofiber membranes were cross-linked to improve the water-resistance ability. In this study, a mixed solution of Acetone:2.5% Glutaraldehyde (GA):HCl (ratio of 9:1:0.01) was used to cross-link the nanofibers membranes. The identical sizes (2 × 2 cm^2) of the sample were exposed to the mixed solution for 4 h in an oven desiccator at room temperature. Samples were taken out from the solution, rinsed with distilled water, wiped off the excess water with paper towels, and weighed as the initial mass of the samples (m_0).

The swelling characteristic of the nanofiber membranes was carried out using a gravimetric technique at room temperature. The samples were soaked in distilled water and weighed at various times (5, 10, 15, 30, 60, 90, 120,150, 180, and 210 minutes (until the mass of samples was constant), and weighed each mass as wet samples (m). The swelling percentage (%) was calculated using Equation (1) [14, 15].

$$\text{Swelling percentage } (\%) = \frac{m-m_0}{m_0} x\ 100\% \qquad (1)$$

3. Results and discussion

3.1 FTIR characterization

The FTIR spectra of P/Ch/C0, P/Ch/C2, P/Ch/C4, and P/Ch/C8 membranes is illustrated in Figure 1. The spectra show that samples have an identical pattern.

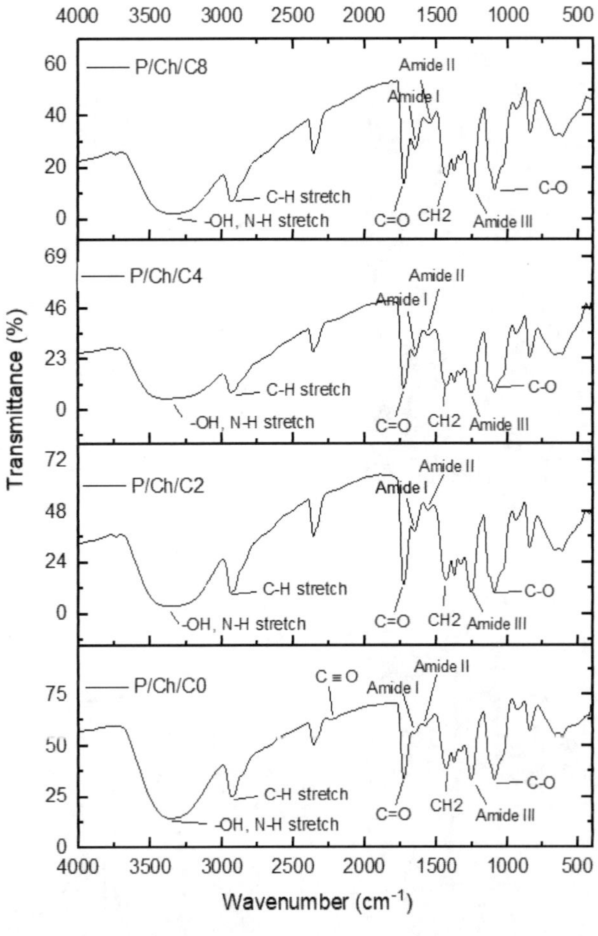

Figure 1. The FTIR spectra of the P/Ch/C0, P/Ch/C2, P/Ch/C4, and P/Ch/C8 nanofiber membranes.

Figure 1 shows that adding collagen changes the percentage of transmittance and shifts in wave numbers in certain functional groups. The -OH and N-H stretching groups in the P/Ch/C2 and P/Ch/C4 membranes experienced a shift to a larger wave number, while in the P/Ch/C8 membrane, it shifted to a smaller wave number. The C-H stretching group and amide I group experienced a shift to a larger wave number in the membrane with the addition of collagen, namely P/Ch/C2, P/Ch/C4, and P/Ch/C8, while the Amide III group experienced a shift to a higher wave number smaller. This shift shows the interaction between functional groups in PVA, chitosan, and collagen [16]. Meanwhile, in the C-O, C=O, CH_2, and Amide III groups in the P/Ch/C2, P/Ch/C4 membranes, and P/Ch/C8 do not shift. Figure 1 also shows that the N-H stretch, C-H stretch, -OH stretch, C-O, C=O, CH_2, Amide I, Amide II, and Amide III in P/Ch/C2, P/Ch/C4, and P/Ch/C8 experienced a decrease in transmittance, showing an increase in the absorption of these functional groups [17].

3.2 Morphological characterization

The development of wound dressing materials that look like skin tissue (extracellular matrix (ECM)) is very important in healing injured tissue [18]. Nanofibers with homogeneous fiber shapes, no beads, and solution droplets, are the main properties of an ideal wound dressing. The SEM images and the diameter distributions of P/Ch/C2, P/Ch/C4, and P/Ch/C8 membranes are shown in Figure 2.

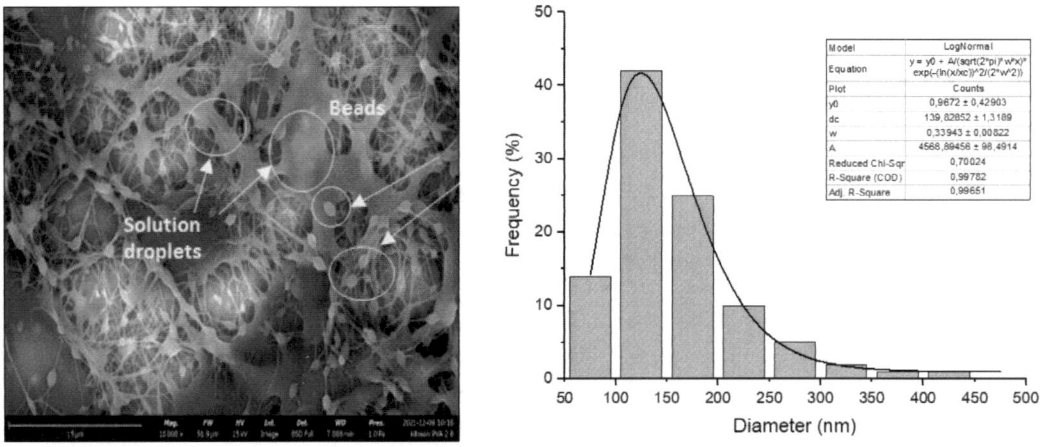

a)

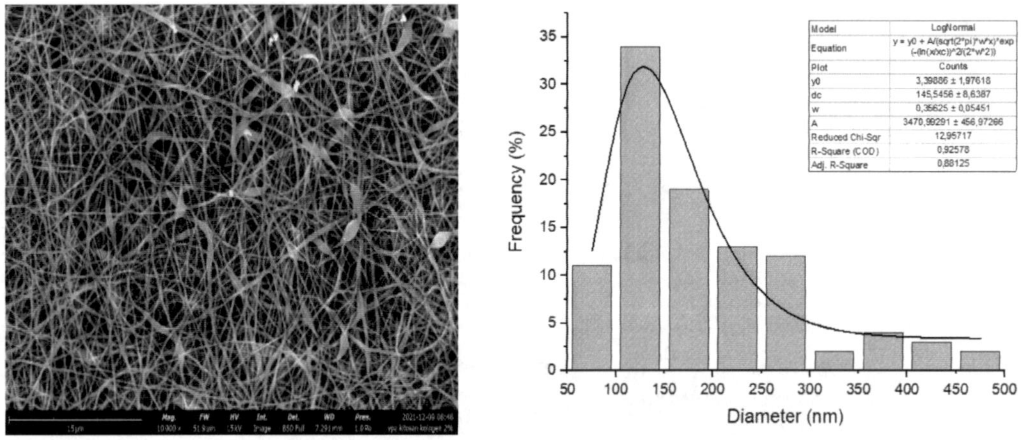

b)

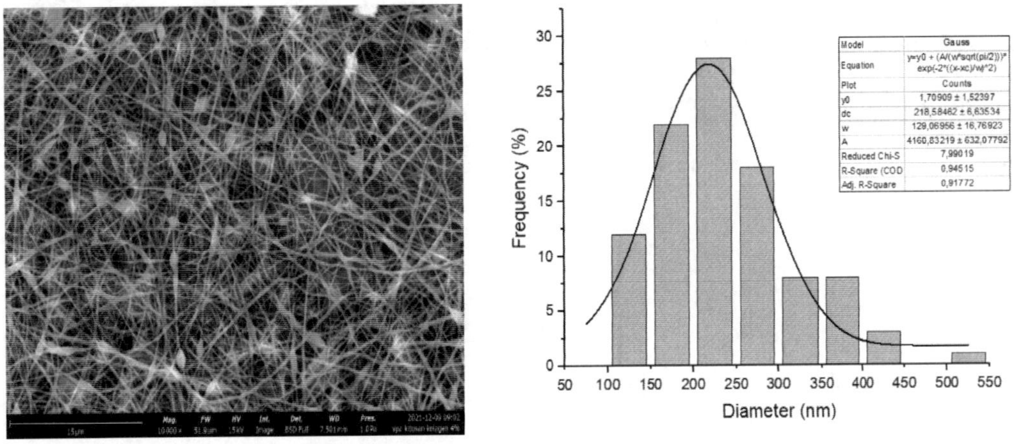

c)

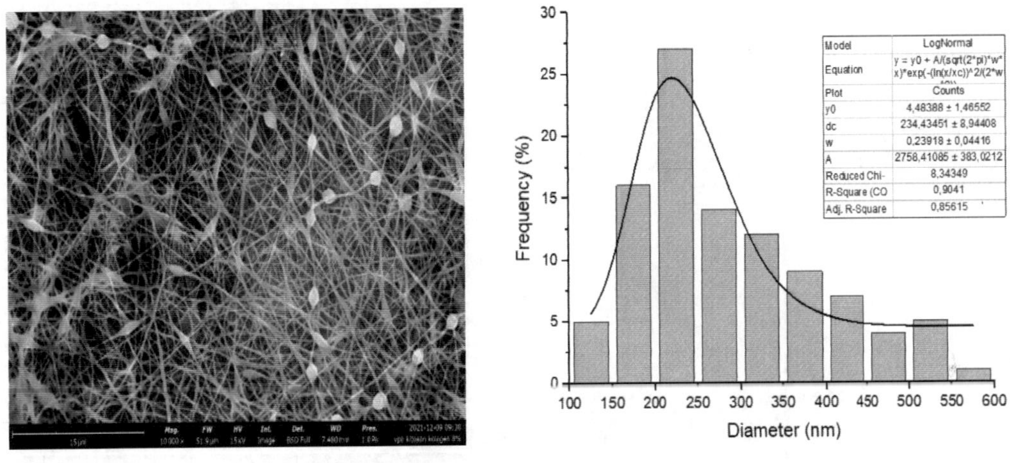

d)

Figure 2. The characterization of nanofiber membranes using SEM: a) P/Ch/C0, b) P/Ch/C2, c) P/Ch/C4, and d) P/Ch/C8.

In Figure 2, it can be seen that the P/Ch/C2 membrane has the best morphology compared to other membranes, where the number of solution droplets and beads observed is the least, while the P/Ch/C0 membrane has the most. In Figure 2(a), unlike other samples, the P/Ch/C0 sample appears to exhibit a fused fiber morphology, which forms a thin film structure. It is also observed that, in the P/Ch/C4 and P/Ch/C8 membranes, the number of beads observed increases with increasing collagen content. So the addition of 2% collagen can improve the morphology of the P/Ch/C0 membrane, in this case, the addition of collagen causes an increase in the strength of the chain bonds in the solution so that it can maintain the continuity of the solution flow during the electrospinning process [18]. Meanwhile, the addition of 4 and 8% collagen resulted in the solution concentration approaching the maximum limit, which affected the results of the formed nanofibers [19].

By using the ImageJ application, it obtained the average diameter (dc) and length (lc) of the nanofiber that are presented in Table 1. This shows that adding collagen causes an increase in the diameter and length of the P/Ch/C0 membrane fibers. As the content of collagen increased, there was an increase in the diameter of the fibers, but the length decreased. In this case, the fiber diameter is still within the range of nanofiber size made for wound dressings, around 50 to 300 nm [20]. So it can be said that collagen can increase the continuity of P/Ch/C0 nanofibers. It is also seen that the P/Ch/C2 membrane has the highest average length, which means it has the highest continuity. Thus, collagen can increase the diameter and improve the continuity of the fiber.

Table 1. The average diameter and average length of P/Ch/C0, P/Ch/C2, P/Ch/C4, and P/Ch/C8 nanofiber membranes.

No.	Nanofibers	The average diameter (nm)	The average length (μm)
1	P/Ch/C0	139.83±1.32	1.69±0.66
2	P/Ch/C2	145.55±8.64	24.19±1.03
3	P/Ch/C4	218.58±6.64	17.36±0.67
4	P/Ch/C8	234.43±8.94	13.39±0.72

3.3 Mechanical characterization

From the mechanical test, a stress-strain graph is obtained as shown in Figure 3. From the stress-strain graph, the values of ultimate tensile strength (UTS), elongation at break (BE) and Young's Modulus (E) are obtained as shown in Table 2.

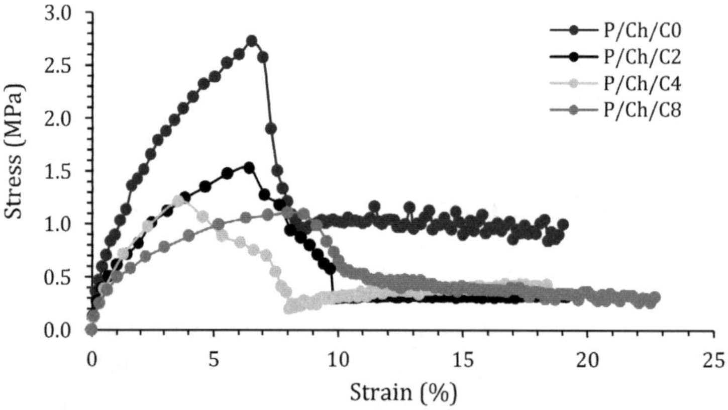

Figure 3. The stress-strain graph of P/Ch/C0, P/Ch/C2, P/Ch/C4, and P/Ch/C8 nanofiber membranes.

Table 2 shows that P/Ch/C0 nanofiber have the highest UTS and E compared to the others. The highest BE was shown by P/Ch/C8 nanofiber, while the UTS was the lowest. It also appears that the lowest BE and E were shown by P/Ch/C4 and P/Ch/C2, respectively. From these results it can be concluded that the added collagen concentration greatly affected the mechanical properties of the P/Ch/C0 nanofiber. The highest E indicates that the P/Ch/C0 nanofiber are stiffer compared to the others, so it can be said that the addition of collagen can increase the elasticity of the P/Ch/C0 nanofiber. The smallest E is shown by the P/Ch/C2 nanofiber, so it is the most elastic one. While the UTS decreases with increasing concentration of collagen added. The decrease in the UTS was due to the reduced amount of amino groups in the chitosan chain which serves to strengthen the nanofiber bonds as collagen is added [21]. However, the resulting UTS and BE are still within the standard range for medical materials, the human skin elongation varies between 17% and 207%, and for tensile strength between 1 and 32 MPa [22, 23]. Therefore, for the membrane with collagen that gives the highest UTS is P/Ch/C2. This is also strengthened by the fiber length of the P/Ch/C2 membrane, that this membrane has the longest fibers among all membranes, Table 1.

Table 2. The values of E, UTS, and BE of P/Ch/C0, P/Ch/C2, P/Ch/C4, and P/Ch/C8 nanofiber membranes.

Nanofiber Membranes	E (MPa)	UTS (MPa)	BE (%)
P/Ch/C0	106.19	2.73	19.03
P/Ch/C2	39.67	1.52	19.16
P/Ch/C4	53.71	1.22	18.38
P/Ch/C8	44.97	1.10	22.68

3.4 Cross-link and swelling characterizations

Figure 4 shows the swelling percentage as a function of time of the nanofiber membranes. The graph of the maximum swelling percentage of the nanofiber membranes is shown in Figure 5. Figure 4 shows that the percentage rate of swelling increased sharply in the first 5 minutes. This sharp increase can occur due to the influence of uneven collagen surface. Uneven and wavy surface areas can expand the contact surface with water and allow rapid hydration to occur [24]. The rate of swelling fluctuated until 90 minutes and stabilized slowly. After 120 minutes, it reached saturation, which means the nanofiber membranes could no longer absorb water. This is because the hydrogen interaction between collagen and chitosan has exceeded the maximum limit for binding water [25].

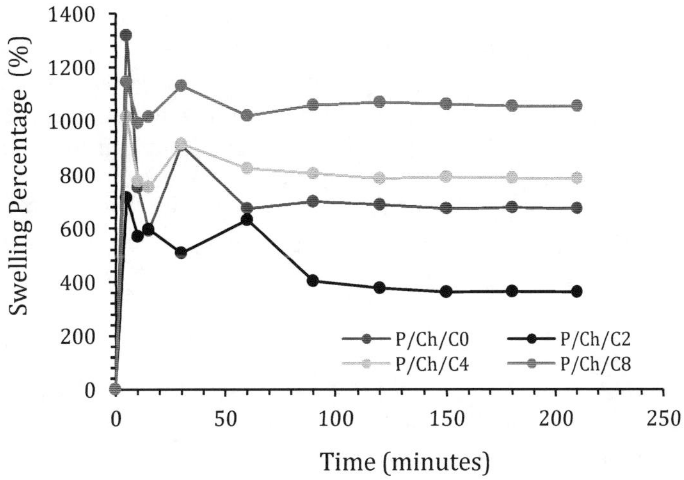

Figure 4. The graph of swelling percentage vs. time of P/Ch/C0, P/Ch/C2, P/Ch/C4, and P/Ch/C8 nanofiber membranes.

Journal of Physics: Conference Series **2980** (2025) 012005 doi:10.1088/1742-6596/2980/1/012005

Figure 5 shows that the 2% collagen (P/Ch/C2) reduced the swelling percentage of P/Ch/C0, where P/Ch/C2 had the smallest swelling percentage. Meanwhile, the tensile strength test showed that the P/Ch/C2 membrane had the smallest E value, which means it is the most elastic, so the swelling percentage should be high. This condition occurs due to the presence of collagen that inhibits the water absorption in the nanofiber membrane. Collagen can act as a modifier on the chitosan surface where the triple helix of collagen binds amino groups to the long chains of chitosan, thus the chitosan surface changing to be smaller. This causes the swelling of P/Ch/C2 smaller than that of P/Ch/C0 [21]. The addition of collagen concentrations of 4% and 8% can increase the percentage of swelling due to the presence of hydrophilic groups in collagen, such as -COOH, $-NH_2$, and -OH which have the property to bind water through hydrogen bonds [21].

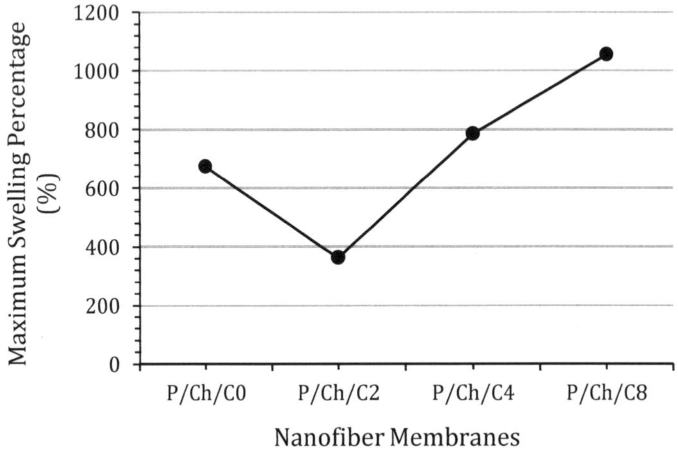

Figure 5. The maximum swelling percentage graph as a function of time of P/Ch/C0, P/Ch/C2, P/Ch/C4, and P/Ch/C8 nanofiber membranes.

4 Conclusion

In this study, the role of the addition of collagen on the mechanical properties of PVA/chitosan/collagen composite nanofibers as wound dressings has been investigated. It was found that the addition of collagen caused a decrease in maximum tensile strength and an increase in fiber elasticity (E). The ratio of PVA/chitosan/collagen that produces the best nanofiber is 8:2:2 (name P/Ch/C2) with the fiber morphology as follows, the number of beads and solution droplets is the least, the average diameter and length of the fiber is 145.55 ± 8.64 nm and 24.19 ± 1.03 μm respectively. The mechanical properties are UTS = 1.52 MPa, Young's modulus (E) = 39.67 MPa, percentage of elongation at break (BE) = 19.16%, and swelling percentage of 362%. It fulfills the standard of medical materials i.e. as a wound dressing application.

Acknowledgments

This study was financially supported by DIPA PNBP Udayana University budget-2024 by the Research Implementation Assignment Agreement B/255.517/UN14.4.A/PT.01.03/2024, dated April 17, 2024, highly acknowledged.

References

[1] Peng-Yu Zhuang, You-Liang Li, Li Fan, Jun Lin, Qiao-Ling Hu, Modification of chitosan membrane with poly(vinyl alcohol) and biocompatibility evaluation, International Journal of Biological Macromolecules 50 (2012) 658-663.

[2] Qiong Wu, Wei-shou Miao, Yi-du Zhang, Han-jun Gao, David Hui, Mechanical properties of nanomaterials: A review, Nanotechnol Rev. 9 (2020) 259-273.

[3] L.C. Henríquez, R.V. Zúñiga, J.P. Molina, J.V. Baudrit, Electrospun nanofibers: A nanotechnological approach for drug delivery and dissolution optimization in poorly water-soluble drugs, ADMET & DMPK. 8 (2020) 325-353.

[4] W.P. Lu, Y. Guo, Electrospinning of Collagen and Its Derivatives for Biomedical Applications, in: Tong Lin (Eds.), Novel Aspects of Nanofibers, IntechOpen, 2018, pp. 39-65.

[5] C. Cui, S. Sun, S. Wu, S. Chen, J. Ma, F. Zhou, Electrospun chitosan nanofibers for wound healing application, Engineered Regeneration 2 (2021) 82-90.

[6] T.Y. Kuo, C.F. Jhang, C.M. Lin, T.Y. Hsien, H.J. Hsieh, Fabrication and application of coaxial polyvinyl alcohol/chitosan nanofiber membranes, De Gruyter Open 15 (2017) 1004-1014.

[7] M.M. Gonçalvesi, J. Carneiro, B. Justus, J.T. Espinoza, J.M. Budel, P.V. Farago, J.P. de Paula, Preparation and characterization of a novel antimicrobial film dressing for wound healing application, Brazilian Journal of Pharmaceutical Sciences 56 (2020) e18784.

[8] Lele Chen, Zhao Liu, Jipeng Shi, Chengbo Wang, Ling Ding, Xu Ding, Guixiang Teng, Jianping Wu, Ji Zhang, Preparation and antibacterial properties of chitosan/polyvinyl alcohol nanofibrous mats using different organic acids as solvents, Process Biochemistry 122 (1) (2022) 13-28.

[9] S. Mirzaeei, S. Taghe, K.A. Addo, A. Nokhodchi, Polyvinyl Alcohol/Chitosan Single-Layered and Polyvinyl Alcohol/Chitosan/Eudragit RL100 Multi-layered Electrospun Nanofibers as an Ocular Matrix for the Controlled Release of Ofloxacin: an In Vitro and In Vivo Evaluation, AAPS PharmSciTech (2021) AAPS PharmSciTech 22 (2021) 170.

[10] D.J. Indrani, F. Lukitowati, Y. Yulizar, Preparation of Chitosan/Collagen Blend Membranes for Wound Dressing: A Study on FTIR Spectroscopy and Mechanical Properties, *IOP Conf. Ser.: Mater. Sci. Eng.* 202 (2017) 012020.

[11] N. T. Chinh, V. Q. Manh, V. Q. Trung, T.D. Lam, M.D. Huynh, N.Q. Tung, N. D. Trinh, T.Hoang, Characterization of collagen derived from tropical freshwater carp fish scale wastes and its amino acid sequence, Natural Product Communications 14 (2019) 1-12.

[12] Junde Chen, Jianying Li, Zhongbao Li, Ruizao Yi, Shenjia Shi, Kunyuan Wu, Yushuang Li, Sijia Wu, Physicochemical and Functional Properties of Type I Collagens in Red Stingray (Dasyatis akajei) Skin, Mar Drugs 17 (2019) 558.

[13] K. Thongchai, P. Chuysinuan, T. Thanyacharoen, S. Techasakul, S. Ummartyotin, Integration of collagen into chitosan blend film composites: physicochemical property aspects for pharmaceutical materials, Springer Nature Journal Applied Sciences 2 (2020) 255

[14] Romeo, A.; Kazsoki, A.; Omer, S.; Pinke, B.; Mészáros, L.; Musumeci, T.; Zelkó, R. Formulation and Characterization of Electrospun Nanofibers for Melatonin Ocular Delivery, Pharmaceutics 15 (2023) 1296. https://doi.org/10.3390/pharmaceutics15041296.

[15] Sung WonKo, JiYeonLee, Joshua Lee, Byeong Cheol Son, Se Rim Jang, Ludwig Erik Aguilar, Young Min Oh, Chan Hee Park, and Cheol Sang Kim. Analysis of Drug Release Behavior Utilizing the Swelling Characteristics of Cellulosic Nanofibers, Polymers 11 (2019) 1376.

[16] Sudirman, A.K. Karo, S.G. Sukaryo, K.D. Adistiana, K. Dahlan. Synthesis of Nanofiber from Poly Vinyl Alcohol (PVA)-Collagen using Electrospinning Methods, *Indonesian Journal of Applied Chemistry* 21 (2020) 55-65.

[17] D.L. Pavia, G.M. Lampman, G.S. Kriz, Introduction To Spectroscopy A Guide for Students of Organic Chemistry, third ed., Brooks/Cole Thomson Learning, United States, 2001.

[18] Li T, Sun M, Wu S. State-of-the-Art Review of Electrospun Gelatin-Based Nanofiber Dressings for Wound Healing Applications, Nanomaterials (Basel) 12(5) (2022):784.

[19] N.N. Rupiasih, M. Sumadiyasa, H. Suyanto, P. Windari, Utilization of Shrimp Skin Waste (Sea Lobster) As Raw Material for the Membrane Filtration, Journal of Physics: Conf. Series 846 (2017) 012003.

[20] Qi Q, Li R, Wang C, Hou G, Li C, Embryonic perfect repair inspired electrospun nanofibers dressing with temperature-sensitive and antibacterial properties for wound healing, Front. Microbiol. 14 (2023)1233559.

[21] C.O. García-Sifuentes, J.C. Zamorano-Apodaca, M. Martínez-Porchas, S.M. Scheuren-Acevedo, M.A. Mazorra-Manzano, Isolation and properties of collagen extracted from mixed by-products obtained from different fish species, Biotecnia 23 (2021) 109-116.

[22] Yuqing Tan, Sam K C Chang, Isolation and characterization of collagen extracted from channel catfish (Ictalurus punctatus) skin, Food Chemistry 242 (2018) 147-155.

[23] Tobin, D.J. Introduction to skin aging, J. Tissue Viability 26 (2017) 37–46.

[24] Putri Nurfriana, Prihartini Widiyanti, Djoni Izak Rudyardjo, Effect of Collagen-Chitosan-Glycerol Composition in Scaffold for Gingival Recession Therapy, Journal of Biomimetics, Biomaterials and Biomedical Engineering 40 (2019) 101-108.

[25] Jing Wang, Xinli Pei, Haiying Liu, Dan Zhou, Extraction and characterization of acid-soluble and pepsin-soluble collagen from skin of loach (Misgurnus anguillicaudatus), Int J Biol Macromol 106 (2018) 544-550.

Journal of Physics: Conference Series **2980** (2025) 012006
doi:10.1088/1742-6596/2980/1/012006

Interfacial properties improvement of AEC surfactant via Titanium dioxide nanoparticles

Taufan Marhaendrajana[1, 2, 3 *], Miftah Hidayat[1, 2, 3], Rima Megayanti[2], Ndaru Cahyaningtyas[2, 4], Mahruri Sanmurjana[2], Zeta Nur Muhammad Yahya[2]

[1] Department of Petroleum Engineering, Faculty of Mining and Petroleum Engineering, Institut Teknologi Bandung, Bandung, Indonesia
[2] Enhanced Oil Recovery Laboratory, Faculty of Mining and Petroleum Engineering, Institut Teknologi Bandung, Bandung, Indonesia
[3] Research Center for CO_2 and Flare Gas Utilization, Institut Teknologi Bandung, Bandung, Indonesia
[4] Department of Petroleum Engineering, Universitas Pembangunan Nasional "Veteran" Yogyakarta, Yogyakarta, Indonesia

*E-mail: tmarhaendrajana@itb.ac.id

Abstract. In this study, we explore the improvement of interfacial properties, specifically interfacial tension (IFT) and contact angle, of a surfactant solution composed of Alkyl Ethoxy Carboxylate (AEC) surfactant with the addition of titanium dioxide (TiO_2) nanoparticles. The integration of TiO_2 nanoparticles into the AEC surfactant is hypothesized to significantly improve the performance of the AEC surfactant in reducing IFT and altering wettability, which are critical parameters in Enhanced Oil Recovery (EOR) applications. Experimental investigations were conducted to measure the IFT between the AEC surfactant solution and crude oil, as well as the contact angle on a thin section of Berea sandstone before and after the addition of TiO_2 nanoparticles. The results show that the addition of TiO_2 nanoparticles had the potential to significantly reduce the IFT approximately by two orders of magnitude compared to the use of AEC surfactant only. On the other hand, incorporating the TiO_2 nanoparticles into the AEC surfactant only slightly changed the measured angle within the experimental uncertainty. These improvements suggest that the AEC surfactant, when combined with TiO_2 nanoparticles, can more effectively mobilize trapped oil within the sandstone reservoir by facilitating the detachment of oil droplets from rock surfaces and promoting their displacement. The findings imply that this modified surfactant system has the potential to enhance the efficiency of EOR processes, leading to increased oil recovery from mature reservoirs.

1. Introduction

Despite the shift towards renewable energy, fossil fuels, especially oil, still play an essential role in meeting Indonesia's energy needs. As most of the oil fields in Indonesia have already matured, the oil industry has turned to enhanced oil recovery (EOR) techniques to increase the oil production from the existing fields. Conventional chemical EOR methods, including surfactant [1-3] and polymer [4-6], become an alternative solution due to their ability to change the interfacial properties between crude oil and formation brine [7-9] improve sweep efficiency

[10,11] and hence, resulting in to recover the remaining trapped oil to secure domestic energy demands.

In recent years, incorporating nanoparticles, particularly titanium dioxide (TiO_2) nanoparticles, into surfactant solutions have been found to have the potential to enhance the performance of surfactants [12-14]. Therefore, this study investigates the enhancement of interfacial properties, particularly interfacial tension (IFT) and contact angle, in a surfactant solution consisting of alkyl ethoxy carboxylate (AEC) surfactant with various added concentrations of TiO_2 nanoparticles. The results obtained from this study are important as they become a foundation for the first step toward applying combined TiO_2 nanoparticles with AEC surfactant in existing matured oil fields.

2. Materials and methods.

2.1 Materials

Titanium dioxide (TiO_2) nanoparticles (a purity of ≥ 99% wt.) used in this study were sourced from XFNANO Material Tech, China. The internal structure of TiO_2 nanoparticles had already been characterized by a previous study by Megayanti et al. [15] as a spherical shape. Referring to the manufacture information, the average diameter of TiO_2 nanoparticles is 13.27 nm with surface area of 52.32 m^2/g.

An oil sample used in this study was obtained from one of the oil fields in Indonesia. The crude oil's API gravity is above 40 °API, which can be classified as light oil. A SARA (saturated-aromatics-resins-asphaltenes) analysis also shows that the crude oil sample is approximately composed of a saturated fraction (72 %), followed by aromatics (25 %), and the remaining 3% is a mixed composition of resins and asphaltenes.

An alkyl ethoxy carboxylate (AEC) surfactant was selected for its characteristics, which can stand up to a wide range of salinity and high temperatures [16,17]. This surfactant is classified as an anionic surfactant due to having the negative polar head component. In order to simulate the actual condition of subsurface formation brine, a synthetic brine of 8000 ppm NaCl was dissolved in demineralized water, which mimicked the actual salinity of the formation brine where the crude oil sample was sourced. The critical micelle concentration (CMC) of AEC surfactant on 8000 ppm NaCl brine and crude oil sample was obtained at the concentration of 1.25% wt. corresponding to the interfacial tension (IFT) value of 2.7×10^{-3} mN/m [15].

2.2 Methods

2.2.1 Interfacial tension measurements

A spinning drop tensiometer (TX-500D, supplied by KINO Scientific Instrument) was used to conduct the interfacial tension (IFT) measurement between crude oil and AEC surfactant with and without the addition of TiO_2 nanoparticles. The measurement was conducted at an elevated temperature of 68.3°C, similar to the actual temperature of subsurface formations. The concentrations of TiO_2 nanoparticles tested in this study varied from 0 ppm – 500 ppm. Prior to measuring the IFT and contact angle, a solution of 1.25% wt. AEC surfactant on 8000 ppm NaCl was first prepared, and the tested concentration of TiO_2 nanoparticles was added after that. To create a homogeneous solution between the tested TiO_2 nanoparticles in the AEC surfactant solution, an ultrasonic homogenizer was used to homogenize the combined solution for 30 minutes of sonication time.

2.2.2 Contact angle measurements

In order to understand the impact of TiO_2 addition on AEC surfactant related to the wettability, a contact angle method was employed using a Theta lite tensiometer (supplied by Biolin Scientific) at the temperature of 25°C. Improving from the previous measurements by Megayanti et al. [15], in this study, the drop of crude oil was placed on top of a thin section of Berea sandstone surrounded by the immiscible phase of AEC surfactant with various tested concentrations of added TiO_2 nanoparticles. The whole process was recorded during the measurement, and the measured angle was later analyzed using Image J software. To ensure the experimental repeatability and determine the uncertainty, this measurement was repeated at least three times.

3. Results and discussions

3.1 Impact of the addition of TiO_2 nanoparticles to AEC surfactant on the interfacial tension

Figure 1 shows the results of the interfacial tension between crude oil and AEC surfactant before and after the addition of the TiO_2 nanoparticles, from 0 ppm – 500 ppm (see details on Megayanti et al. [15]). The addition of TiO_2 nanoparticles to AEC surfactant can significantly enhance the performance of AEC surfactant by lowering the IFT at the lowest value by almost two orders of magnitudes. We hypothesize that this phenomenon is due to the interaction of TiO_2 nanoparticles with AEC surfactants, in which surfactant molecules can adsorb on TiO_2 nanoparticles using the electrostatic interaction mechanism. Therefore, when TiO_2 nanoparticles move to the interface due to their tiny size, some surfactant molecules will partially desorb from nanoparticles and reallocate on the interface, thus contributing to reducing the IFT [18].

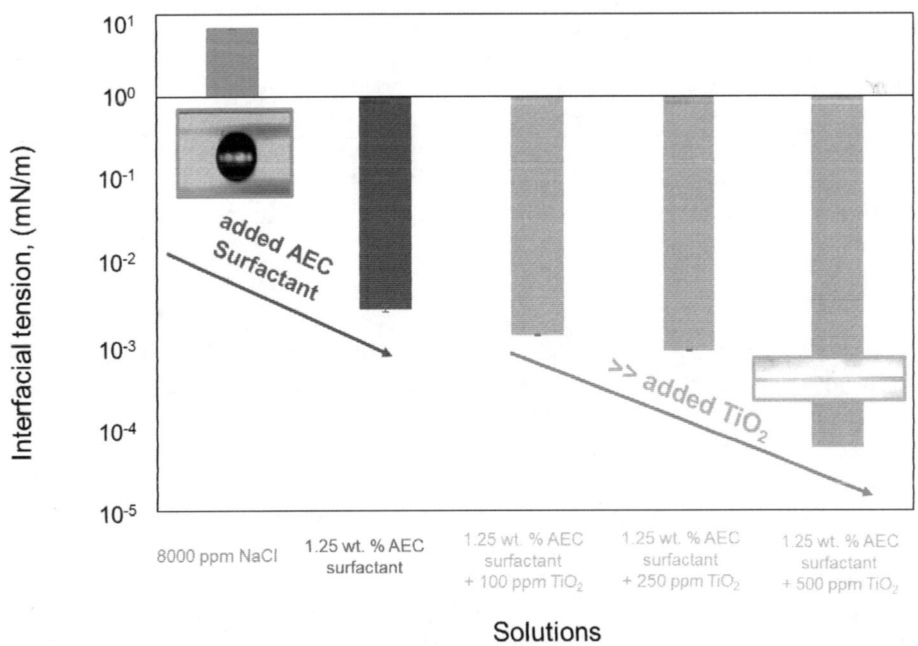

Figure 1. The interfacial tension between crude oil and 8000 ppm NaCl brine in combination with AEC surfactant and TiO_2 nanoparticles. The experimental data presented here was adapted from Megayanti et al. [15].

Journal of Physics: Conference Series **2980** (2025) 012006 doi:10.1088/1742-6596/2980/1/012006

3.2 Impact of the addition of TiO_2 nanoparticles to AEC surfactant on the wettability

The results of the measured contact angle on a thin section of Berea sandstone with various concentrations of TiO_2 nanoparticles on 1.25% wt. AEC surfactant is demonstrated in Figure 2. Employing 1.25% wt. AEC surfactant makes the wetting state more water-wet than only 8000 ppm NaCl brine, as shown in Figure 2. Further, after adding TiO_2 nanoparticles to the AEC surfactant, the contact angle slightly increases with the increase in the concentration of TiO_2, from 46° to 55°, within the experimental uncertainty. Therefore, in this study, adding TiO_2 to AEC surfactant does not significantly improve wettability to be more water wet, but instead, shifting the wetting state to be more intermediate-wet condition. We hypothesize that this phenomenon is due to the instability of TiO_2 nanoparticles on AEC surfactant. Therefore, future experimental works will focus on the stability of various added TiO_2 nanoparticles to AEC surfactant.

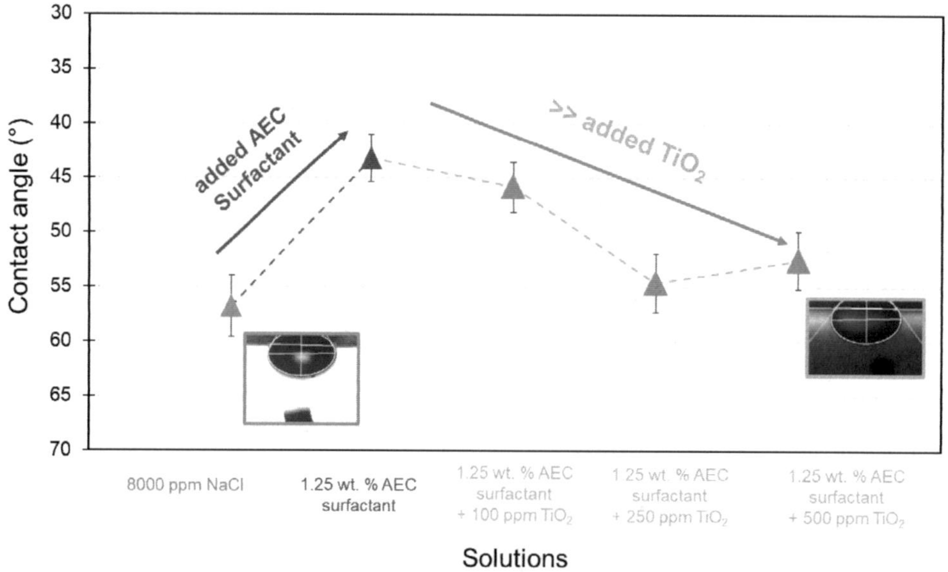

Figure 2. The contact angle of crude oil-rock-brine (AEC surfactant/TiO_2 nanoparticles) system.

4. Conclusion

In this study, we explore the possibility of combining titanium dioxide nanoparticles with AEC surfactant to improve the interfacial properties of surfactant when in contact with crude oil and on a surface of Berea sandstone. Our results suggest that the addition of TiO_2 nanoparticles to AEC surfactant can significantly reduce the IFT, approximately by two orders of magnitude, compared to the use of AEC surfactant only. This also indicates that the application of combining TiO_2 nanoparticles with AEC surfactant can be extended to other light oil samples to reduce the IFT. Conversely, adding the TiO_2 nanoparticles to the AEC surfactant slightly increased the measured contact angle on a thin section of the Berea sandstone sample compared with the AEC surfactant alone. Therefore, future works will be planned to understand this phenomenon to optimize the wettability alteration mechanism by combining TiO_2 nanoparticles with AEC surfactant on sandstone formations.

References

[1] Madani, M., Zargar, G., Takassi, M.A., Daryasafar, A., Wood, D.A. and Zhang, Z., 2019. Fundamental investigation of an environmentally-friendly surfactant agent for chemical enhanced oil recovery. *Fuel, 238*, pp.186-197.

[2] Marhaendrajana, T., Kurnia, R., Irfana, D., Abdassah, D. and Wahyuningrum, D., 2019. Study to improve an amphoteric sulfonate alkyl ester surfactant by mixing with nonionic surfactant to reduce brine–waxy oil interfacial tension and to increase oil recovery in sandstone reservoir: T-KS field, Indonesia. *Journal of Petroleum Exploration and Production Technology*, *9*, pp.675-683.

[3] Massarweh, O. and Abushaikha, A.S., 2020. The use of surfactants in enhanced oil recovery: A review of recent advances. *Energy Reports*, *6*, pp.3150-3178.

[4] Rellegadla, S., Prajapat, G. and Agrawal, A., 2017. Polymers for enhanced oil recovery: fundamentals and selection criteria. *Applied microbiology and biotechnology*, *101*, pp.4387-4402.

[5] Santoso, R.K., Fauzi, I., Hidayat, M., Swadesi, B., Aslam, B.M. and Marhaendrajana, T., 2018. Study of Non-Newtonian fluid flow in porous media at core scale using analytical approach. *Geosystem Engineering*, *21*(1), pp.21-30.

[6] Gbadamosi, A., Patil, S., Kamal, M.S., Adewunmi, A.A., Yusuff, A.S., Agi, A. and Oseh, J., 2022. Application of polymers for chemical enhanced oil recovery: a review. *Polymers*, *14*(7), p.1433.

[7] Swadesi, B., Marhaendrajana, T., Siregar, H.P. and Mucharam, L., 2015. The Effect of Surfactant Characteristics on IFT to Improve Oil Recovery in Tempino Light Oil Field Indonesia. *Journal of Engineering & Technological Sciences*, *47*(3).

[8] Hou, B., Wang, Y., Cao, X., Zhang, J., Song, X., Ding, M. and Chen, W., 2016. Surfactant-induced wettability alteration of oil-wet sandstone surface: mechanisms and its effect on oil recovery. *Journal of Surfactants and Detergents*, *19*, pp.315-324.

[9] Qi, Z., Han, M., Chen, S. and Wang, J., 2022. Surfactant enhanced imbibition in carbonate reservoirs: Effect of IFT reduction and surfactant partitioning. *JCIS Open*, *5*, p.100045.

[10] Wang, S., Shi, L., Ye, Z., Wang, Y., Liu, C. and Xue, X., 2021. Microscopic experimental study on the sweep and displacement efficiencies in heterogeneous heavy oil reservoirs. *Energy Reports*, *7*, pp.1627-1635.

[11] Aslam, B., Yan, B., Hidayat, M., Husenido, S., Swadesi, B., Permadi, P. and Marhaendrajana, T., 2023. Data driven approach using capacitance resistance model to determine polymer in-situ retention level. *Geoenergy Science and Engineering*, *229*, p.212043.

[12] Cheraghian, G., 2016. Effects of titanium dioxide nanoparticles on the efficiency of surfactant flooding of heavy oil in a glass micromodel. *Petroleum Science and Technology*, *34*(3), pp.260-267.

[13] Negi, G.S., Anirbid, S. and Sivakumar, P., 2021. Applications of silica and titanium dioxide nanoparticles in enhanced oil recovery: Promises and challenges. *Petroleum Research*, *6*(3), pp.224-246.

[14] Shahrabadi, A., Daghbandan, A. and Arabiyoun, M., 2022. Experimental investigation of the adsorption process of the surfactant-nanoparticle combination onto the carbonate reservoir rock surface in the enhanced oil recovery (EOR) process. *Chemical Thermodynamics and Thermal Analysis*, *6*, p.100036.

[15] Megayanti, R., Hidayat, M., Cahyaningtyas, N., Sanmurjana, M., Nur Muhammad Yahya, Z., Sagita, F., Kadja, G.T.M. and Marhaendrajana, T., 2023. Effect of Titanium Dioxide

Nanoparticles on Surfactants and Their Impact on the Interfacial Properties of the Oil–Water–Rock System. *ACS omega*, *8*(41), pp.38539-38545.

[16] Lu, J., Liyanage, P.J., Solairaj, S., Adkins, S., Arachchilage, G.P., Kim, D.H., Britton, C., Weerasooriya, U. and Pope, G.A., 2014. New surfactant developments for chemical enhanced oil recovery. *Journal of Petroleum Science and Engineering*, *120*, pp.94-101.

[17] Marhaendrajana, T., Kurnia, R., Wahyuningrum, D. and Fauzi, I., 2015. A Novel Sulfonated Alkyl Ester Surfactant to Reduce Interfacial Tensions in a Wide Range Salinity with Monovalent and Divalent Ions. *Modern Applied Science*, *10*(1), pp.1-93.

[18] Xu, F., Zhong, X., Li, Z., Cao, W., Yang, Y. and Liu, M., 2022. Synergistic mechanisms between nanoparticles and surfactants: insight into NP–surfactant interactions. *Frontiers in Energy Research*, *10*, p.913360.

The Effect of Substrate Temperature on CNT Deposition with Ni Catalyst as Precursor

Ajeng Eliyana[1,2], Rustan Ruslan[1], Aveni Christy Keintjem[1], Jasruddin Daud Malago[3], and Toto Winata[1*]

[1] Department of Physics, Institut Teknologi Bandung, Bandung, Indonesia.
[2] Department of Physics, Institut Teknologi Sumatera, South Lampung, Indonesia.
[3] Department of Physics, State University of Makassar, Makassar, Indonesia.

*E-mail: twinata@itb.ac.id

Abstract. The effect of temperature substrat on the growth of carbon nanotubes (CNTs) with a Ni catalyst precursor were studied, where the CNTs were grown by HWC in plasma-VHF-PECVD method. The Ni catalyst was grown on a glass substrate utilizing the evaporation method, with parameters such as deposition time 50 seconds, annealing temperature 500 °C for 4.5 hours. Morphological characterization of the Ni catalyst was done using a Scanning Electron Microscope (SEM). Subsequently, the Ni catalyst served as a growth precursor for CNTs with parameters such as CH4 gas flow rate of 80 sccm and a pressure of 300 mTorr for 60 minutes. Substrate temperature was varied at 250, 275 and 300 °C. Characterization of the CNT thin layer using SEM, and Raman spectroscopy produced ID/IG values of 0.94, 1.14, and 1.06, respectively. From the comparison of the intensity ratio of the D-band and G-band, the optimum temperature for CNT growth was 250 °C.

Keywords: Ni Catalyst, Substrate Temperature, Carbon Nanotubes, HWC in plasma-VHF-PECVD

1. Introduction

Carbon nanotubes (CNT) have become one of the most intriguing research topics in the field of nanotechnology over the past few decades, since their discovery by Iijima in 1991 [1]. CNT is known for its extraordinary properties, including high elastic modulus and superior electrical and thermal conductivity. These properties make CNT a potential candidate material for various technological applications, such as nanoelectronics, high-strength composite materials, high-energy batteries, and photovoltaic applications [2-5].

The process of CNT deposition is carried out through several methods, namely chemical vapor deposition (CVD), laser ablation, and arc discharge. Of the three methods, CVD is one of the most commonly used due to its advantages in controlling the morphology of CNTs, cost efficiency, and ease of scaling up. Based on the type of energy used to trigger chemical reactions during the deposition process, CVD (Chemical Vapor Deposition) can be categorized into the following types: Thermal CVD, Plasma-Enhanced CVD (PECVD), Photon-Enhanced CVD (Photo-CVD), Laser-Enhanced CVD (LCVD), Plasma-Assisted CVD (PACVD), Hot Filament CVD (HFCVD), Direct Current

CVD (DC-CVD), and Microwave-Enhanced CVD (MWCVD). According to Lee et al., the PECVD method has demonstrated a greater ability to produce vertically aligned nanotubes, which are often used in the fabrication of CNT-based devices, such as solar cells and field emitters [6]. In the PECVD process, the use of transition metal catalysts, such as nickel (Ni), iron (Fe), or cobalt (Co), is crucial for initiating and facilitating the growth of CNT [7]. This catalyst aids in the process of decomposing carbon-carrying gases (such as methane, ethylene, or acetylene) into carbon atoms, which will then combine to form nanotube structures [8].

For this study, the growth of CNT was developed using the PECVD method with very high frequency and a hot wire cell. This combination is expected to enhance the deposition rate and facilitate the dissociation of precursor gas before plasma formation.

One important parameter in the synthesis of CNT using the PECVD method is the substrate temperature. The substrate temperature has a direct influence on the deposition rate, morphology, and quality of the produced CNT [9-10]. Temperatures that are too low can hinder the decomposition of carbon-carrying gases, while temperatures that are too high can lead to catalyst agglomeration, negatively affecting CNT growth. Therefore, the optimal substrate temperature needs to be identified to achieve efficient and high-quality CNT growth. Several studies have shown that variations in substrate temperature can produce CNTs with different dimensions, densities, and purities, but the specific effects of temperature on the deposition process still require further research, especially when using Ni catalysts.

In this study, the material used as a catalyst source was nickel (Ni), which was grown using the thermal evaporation method. Ni catalyst was chosen as a precursor due to its good ability to break down carbon molecules at high temperatures and its suitable characteristics that allow for the formation of a stable catalytic core for CNT growth. Nickel (Ni) has a high carbon solubility and a high carbon diffusion value [11-13]. Previous studies have shown that the use of Ni as a catalyst can produce CNTs with good structural properties; however, variations in substrate temperature can lead to significant differences in the size and shape of the resulting CNTs [14]. Therefore, a more in-depth study on the effects of substrate temperature on the CNT deposition process using Ni catalyst becomes very important.

This research aims to explore the effects of substrate temperature on the CNT deposition process using Ni catalyst as a precursor [15]. The main objective is to understand how this temperature variation affects the morphology, size, and quality of the produced CNTs. We have conducted an in-depth analysis using Raman spectroscopy characterization and scanning electron microscopy (SEM) to identify the structural changes occurring in CNT. The results of this research are expected to provide guidance for the optimization of the CNT synthesis process using Ni catalysts, as well as to open new opportunities for the application of CNT in various future technologies.

2. Experimental Methods

In this research, the growth of Ni catalyst on SiO_2 glass substrate has been carried out using an thermal evaporation method, with evaporation times varied at 12, 25, and 50 seconds, while maintaining a constant Ni source mass of 2.2 mg. The evaporation was conducted at BATAN, Yogyakarta, Indonesia. The initial process involves cleaning the substrate using an ultrasonic

Journal of Physics: Conference Series **2980** (2025) 012007 doi:10.1088/1742-6596/2980/1/012007

cleaner for 15 minutes in deionized water (DI), acetone, methanol, and finally deionized water again.

Table 1. Parameters for the growth of carbon nanotubes.

Parameters	Value
Gas flow rate CH_4	80 sccm
Deposition time	60 minutes
Growth pressure	300 mTorr
RF	8 watt
electrode distance	2.562 cm

The substrate is then dried by spraying the glass surface with nitrogen gas until no traces of water remain. The next process is annealing, at a temperature of 500 °C for 4.5 hours. The final process is the deposition of CNT with optimized growth temperatures varied at 250, 275, and 300 °C. Another parameter used in this optimization is shown in Table 1. Nickel catalysts were then characterized using scanning electron microscopy (SEM) to observe the morphology of the catalyst layer composition and Raman spectroscopy to identify the electrical characteristics of CNT.

3. Result and Discussion

3.1 Optimization of deposition Ni

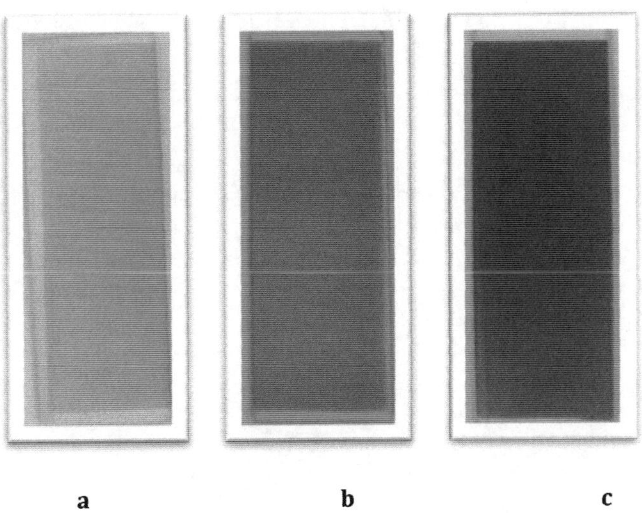

Figure 1. Results of the evaporation of Ni catalyst grown on glass substrate (25.4 x 76.2 mm) with deposition times of a) 12 s, b) 25 s, and c) 50 s.

Growth of the catalyst layer on SiO_2 glass substrate has been carried out using the evaporation method, with evaporation times varying for 12, 25, and 50 seconds. The results of evaporation on the glass substrate are shown in figure 1. From figure 1, it can be seen that the longer the time used for the deposition of the Ni catalyst, the thicker (darker) the surface of the glass becomes. The analysis of the influence of evaporation time on the Ni layer formed on the substrate before the annealing process through SEM characterization is shown in Figure 2. From the figure, the morphological condition for the thin film sample grown for 12 seconds shows very little visible Ni (the parts that are lighter in color, silver). In contrast, the sample deposited with a longer evaporation time shows more areas that are lighter in color or Ni, the substrate surface appears denser, and the size of the formed clusters is clearer and larger.

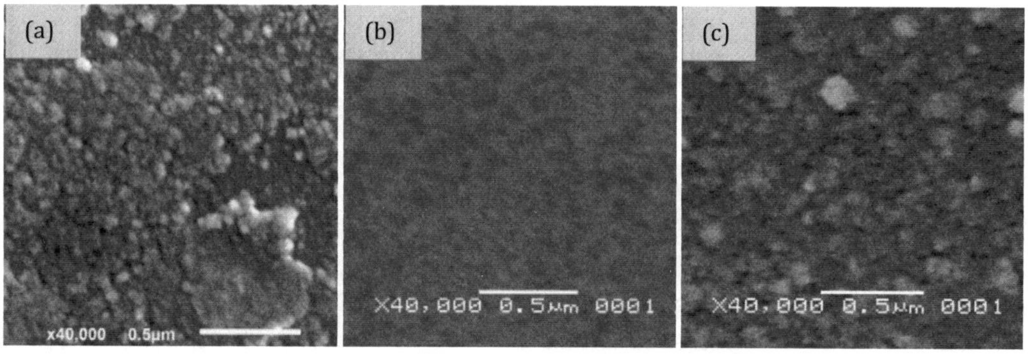

Figure 2. SEM results of the Ni layer samples evaporated for a) 12 s, b) 25 s, and c) 50 s.

The increase in the size of the clusters formed and the growth in diameter over time of deposition is also stated by Wu et al. (2012) [16]. This enlargement of cluster size is caused by the process of merging previously formed clusters, known as coalescence. The arrangement of the constituent atoms still appears denser for evaporation times of 12 seconds and 25 seconds, unlike the sample deposited for 50 seconds, which shows that Ni catalyst particles are more formed, although the particles tend to have an irregular shape. With the increase in evaporation time, it means that the number of deposited Ni atoms is also greater.

Table 2. EDS results of the effect of evaporation time on the atomic content of Ni after annealing.

Sample	T_{ann} (°C)	t_{ann} (hours)	t_{eva} (s)	% Atom Ni
1	400	4	12	0.84
2	400	4	25	1.24
3	400	4	50	2.30

The increase in the number of Ni atoms makes the formed layer thicker and initiates grain agglomeration to create a more continuous layer [6]. This is reinforced by the EDS results shown in Table 2, where the number of Ni atoms increases with the longer evaporation time. The deposition time parameter of 50

The next process is annealing, where this annealing process helps in the formation of clusters with smaller grain sizes [17]. This cluster is formed due to surface tension and pressure between the substrate and the deposited catalyst as a result of the difference in thermal expansion coefficients at high temperatures.

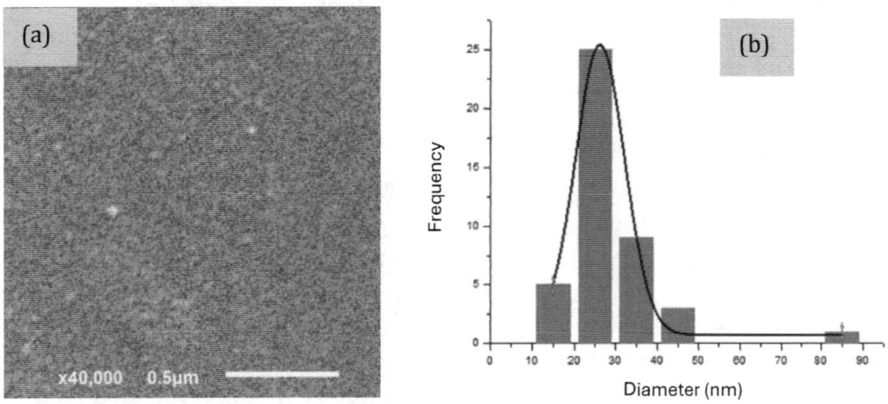

Figure 3. a) SEM results of the Ni catalyst surface, b) Histogram of the Ni catalyst that was evaporated for 50 seconds and annealed for 4.5 hours at a temperature of 500 °C.

The SEM results from the sample deposited for 50 seconds and annealed at a temperature of 500 °C for 4.5 hours are shown in Figure 3 a), where the figure reveals a more uniform distribution of shape changes [18-19]. By using image-j and distribution calculations in Origin Lab, an average diameter of around 26.18 nm was obtained, as shown in Figure 3 b) in the form of a histogram. The results obtained are used as fixed variables in the subsequent optimization analysis, namely the growth of CNT.

3.2 The Influence of Substrate Temperature

The results of CNT deposition with substrate temperature optimization were analyzed using SEM characterization and Raman spectroscopy. Figure 4 shows the SEM characterization results of CNT for the three variations of substrate temperature, which are similar for the 275 °C and 300 °C variations. At a temperature of 275 °C, the presence of CNTs is observed with the longest diameter and length values of 138.69 nm and 1651.62 nm, respectively. Meanwhile, at a temperature of 300 °C, there is an uneven distribution and the emergence of clumped areas. In contrast, at 250 °C, there is a kind of accumulation in some areas. Looking at the Raman characterization results shown in Figure 5, at 300 °C, the second CNT band, the G'-band, appears at a position of 2657.03, while at 250 °C, it does not appear.

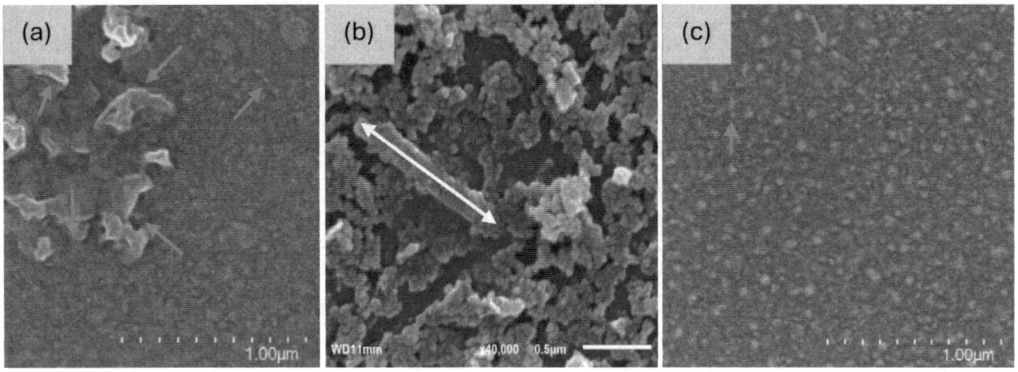

Figure 4. Results of SEM CNT at substrate temperatures a) 250 °C, b) 275 °C, dan c) 300 °C.

Likewise, the RBM ribbon essentially appears and its peak value can be determined; this result can be seen in Table 3. From Table 3, at a temperature of 250 °C, the comparison of the intensity ratio of the D-band and G-band (I_D/I_G) is also less than 1 [20], specifically 0.94, which indicates better quality of CNT, where the D-band is lower than the G-band. This can be interpreted that at a temperature of 250 °C, it may be possible to become the next parameter in CNT deposition.

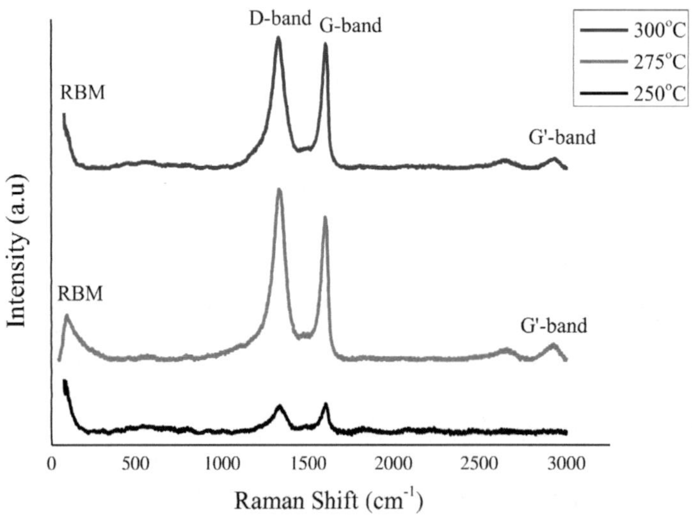

Figure 5. Results of Raman spectroscopy characterization with substrate temperatures at 250 °C, 275 °C, and 300 °C.

Table 3. Peak position of the band (Raman shift) from Raman characterization results for substrate temperature optimization.

No	T (°C)	Raman shift (cm^{-1})				I_D/I_G
		RBM	D-Band	G-Band	G'-Band	
1	250	85.09	1339.03	1606.88	-	0.94
2	275	88	1323.53	1606.17	2921.3	1.14
3	300	81.48	1326.83	1604	2657.03	1.06

4. Conclusions

In this research, the optimization of the deposition time of Ni catalyst using the thermal evaporation method has been investigated, with CNT successfully grown on the Ni metal catalyst using the HWC method in Plasma-VHF-PECVD. The results show that a longer evaporation time leads to an increased number of Ni atoms, the substrate surface appears denser, and the size of the clusters formed becomes clearer and larger. Additionally, a longer annealing process is directly correlated with an increase in the average diameter of the Ni islands, which simultaneously enhances the homogeneity of the cluster diameters, as evidenced by the reduced polydispersity values. The optimal sample of Ni island was obtained with an evaporation time of 50 seconds, an annealing temperature of 500°C, and an annealing time of 4.5 hours. For the deposition of CNT with variations in substrate temperature, observations from SEM at a temperature of 275 °C reveal CNT with a diameter of 138.69 nm and a length of 1651.62 nm. However, unlike the Raman spectroscopy results, which show that CNT has better quality at a temperature of 250 °C. Where the comparison of the D-band and G band is less than 1, specifically 0.94, and the values of RBM, D-band, and G'-band are respectively 85.09/cm, 1339.03/cm, and 1606.88/cm.

5. Acknowledgment

This research is funded by "Penelitian, Pengabdian kepada Masyarakat, dan Inovasi (PPMI)" in the fiscal year 2024. RR acknowledges the support from the Indonesia Endowment Funds for Education (LPDP), the Ministry of Finance of the Republic of Indonesia.

References

[1] Iijima, S., 1991, Helical Microtubules of Graphitic Carbon, Nature (London), 354, 56-58.
[2] Sibiński M, Jakubowska M, Znajdek K, Słoma M and Guzowski B 2011 Carbon Nanotube Transparent conductive layers for solar cells applications Opt. Appl. 41 375–81.
[3] Ferguson A J, Blackburn J L and Kopidakis N 2013, Fullerenes and carbon nanotubes as acceptor materials in organic photovoltaics Mater. Lett. 90 115–25
[4] Endo M, Strano M S and Ajayan P M, 2008, Potential applications of carbon nanotubes Top. Appl. Phys. 111 13–62

Journal of Physics: Conference Series **2980** (2025) 012007 doi:10.1088/1742-6596/2980/1/012007

[5] Venkataraman A, Amadi E V, Chen Y and Papadopoulos C, 2019, Carbon Nanotube Assembly and Integration for Applications Nanoscale Res. Lett. 14

[6] Lee, M. W.; Haniff, M. A. S. M.; Teh, A. S.; Bien, D. C. S.; Chen, S. K., 2015, Effect of Co and Ni Nanoparticles Formation on Carbon Nanotubes Growth via PECVD. J Exp Nanosci, 10 (16), 1232–1241.

[7] Saito, R., Dresselhaus, G. & Dresselhaus, M.S., 1998, Physical Properties of Carbon Nanotubes. Imperial College Press.

[8] Meyyappan, M., Delzet, L., Cassell, A. & Hash, D., 2003, Carbon Nanotube Growth by PECVD: a Review, Plasma Sources Sci. Technol, 12, 205-216, 2003.

[9] Pargellis, A. N. 1989, Evaporating and Sputtering: Substrate Heating Dependence on Deposition Rate. Journal of Vacuum Science & Technology A: Vacuum, Surfaces, and Films., 7 (1), 27–30.

[10] Baptista, A.; Silva, F.; Porteiro, J.; Míguez, J.; Pinto, G. 2018, Sputtering Physical Vapour Deposition (PVD) Coatings: A Critical Review on Process Improvement and Market Trend Demands. Coatings, 8 (11), 402.

[11] Lim Y D, Avramchuck A V, Grapov D, Tan C W, Tay B K, Aditya S and Labunov V, 2017, Enhanced Carbon Nanotubes Growth Using Nickel/Ferrocene-Hybridized Catalyst ACS Omega 2 6063–71

[12] Huang Z P, Wang D Z, Wen J G, Sennett M, Gibson H and Ren Z F, 2002, Effect of nickel, iron and cobalt on growth of aligned carbon nanotubes Appl. Phys. A Mater. Sci. Process. 74 387–91

[13] Rao R, Sharma R, Abild-Pedersen F, Nørskov J K and Harutyunyan A R, 2014, Insights into Carbon Nanotube Nucleation: Cap formation governed by catalyst interfacial step flow Sci. Rep. 4 1–6

[14] Krishna, A.; Aravinda, L. S.; Murugan, A.; Kumar, N. S.; Sankar, M. R.; Nagahanumaiah; Reddy, K. N.; Balashanmugam, N. 2022, A Study on Wafer Scalable, Industrially Applicable CNT Based Nanocomposites of Al-CNT, Cu-CNT, Ti-CNT, and Ni-CNT as Thermal Interface Materials Synthesised by Thin Film Techniques. Surf Coat Technol,429, 127926.

[15] Xiao, Y., Ahmed, Z., Ma, Z., Zhou, C., Zhang, L., and Chan, M., 2019, Low Temperature Synthesis of High-Density Carbon nanotubes on Insulating Substrate. Nanomaterials, 9(3), 473.

[16] Wu, B.R., Lo, S.Y., Wuu, D.S., Ou, S.L., Mao, H.Y., Wang, J.H., Horng, R.H. 2012, Direct growth of large grain polycrystalline silicon films on aluminum-induced crystallization seed layer using hot-wire chemical vapor deposition. Thin Solid Films, 520(18), 5860–5866.

[17] Carey, J. D.; Ong, L. L.; Silva, S. R. P. 2003, Formation of Low-Temperature Self-Organized Nanoscale Nickel Metal Islands. Nanotechnology. 14 (11), 1223–1227.

[18] Min Shin, Y.; Yol Jeong, S.; Jin Jeong, H.; Jin Eum, S.; Woong Yang, C.; Yun Park, C.; Hee Lee, Y, 2004, Influence of Morphology of Catalyst Thin Film on Vertically Aligned Carbon Nanotube Growth. J Cryst Growth, 271 (1–2), 81–89.

[19] Abidin, K.; Keintjem, A. C.; Yusuf, A. M.; Eliyana, A.; Malago, J. D.; Noor, F. A.; Winata, T. 2021, Optimization of Ag Catalyst Growth by Vacuum Thermal Evaporation. Fuller. Nanotube. Carbon. Nanostruct, 29 (9).

[20] Childres, I., Jauregui, L.A., Park, W., Cao, H., Chen, Y.P., Raman Spectroscopy of Graphene and Related materials, Chapter.ID_13194_6x9_os (1).

Journal of Physics: Conference Series **2980** (2025) 012008 doi:10.1088/1742-6596/2980/1/012008

Study on Gamma-Ray Buildup Factor of TeO_2-B_2O_3-TiO_2-ZnO Glasses using Py-MLBUF Simulation

Ahmad Marzuki[1*], Ari Handono Ramelan[1], Devara Ega Fausta[1], Rivina Oktafiani[1], and Lita Rahmasari[2]

[1] Physics Department, Faculty of Mathematics and Natural Science, Sebelas Maret University , 57126, Surakarta, Indonesia
[2] Physics Education Department, Faculty of Teacher Training and Education, Sebelas Maret University, 57126, Surakarta, Indonesia

*e-mail: amarzuki@mipa.uns.ac.id

Abstract. One of the factors that characterized gamma shielding material is Buildup Factor. Its value determines the appropriate thickness of the shielding material. In the present work, tellurite glasses having compositions (in mol%): $60TeO_2$ - (30-x)B_2O_3- $7TiO_2$-(3+x) ZnO (TBTZ-x), with x = 0, 1, 2, 3, 4, 5, 6, 7, and 8 has been successfully fabricated using the traditional melt-quenching procedure. The density of the glasses was measured by applying the Arhcimedes law and found to be increased from 3.667 to 4.736 g cm^{-3} due to partial substitution between B_2O_3 and ZnO compounds with higher molecular weight. The exposure buildup factor (EBF) of the glasses was computed by utilizing Py-MLBUF simulation software. The simulation was carried out by varying penetration depth value (PD) from 0.5 to 40 mfp within the energy range between 0.015 and 15 MeV. Evaluation of the EBF was carried out at three different regions, i.e., 0.015 MeV, 1.5 MeV, and 15 MeV that represent Photoelectric (PE) mechanism, Compton Scattering (CS) and Pair-Production (PP) effect. Based on the simulation, it is clearly seen from 0 to 40 mfp, at 0.03 MeV show maximum EBF value around 10^9, which interpreted as K-shell binding energy of Te atom. while the TBTZ-0 and TBTZ-8 shows the highest and lowest EBF value.

1. Introduction

In the recent era, demand for gamma radiation shielding material has arisen with the growth of radiation chambers for medical examination [1]. It was acknowledged that radiation chambers have more than one radiation shielding type material. For example, the wall of the chamber is constructed of concrete which is layered by a lead sheet while the visible part of the chamber is usually built by transparent material like glass. The choice of glass is due to its unique properties such as high transparency, good chemical durability, excellent mechanical properties, and has a high radiation absorption cross-section [2]. Nowadays, existing gamma radiation shielding glass in radiological chambers is formed by high concentrations of Lead (Pb) to create lead-based glass [3]. However, although provide good shielding characterics, lead has serious disadvantageous effects on human health and environmental if continously exposured in long time period [4]. Since 70's, it was reported the lead usage in any applied materials has significantly reduced [5] [6].

Journal of Physics: Conference Series **2980** (2025) 012008 doi:10.1088/1742-6596/2980/1/012008

Then, seeking alternative compounds that are used for major or additional components to replace the importance of lead, especially in lead-based glass becomes an interesting topic.

The use of more friendly compounds that possess similar properties with Pb like bismuth trioxide (Bi_2O_3), Tungsten trioxide (WO_3), Aluminium trioxide (Al_2O_3), and zinc oxide (ZnO), have been reported [7] [8] [9] [10]. However, these compounds are limited by their high crystallization rate which prevents them from being reinforced into a glass by themselves [11]. So, It would need a stable glass system as its host, one of the glass hosts which can accommodate that requirement is the boro-telurite glass system.

Boro-tellurite is a glass system that is dominated by Boron trioxide (B_2O_3) and tellurium dioxide (TeO_2) proportion in glass networks [12]. The development of boro-tellurite glass itself as a gamma shielding material has also been intensively studied in recent years [13] [14] [15] [16]. In our previous report, we developed boro-tellurite glass which consisted of B_2O_3, TeO_2, TiO_2, and ZnO as its main components [17]. We used the computational approach to achieve glass density, accompanied by its shielding parameters in terms of Linear Attenuation Coefficient (LAC) and Half Value Layer (HVL). According to the results, partial substitution between B_2O_3/ZnO can alter the calculated density value and directly improve the gamma shielding performance of glass.

Therefore, in this present work, we continue the investigation by experimentally measuring the density of glass and using it to compute the buildup factor parameter. The buildup factor is one of the essential parameters in designing gamma radiation shielding glass because it can predict the gamma radiation dose behind the shielding material by calculating the secondary effects, for example, the scattered gamma radiation [18]. The Exposure Buildup Factor (EBF) of the present glass was later analysed and discussed in the section below.

2. Experimental

2.1 Glass synthesis

In this present work, nine boro-tellurite glasses with general composition $60TeO_2$ - $(30\text{-}x)B_2O_3$-$7TiO_2$-$(3+x)$ ZnO: x=0-8 mol% in increments each 1 mol% are synthesized using traditional melt quenching method. Raw materials that consist of TeO_2, B_2O_3, TiO_2, and ZnO were prepared for a total of 10 grams masses. The raw glass batch was mixed inside a ceramic mortar and subsequently poured into the alumina crucible. The mixture was then heated at 950°C for 45 minutes using a muffle furnace and followed by an annealing process at 285°C for twelve hours until reached room temperature.

2.2 Density and Buildup factor estimation

The density of glasses was measured using the pycnometer method by obeying Archimedes law. After the measurement, several parameters such as the mass of empty pycnometer (m_p), pycnometer plus glass (m_{p+s}), pycnometer plus immersion liquid (m_{p+a}), and pycnometer with immersion liquid plus glass ((m_{p+a+s}) were obtained then, these parameters processed into Equation (1)

$$\rho_g = \frac{[(m_{p+s})-(m_p)]}{[(m_{p+s})-(m_p)-(m_{p+s+a}-m_{p+a})]}\rho_s \quad (1)$$

Where ρ_g and ρ_s are glass density and immersion liquid density. The measurement process was performed at room temperature (25°C). Then, the density can be used to estimate the EBF within the energy range between 0.015-15 MeV and varied penetration depth (PD) from 0.5 to 40 mfp. The energy range was chosen from 0.015 to 15 MeV due to its coverage to majority radiation that encountered in real shielding material application such in industrial, medical, or nuclear application. As example for low to medium energy that ranging from 0.015 to 1.5 MeV, it commonly meets in medical imaging (i.e. CT-Scans and X-ray) [19]. While at high energy range (1.5-15 MeV) can be found in nuclear reactor application [20]. The simulation was performed by using the Python program for Multi-Layered Buildup Factors (Py-MLBUF), a fast free interactive online software invented by Mann and Mann [21].

3. Results and Discussion

According to the measurement result, the density of TBTZ glass arises from 3.667 to 4.736 g cm^{-3} with increasing of ZnO contents. The main reason for this enhancement is because of partial substitution between B_2O_3 and ZnO. B_2O_3 has a smaller molecular weight compared to the ZnO. An increase in density leads to the formation of bridging oxygen (BO) which creates a stronger bond between the constituent compound inside the glass networks [22]. It also means the ability to reduce the travelled radiation inside the material enhanced.

EBF is a shielding parameter that captures the response of non-colliding photons to add the participation of scattered photons with the amount of energy stored in the air [23]. It can be used to help determine and control the thickness of the created radiation shielding material [24]. **Figure 1**. shows the relationship between the EBF of each TBTZ glass in the energy range from 0.015 to 15 MeV.

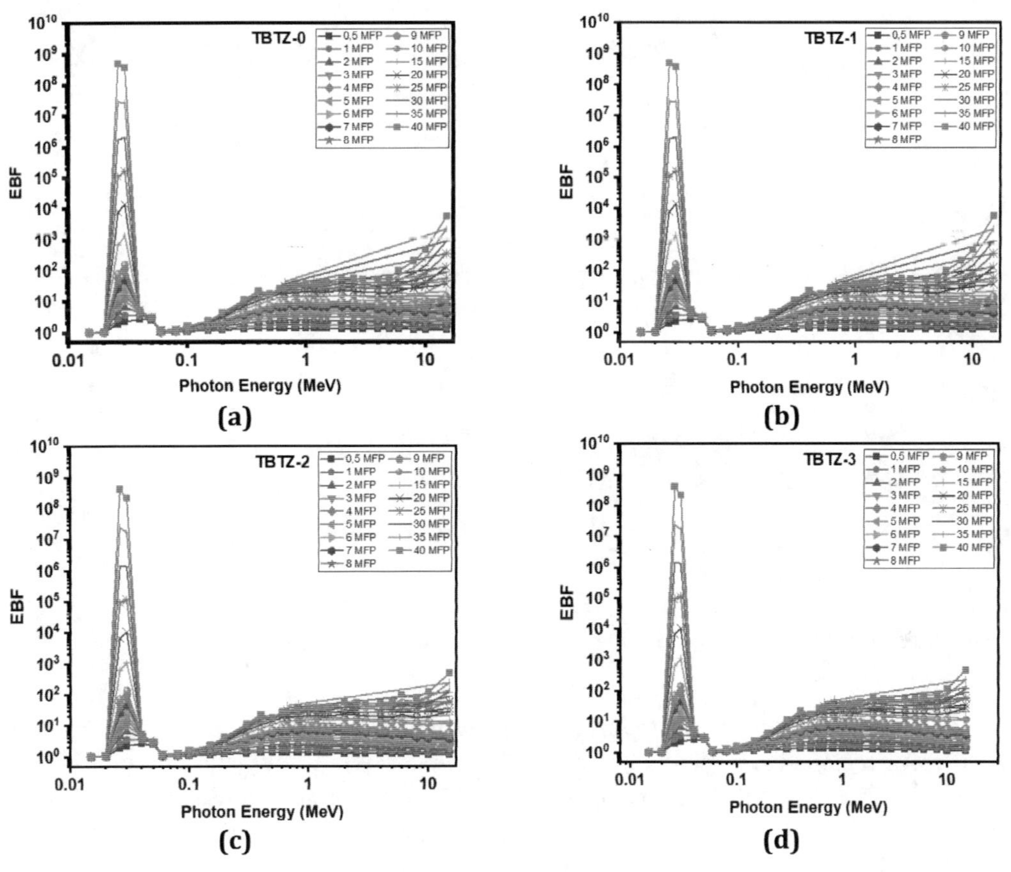

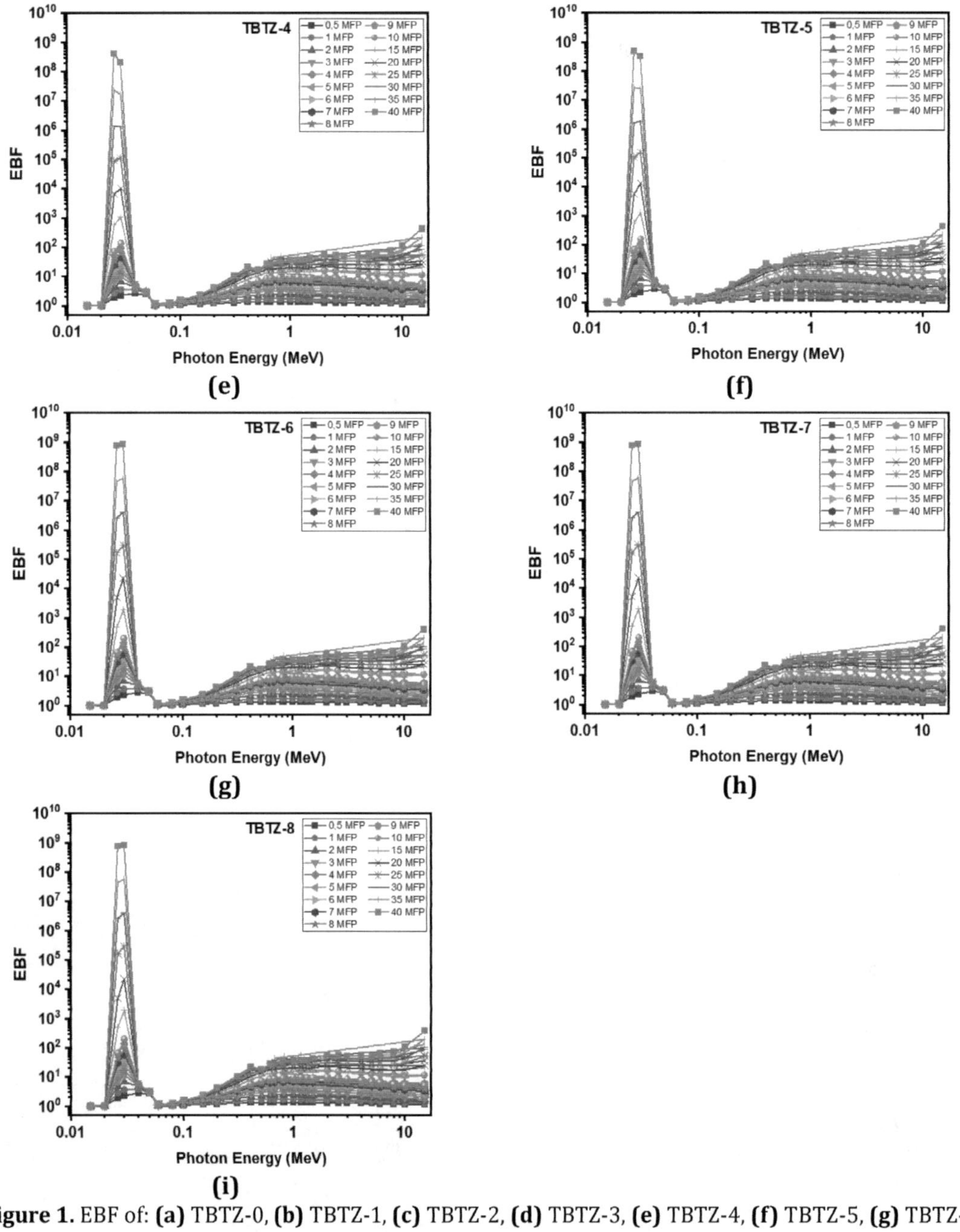

Figure 1. EBF of: **(a)** TBTZ-0, **(b)** TBTZ-1, **(c)** TBTZ-2, **(d)** TBTZ-3, **(e)** TBTZ-4, **(f)** TBTZ-5, **(g)** TBTZ-6, **(h)** TBTZ-7, and **(i)** TBTZ-8 glasses in range 0.015-15MeV with variation in penetration depth.

It can be seen from **Figure 1. (a)** to **(i)** the EBF regarding the photon energy exhibits similar results for all glass samples. Evaluated from the photon energy, EBF itself can be categorized into three different regions (low, medium, and high energy levels). At low energy around 0.015-0.1 MeV, the Photo Electrics (PE) mechanism occurred [25]. At this region, EBF also exhibits a sudden alteration to its maximum magnitude around 0.03 MeV, which relates to the K-shell binding energy of Te (0.0318 MeV) [26], thus it was slightly decayed until 0.06 MeV. At medium energy from 0.1 to 1 MeV, the Compton Scattering (CS) process was detected and the EBF increased [27]. At the same time, at a high energy level around 1-15 MeV the Pair Production (PP) process was discovered [28]. In terms of penetration depth (PD), the EBF of all glass samples increased due to the rise of photon scattering. It appears that the EBF of TBTZ glasses exhibit the same value and the increasing ZnO concentration does not have a significant effect. It can be suggested due to TeO_2 contents that retain the highest concentration among the others oxide [29].

For further investigation, the correlation between EBF and penetration depth in various energy levels is illustrated in **Figure 2. (a)** and **(b)**. Three different energies were chosen (0.015, 1.5, and 15 MeV) to represent low, medium, and high energy levels. It can be noticed that EBF enhances with an increase in penetration depth value. EBF increase gradually from 0.5 to 10 mfp and continues to rise drastically at 10-40 mfp. From these results, it is clear that the TBTZ-0 glass has a higher EBF magnitude compared to the TBTZ-8 glass in all energy regions. Substitution between B_2O_3 and ZnO gives an indication that ZnO can reduce the energy buildup and enhance glass structure which is related to the ability of TBTZ glass to absorb or scatter the gamma radiation [30].

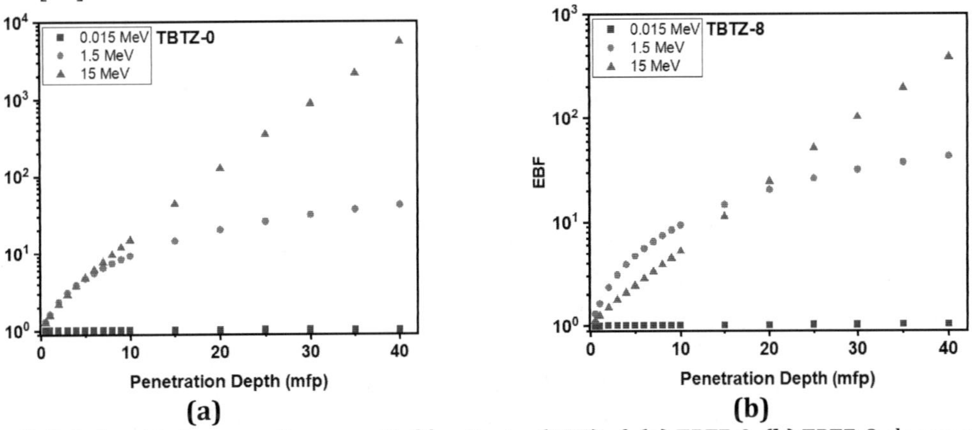

Figure 2. Relationship between Exposure Buildup Factor (EBF) of: **(a)** TBTZ-0, **(b)** TBTZ-8 glasses with penetration depth at 0.015 MeV, 1.5 MeV, and 15 MeV

ZnO is well known for its glass modifier role [31]. In present study, when ZnO concentration increasing from 3 to 11 mol%, it enhances the glass rigidity. This proved by an alteration in glass density from 3.667 to 4.736 g cm^{-3}. The glass density improves due to the enrichment of BO bonds inside the glass networks with increasing ZnO concentration. Therefore, it seems lower the ability of the present glass to trap the radiation energy [32] that causing the magnitude of EBF to decline. It also can be evaluated that Zn (65.38 amu) has a relatively high atomic mass compared to B (10.81 amu). So, when ZnO concentration increases it leading to more compact glass structure that enlarge the number of colliding photon and lessen the EBF [33]. The lessen in EBF value weakens energy retention and thereby creates a glass with less effective ability to buildup the energy [34]. Hence, it displays the potential application of the TBTZ glass series as a gamma radiation shielding material.

4. Conclusion

Synthesis of boro-tellurite glass series with main formula consisting of $60TeO_2$ - $(30\text{-}x)B_2O_3$-$7TiO_2$-$(3+x)$ ZnO: x=0-8 mol% has completely done. The density of glass was measured using the pycnometer method and obtained that an increase in the ZnO concentration altered the density from 3.667 to 4.736 g cm^{-3}. The buildup parameter in terms of EBF was computed and it described the interaction of glass with incoming photons at low, medium, and high energy levels. The sudden escalation of EBF magnitude at 0.03 MeV is indicated as the absorption edge of Te due to its highest concentration among the other compounds. Additional investigation to analyze the relation between EBF and MFP shows that EBF increases in the same manner with increasing depth penetration. From this evaluation, the TBTZ boro-tellurite glass series has possible applications as gamma radiation shielding glass.

Acknowledgments

The authors acknowledged with gratitude to Ministry of Education, Culture, Research and Technology Indonesia for Penelitian Fundamental Regular Research Scheme with Contract Number: 1076.1/UN27.22/PT.01.03/2024.

References

[1] Zarei, M., Sina, S., & Hashemi, S. A. (2021). Radiation Physics and Chemistry, **188**, 109588
[2] Kaur, P., Singh, D., & Singh, T. (2016). Nuclear Engineering and Design, **307**, 364–376.
[3] AbuAlRoos, N. J., Baharul Amin, N. A., & Zainon, R. (2019). Radiation Physics and Chemistry, **165**, 108439
[4] Sigel, A., Sigel, H., & Sigel, R. K. O. (2017). De Gruyter.
[5] Nicolas, E., Ruiz-Pino, D., Buat-Ménard, P., & Bethoux, J. P. (1994). Geophysical Research Letters, **21**, 19, 2119–2122
[6] B L Johnson and R W Mason (1984). Neurotoxicology, **5**, 3, 1-22
[7] Alrowaili, Z. A., Yılmaz, E., Çalışkan, F., Öztürk, B., Olarinoye, I. O., Arslan, H., & Al-Buriahi, M. S. (2023). Radiation Physics and Chemistry, **204**, 110711.
[8] Acikgoz, A., Demircan, G., Yılmaz, D., Aktas, B., Yalcin, S., & Yorulmaz, N. (2022). Materials Science and Engineering: B, **276**, 115519.
[9] Esawii, H. A., Salama, E., Sayed El-ahll, L., Moustafa, M., & Saleh, H. M. (2022). Progress in Nuclear Energy, **150**, 104321
[10] Al-Buriahi, M. S., Alrowaili, Z. A., Eke, C., Alzahrani, J. S., Olarinoye, I. O., & Sriwunkum, C. (2022). Optik, **257**, 168821.
[11] Lezal, D., Pedlikova, J., Kostka, P., Bludska, J., Poulain, M., & Zavadil, J. (2001). Journal of Non-Crystalline Solids, **284**, 1–3, 288–295.
[12] Alajerami, Y. S. M., Mhareb, M. H. A., Sayyed, M. I., Hamad, M. Kh., Abuelhia, E., Alshamari, A., Alqahtani, M. Y., & Imheidat, M. A. (2024). Optical Materials, **147**, 114558.
[13] Lakshminarayana, G., Sayyed, M. I., Baki, S. O., Lira, A., Dong, M. G., Bashar, Kh. A., Kityk, I. V., & Mahdi, M. A. (2018). Journal of Electronic Materials, **48**, 2, 930–941
[14] Tamam, N., Al Huwayz, M., Alrowaili, Z. A., Alwadai, N., Katubi, K. M., Alqahtani, M. S., Olarinoye, I. O., & Al-Buriahi, M. S. (2024). Applied Radiation and Isotopes, **203**, 111080.
[15] Azuraida, A., Halimah, M. K., Ishak, M., Hasnimulyati, L., & Ahmad, S. I. (2020). Chalcogenide Letters, **17**, 4, 187–196.
[16] Halimah, M. K., Azuraida, A., Ishak, M., & Hasnimulyati, L. (2019). Journal of Non-Crystalline Solids, **512**, 140–147.
[17] Marzuki, A., Fausta, D. E., Purnomo, A. T., Yannaf, N. H., Istiqomah, H. D., & Astuti, R. W. (2024). AIP Conference Proceedings, **3165**, 070008.
[18] Rafiei, M. M., Tavakoli-Anbaran, H., & Kurudirek, M. (2020). Radiation Physics and Chemistry. **177**, 109118.

[19] Alqahtani, M. S., Alqahtani, F., Almarhaby, A. M., Hussain, K. I., Khalid, Y., Almohiy, H., Yaha, I. S., & Yousef, E. S. (2023). Chalcogenide Letters, **20**, 1, 43–54.

[20] Murthy, K. I. & Charit, I. (2013). *An Introduction to Nuclear Materials: Fundamentals and Applications.* Wiley

[21] Mann, K. S., & Mann, S. S. (2021). Annals of Nuclear Energy, **150**, 107845.

[22] Bixmaiah, Ch., Naresh, P., Laxman, A., Bhavani, S., Siva Kumar, K., & Sreenivasu, D. (2023). Materials Today: Proceedings.

[23] Sharaf, J. M., & Saleh, H. (2015). Radiation Physics and Chemistry, **110**, 87–95.

[24] Kaur, P., Singh, D., & Singh, T. (2018). Physics and Chemistry, **144**, 336–343

[25] D'Souza, A. N., Murari, M. S., Sayyed, M. I., Al-Ghamdi, H., Almuqrin, A. H., & Kamath, S. D. (2024). Chemical Papers, **78**, 10, 5773–5791.

[26] Bearden, J. A., & Burr, A. F. (1967). Reviews of Modern Physics, **39**, 1, 125–142.

[27] Sriwongsa K., Glumglomchit, P., Sualang B., Arunoros P., Pansuay, M., Ravangvong, S., Khobkham C. (2022). Science and Technology, **30**, 1, 109-119.

[28] Srinivasa Rao, L., Hussain, S., Navalika, A., Chennakesava Rao, B., Venkatappa Rao, T., & Hila, F. C. (2023). Results in Optics, **12**, 100491.

[29] Rammah, Y. S., Al-Buriahi, M. S., Sriwunkum, C., Shams, M. S., & Yousef, E. S. (2020). Journal of Materials Science: Materials in Electronics, **31**, 23, 21431–21443.

[30] Susoy Dogan, G. (2020). European Journal of Science and Technology, 531–544.

[31] Effendy, N., Sidek, H. A. A., Halimah, M. K., & Zaid, M. H. M. (2021). Materials Chemistry and Physics, **273**, 125156.

[32] Kartika, S., Marimuthu, K., Evangelin T.P., Meena, RC, Sulania, I., Asokan K., Shanmuga, S,S. (2023). Radiation Physics and Chemistry, **202**, 110560.

[33] ALMisned, G., Elshami, W., Issa, S.A.M., Susoy, G., Zakaly, H.M.H., Alegtani, M., Rammah Y.S., Ene, A., Al-Ghamdi, S.A., Ibraheem A.A., and Tekin H.O. (2021). Materials, **14**, 7703.

[34] Mahmoud, I. S., Issa, S. A. M., Saddeek, Y. B., Tekin, H. O., Kilicoglu, O., Alharbi, T., Sayyed, M. I., Erguzel, T. T., & Elsaman, R. (2019). Ceramics International, **45**, 11, 14058–14072.

Journal of Physics: Conference Series **2980** (2025) 012009 doi:10.1088/1742-6596/2980/1/012009

The effect of deposition bead pattern and multilayer hardfacing on wear resistance of Q55 rail using SMAW

Defi Pramesti[1*], Winarto Winarto[1], Muhammad Athaila Ramadhanu[1]

[1] Department of Metallurgical and Material Engineering, Faculty of Engineering, Universitas Indonesia, Depok 16424, Indonesia

*E-mail: defi.pramesti31@ui.ac.id

Abstract. This study aims to analyze the effect of weld bead pattern and multilayer hardfacing on the hardness, toughness, and wear resistance values of Q55 using the Shielded Metal Arc Welding (SMAW) process with Panalloy 312 MA and 307 MA for the buffer layer and HiMn for the surface. The welding technique used bead pattern string and weaving on a railroad substrate Q55. Tests included hardness testing using the Vickers Micro Hardness, Charpy impact testing, the Ogoshi wear testing, and metallographic analysis using Optical Microscopy and Scanning Electron Microscopy (SEM-EDS). The results show that the weld bead pattern significantly affects the material's hardness, wear, and toughness values. String patterns produce higher hardness, with the most considerable hardness value of 467 HV, compared to weaving, which obtains the most considerable hardness value of 355 HV. Impact testing showed that the sample with the string had better toughness, with an average impact test value of 46.05 J, while the weaving obtained an average impact value of 44.30 J. The wear test results show that the string application has better wear resistance with an average value of the abraded volume of 0.01667 mm^3 compared to the weaving, obtaining a wear test result value of 0.04306 mm^3. SEM-EDS analysis showed homogeneous chemical distribution in each weld seam and the formation of ductile fracture mechanism in the test samples after impact testing. This study contributes to selecting the optimal weld bead pattern to improve the quality and service life of high-carbon steel in rail.

1. Introduction

Hardfacing is a process that improves the hardness and wear and erosion resistance of necessary metal substrates by applying hard metals as a filler on their surface [1]. One such class of alloys is the Fe-Cr-C (Iron-Chromium-Carbon) hardfacing alloys, commonly used on structural steel surfaces to improve their resistance to wear, abrasion, and corrosion. The microstructure of these alloys plays a crucial role in determining their performance, particularly in terms of wear resistance, which is critical for prolonging the lifespan of machinery and structures. The interaction between the alloy's microstructure and base steel affects the mechanical characteristics and the wear behavior under operating conditions when Fe-Cr-C hardfacing alloys are welded onto structural steel. Therefore, optimizing hardfacing procedures requires understanding how the microstructure changes in these welded locations. Because of the strong wear resistance from a combination of hard carbides (primary carbides) and a moderately ferrous matrix, Fe-Cr-C alloy is primarily employed as a hard-facing filler [2, 3]. Fe-Cr-C (iron-chromium-carbon) alloys can form eutectic carbide phases, especially in specific compositions with high concentrations of chromium and carbon. In these alloys, the carbide phases that are often formed are Fe_3C (iron carbide) or Cr_7C_3 (chrome carbide).

The eutectic carbides in the Fe-Cr-C alloy produced high hardness and appropriate abrasive wear resistance. However, the impact resistance and hardfacing quality are still inadequate and must be improved [4-6]. It is because carbides formed in Fe-Cr-C alloys, such as Cr_7C_3, although very hard, have high brittleness. The carbide phase can crack or break when these alloys are exposed to wear conditions involving severe impact or friction. This cracking reduces the material's ability to maintain long-term wear resistance, as carbide fragments can cause faster material loss. The welding process on structural steel with Fe-Cr-C alloy can cause microstructural changes that affect wear resistance. At high temperatures during welding, the carbides formed can undergo dissolution or structural changes so that the complex carbide phase can be eroded or unstable, which reduces wear resistance. In addition, the rapid cooling process during welding can produce martensite or unstable microstructures, reducing wear resistance.

Fe-Cr-C alloys have significant potential for improving wear resistance; certain factors, such as carbide brittleness, uneven carbide distribution, and microstructure changes during welding, can present challenges affecting overall wear resistance. Therefore, for hardfacing applications or materials requiring high wear resistance, attention must be given to optimizing the alloy composition and ensuring proper welding conditions to achieve the best mechanical properties and wear resistance.

Moreover, it is crucial to explore processes to enhance the structure, properties, and wear resistance of hardfacing through various techniques, including matrix fractionation, alloying additions, cooling rate control, carbide refinement, and heat treatment [7-9]. When used in wear environments, undesirable properties such as incompatibility, poor weldability, and high stress may lead to defects such as stress cracking in the hardfacing welds and subsequent failure [10-14]. Previous studies have suggested that preheating and post-welding heating techniques can improve dilution and weldability while reducing flaws caused by moisture and hydrogen gas [15, 16]. Additionally, controlling the cooling rate during welding can influence the microstructure and wear resistance of the weld, while buffering or buttering processes can improve the elemental compatibility between the base substrate and the weld, reducing discontinuities [17, 18]. The choice of shielding gas also protects the weld from oxidation and maintains arc stability, puddle fluidity, and metal transfer [19, 20].

However, no reports have been made on the microstructure and wear resistance of Fe-Cr-C hard facing alloys welded onto structural steel using Shielded Metal Arc Welding (SMAW) under various conditions, such as preheating, welding technique, and buffering. Therefore, this study investigates the impacts of SMAW on the properties and wear resistance of Fe-Cr-C welded rail steel under these conditions [21, 22]. The findings from this research are expected to offer strategies for improving welding quality, enhance the understanding of hardfacing weld characteristics, and ultimately improve the wear resistance of future hardfacing applications.

2. Experimental procedure

2.1 Material and welding process

Material rail Q55 has dimensions 95 mm x 90 mm x 150 mm. This research uses a Shielded Metal Arc Welding (SMAW) welding machine to apply hardfacing layers to carbon steel-based rail track

Journal of Physics: Conference Series **2980** (2025) 012009 doi:10.1088/1742-6596/2980/1/012009

substrates—electrode for layer 1 Panalloy 312MA, layer 2 Panalloy 307MA, and layer 3 Panalloy HiMn. The chemical composition of the electrodes is listed in Table 1 . Welding was performed on a rail surface with a 33-35 mm thickness using ampere 95-110 A, voltage 21-22 Volt, and polarity direct current electrode positive (DCEN). The layering technique involves two welding patterns, as shown in Figure 1. string at sample a) and weaving at sample b), arranged in horizontal and vertical orientations. Before carrying out the welding process, a further sample preparation process is carried out formerly. The sample preparation process involves cutting the sample using a metal cutting band saw machine into two parts with a length of 15 cm each. Next, the sample is smoothed and cleaned through sanding with sandpaper of various grits. The next stage is to carry out the preheating process with the burner method at a temperature of 200°C to achieve an even temperature throughout the sample surface.

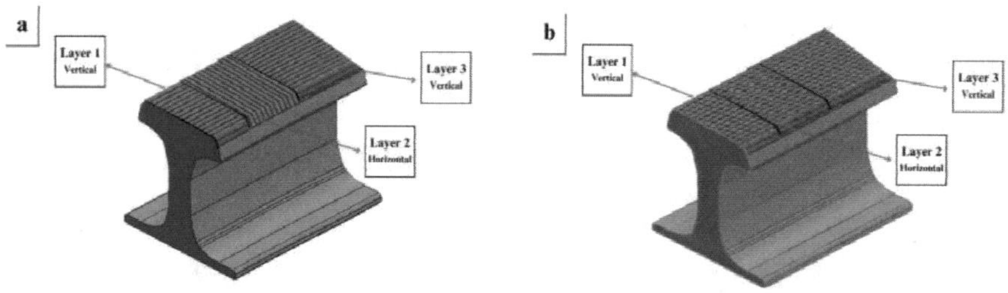

Figure 1. Hardfacing process on Q55 railroads a) String b) weaving

Table 1. Chemical composition of material

Elements (%)	Chemical compounds					
	C	Si	Mn	Cr	Ni	Fe
Q55 - Rail	0.75	0.24	0.75	0.016	0.005	Bal.
Panalloy 312 MA	0.12	0.5	1.35	28.5	8.95	Bal.
Panalloy 307 MA	0.09	0.3	5.8	21.5	8.4	Bal.
Panalloy Hi Mn	0.68	-	13.5	-	3.7	Bal.

2.2 Mechanical properties

The mechanical properties of microhardness impact Charpy and wear testing. The tool used in the microhardness Vickers test is Nobel, type MHVS-50AT, No—series 17043. Hardness testing is carried out referring to the ASTM E92-17 standard. Hardness testing using the Vickers method aims to determine the hardness value of each layer and the Heat Affected Zone (HAZ) of each layer due to the influence of the hardfacing process. Impact testing used HungTa/Taiwan type HT-8041A, Serial No. 7472, Capacity 300 Joules. Charpy impact testing was carried out at a temperature of 25°C, referring to the ASTM E23-23a standard. The toughness test is used to determine the brittleness or ductility of a material by suddenly loading the specimen. The test specimen has a V-shaped notch with a notch depth of 2 mm. Wear testing using Riken-Ogoshi's Universal with reference standard ASTM G99.

2.3 Microstructure investigation

The sample of the cross-section of joints was prepared by grinding and polishing as per the standard of ASTM E407-07:2015 and etched under ASTM E1 for metallographic and microstructural studies. Microstructure observation was taken using optical microscopy (OM) and scanning electron microscopy (SEM). Microstructure observation aims to get a detailed picture of the microstructure in the areas of the base metal, heat-affected areas (HAZ) at each layer. Microstructural observations were carried out using an optical microscope after the surface preparation of the specimen was carried out by cutting the specimen, grinding from coarse-grit sandpaper to fine-grit, and etching. An optical microscope was conducted using Nikkon MA 100N and a scanning electron microscope with energy-dispersive X-ray spectroscopy (SEM-EDS) using JEOL JSM – IT 200.

3. Result and discussion

3.1 Macro and microstructure analysis

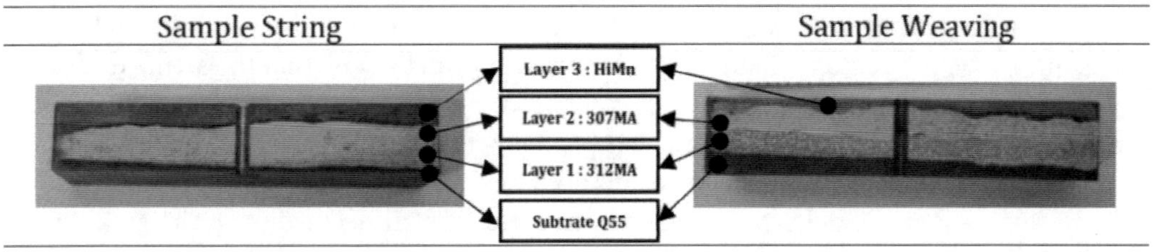

Figure 2. Macro photograph of hardfacing

Table 2 shows the chemical composition of filler metal and rail as a substrate. Rail material contains high carbon. On the surface, layer 3 must have high wear resistance properties to accommodate frictional loads from train wheels. The type of material suitable for train rails is one that contains high manganese. Manganese steel has hard and high wear resistance properties; this is because manganese steel has a very prominent work hardening ability, which is one of the main characteristics of this steel. This ability occurs due to the plastic beating process on the steel surface, which results in increased hardness in the surface layer without damaging the steel core, which remains tough. This process can occur under high conditions and pressure, such as applications on railroad tracks. On layers 1 and 2 are buffer layers that function to accommodate sudden loads so that the rails do not break and also so that layer 3 with the Q55 rail subtract can fusion without delamination. The selection of filler is based on the characteristics required for railroad track repairs. Layer 1 and layer 2 as buffer layers to connect layer 3, which has high manganese content with its substrate Q55, while in layer 3, fillers are selected that have hard, tough, and wear-resistant characteristics. The fillers selected for the buffer layer are those with high Ni content so that they can produce an austenite matrix, namely Panalloy 312 MA and Panalloy 307 MA, while the filler for layer 3, which has a high Nickel composition is Panalloy HiMn.

Figure 2 shows a macro photo cross-section of Layer 1, Layer 2, and Layer 3, as well as the substrate with different welding patterns, strings, and weaving. The hardfacing process was successfully applied, and the fusion was completed. In Figure 3 (a) and (b), the microstructure formed in the railway track substrate shows that the dominating phase is pearlite, a mixture of ferrite (α-Fe) and cementite (Fe_3C). The pearlite formed can be seen by globular pearlite and lamellar pearlite morphology as dominant with inter-distance lamellar pearlite is so close that it can be concluded that the mechanical properties it has by the substrate are that it has good strength, hardness, and machinability properties [23]. In the hardfacing welding section, layer 1 is the welding part that is a buffer layer to connect the hardfacing layer to the substrate. Application This layer aims to avoid shrinkage cracking, which can occur on the sample surface. The metal filler used can withstand high temperatures during the welding process. This is due to the chemical composition of Cr of 28.5% and Ni of 8.95%, which is contained in the filler metal, increasing heat resistance and forming an austenitic microstructure [24].

Previously, the chrome and nickel content played a significant role in forming an austenite phase stabilizer. This follows the literature stating that if the nickel composition exceeds 7% and is combined with a high chromium composition, the austenite microstructure will be formed [25]. Apart from identifying the austenite phase, there is also a visible distribution of delta ferrite in darker colors, which is quite dominant with various morphologies, such as lathy ferrite and vermicular ferrite. The process of delta ferrite formation in this study occurred during the solidification process of the hardfacing weld layer due to the filler metal used having a high chrome and nickel content, which acts as a delta ferrite formation. It can be seen that the string test sample obtained a higher delta ferrite fraction when compared with the weaving test sample, which is caused by the speed cooling of the string test sample being faster. The presence of delta ferrite can prevent the occurrence of hot cracks because delta ferrite can remain stable at high temperatures. Besides that, the presence of delta ferrite can prevent cracks that occur during the solidification process [26]. Delta ferrite can inhibit crack propagation through several mechanisms related to its unique structure and the interactions between phases within the material. Although delta ferrite is more brittle than austenite, its presence in the material matrix serves to distribute stress more evenly, reducing stress concentrations that can lead to crack formation. Delta ferrite also acts as a physical barrier that slows or even blocks the path of crack propagation, forcing cracks to pass through harder phase boundaries. In addition, delta ferrite helps stabilize the tougher austenite phase at high temperatures, thereby reducing the formation of the more brittle martensite phase that is prone to crack propagation. The presence of delta ferrite also reduces subsequent crack formation by reducing local stress concentrations that can exacerbate damage. Overall, delta ferrite improves the material's resistance to crack propagation, especially at high temperatures, by slowing crack growth and increasing structural stability under thermally stressed conditions.

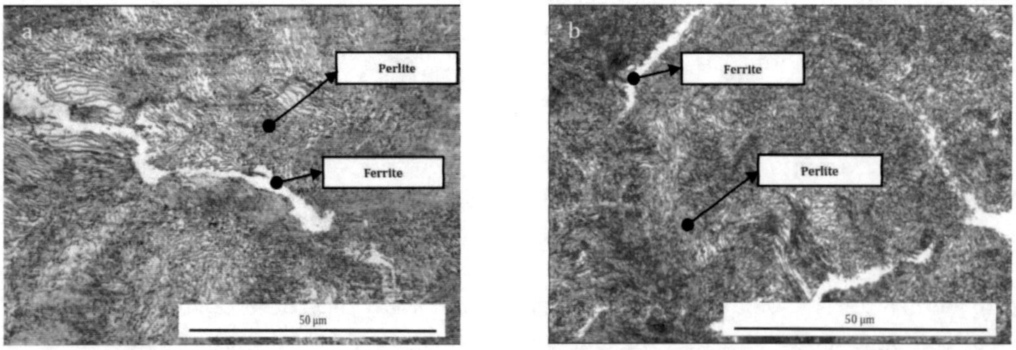

Figure 3. Microstructure of Q55 Rail with (a) string (b) weaving

As shown in Figure 4 (e) and (f), because it has a composition high chrome and nickel exceeds 7%, the phase that will be formed will be similar to the phase formed in layer 1, where the phase formed is austenite as matrix followed by the formation of delta ferrite because there are several alloy components contained in filler metal. However, due to the chrome composition and the nickel filler metal used in layer two welding is not as much as in layer 1, then the delta ferrite produced is also not as much as the microstructure is in layer 1, so it can be concluded that the values of hardness, toughness, and the ductility of the second layer is better than the third layer. It can be seen that the delta ferrite fraction contained in the string test sample is more dominant when compared with the weaving test sample; this is similar to the formation of delta ferrite in layer 1, which is influenced by the cooling rate. Besides that, delta ferrite morphology is formed, like vermicular ferrite, as seen in Figure 4. As in layer one conditions, delta ferrite in the microstructure austenite can inhibit the propagation of cracks that might form when the hardfacing layer two welding process is carried out on the test sample [27]. So that it can, It was concluded that the microstructure formed in layer 2 had conditions similar to layer one but with a smaller fraction, and the string test sample had a fraction of the larger delta ferrite due to the faster cooling rate compared with the weaving test sample.

As shown in Figure 4, an austenite phase is formed as a matrix phase due to the content of Mn and Ni, which play a role in creating and stabilizing the austenite phase [28]. Apart from the formation of the austenite phase, it can be seen that carbide distribution is formed in the form of fine particles and is also present at the boundary austenite grains caused by the carbon content of the filler metal. It can be seen that the grain boundaries of the string test samples look darker when compared with the weaving test sample, so the carbide precipitates that occur are more dominant when compared to the weaving test sample. This is due to the cooling rate produced during the welding process. The string test sample is faster than the weaving test samples. Therefore, it can be concluded that the mechanical properties created in the first layer include high hardness and wear resistance values, and because carbide precipitates occur more dominantly in the string test samples, they obtained a higher hardness value when compared with the test sample weaving [29].

3.2 Mechanical properties

Analysis of the data shown in Table 2 and Figure 5 shows that the parts with the highest hardness value up to the lowest are Layer 3, Layer 2, Layer 1, and substrate. This is influenced by the composition of the chemistry in each part and the microstructure that forms after the hardfacing welding process.

Besides the weld part, the bead deposition pattern influences the hardness value. The string technique produces higher hardness than weaving techniques due to its significant role in the heat input obtained, which can affect different hardness values; the higher the heat input obtained, the higher the hardness value will be. These results also represent optimal values when applied to rails trains because the surface layer obtains a high hardness value [30]. As shown in Figure 6, a graph of the impact test results can be seen in the results of hardfacing welds with bead pattern strings and weaving, indicating that the first test obtained the same, but existing impact toughness values were reduced for the second test on the test sample using the bead pattern deposition technique weaving, after visually analyzing the fracture section of the test sample, cracks were found which has quite a large volume. Applying a hardfacing weld layer with a bead weaving pattern produces an increased heat input value and a wider weld bead, making it possible for slag inclusion to occur.

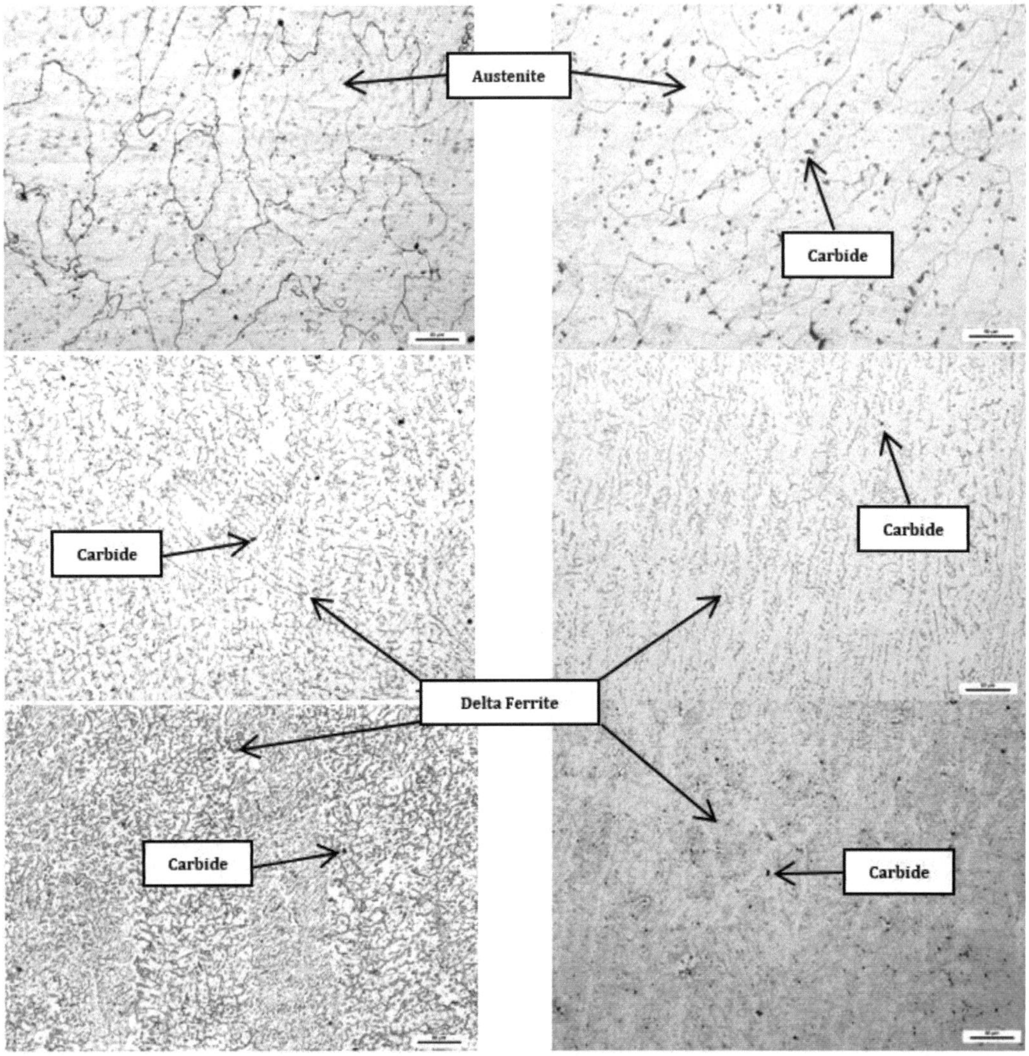

Figure 4. Microstructure of Layer 1, layer 2 and layer 3 (a) string (b) weaving

Figure 6 shows the results of the wear test in the form of an abrasion volume on the string, weaving, and substrate test samples so that it can be seen that the results are obtained for the application of bead pattern string and weaving produce values that are in contrast to the string technique, the results of the abrasion volume are slightly smaller if compared to weaving techniques. This is because there are significant differences in the hardness values obtained by the two test samples. The higher the hardness value obtained by a material, the more value it will produce; conversely, better wear resistance if a material has a low hardness value will result in lower wear resistance. However, for the long-term service life of the material, using strings produces dimensions, and the weld results in a higher thickness while the weaving is more comprehensive laterally. Therefore, hardfacing welding with bead pattern string is better than weaving because it has a greater thickness, so it can withstand wear and tear longer.

Data obtained in wear testing for substrate test samples show results the abrasion volume is much greater than the weld result; this is due to the content the chemistry of the railroad track material is high carbon steel, so the properties of the mechanical wear resistance obtained are lower when compared to coating third. Therefore, it can be concluded that there is an influence of bead pattern on the wear resistance of the material is due to the hardness value obtained on the string test sample higher than the weaving test samples and the application of the hardfacing layer using filler metal with manganese steel material can increase wear resistance and the service life of railway track substrates.

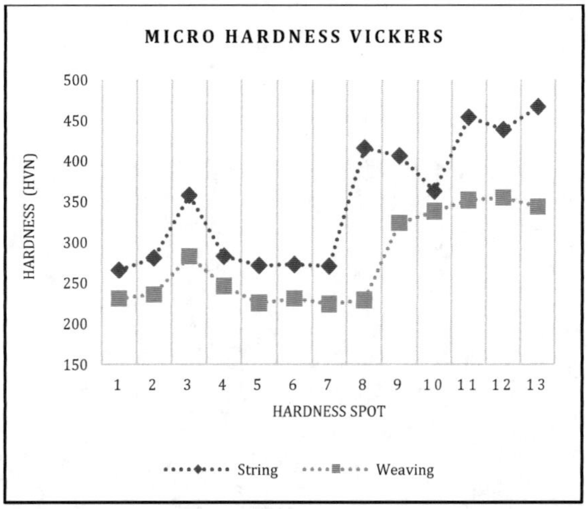

Figure 5. Hardness testing results based on the hardness sample testing tracking point

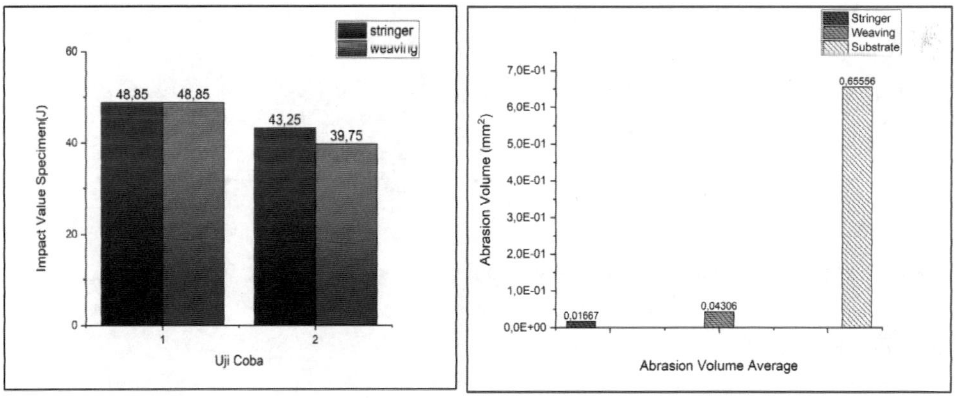

Figure 6. Impact testing and Wear testing result

3.3 SEM EDS

As shown in Figure 8 and Table 2 show that the results the chemical composition of each part of the weld deposited by filler metal produces the same results quite good for both bead patterns when compared with the specifications available by each filler metal used so that it can be concluded that the role of each layer can be done well. When comparing the chemical composition shown in the data obtained above, there are differences in the oxygen composition contained in the test sample by applying hardfacing welding with bead pattern weaving, so there may be defects or empty spaces in the weld layer. The discovery of cracks in the fracture section from the sample impact test results weaving test strengthened that matter. After finding the oxygen content in the weaving test sample Next, fractographic observations are carried out by focusing on the area on the surface fracture obtained from impact test results so that the mechanism can identify fractures that occurred in the weaving test sample.

Fractographic observations show the results of the data obtained using SEM-EDS, which is shown in Figure 7. It can be concluded from these data that the fracture mechanism that occurred in the string test samples and weaving is a very ductile fracture that can be identified by the appearance of sufficient dimples in the results of the data obtained. Therefore, according to the results of the data obtained after SEM-EDS observations on the fracture surface of the test sample string and weaving after impact testing, it can be concluded that the test samples obtained can absorb energy when given a load and are ductile.

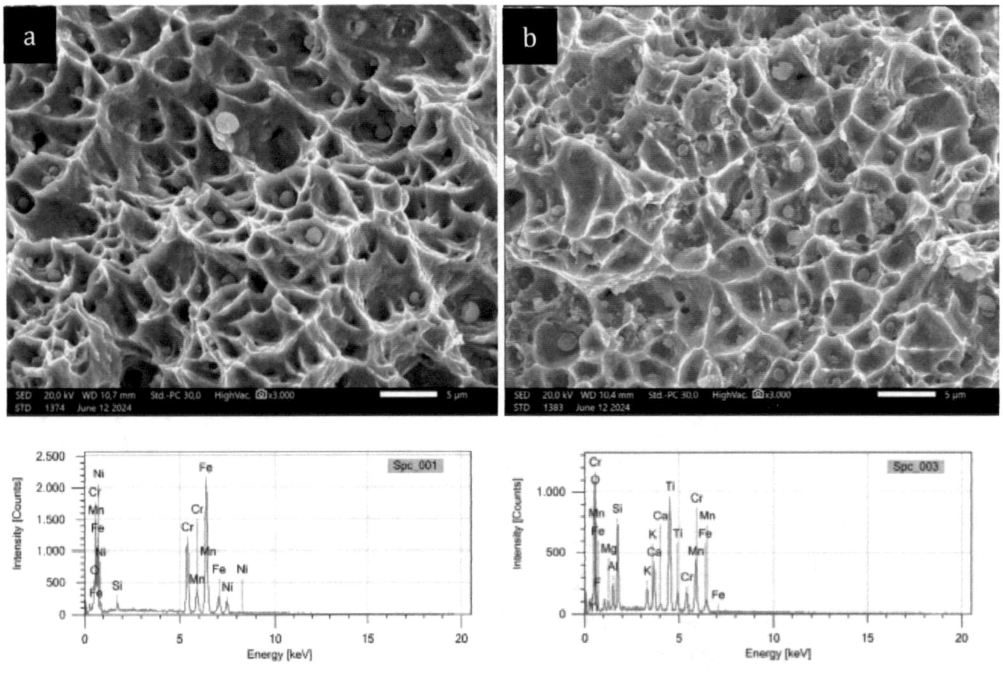

Figure 7. SEM and EDS at surface fracture of impact testing sample a) String b) weaving

Table 2. EDS Result

Bead pattern	Weld zone	Chemical composition (wt%)						
		Mn	Cr	Ni	Si	C	O	Fe
String	Layer 1	1.76	26.57	7.9	0.98	-	-	62.79
	Layer 2	4.41	21.62	8.75	0.96	-	-	64.26
	Layer 3	12.82	4.64	1.87	0.55	4.88	-	75.24
Weaving	Layer 1	1.27	22.88	6.49	9.93	8.76	1.24	58.33
	Layer 2	4.61	20.31	8.35	0.82	5.41	-	60.51
	Layer 3	13.56	3.37	1.73	0.66	7.91	1.43	71.33

Journal of Physics: Conference Series **2980** (2025) 012009 doi:10.1088/1742-6596/2980/1/012009

The Cr content encourages the formation of delta ferrite, especially at high temperatures, while Nickel tends to stabilize austenite. Materials with higher Delta ferrite phase content can reduce the toughness value. The string pattern contains an average of 24% Cr while the weaving pattern contains an average of 21.5% Cr, so in the string, the delta ferrite phase is high, and also Ni in the string is higher, and the austenite is more stable than in weaving. This is evident in the toughness value of the string pattern, which is 3.95% higher than the weaving pattern. In the microstructure produced in Fig. 4 in images a and b, the phase shape and phase differences produced are not visible significantly. This is because the filler composition used is the same. In contrast, in the SEM results Table 2, after welding, using different patterns will produce a slight difference in the formation of its composition. The controlled proportion of Cr and Ni can achieve a balanced microstructure and can improve material properties such as hardness, toughness, and wear resistance in hardfacing applications. The proportion of Cr and Ni in the string pattern is 2.8, while in the weaving pattern is 2.9. The smaller the value of this proportion, the greater the wear value. The wear value on the string pattern produces an abrasion volume 61.27% lower than the weaving pattern; this indicates that the wear value on the string pattern is the highest and more wear-resistant.

Manganese steel is in the austenite phase at high temperatures, where the crystal structure is face-centered cubic (FCC). This structure is more ductile and plastic. When this steel is subjected to stress or friction, the austenite structure approaches its elastic limit and begins to deform plastically, forming dislocations. Austenite is a more ductile phase and is resistant to plastic deformation. When austenite is deformed, it can accumulate dislocations in its crystal structure, which increases the surface hardness of the steel, a process known as work hardening. Plastic deformation and dislocation formation when this steel is exposed to high stress or friction, the surface layer of the steel undergoes plastic deformation, a change in shape that cannot be returned to its original shape after the load is removed. This deformation causes the movement of dislocations in the steel's crystal lattice. Dislocations are crystal defects that affect the ability of atoms to move in the crystal structure. The more dislocations that are formed, the harder the material is because the movement of dislocations becomes more difficult. In manganese steel, dislocations formed during plastic deformation increase the dislocation density on the steel surface, thereby increasing the surface hardness. Dislocations accumulated in the austenite structure inhibit further atomic movement, which increases the hardness and resistance to further deformation.

4. Conclusion

In conclusion, this research demonstrates that using the SMAW method, the weld bead pattern and hardfacing technique significantly influence the hardness, toughness, and wear resistance of welded carbon steel. The string weld bead deposition technique outperforms the weaving technique in all tested properties, showing a higher hardness value (467 HV compared to 355 HV), superior toughness with an average impact value of 46.05 J versus 44.30 J, and better wear resistance, evidenced by a 61.27% lower abrasion volume (0.01667 mm^2 compared to 0.04306 mm^2). These findings highlight the importance of selecting the appropriate welding technique to optimize material performance for applications requiring high hardness, toughness, and wear resistance.

Acknowledgment

This research was funded by PT. Pusat Pelatihan Kompetensi Mandiri (Welding Study Center).

References

[1] C. Tippayasam, C. Taengwa, J. Palomas, T. Siripongsakul, T. Thaweechai, and A. Kaewvilai, "Effects of flux-cored arc welding technology on microstructure and wear resistance of Fe-Cr-C hardfacing alloy," *Materials Today Communications,* vol. 35, p. 105569, 2023/06/01/ 2023, doi: https://doi.org/10.1016/j.mtcomm.2023.105569.

[2] N. ÇÖMez and S. Şahin, "Wear behavior–hardness–microstructure relation of Fe–Cr–C and Fe–Cr–C–B based hardfacing alloys," *Materials and Design,* vol. 58, pp. 491–498, 02/20 2014, doi: 10.1016/j.matdes.2014.02.032.

[3] M. Balakrishnan, V. Balasubramanian, and G. Madhusudhan Reddy, "Effect of Hardfacing Consumables on Ballistic Performance of Q&T Steel Joints," *Defence Technology,* vol. 9, no. 4, pp. 249-258, 2013/12/01/ 2013, doi: https://doi.org/10.1016/j.dt.2013.12.007.

[4] A. Appiah, O. Bialas, M. Żuk, A. Czupryński, D. Konadu, and M. Adamiak, "Hardfacing of mild steel with wear-resistant Ni-based powders containing tungsten carbide particles using powder plasma transferred arc welding technology," *Materials Science-Poland,* vol. 40, pp. 42-63, 12/01 2022, doi: 10.2478/msp-2022-0033.

[5] R. J. Chung, X. Tang, D. Li, B. Hinckley, and K. Dolman, "Microstructure refinement of hypereutectic high Cr cast irons using hard carbide-forming elements for improved wear resistance," *Wear,* vol. 301, pp. 695–706, 04/01 2013, doi: 10.1016/j.wear.2013.01.079.

[6] K. Wieczerzak, P. Bała, M. Stępień, G. Cios, and T. Kozieł, "Formation of eutectic carbides in Fe–Cr–Mo–C alloy during non-equilibrium crystallization," *Materials & design,* vol. 94, 03/15 2016, doi: 10.1016/j.matdes.2016.01.028.

[7] Y. Singla, M. Maughan, N. Arora, and D. Dwivedi, "Enhancing the wear resistance of iron-based alloys: A comprehensive review of alloying element effects," *Journal of Manufacturing Processes,* vol. 120, pp. 135-160, 06/01 2024, doi: 10.1016/j.jmapro.2024.04.038.

[8] D. Liu, R. Liu, Y. Wei, Y. Ma, and K. Zhu, "Microstructure and wear properties of Fe–15Cr–2.5Ti–2C–xB wt.% hardfacing alloys," *Applied Surface Science,* vol. 271, pp. 253–259, 04/01 2013, doi: 10.1016/j.apsusc.2013.01.169.

[9] R. M. Triani, J. B. T. D. R. Neto, P. G. B. De Oliveira, G. C. Rêgo, A. L. Neto, and L. C. Casteletti, "In-Situ Production of Metal Matrix Composites Layers by TIG Surface Alloying to Improve Wear Resistance of Ductile Cast Iron Using a Buffer-Layer and Post Weld Heat Treatment," *Coatings,* vol. 13, no. 7, p. 1137, 2023. [Online]. Available: https://www.mdpi.com/2079-6412/13/7/1137.

[10] S. Balaguru and M. Gupta, "Hardfacing studies of Ni alloys: a critical review," *Journal of Materials Research and Technology,* vol. 10, pp. 1210-1242, 2021/01/01/ 2021, doi: https://doi.org/10.1016/j.jmrt.2020.12.026.

[11] N. Srisuwan, N. Kumsri, T. Yingsamphancharoen, and A. Kaewvilai, "Hardfacing Welded ASTM A572-Based, High-Strength, Low-Alloy Steel: Welding, Characterization, and Surface Properties Related to the Wear Resistance," *Metals,* vol. 9, no. 2, p. 244, 2019. [Online]. Available: https://www.mdpi.com/2075-4701/9/2/244.

[12] J. Coronado, H. Caicedo, and A. Gomez, "The effects of welding processes on abrasive wear resistance for hardfacing deposits," *Tribology International - TRIBOL INT,* vol. 42, pp. 745-749, 05/01 2009, doi: 10.1016/j.triboint.2008.10.012.

[13] A. Gómez, C. Cañón, and D. Ramírez, "Hard faced welded tips in shredder hammers: technical and economical performance," 2008.

[14] C. Taengwa and A. Kaewvilai, "Effects of preheating and dual shielding on flux-cored arc welded high-strength carbon steel for hardfacing application," in *Key Engineering Materials*, 2020, vol. 856: Trans Tech Publ, pp. 112-118.

[15] B. Srivastava, D. Tewari, and P. Jyoti, "A Review on Effect of Preheating and/or Post Weld Heat Treatmemt (Pwht) on Mechanical Behaviour of Ferrous Metals," *International Journal of Engineering Science and Technology,* vol. 2, 04/01 2010.

[16] M. Franceschi *et al.*, "Effect of the welding technique and post-welding heat treatments on the microstructure and mechanical properties of a high silicon nanostructured carbide-free bainitic steel," *Journal of Materials Research and Technology,* vol. 31, pp. 718-732, 2024/07/01/ 2024, doi: https://doi.org/10.1016/j.jmrt.2024.06.075.

[17] D. Rathod, S. Pandey, S. Aravindan, and P. Singh, "Diffusion Control and Metallurgical Behavior of Successive Buttering on SA508 Steel Using Ni-Fe Alloy and Inconel 182," *Metallography Microstructure and Analysis - ASM International,* vol. Vol-5, pp. 450-460, 10/18 2016, doi: 10.1007/s13632-016-0311-z.

[18] W. Winarto, M. Anis, and T. P. Hertanto, "Mechanical Properties and Microstructure of Welded Dissimilar Metals Using Buttering & Non-Buttering Layer," *Advanced Materials Research,* vol. 789, pp. 341-346, 2013, doi: 10.4028/www.scientific.net/AMR.789.341.

[19] S. Chidambaresh, N. Radhika, D. k. N H, and S. Bal, "A review on welding techniques: properties, characterisations and engineering applications," *Advances in Materials and Processing Technologies,* vol. 10, pp. 1-56, 03/05 2023, doi: 10.1080/2374068X.2023.2186638.

[20] M. Ťavodová, M. Hnilicová, P. Beno, and N. Náprstková, "Quality Evaluation of Welding Joints by Different Methods," *FME Transactions,* vol. 48, pp. 816-824, 08/15 2020, doi: 10.5937/fme2004816T.

[21] B. Srikarun, H. Oo, and P. Muangjunburee, "Influence of Different Welding Processes on Microstructure, Hardness, and Wear Behavior of Martensitic Hardfaced Cladding," *Journal of Materials Engineering and Performance,* vol. 30, 08/19 2021, doi: 10.1007/s11665-021-06109-0.

[22] J. Zhang, J. Wei, S. Wei, Z. Huang, W. Wei, and L. Xu, "Microstructure and abrasive wear resistance evolution of water and liquid nitrogen forced cooling Fe-Cr-C deposit," *Tribology International,* vol. 174, p. 107732, 10/01 2022, doi: 10.1016/j.triboint.2022.107732.

[23] S. Allain, A. Roth, O. Bouaziz, and E. D'Eramo, "Microstructure-based behavior law for globular pearlitic steels," *Journal of Materials Research and Technology,* vol. 8, 05/01 2019, doi: 10.1016/j.jmrt.2019.03.014.

[24] H. Alamdari, "The effect of nickel, silicon and austenite deep-freezing treatment on the microstructure and wear properties of high chromium white cast iron containing niobium," *International Journal of Cast Metals Research,* vol. 11, pp. 89-95, 09/01 1998, doi: 10.1080/13640461.1998.11819262.

[25] V. Tandon, M. A. Thombre, A. P. Patil, R. V. Taiwade, and H. Vashishtha, "Effect of Heat Input on the Microstructural, Mechanical, and Corrosion Properties of Dissimilar Weldment of Conventional Austenitic Stainless Steel and Low-Nickel Stainless Steel," *Metallography, Microstructure, and Analysis,* vol. 9, no. 5, pp. 668-677, 2020/10/01 2020, doi: 10.1007/s13632-020-00681-y.

[26] H. Inoue, T. Koseki, S. Ohkita, and M. Fuji, "Formation mechanism of vermicular and lacy ferrite in austenitic stainless steel weld metals," *Science and Technology of Welding & Joining,* vol. 5, pp. 385-396, 12/01 2000, doi: 10.1179/136217100101538452.

[27] H. Hong, B. S. Rho, and S. Nam, "A study on the crack initiation and growth from δ-ferrite/γ phase interface under continuous fatigue and creep-fatigue conditions in type 304L stainless steels," *International Journal of Fatigue,* vol. 24, pp. 1063-1070, 10/01 2002, doi: 10.1016/S0142-1123(02)00019-1.

[28] S. Godin, J. Hamel-Akré, D. Thibault, A.-M. Serventi, and P. Bocher, "Ni and Mn enrichment effects on reformed austenite: thermodynamical and low cycle fatigue stability of 13%Cr–4%Ni and 13%Cr–6%Ni stainless steels," *SN Applied Sciences,* vol. 2, no. 3, p. 382, 2020/02/11 2020, doi: 10.1007/s42452-020-2180-y.

[29] L. Hao, X. Ji, G. Zhang, W. Zhao, M. Sun, and Y. Peng, "Carbide precipitation behavior and mechanical properties of micro-alloyed medium Mn steel," *Journal of Materials Science & Technology,* vol. 47, pp. 122-130, 2020/06/15/ 2020, doi: https://doi.org/10.1016/j.jmst.2019.12.024.

[30] O. Popović, R. P. Cvetković, L. Radović, Z. Burzić, and D. Arsić, "The influence of heat input on the toughness and fracture mechanism of surface weld metal," *Procedia Structural Integrity,* vol. 13, pp. 2216-2220, 2018/01/01/ 2018, doi: https://doi.org/10.1016/j.prostr.2018.12.138.

Influence of Annealing Treatment on Electrical Characteristics of Ce-Doped MoS_2 Thin Film Grown by DC-Unbalanced Magnetron Sputtering

Aljufri Hadju, Indris S. Sianturi, Kholik Hidayatullah, Idawati Supu, and Yudi Darma*

Department of Physics, Institut Teknologi Bandung, Bandung, Indonesia

*E-mail: yudi@itb.ac.id

Abstract We study the annealing effect on the electrical characteristics of Ce-doped MoS_2 thin films. In the present study, we successfully synthesized thin film Ce-doped MoS_2 and the treatment of annealing Ce-doped MoS_2 at 500 ^{0}C using the DC-unbalanced magnetron sputtering method. The thin film is characterized by SEM, X-ray diffraction, and current-voltage (I-V). We find the XRD results revealed changes in crystallinity and phase composition, indicating improved crystal quality with increased annealing temperature. SEM analysis demonstrated variations in surface morphology and grain size, which correlated with the annealing conditions. The I-V measurements showed responsibility and sensitivity, highlighting the role of annealing in enhancing the electrical performance of Ce-doped MoS_2 thin films.

Keywords: Ce-doped MoS_2, DC-Unbalanced Magentron Sputtering, electrical characteristics.

1. Introduction

Transition metal dichalcogenides (TMD) materials with graphene-like layer structures have attracted the attention of scientists in recent decades because they have several advantages such as good optical, structural, and electronic properties, efficient co-catalytic, suitable band gaps [1] and can facilitate charge generation and separation [2,3]. As a result, TMD materials are widely used in many different sectors, including physicochemical, electronics, photonics, and energy. Molybdenum disulfide (MoS_2) is one kind of TMD material that has gained attention. Materials with such two-dimensional structures have found widespread application in photocatalytic degradation [4], photonics and energy [5], catalysts [6], micro-electronics [7], super solid lubricants, photocatalytic dye degradation, lithium-ion batteries, dye-sensitized solar cells, and materials science and biosensors [8]. MoS_2 is widely used because of its excellent biocompatibility and low toxicity. Moreover, MoS_2 possesses special qualities and attributes as a big surface area, chemical inertness, and anisotropic [9], good electronic properties and thermal stability [10] high active sites of the layered nanosheets, chemical stability, ability to absorb visible light, and good carrier mobility to facilitate the transport of electrons produced by light [11]. In some research, MoS_2 is also used as a co-catalyst or as a constituent component of composites/heterojunctions.

Journal of Physics: Conference Series **2980** (2025) 012010 doi:10.1088/1742-6596/2980/1/012010

Heterojunction has good potential for fast charge transfer, generation and separation of photogenerated charge carriers, and better interface contact [12,13]

Rare-earth materials are still not widely used as dopants. Some of these are: europium-doped $BiVO_4$ photocatalytic degradation of methyl orange (MO) under visible light irradiation [14], Er-doped ZnO for degradation of methylene blue under UV illumination [15], and dysprosium-doped ZnO nanoparticles for photocatalytic degrading of a textile dye under visible light irradiation [16]. Nonetheless, this kind of material can be used in optoelectronic devices due to its special qualities that increase photoelectric conversion efficiency. Cerium has been widely applied in degradation of dibenzothiophene in gasoline under visible light irradiation [17], catalytic activity toward H_2 production [18], symmetric supercapacitors [19], and photocatalytic degrade of methyl orange under visible light illumination [20]. The ability to adjust the adsorption edge makes cerium widely used in increasing photocatalytic activity [21,22]. Ce ions have Ce^{3+} and Ce^{4+} oxidation states used in redox potential and oxygen storage capacity [23]. Therefore, cerium becomes an ideal candidate material to be combined with MoS_2 to produce more active sites that can enhance the photocatalytic activity of MoS_2 under visible light irradiation.

One of the key processes that influence the properties of thin films, including doped MoS_2, is the annealing treatment. Annealing involves heating the material at elevated temperatures, which can lead to changes in the structural, optical, and electrical characteristics of the thin films. In the case of Ce-doped MoS_2, annealing is expected to improve the crystal quality by reducing defects, redistributing dopants, and potentially modifying the material's electronic band structure. These changes can significantly impact the electrical characteristics of the thin film, including its current-voltage (I-V) behaviour, which is crucial for evaluating its performance in device applications.

2. Method

In this experiment, all the chemicals were purchased from Sigma-Aldrich and used without further purification. Molybdenum disulfide doping cerium (Ce-MoS_2) target was produced from pressed Ce-MoS_2 (Ce 0.3 g and MoS_2 g) mixing powder by using a mold with diameter of 2.8 cm. silicon. Ce-doped MoS_2 were deposited on Si (100) substrates using DC-unbalanced magnetron sputtering process. The deposition was performed using chamber pressure of 3,5 x 10^{-2} mbar, deposition power 12 W. After deposition, the RT samples were annealed for 30 min under N_2 environment at temperatures 500°C.

The morphologies and elemental compositions of the samples were systematically assessed through the utilization of scanning electron microscopy (SEM) and energy-dispersive X-ray spectroscopy (EDX). Additionally, the structural characteristics of the specimens were analyzed by means of X-ray diffraction (XRD). The electrical properties of the samples were delineated using current-voltage (I-V) measurements, which were conducted utilizing the Keithley 617 programmable electrometer. The fundamental principle underlying I-V measurement involves the application of a voltage across the source and drain (-5 to 5 volts), followed by the recording of the resultant current traversing the material. Silver metal (Ag) was employed as the electrode and configured in a metal-semiconductor-metal (MSM) arrangement.

Journal of Physics: Conference Series **2980** (2025) 012010
doi:10.1088/1742-6596/2980/1/012010

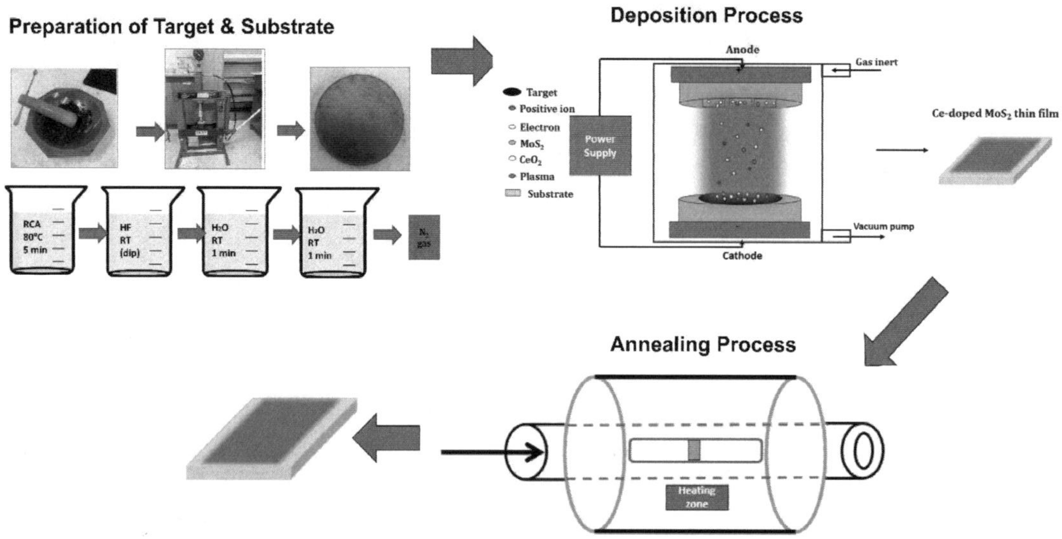

Figure 1. Schematic illustration of the experiment method

3. Result and discussions

The crystal planes of Ce-MoS_2 film were analysed by XRD, which utilized Hitachi SU3500-A at 15.0 kV. The patterns of Ce-MoS_2 film are exhibited in figure 1. The diffraction angle of above sample was scanned in the range 35^0 - 65^0. As displayed in the XRD pattern of Ce-MoS_2, the planes of (100), (220), (106) and (008) are respectively located at $32{,}97^0$, $48{,}84^0$, $57{,}61^0$, $61{,}87^0$. The XRD pattern of Ce-MoS_2 film illustrates the existence of cubic-typed CeO_2 and hexagonal MoS_2 in sample. The matching of standard and observed d value of MoS_2 hexagonal crystal structure of MoS_2 [24]and the cubic crystal structure of CeO_2 [25]. Cerium (Ce) is classified as a rare-earth element that is frequently utilized as a dopant owing to its dual oxidation states, namely Ce^{3+} and Ce^{4+}. Within the MoS_2 lattice, cerium generally substitutes a portion of the molybdenum (Mo) atoms or engages with lattice sites. This doping mechanism alters the electronic band structure by the introduction of localized states. Another treatment was performed on the Ce-MoS_2 film where the Ce-MoS_2 film was annealed at a temperature of 500^0C. The annealing treatment of Ce-MoS_2 film the crystallinity of Ce-MoS_2 film increase due to the recrystallization proses during annealing, as indicated by intensity of the diffraction peak increases. Ce doping increases the interlayer distance in MoS2, facilitating better substrate and electron transport, which contributes to enhanced crystallinity[26]. The annealing process under nitrogen (N_2) gas plays a significant role in improving the lattice structure and electronic properties of Ce-doped MoS_2. The presence of nitrogen contributes not only to the stabilization of the lattice but also exerts a significant influence on the electronic states, thereby augmenting the material's efficacy in various applications, including electronic systems and energy conversion devices. The presence of air can facilitate the formation of larger crystal dimensions in contrast to nitrogen atmospheres, which tend to stabilize smaller crystal structures, thereby influencing the overall morphology and characteristics of the film. Nitrogen functions as a pinning agent, markedly diminishing grain size and inhibiting grain growth throughout the deposition process. These changes are evident in the observed improvements in diffraction intensity and surface morphology, as demonstrated in the XRD and SEM results.

Journal of Physics: Conference Series **2980** (2025) 012010 doi:10.1088/1742-6596/2980/1/012010

Table 1. Crystal structure of Ce-MoS_2 RT and annealing

Sample	2θ (°)	β (°)	D (nm)	d (Å)	a (Å)	c (Å)	ε	δ (nm^{-2})
Ce-MoS_2 RT	61.87	5.31	1.74	1.498	3.12	11.98	0.04	0.33
Ce-MoS_2 Annealling	61.83	0.44	10.5	1.499	3.11	11.99	0.03	0.002

The table provides a comparison of various crystal structure parameters for Ce-MoS_2 under two conditions: at room temperature (RT) and after annealing. The annealing process induces notable improvements in the structural properties of the material. The diffraction angle (2θ) remains nearly unchanged, from 61.87° at room temperature (RT) to 61.83° after annealing, indicating preserved crystal plane orientation and no major phase transformations. The full width at half maximum (FWHM), β, narrows significantly from 5.31° to 0.44°, reflecting enhanced crystallinity due to reduced defect density and larger crystallite size, which increases from 1.74 nm to 10.5 nm. This growth is facilitated by atomic diffusion during annealing, leading to more ordered crystalline domains. The interplanar spacing (d) shows minimal expansion from 1.498 Å to 1.499 Å, while lattice parameters reveal a slight decrease in a (3.12 Å to 3.11 Å) and a minor increase in c (11.98 Å to 11.99 Å), indicating subtle restructuring and stress relaxation. Lattice strain (ε) reduces from 0.04 to 0.03, and dislocation density (δ) drops significantly from 0.33 nm^{-2} to 0.002 nm^{-2}, demonstrating the elimination of defects and the formation of a more stable, defect-free lattice. The annealing process markedly improves the crystalline quality of Ce-MoS_2 by enhancing the crystal size, reducing strain, and significantly lowering defect densities. These improvements are critical for applications where high crystallinity and minimal defects are essential, such as in electronic devices, catalytic processes, and energy storage materials.

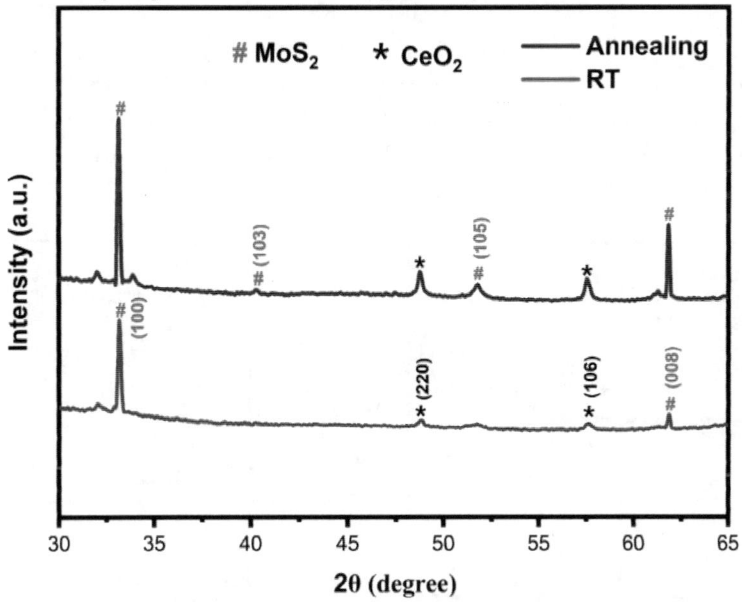

Figure 2. Diffraction pattern of the Ce-MoS_2 RT (black-line) and Annealing sample at 500°C (blue-line).

Journal of Physics: Conference Series **2980** (2025) 012010 doi:10.1088/1742-6596/2980/1/012010

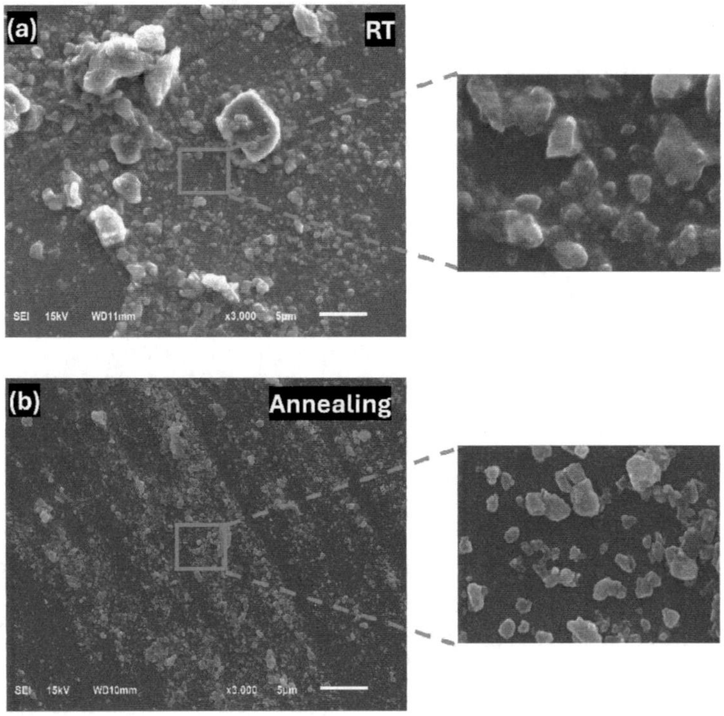

Figure 3. Scanning electron microscope (SEM) image of the Ce-MoS_2 a) RT and b) annealing 500^{0}C.

Figure 3(a)-(b), shows the SEM images of the Ce-MoS_2 RT and annealing treatment. From the SEM images, it is observed that Ce-MoS_2 RT is in the form of flakes. The addition of annealing treatment can change the flake structure of Ce-MoS_2 into structured grains. High-temperature annealing can lead to grain growth, which increases the size of the Ce-MoS_2 crystal grains. This is due to the increased atomic mobility at higher temperatures, resulting in recrystallization and grain growth, which can alter the size and surface morphology.

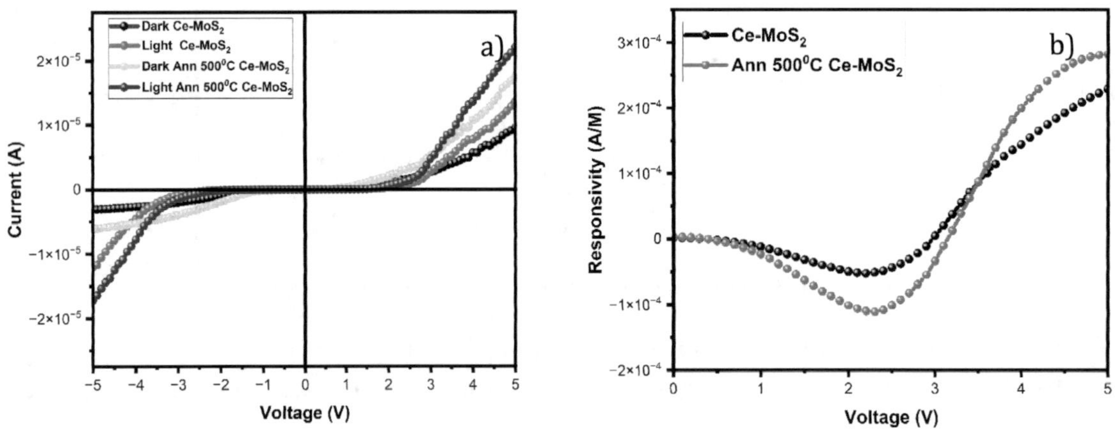

Figure 4. *I-V* characteristic a) Ce-MoS_2 and Ann Ce-MoS_2, b) responsivity Ce-MoS_2 and Ann Ce-MoS_2

Figure 4. a) shows *I-V* characteristic at room temperature of the Ce-MoS_2 and annealing 500°C Ce-MoS_2 thin films in the dark and light conditions. The input voltage is set in reverse-forward bias ranging from -5 V to 5 V. The photocurrent generated being higher in the annealed thin film, which is 1.36 × 10-5 A for RT Ce-doped MoS_2 thin film and 2.21 × 10-5 A for annealed Ce-doped MoS_2 thin film at voltage of 5 V, indicating a commendable sensitivity of both samples to the incident photons. Moreover, the I-V curve shows non-linear behaviour due to the Schottky barrier between metal semiconductor interfaces. The current exhibited a gradual increase from 0 to ±1.5 V, followed by a pronounced escalation once the voltage surpassed ±1.5 V. Additionally, it was noted that the dark current escalated in conjunction with the annealing treatment of Ce-MoS_2. Fig. 4 b) shows responsivity of the Ce-MoS_2 and annealing 500°C Ce-MoS_2 thin films. As a result, we report that the annealing treatment enhances the responsivity of the material. The responsivity value increased from 9.12 × 10-7 A/W to 1.54 × 10-6 A/W on annealed Ce-doped thin films. This improvement is attributed the annealing process reduces structural defects, optimizes the crystal structure. the responsivity-voltage correlation demonstrates that annealing markedly enhances the responsivity of Ce-MoS_2 towards light stimuli. This increased sensitivity can be attributed to the passivation of defects and the augmentation of carrier generation, culminating in enhanced transport of photogenerated charges. The improved responsivity observed at elevated positive voltages suggests alterations in charge transport dynamics, likely instigated by modifications in the band structure following annealing. Collectively, annealing at 500°C serves to enhance both the electrical conductivity and optoelectronic efficacy of Ce-MoS_2, thereby rendering it more advantageous for utilization in photodetectors and optoelectronic apparatuses.

The observed current exhibits an augmentation concomitant with the elevation of incident light intensity for both films, thereby signifying heightened photosensitive behavior. Given that the energy of an incident photon surpasses the band gap energy of the film, electrons residing in the valence band (Ev) are consequently elevated to the conduction band (Ec).

4. Conclusions

The Ce-doped MoS_2 thin films with annealing treatment have been successfully grown on the surface of Si (100) substrate using the DC unbalanced magnetron sputtering method. The SEM analysis showed a change in morphology from a flake shape to a more structured grain after annealing. In addition, the XRD analysis demonstrated that the obtained MoS_2 has a hexagonal structure, with diffraction peaks observed at 2θ = 32.97°, 48.84°, 57.61°, 61.87°, which corresponding with (100), (220), (106) and (008) planes, respectively. The annealing treatment affects the crystallinity of the thin film, where several parameters such as crystal size increase, lattice strain and dislocation density decrease, indicating low defect density and good crystallinity. In addition, the I-V analysis revealed an excellent response of current-voltage, with the photocurrent generated being higher in the annealed thin film, which is 1.36 × 10-5 A for RT Ce-doped MoS_2 thin film and 2.21 × 10-5 A for annealed Ce-doped MoS_2 thin film at voltage of 5 V. The responsivity value also increased from 9.12 × 10-7 A/W to 1.54 × 10-6 A/W on annealed Ce-doped thin films. The results of this study indicate that annealing treatment can improve the electrical properties of Ce-doped MoS_2 thin films, which are expected to be new insights for future research and showed the potential of Ce-doped MoS_2 thin films with annealing treatment can be applied for optoelectronic devices applications.

Acknowledgements

This work is partly supported by the Ministry of Education, Culture, Research and Technology of the Republic of Indonesia through the 2024 research program and ITB research program year 2024. A.H, I.S.S, K.H., thanks to the Indonesian Endowment Fund for Education (LPDP) for their scholarship and fund for this research. I. S. acknowledges *Beasiswa Pendidikan Indonesia (BPI)* scholarship. All lab works were conducted at the Quantum Semiconductor and Device (QSD) laboratory of ITB for the DC-unbalanced magnetron sputtering.

References

[1] Y.M. Hunge, A.A. Yadav, S.-W. Kang, H. Kim, Photocatalytic degradation of tetracycline antibiotics using hydrothermally synthesized two-dimensional molybdenum disulfide/titanium dioxide composites, Journal of Colloid and Interface Science 606 (2022) 454–463. https://doi.org/10.1016/j.jcis.2021.07.151.

[2] A.A. Yadav, Y.M. Hunge, S.-W. Kang, Ultrasound assisted synthesis of highly active nanoflower-like $CoMoS_4$ electrocatalyst for oxygen and hydrogen evolution reactions, Ultrasonics Sonochemistry 72 (2021) 105454. https://doi.org/10.1016/j.ultsonch.2020.105454.

[3] D. You, B. Pan, F. Jiang, Y. Zhou, W. Su, CdS nanoparticles/CeO_2 nanorods composite with high-efficiency visible-light-driven photocatalytic activity, Applied Surface Science 363 (2016) 154–160. https://doi.org/10.1016/j.apsusc.2015.12.021.

[4] Z.-Q. Liu, P.-Y. Kuang, R.-B. Wei, N. Li, Y.-B. Chen, Y.-Z. Su, BiOBr nanoplate-wrapped ZnO nanorod arrays for high performance photoelectrocatalytic application, RSC Advances 6 (2016) 16122–16130. https://doi.org/10.1039/C5RA27310B.

[5] M. Wu, L. Li, N. Liu, D. Wang, Y. Xue, L. Tang, Molybdenum disulfide (MoS_2) as a co-catalyst for photocatalytic degradation of organic contaminants: A review, Process Safety and Environmental Protection 118 (2018) 40–58. https://doi.org/10.1016/j.psep.2018.06.025.

[6] H. Hwang, H. Kim, J. Cho, MoS_2 Nanoplates Consisting of Disordered Graphene-like Layers for High Rate Lithium Battery Anode Materials, Nano Letters 11 (2011) 4826–4830. https://doi.org/10.1021/nl202675f.

[7] D. Voiry, M. Salehi, R. Silva, T. Fujita, M. Chen, T. Asefa, V.B. Shenoy, G. Eda, M. Chhowalla, Conducting MoS_2 Nanosheets as Catalysts for Hydrogen Evolution Reaction, Nano Letters 13 (2013) 6222–6227. https://doi.org/10.1021/nl403661s.

[8] Z. Wang, B. Mi, Environmental Applications of 2D Molybdenum Disulfide (MoS_2) Nanosheets, Environmental Science & Technology 51 (2017) 8229–8244. https://doi.org/10.1021/acs.est.7b01466.

[9] X. Gan, H. Zhao, X. Quan, Two-dimensional MoS_2: A promising building block for biosensors, Biosensors and Bioelectronics 89 (2017) 56–71. https://doi.org/10.1016/j.bios.2016.03.042.

[10] Y. Sharma, M.A. Khan, T. Chellapandi, M.T. Sukumar, G. Madhumitha, M.M.R. Khan, M. Danish, S. Velu, S.M. Roopan, Biosynthesis of 3D/2D CeO_2/MoS_2 Nanocomposites with Enhanced Photocatalytic Activity to Degrade Organic Dye in Wastewater and Statistical Optimization of Reaction Parameters, Research Square (2021) 1–21. https://doi.org/https://doi.org/10.21203/rs.3.rs-853778/v1.

[11] P. Kuang, J. Ran, Z. Liu, H. Wang, N. Li, Y. Su, Y. Jin, S. Qiao, Enhanced Photoelectrocatalytic Activity of BiOI Nanoplate–Zinc Oxide Nanorod p–n Heterojunction, Chemistry – A European Journal 21 (2015) 15360–15368. https://doi.org/10.1002/chem.201501183.

[12] Q. Xiang, J. Yu, M. Jaroniec, Synergetic Effect of MoS_2 and Graphene as Cocatalysts for Enhanced Photocatalytic H_2 Production Activity of TiO_2 Nanoparticles, Journal of the American Chemical Society 134 (2012) 6575–6578. https://doi.org/10.1021/ja302846n.

[13] H.S.S. Ramakrishna Matte, A. Gomathi, A.K. Manna, D.J. Late, R. Datta, S.K. Pati, C.N.R. Rao, MoS_2 and WS_2 Analogues of Graphene, Angewandte Chemie 122 (2010) 4153–4156. https://doi.org/10.1002/ange.201000009.

[14] G. Ye, Y. Gong, J. Lin, B. Li, Y. He, S.T. Pantelides, W. Zhou, R. Vajtai, P.M. Ajayan, Defects Engineered Monolayer MoS_2 for Improved Hydrogen Evolution Reaction, Nano Letters 16 (2016) 1097–1103. https://doi.org/10.1021/acs.nanolett.5b04331.

[15] W. Zhang, H. He, Y. Tian, H. Li, K. Lan, L. Zu, Y. Xia, L. Duan, W. Li, D. Zhao, Defect-engineering of mesoporous TiO_2 microspheres with phase junctions for efficient visible-light driven fuel production, Nano Energy 66 (2019) 104113. https://doi.org/10.1016/j.nanoen.2019.104113.

[16] A. Zhang, J. Zhang, Effects of europium doping on the photocatalytic behavior of BiVO4, Journal of Hazardous Materials 173 (2010) 265–272. https://doi.org/10.1016/j.jhazmat.2009.08.079.

[17] K. Yu, J. Shi, Z. Zhang, Y. Liang, W. Liu, Synthesis, Characterization, and Photocatalysis of ZnO and Er-Doped ZnO, Journal of Nanomaterials 2013 (2013). https://doi.org/10.1155/2013/372951.

[18] A. Khataee, R. Darvishi Cheshmeh Soltani, Y. Hanifehpour, M. Safarpour, H. Gholipour Ranjbar, S.W. Joo, Synthesis and Characterization of Dysprosium-Doped ZnO Nanoparticles for Photocatalysis of a Textile Dye under Visible Light Irradiation, Industrial & Engineering Chemistry Research 53 (2014) 1924–1932. https://doi.org/10.1021/ie402743u.

[19] Z.M. El-Bahy, A.A. Ismail, R.M. Mohamed, Enhancement of titania by doping rare earth for photodegradation of organic dye (Direct Blue), Journal of Hazardous Materials 166 (2009) 138–143. https://doi.org/10.1016/j.jhazmat.2008.11.022.

[20] X. Zhang, N. Zhang, Y.-J. Xu, Z.-R. Tang, One-dimensional CdS nanowires–CeO_2 nanoparticles composites with boosted photocatalytic activity, New Journal of Chemistry 39 (2015) 6756–6764. https://doi.org/10.1039/C5NJ00976F.

[21] X. Li, Z. Zhang, C. Yao, X. Lu, X. Zhao, C. Ni, Attapulgite-CeO_2/MoS_2 ternary nanocomposite for photocatalytic oxidative desulfurization, Applied Surface Science 364 (2016) 589–596. https://doi.org/10.1016/j.apsusc.2015.12.196.

[22] X. Gong, Y.-Q. Gu, N. Li, H. Zhao, C.-J. Jia, Y. Du, Thermally Stable Hierarchical Nanostructures of Ultrathin MoS_2 Nanosheet-Coated CeO_2 Hollow Spheres as Catalyst for Ammonia Decomposition, Inorganic Chemistry 55 (2016) 3992–3999. https://doi.org/10.1021/acs.inorgchem.6b00265.

[23] N. Li, H. Zhao, Y. Zhang, Z. Liu, X. Gong, Y. Du, Core–shell structured CeO_2 @MoS_2 nanocomposites for high performance symmetric supercapacitors, CrystEngComm 18 (2016) 4158–4164. https://doi.org/10.1039/C5CE02466H.

[24] S.J. Adilla, E. Nurfani, R. Kurniawan, C.D. Satrya, Y. Darma, Structural and optical properties analysis of MoS_2 nanoflakes on quartz substrate as prepared by mechanical exfoliation, in: J Phys Conf Ser, Institute of Physics Publishing, 2017. https://doi.org/10.1088/1742-6596/877/1/012036.

[25] T. Taniguchi, Y. Sonoda, M. Echikawa, Y. Watanabe, K. Hatakeyama, S. Ida, M. Koinuma, Y. Matsumoto, Intense Photoluminescence from Ceria-Based Nanoscale Lamellar Hybrid, ACS Appl Mater Interfaces 4 (2012) 1010–1015. https://doi.org/10.1021/am201613z.

[26] Zhang, X., Wang, C. & Gao, Y. Cerium(III)-doped MoS_2 nanosheets with expanded interlayer spacing and peroxidase-mimicking properties for colorimetric determination of hydrogen peroxide. *Microchim Acta* **187**, 111 (2020). https://doi.org/10.1007/s00604-019-4078-1

Journal of Physics: Conference Series **2980** (2025) 012011 doi:10.1088/1742-6596/2980/1/012011

Graphene growth by HW-PECVD method and its suitable characteristics as p-type layer in p-i-n amorphous silicon-based solar cells

A. Momang Yusuf* **and Jasruddin**

Laboratorium Fisika Material Program Studi Fisika Jurusan Fisika FMIPA, Universitas Negeri Makassar, Jl. Mallengkeri, Makassar, Indonesia

*E-mail: a.momang.yusuf@unm.ac.id

Abstract. Application of graphene, either as an active layer or as a transparent conductive layer in crystalline silicon-based solar cells has been widely carried out. In contrast, for amorphous silicon-based solar cells, the application of graphene has not been widely performed. In fact, amorphous silicon-based solar cells can be an alternative photovoltaic device that is more economical and flexible than solar cells from crystalline silicon materials. In this study, we grew graphene by using PECVD method for amorphous/microcrystalline silicon-based solar cells application. Based on our results, it was found that graphene could be obtained at temperature of 300°C, pressure of 500 mTorr, and gas flow rate of methane of 30 sccm. The graphene was grown on a glass substrate that had been previously coated with nickel layer as catalyst for graphene growth. The results of UV-Vis measurements showed that the transmittance of the graphene sample was 65% at a wavelength of between 440 nm to 800 nm. By using Tauc method, we estimated the bandgap of the sample was of 2.92 eV. Based on the analysis of the Raman spectrum of the sample, the graphene sample tends to be of the p-type. With these characteristics, the resulting graphene layer has the potential to be applied as a p-type layer in amorphous/microcrystalline silicon-based solar cells.

1. Introduction

Since it was first successfully synthesized by Novoselov et al. [1], graphene immediately attracted attention of researchers all around the world. Its characteristics, especially the high conductivity and high layer transparency, have the potential to be applied in photovoltaic technology.

The use of graphene in solar cells was first reported by Li et al. by fabricating Schottky junction type solar cells based on crystalline silicon [2]. In this device, graphene acts as an antireflective coating and reduces the occurrence of reflection by up to 70% in the visible light region and about 80% in the infrared region. The conversion efficiency achieved with this device structure was about 1.5%. Following the results achieved by Li, several other researchers then reported the utilization of graphene in crystalline silicon-based solar cells with increasing efficiency [3–5].

On the other hand, research on application of graphene in amorphous silicon-based solar cells has not been widely conducted. In fact, the amorphous silicon-based solar cells have long been used as a cheaper alternative to crystalline silicon-based solar cells. Amorphous silicon materials'

Journal of Physics: Conference Series **2980** (2025) 012011 doi:10.1088/1742-6596/2980/1/012011

production cost is much cheaper. It also has more efficient photon absorption ability so that it can be made in the form of a thin layer.

Amorphous silicon-based solar cells are usually in p-i-n or n-i-p structures grown on a glass substrate coated with transparent conducting oxides (TCO). In addition to being transparent, TCO must be conductive to prevent ohmic losses at the junction with the p-type or n-type layer. In general, the research on graphene applications for solar cell was directed to be able to act as a TCO layer to substitute the more expensive material like indium tin oxide (ITO) [6–9].

In amorphous silicon-based solar cells with a p-i-n structure, light (photons) enters the material through the p-layer after passing through the TCO. The electron-hole pairs formed in this layer when exposed to light almost has no contribution to the photocurrent, because they are quickly lost through the recombination mechanism. Therefore, the p-layer must be as thin as possible and has a wide enough energy gap to avoid absorption losses but must be thick enough to be able to generate an electric field through the i-type. These criteria are usually met by hydrogenated amorphous silicon carbide (a-SiC:H) or microcrystalline silicon films. Because of graphene is carbon-based material, and there is an important role of carbon in the suitable application of a-SiC:H as p-type in amorphous silicon solar cells, we therefore study the growth of graphene to be used as type-p layer in amorphous silicon based solar cells.

2. Materials and Methods

In this study, the graphene was grown by using HW-PECVD method on a microscope slide glass as substrate. The microscope slide previously was coated with 30 – 70 nm Ni by thermal evaporation. The HW-PECVD system that we used in this study was equipped with hot-wire, placed in the plasma area. The detail construction of our HW-PECVD can be found in our published article [10].

As carbon source, we use methane gas with 99,9% purity. The graphene growth was conducted with growth parameters mentioned in Table 1.

Table 1. Parameters of HW-PECVD for graphene growth

Growth parameters	Value
Pressure in chamber	500 mTorr
Temperature of substrate	300 °C
Effective RF power	(10 ± 2) watt
Gas flow of methane	30 sccm

The graphene growth was carried out using the following procedure. Firstly, the Ni-precoated glass substrate was inserted into chamber. The chamber was vacuumed by pumping gas out until it reaches the base pressure. After the base pressure was reached, nitrogen gas was flowed into the chamber for approximately 5 minutes to ensure that there was no other gas or water vapor trapped in the chamber or gas pipe. After that, the growth temperature was set according to the specified value, which was of 300 °C. When the specified temperature had been reached, methane gas was flowed into the chamber at flow rate of 30 sccm until the pressure in the chamber reached the growth pressure as given in Table 1.

Subsequent to all specified parameters has been reached, the methane gas in the chamber was excited using 70 MHz very high frequency (VHF) radio frequency and power of 12 watts. The graphene growth process was carried out for 10 minutes. Following the growth process was complete, the flow of methane gas was stopped by closing the gas source and turning off the VHF radio wave generator. The sample was left in the chamber until the temperature reach the room temperature. The obtained sample was then analysed by using SEM, Raman spectroscopy, and UV-Vis spectroscopy.

3. Results and Discussion

Raman spectroscopy of the obtained sample was shown in Figure 1. Refers to the figure, it appears that the sample shows a Raman spectrum with bands that are typical for carbon allotropes, namely the G and D bands [11]. The G-band of sample is strong with narrow width and locate around 1601 cm^{-1}. The G-band is related to phonon vibrations in sp^2 carbon materials and a direct indication that a sample contain sp^2 structure of carbon. On the other hand, a D-band which also shows strong peak comparable with G-band located around 1350 cm^{-1} and is indication that the sample contain large number of defect sites and impurities.

Figure 1 also shows the presence of 2D bands, which is a characteristic band of graphene. The 2D bands in this sample appear weak and are allegedly to be a consequence of the large content of defects and impurities in the sample, especially the presence of oxygen functional groups. In addition, we also noted the shift in the G-band of Raman spectra of the sample toward a wavelength around 1601 cm^{-1}. This shift indicates a tendency for the formation of nanocrystalline structure of graphene rather than crystalline graphene [12]. In addition to indication of crystalline size of the graphene sample, the shift in the G-band towards a higher wavelength also indicates the occurrence of p-type doping in the sample [13,14].

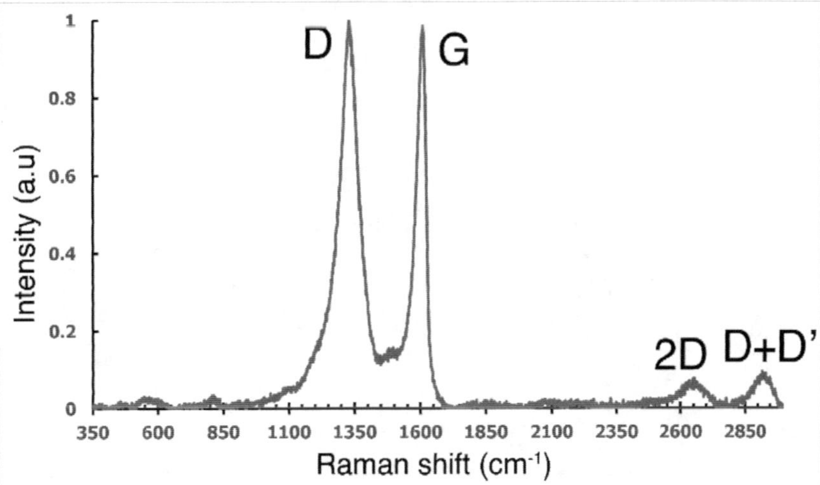

Figure 1. Raman spectrum of the obtained graphene. The sample was grown with growth parameters given in Table 1.

The SEM image of the sample is given in Figure 2. This image shows the surface of the nanocrystalline graphene sample. This kind of graphene surface image has been widely reported by several researchers [15–17].

Figure 2. SEM image of the obtained graphene

In addition to SEM and Raman characterization, the analysis of the obtained sample was also carried out through UV-Vis measurements. The results of UV-Vis measurements showed that the obtained graphene had an optical transmittance of 68% at a wavelength of 440 - 800 nm. This value was smaller than the transmittance of nanocrystalline graphene obtained by other researchers which can reach around 97% at the same wavelength [18]. The low optical transmittance value of nanocrystalline graphene obtained in this study can be suspected because of layer structure of the sample that consist of several layers and the presence of a nickel catalyst layer. Roughly, by using the graphene transmittance value at a wavelength of 550 nm obtained from several research [19], the number of graphene layers in this sample can be extrapolated as shown in Figure 3. Based on this figure, with an optical transmittance value of 68% at a wavelength of 550 nm, it was estimated that the number of graphene layers in the sample was between 9 - 13 layers.

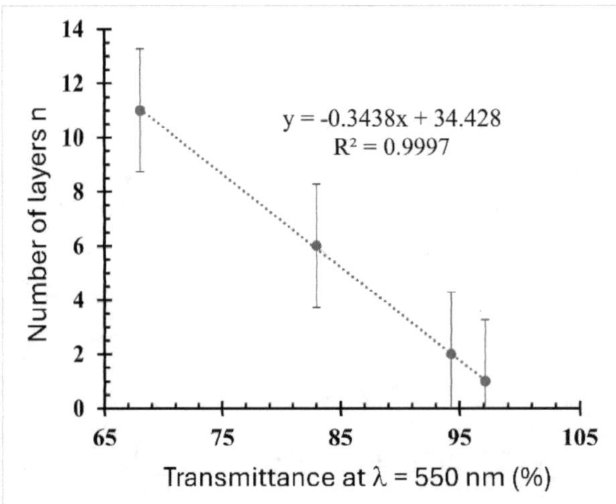

Figure 3. Extrapolation of number of layers in the obtained graphene.

Based on the UV-Vis absorbance data, the energy gap of graphene can be determined using Tauc method as shown in Figure 4. According to the figure, it was estimated that the energy gap of the obtained graphene was around 2.92 eV. With this energy gap value, the graphene sample has the potential to be applied as a window layer, namely as a p-type layer in amorphous silicon-based solar cells with a p-i-n structure.

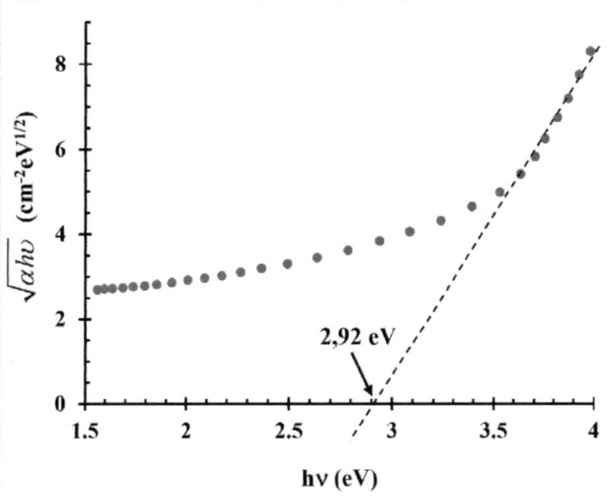

Figure 4. Using Tauc method to determine graphene's bandgap showed that the value of the bandgap was of 2,92 eV.

4. Conclusions

In summary, we have grown multilayer nanocrystalline graphene on Ni-coated microscope slide glass by using HW-PECVD method. The Raman spectrum of graphene shows the shift of G-band to higher wavelength, that is to around 1601 cm^{-1} than the usual at around 1598 cm^{-1}. This shift indicated p-type doping by impurities contain in the graphene. By using Tauc method we estimate the bandgap of the graphene was of 2,92 eV. With these characteristics, the nanocrystalline graphene has the potential to be applied as p-type layer in p-i-n amorphous silicon based solar cells.

Acknowledgments

The authors acknowledge financial support from PNBP of Universitas Negeri Makassar, Indonesia. A. M. Y. acknowledges facilities from PECVD Laboratory FMIPA Institut Teknologi Bandung, Indonesia.

References

1. Novoselov K S 2004 Electric field effect in atomically thin carbon films *Science* **306**(5696):666–9.
2. Li X, Zhu H, Wang K, Cao A, Wei J, Li C, et al. 2010 Graphene-on-silicon Schottky junction solar cells *Advanced Materials* **22**(25): 2743–8.
3. Feng T, Xie D, Lin Y, Zhao H, Chen Y, Tian H, et al. 2012 Efficiency enhancement of graphene/silicon-pillar-array solar cells by HNO_3 and PEDOT-PSS *Nanoscale* **4**(6):2130–3.
4. Fan G, Zhu H, Wang K, Wei J, Li X, Shu Q, et al. 2011 Graphene/silicon nanowire Schottky junction for enhanced light harvesting *ACS Applied Materials & Interfaces* **3**(3):721–5.
5. Miao X, Tongay S, Petterson M K, Berke K, Rinzler A G, Appleton B R, et al. 2012 High efficiency graphene solar cells by chemical doping. *Nano Letters* **12**(6):2745–50.
6. Bae S, Kim H, Lee Y, Xu X, Park J S, Zheng Y, et al. 2010 Roll-to-roll production of 30-inch graphene films for transparent electrodes *Nature Nanotechnology* **5**(8):574–8.

7. Diao S, Zhang X, Shao Z, Ding K, Jie J and Zhang X 2017 12.35% efficient graphene quantum dots/silicon heterojunction solar cells using graphene transparent electrode *Nano Energy* **31**:359–66.
8. Kalita G, Wakita K and Umeno M 2012 Low temperature growth of graphene film by microwave assisted surface wave plasma CVD for transparent electrode application *RSC Advances* **2**(7):2815.
9. Li X, Zhu Y, Cai W, Borysiak M, Han B, Chen D, et al. 2009 Transfer of large-area graphene films for high-performance transparent conductive electrodes *Nano Letters* **9**(12):4359–63.
10. Yusuf A M, Abidin K, Eliyana A, Usman I, Malago J D, Noor F A, et al. 2022 Effect of precursor gas inlet position relative to hot wire cells in HWC-IP-PECVD systems for low-temperature graphene growth *Materials Research Innovations* 1–7.
11. Ferrari A C 2007 Raman spectroscopy of graphene and graphite: disorder, electron–phonon coupling, doping and nonadiabatic effects. *Solid State Communications* **143**(1):47–57.
12. Ferrari A C and Robertson J 2000 Interpretation of Raman spectra of disordered and amorphous carbon *Phys Rev B* **61**(20):14095–107.
13. Das A, Pisana S, Chakraborty B, Piscanec S, Saha S K, Waghmare U V, et al. 2008 Monitoring dopants by Raman scattering in an electrochemically top-gated graphene transistor *Nature Nanotech* **3**(4):210–5.
14. Bong J H, Sul O, Yoon A, Choi S Y and Cho B J 2014 Facile graphene n-doping by wet chemical treatment for electronic applications *Nanoscale* **6**(15):8503.
15. Deokar G, Avila J, Razado-Colambo I, Codron J L, Boyaval C, Galopin E, et al. Towards high quality CVD graphene growth and transfer *Carbon* 89:82–92.
16. Yao Y, Li Z, Lin Z, Moon K S, Agar J and Wong C 2011 Controlled growth of multilayer, few-layer, and single-layer graphene on metal substrates. *J Phys Chem C* **115**(13):5232–8.
17. Sun J(孙捷), Lindvall N, Cole M T, Wang T(王腾), Booth T J, Bøggild P, et al. 2012 Controllable chemical vapor deposition of large area uniform nanocrystalline graphene directly on silicon dioxide *Journal of Applied Physics* **111**(4):044103.
18. Zhao T, Xu C, Ma W, Liu Z, Zhou T, Liu Z, et al. 2019 Ultrafast growth of nanocrystalline graphene films by quenching and grain-size-dependent strength and bandgap opening *Nat Commun* **10**(1):4854.
19. Sun Z, Yan Z, Yao J, Beitler E, Zhu Y and Tour J M 2010 Growth of graphene from solid carbon sources *Nature* **468** (7323):549–52.

Thermodynamic Assessment of Phase Transformation during Carbothermic Ilmenite Using Non-Carbonization Palm Kernel Shell Biomass

Boy Attaurrazaq[1], Sri Harjanto[1*], and Reza M. Ulum[1]

[1]Department of Metallurgical and Materials Engineering, Faculty of Engineering, Universitas Indonesia, Kampus, Depok 16424, Indonesia.

*E-mail: sri.harjanto@ui.ac.id

Abstract. Based on the thermodynamic assessment, this study aims to investigate the reduction behavior of carbothermic ilmenite using Non-Carbonization Palm Kernel Shell biomass (PKS-B). Phase transformation, equilibrium phases, and spontaneity reactions during reduction were systematically evaluated. The effect of PKS-B addition in metallic iron and TiO_2-rutile formation temperature are investigated, respectively. Based on thermodynamic results, ilmenite was dissociated indirectly by volatile matter of biomass to FeO and TiO_2 at 400°C to 600°C. Subsequently, biomass carbon took the reduction directly to Fe above 600°C. The metallization degree of Fe was almost close to 100% at 1200°C and 1000°C when adding half and one-time mol ratio of the stoichiometry of PKS-B. In addition, the Fe phase was consistent with the reduction of TiO_2 to form Ti-suboxides such as Ti_3O_5 and Ti_2O_3 provided by adding PKS-B above half along with reduction temperature above 1100°C. The formation of TiN from Ti-suboxides was more spontaneous than via TiO_2 due to the presence of N_2 as confirmed by Gibbs free energy change. Thus, the ilmenite reduction pathway using PKS-B was dissociation and reduction first, then continued by the nitridation process. Furthermore, the nitriding competes with the reduction process, where Fe nitrides are formed at low temperatures while Ti nitrides and metallic Fe are formed at high temperatures.

Keywords: palm kernel shell biomass, ilmenite, carbothermic reduction, thermodynamic simulation, metallization degree.

1. Introduction

The use of secondary ore sources and biomass represents alternative raw materials that can contribute to the goal of achieving zero CO_2 emissions by 2050 [1]. Rutile and Ilmenite are crucial strategic minerals that serve as sources of iron and titanium. Titanium sand, a secondary source of the mineral ilmenite, is rich in iron and titanium, with an abundance of nearly 600 million tons, particularly prevalent in Indonesia [2,3]. However, only 30% of ore reserves have been successfully extracted into economically valuable products [2]. This indicates that titanium mineral resources in Indonesia still offer significant opportunities for utilization as raw materials in the production of advanced materials, thereby increasing added value.

The carbothermic reduction of ilmenite provides a method for extracting iron and rutile from ilmenite. The varying volatile matter and fixed carbon content in carbonaceous reductants influence phase transformations and reduction pathways. Palm kernel shell biomass (PKS-B), characterized by its high volatile matter content, presents opportunities for its use as a reducing agent. Moreover, syngas (CO, CH_4, and C_xH_y) produced from biomass show reduced activity toward metal oxides at temperatures ranging from 300 to 700°C [4]. Additionally, emissions from biomass pyrolysis can be naturally reabsorbed by plants, classifying biomass as a neutral carbon source, capable of reducing CO_2 emissions by at least 50% [5,6].

Previous studies on the thermodynamic simulation of ilmenite reduction have examined the effects of temperature, type of reductant, ratio, and specific conditions. Yunos et al. [7] found that carbonaceous reductants, such as biochar with highly amorphous structures, can improve reduction performance compared to graphite. Furthermore, thermodynamic simulations suggest that the reduction process is limited by gasification, as the phases of $FeTiO_3$, TiO_2, and FeO cannot be reduced by CO at temperatures below 1000°C. Setiawan et al. [8] reported that solid metallic Fe is predicted to form at 900°C with the addition of 6.9 .wt% PKS-charcoal, which is then converted to liquid metallic Fe when the PKS-charcoal content reaches 13.9 .wt%. However, cementite (Fe_3C) reappears when the PKS-charcoal addition reaches 20.8 .wt%, alongside a decrease in $FeTi_2O_5$. Subasinghe et al. [9] noted that ilmenite is significantly reduced by carbon at 1000°C to TiO_2, $Fe_2Ti_3O_9$, and Fe_2TiO_5 as intermediate phases before being fully converted into Fe and TiO_2. Additionally, the efficiency of ilmenite reduction follows the order: biochar > activated carbon > graphite.

This study investigated the reduction behavior during the carbothermic reduction of ilmenite using palm kernel shell biomass (PKS-B) based on thermodynamic simulation. The reduction behavior highlighted phase transformation, equilibrium phases, and spontaneous reactions at the formation temperature of metallic iron and TiO_2-rutile.

2. Materials and Methods

2.1 Materials

In this study, the raw material used was ilmenite obtained from Katingan, Central Kalimantan, Indonesia with the oxide composition shown in Table. 1. Additionally, PKS biomass was obtained from Medan, North Sumatra, Indonesia. The proximate content of biomass is shown in Table. 2. The information on oxide content from ilmenite and the proximate composition of biomass is required as a reference for input into the simulation software.

Table 1. Element composition of raw ilmenite.

Oxide (.wt%)						
Fe_2O_3	TiO_2	Cr_2O_3	Al_2O_3	MnO	MgO	SnO_2
34.45	53.97	4.68	2.74	1.94	1.42	0.19
ZnO	CoO	CuO	NiO	SiO_2	P_2O_5	
0.06	0.07	0.01	0.01	0.03	0.24	

Table 2. Proximate analysis of PKS-B in .wt%.

Samples	Proximate Analysis			
	Fixed Carbon	Volatile Matter	Ash	Moisture
PKS-B	18.4	70.9	3	7.7

2.2 Thermodynamic Simulation Set-up

The phase transformation, equilibrium phases, and spontaneity reactions in the carbothermic reduction of ilmenite were conducted using HSC Chemistry 9.1.5 combined with FactSage 8.3 software student licenses. The thermodynamics simulation was performed under the assumption of a closed system, with a total input of 100 g.

Furthermore, the oxide composition of ilmenite and the proximate PKS listed in Tables 1 and 2 were input into the simulation software for phase transformation, equilibrium, and Gibbs free energy calculations. The reducing agent was PKS biomass, which was not carbonized. In the initial stage of reduction, PKS was decomposed, resulting in carbon and volatile matter, both of which were considered during the reduction process. The details of the amounts input for calculation are shown in Table 3.

Table 3. Thermodynamic assessment set-up.

Reducing Agent (stoichiometry)	**Initial Amount (%)**		
	Oxide	**Carbon fixed**	**Volatile Matter Biomass**
$O_{(in\ ore)}$ to $C_{(in\ PKS\text{-}B)}$ was 1: 1 PKS-B as much as 60 .wt% (PKS-B consisted of 18.42% fixed carbon, 70.86% volatile matter, specifically 30% *CH_4*, 20% *C_2H_6*, 25% *CO*, and 25% *H_2 Above 300°C).*	*32.85 $FeTiO_3$ + 0.03 CoO + 0.005 NiO + 0.009 CuO + 0.005 SnO_2 + 0.007 PbO_2 + 0.43 ZnO + 0.03 MgO + 0.76 Al_2O_3 + 0.13 P_2O_5 + 0.001 SiO_2 + 2.51 Cr_2O_3 + 1.09 MnO_2*	*11.05 C*	*0.005 SO_2 + 21.25 CH_4 +14.17 C_2H_6 + 17.71 H_2 +17.71 CO*

2.3 Characterization

The elemental identification of ilmenite was obtained with X-ray fluorescence (XRF) using Spectrometer EDXRF Epsilon 4 with silicon drift high-resolution detector. XRF analysis was conducted where the sample was micro-sized and then placed into a 95% Pt/Au crucible with lithium tetraborate-metaborate as flux. Furthermore, the proximate and ultimate analysis of PKS-B was determined by referring to ASTM standards D3173 (moisture), D3174 (ash), and D3175 (volatile matter).

3. Results and Discussions

3.1 Phases Transformation during reactions

The Fe-Ti-O_2 system, as illustrated in Fig. 1(a), represents the phase transformation during the carbothermic reduction of ilmenite at various oxygen partial pressures (pO_2) and temperatures, as obtained using FactSage. The pO_2 decreases to 10^{-22} atm as the temperature rises to 1400°C, facilitating an understanding of ilmenite's reduction behavior under increasingly reductive conditions.

Ilmenite begins to dissociate into pseudobrookite (Fe_2TiO_5) and rutile (TiO_2) at $pO_2 > 10^{-3}$ atm and will completely dissociate into FeO and TiO_2 at pO_2 values between 10^{-15} and 10^{-22} atm at 800°C. At 900°C with $pO_2 < 10^{-20}$ atm, the α-Fe and TiO_2 phases are formed. At the same pO_2, the α-Fe phase transforms into γ-Fe within the temperature range of 900°C to 1100°C, where the Ti_3O_5 and Ti_2O_3 phases are formed at temperatures of 1100°C and 1200°C, respectively. The FeTi phase can only be formed under conditions of $pO_2 < 10^{-22}$ atm at 1400°C. Therefore, it can be concluded that transforming ilmenite into Fe and Ti-suboxides was achievable by creating reductive conditions based on the type of reductant used or by increasing the reduction temperature. The transformation of TiO_2 into Ti-suboxides under various atmospheric and temperature conditions has been reported previously [7,10–12].

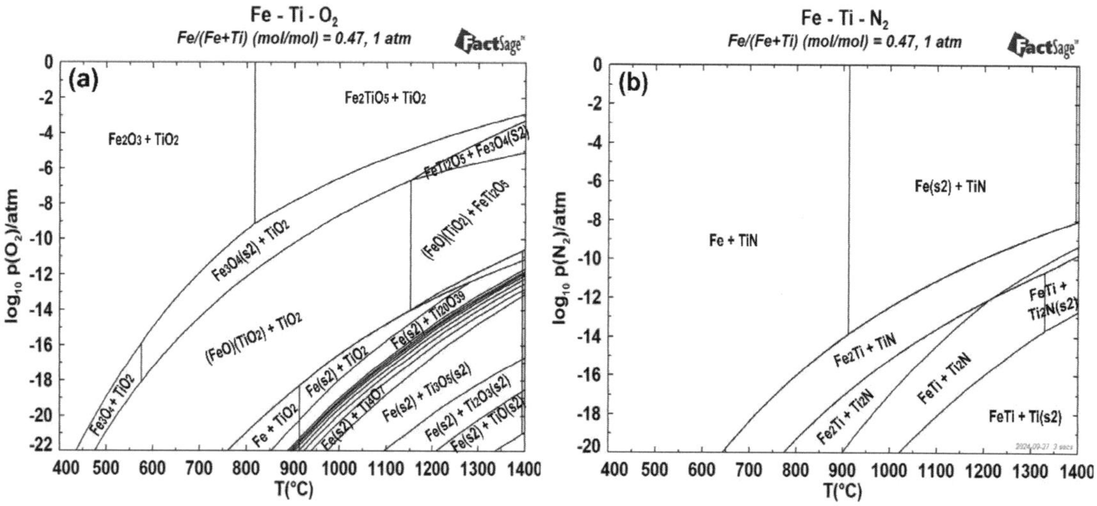

Figure 1. Thermodynamic predictions of stability and phase transformation: **(a)** Fe-Ti oxides in the reduction potential range (pO_2) and **(b)** in the nitriding potential range (pN_2).

The Fe-Ti-N_2 system, as shown in Fig. 1(b), represents the phase transformation during nitriding that occurs simultaneously with the reduction of ilmenite at various N_2 partial pressures (pN_2) and temperatures. In addition to the reduction reaction, the formation of nitride from oxide was possible across all temperature ranges. The α-Fe and TiN phases form at pN_2 = 1 atm and remain stable at temperatures above 400°C, while the α-Fe phase transforms into γ-Fe and TiN remains stable as the primary phase above 900°C. Increasing the temperature up to 900°C at $pN_2 < 10^{-15}$ atm leads to the formation of the Fe_2Ti and TiN phases, with Fe and FeTi forming when the temperature reaches 1200°C. This indicates that while increasing temperature decreases the degree of nitriding, it simultaneously increases the degree of reduction, as supported by Liu et al. [13] and Iwamoto et al. [14].

3.2 Equilibrium phases during reactions

Figure. 2 illustrates the phase equilibrium and metal formation during the carbothermic reduction of ilmenite at varying addition ratios of $C_{(in\ PKS\text{-}B)}$. Generally, ilmenite dissociates into TiO_2 and iron oxides such as Fe_3O_4 and FeO at a temperature of 300°C for each different $C_{(in\ PKS\text{-}B)}$ ratio. The activity of volatile matter capable of reducing iron oxide at low temperatures has been previously reported by Murakami et al. [4].

When the ratio of $O_{(in\ ore)}$ to $C_{(in\ PKS\text{-}B)}$ was 1: 0.5, ilmenite dissociation occurred indirectly through volatile matter (CO, CH_4, and H_2) produced by the carbonization of PKS-B, as illustrated in Fig. 2(a). This is shown by the increasing concentration of H_2 within the temperature range of 300°C to 400°C, while the carbon concentration remains stable. Ilmenite fully dissociates at 600°C, where Fe_3O_4 is reduced to Fe, resulting in Fe, FeO, and TiO_2 as the primary phases formed. At this temperature, reduction occurs both directly and indirectly, as evidenced by carbon consumption alongside rising concentrations of CO and CO_2. In the temperature range of 600°C to 1200°C, FeO begins to be reduced and depleted to form Fe, with direct reduction indicated by a decreasing carbon concentration and a significant increase in CO gas.

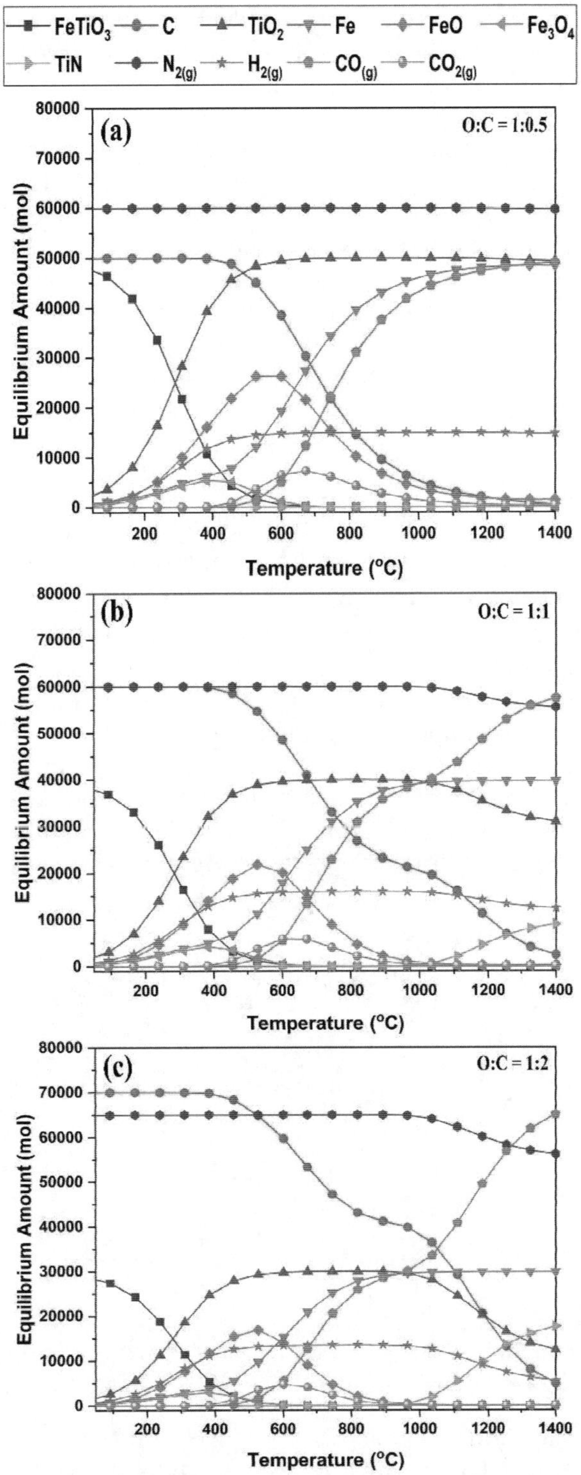

Figure 2. The equilibrium phase formed during the carbothermic reduction of ilmenite at a temperature range of 100°C to 1400°C when the ratio of $O_{(in\ ore)}$ to $C_{(in\ PKS\text{-}B)}$ was **(a)** 1: 0.5; **(b)** 1: 1; and **(c)** 1: 2.

Journal of Physics: Conference Series **2980** (2025) 012012 doi:10.1088/1742-6596/2980/1/012012

When the carbon ratio increases, carbon consumption occurs in two stages. First, during the reduction process and then during the nitriding process at high temperatures. When the ratio of $O_{(in\ ore)}$ to $C_{(in\ PKS\text{-}B)}$ was 1: 1, ilmenite completely dissociates at a temperature of 560°C, where FeO begins to be reduced by carbon and is further reduced to form Fe in the temperature range of 600°C to 1000°C, as shown in Fig. 2(b). When the temperature exceeds 1000°C, TiO_2 is reduced by carbon, and hydrogen gas forms Ti-suboxide, which is simultaneously nitrided to form TiN as the concentrations of H_2, N_2, and TiO_2 decrease.

When the ratio of $O_{(in\ ore)}$ to $C_{(in\ PKS\text{-}B)}$ was 1: 2, the dissociation temperature of ilmenite did not experience a significant shift, while FeO began to be reduced by carbon and is further reduced to form Fe in the temperature range of 600°C to 900°C. The TiO_2 phase is reduced by carbon and hydrogen gas to form Ti-suboxide, which was simultaneously nitrided to form TiN at temperatures above 950°C, as shown in Fig. 2(c). Thus, it can be concluded that by increasing the carbon ratio, reduction can be achieved at lower temperatures, with the metallization degree of Fe approaching nearly 100% at 1200°C and 975°C, when the ratio of $O_{(in\ ore)}$ to $C_{(in\ PKS\text{-}B)}$ was 1: 0.5 and 1: 2. Furthermore, the Fe phase was consistent with the reduction of TiO_2 to form Ti-suboxides such as Ti_3O_5 and Ti_2O_3 occurred by adding PKS-B above half along with reduction temperature above 1100°C.

3.3 Spontaneity reaction

The Gibbs free energy changes is an approach used to determine the possibility of a reaction occurring during reduction. The changes in Gibbs free energy on the Ellingham diagram show the spontaneity during reduction shown in Fig. 3 (a) the solid-solid and Fig. 3 (b) solid-gas reduction of ilmenite using PKS-B as a reducing agent. The Gibbs free energies ($\Delta G°$) of possible chemical reactions are shown in Eq. (1-20) that listed in Table. 4.

PKS-B, as a reducing agent, allowed reduction to occur directly and indirectly. Carbon produced during biomass decomposition facilitates gasification more rapidly than other types of carbon, such as graphite. Spontaneous biomass carbon gasification creates reductive CO even at the decomposition temperature of PKS-B. Initially, PKS-B decomposed at temperatures above 200°C producing volatile matter and carbon. The volatile matter, including tar, CH_4, CO, and H_2 gases, are present at 300°C to 700°C, as reported by Murakami et al. [4].

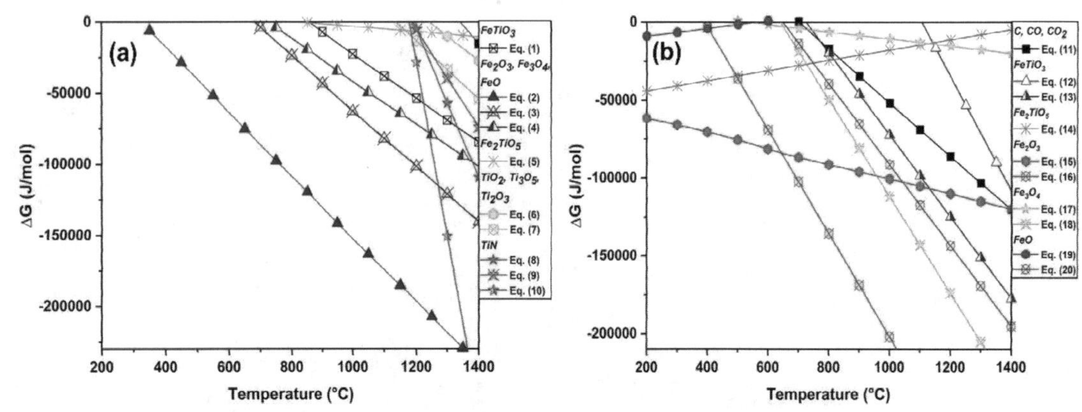

Figure 3. The changes in Gibbs free energy on the Ellingham diagram show the spontaneity of **(a)** the solid-solid and **(b)** solid-gas reduction of ilmenite using PKS-B as a reducing agent.

$FeTiO_3$ will dissociate into FeO and TiO_2 at relatively low temperatures in a reducing volatile environment, as shown in Fig. 2. However, if the conditions are not sufficiently reductive, $Fe_2O_3{\cdot}TiO_2$ is likely to be the phase formed, as illustrated in Fig. 1 [15]. Fe_2O_3 was the first phase to be spontaneously reduced by CO, followed by carbon, to form the Fe_3O_4 phase at temperatures between 300°C and 350°C (Eq. 15 and 2). Furthermore, the formation of FeO from Fe_3O_4 using CH_4 and CO occurs spontaneously at temperatures above 550°C and 650°C, respectively (Eq. 17 and 18). Meanwhile, using carbon, the formation of FeO occurs at 700°C (Eq. 3). Upon the reduction of Fe_3O_4, FeO spontaneously undergoes indirect reduction to form Fe using CH_4 at 700°C (Eq. 20), while carbon is used at 800°C (Eq. 4).

At the same temperature, undissociated ilmenite will spontaneously reduce with CH_4 at 800°C, producing Fe and TiO_2 (Eq. 13). However, ilmenite will be directly reduced with carbon at temperatures above 900°C (Eq. 1), while the presence of CH_4 decreases as the temperature increases. Overall, ilmenite is completely reduced to Fe and TiO_2 at temperatures ranging from 900°C to 1100°C. Conversely, the direct reduction of TiO_2 by carbon occurs spontaneously when the temperature exceeds 1150°C, resulting in Ti-suboxide phases (Ti_3O_5 and Ti_2O_3) (Eq. 6 and 7). Furthermore, when the temperature increases slightly to 1200°C, Ti_3O_5 and Ti_2O_3 will spontaneously react with carbon and N_2 to form TiN (Eq. 8-10).

Table 4. Gibbs free energies (ΔG°) of possible chemical reactions.

Solid-solid Reduction			
Eq		Reaction	Gibbs Free Energy (J/mol); T(K)
(1)	Ilmenite Reduction	$FeTiO_{3(s)} + C_{(s)} \rightarrow Fe_{(s)} + TiO_{2(s)} + CO_{(g)}$	$132589.6 - 155.2T$
(2)	Haematite Reduction	$3Fe_2O_{3(s)} + C_{(s)} \rightarrow 2Fe_3O_{4(s)} + CO_{(g)}$	$71538.5 - 223.7T$
(3)	Magnetite Reduction	$Fe_3O_{4(s)} + C_{(s)} \rightarrow 3FeO_{(s)} + CO_{(g)}$	$140581.6 - 202.8T$
(4)	Wustite Reduction	$FeO_{(s)} + C_{(s)} \rightarrow Fe_{(s)} + CO_{(g)}$	$110179.9 - 151.8T$
(5)	Pseudobrookite Dissociation	$Fe_2TiO_{5(s)} \rightarrow Fe_2O_{3(s)} + TiO_{2(s)}$	$-18722 + 21.8T$
(6)	TiO_2 Reduction/Ti-suboxide formation	$3TiO_{2(s)} + C_{(s)} \rightarrow Ti_3O_{5(s)} + CO_{(g)}$	$211187.5 - 170.4T$
(7)		$2Ti_3O_{5(s)} + C_{(s)} \rightarrow 3Ti_2O_{3(s)} + CO_{(g)}$	$221642.9 - 194.8T$
(8)	TiO_2 Nitridation/TiN Formation	$2TiO_{2(s)} + 4C_{(s)} + N_{2(g)} \rightarrow 2TiN_{(s)} + 4CO_{(g)}$	$620269 - 521.1T$
(9)		$Ti_2O_{3(s)} + 3C_{(s)} + N_{2(g)} \rightarrow 2TiN_{(s)} + 3CO_{(g)}$	$405596.5 - 342.5T$
(10)		$2Ti_3O_{5(s)} + 10C_{(s)} + 3N_{2(g)} \rightarrow 6TiN_{(s)} + 10CO_{(g)}$	$1438432.7 - 1222.4T$
Solid-gas Reduction			
Eq		Reaction	Gibbs Free Energy (J/mol); T(K)
(11)	Carbon Gasification	$C_{(s)} + CO_{2(g)} \rightarrow 2CO_{(g)}$	$123315.1 - 174.9T$
(12)	Ilmenite reduction	$5FeTiO_{3(s)} + 2CH_{4(g)} \rightarrow 3Fe_{(s)} + Fe_2TiO_{5(s)} + 4TiO_{2(s)} + 2CO_{(g)} + 4H_{2(g)}$	$405377.7 - 366.9T$
(13)		$FeTiO_{3(s)} + CH_{4(g)} \rightarrow Fe_{(s)} + TiO_{2(s)} + CO_{(g)} + 2H_{2(g)}$	$190526.7 - 262.9T$
(14)	Pseudobrookite Reduction	$Fe_2TiO_{5(s)} + 3CO_{(g)} \rightarrow 2Fe_{(s)} + TiO_{2(s)} + 3CO_{2(g)}$	$-50740 + 32.9T$

(15)	Haematite Reduction	$3Fe_2O_{3(s)} + CO_{(g)} \rightarrow 2Fe_3O_{4(s)} + CO_{2(g)}$	$-51776.6 - 48.8T$
(16)		$3Fe_2O_{3(s)} + CH_{4(g)} \rightarrow 2Fe_3O_{4(s)} + CO_{(g)} + 2H_{2(g)}$	$129475.6 - 331.35T$
(17)	Magnetite Reduction	$Fe_3O_{4(s)} + CO_{(g)} \rightarrow 3FeO_{(s)} + CO_{2(g)}$	$17266.5 - 27.9T$
(18)		$Fe_3O_{4(s)} + CH_{4(g)} \rightarrow 3FeO_{(s)} + CO_{(g)} + 2H_{2(g)}$	$198518.7 - 310.5T$
(19)	Wustite Reduction	$FeO_{(s)} + CO_{(g)} \rightarrow Fe_{(s)} + CO_{2(g)}$	$-13135.1 + 23T$
(20)		$FeO_{(s)} + CH_{4(g)} \rightarrow Fe_{(s)} + CO_{(g)} + 2H_{2(g)}$	$168117.1 - 259.6T$

Thus, it was concluded that reduction begins with the decomposition of biomass into carbon and volatile matter, where volatile compounds such as CH_4 and CO indirectly reduce oxides while carbon is reduced directly. The CO_2 gas produced from indirect reduction plays a role in carbon gasification, converting it into CO. Direct and indirect reduction occur simultaneously as long as there is a sufficient supply of carbon until the reduction process is complete. An excess supply of carbon at high temperatures significantly reduces TiO_2 to titanium suboxide, ultimately leading to the formation of TiN at 1200°C.

4. Conclusion

The thermodynamic assessment of carbothermic ilmenite using PKS biomass predicted that the reaction pathway for ilmenite begins with dissociation and reduction, followed by the nitridation process. The volatile matter and carbon content in PKS biomass facilitated both direct and indirect reduction. By increasing the carbon ratio, reduction can be achieved at lower temperatures, with the metallization degree of Fe approaching nearly 100% at 1200°C and 975°C, when the ratio of $O_{(\text{in ore})}$ to $C_{(\text{in PKS-B})}$ was 1: 0.5 and 1: 2. The Fe phase was consistent with the reduction of TiO_2 to form Ti-suboxides such as Ti_3O_5 and Ti_2O_3, which only occurred when $C_{(\text{in PKS-B})} \geq 1$ and reduction temperature was above 1100°C. The formation of TiN from Ti-suboxide was more spontaneous than via TiO_2 due to the excess of carbon and N_2 gas. This was evidenced by the absence of TiN product when the ratio of $O_{(\text{in ore})}$ to $C_{(\text{in PKS-B})}$ was 1: 0.5.

References

[1] Cox B, Innis S, Kunz NC, Steen J. The Mining Industry as a net Beneficiary of a Global Tax on Carbon Emissions. Commun Earth Environ. 2022;3(1).
[2] ESDM. Indonesian Minerals , Coal, and Geothermal Resources and Reserves. 2021;
[3] Hararap LML, Idrus A, Sukadana IG, Suwahyudi, Handayani T. Mineralogical Distribution and Characteristics of Fe- , Ti- , and V-Bearing Beach Ironsand Deposit in Adikarto Bay Kulonprogo , Yogyakarta , Indonesia Mineralogical Distribution and Characteristics of Fe- , Ti- , and V-Bearing Beach Ironsand Deposit in Ad. IOP Conf Ser Earth Environ Sci. 2023;
[4] Murakami T, Higashi R, Maruoka D, Kasai E. Reduction of Iron Ore by Uncarbonized Biomass in a Rotary Kiln Type Furnace. ISIJ Int. 2021;61(12):2971–8.
[5] Swennenhuis F, de Gooyert V, de Coninck H. Towards a CO2-neutral Steel Industry: Justice Aspects of CO2 Capture and Storage, Biomass- and Green Hydrogen-based Emission Reductions. Energy Res Soc Sci. 2022;88(April):102598.
[6] Holappa L. A General Vision for Reduction of Energy Consumption and CO2 Emissions from The Steel Industry. Metals (Basel). 2020;10(9):1–20.
[7] Yunos NFM, Idris MA, Nasrun NA, Kurniawan A, Nomura T, Rezan SA. Structural Characterizations and Phase Transition on the Reducibility of Ilmenite Ore with Different Carbon Reductants by Carbothermal Reduction Under Hydrogen Atmosphere. J Sustain Metall. 2023;

[8] Setiawan A, Shaw M, Torpy A, Pownceby MI, Harjanto S, Rhamdhani MA. Solar Carbothermic Reduction of Ilmenite Using Palm Kernel Shell Biomass. Jom. 2020;72(10):3410–21.
[9] Subasinghe HCS, Ratnayake AS. Effect of Different Carbon Sources on The Conversion of Ilmenite into Synthetic Rutile via Ball Milling Induced Carbothermic Reduction. J Alloys Compd. 2023;954:170086.
[10] Adipuri A, Li Y, Zhang G, Ostrovski O. Chlorination of Reduced Ilmenite Concentrates and Synthetic Rutile. Int J Miner Process. 2011;100(3–4):166–71.
[11] Lv W, Bai C, Lv X, Hu K, Lv X, Xiang J, et al. Carbothermic Reduction of Ilmenite Concentrate in Semi-molten State by Adding Sodium Sulfate. Powder Technol. 2018;340:354–61.
[12] Crespo-Monteiro N, Jourlin Y, Valour A, Usuga Higuita MA, Reynaud S, Hochedel M, et al. Micro–Nanostructured TiN Thin Film: Synthesis from a Photo-Patternable TiO2 Sol–Gel Coating and Rapid Thermal Nitridation. J Phys Chem C. 2020;124(46):25480–8.
[13] Liu Y, He F, Hu Q, Huang Q, You Z, Qiu G. Isothermal reduction and nitridation kinetics of ilmenite concentrate in ammonia gas. Miner Eng. 2023;203(174):108319.
[14] Iwamoto I, Kurniawan A, Hasegawa H, Kashiwaya Y, Nomura T, Akiyama T. Reduction Behaviors and Generated Phases of Iron Ores using Ammonia as Reducing Agent. ISIJ Int. 2022;62(12):2483–90.
[15] Ri V, Nersisyan H, Kwon SC, Lee JH, Suh H, Kim JG. A Thermochemical and Experimental Study for The Conversion of Ilmenite Sand into Fine Powders of Titanium Compounds. Mater Chem Phys. 2019;221(May 2018):1–10.

Journal of Physics: Conference Series **2980** (2025) 012013 doi:10.1088/1742-6596/2980/1/012013

Development of Bentonite-Cellulose-Lignin Composite for Fire-Retardant Roof Tile Applications

Arie Josua Franto Pratama Siahaan[1], Deni Purnomo[2], Ismadi[2], Dian Burhani[2], Ria Yolanda Arundina[3], Ester Rimma Suryani Togatorop[4], Sukma Surya Kusumah[2], Dede Hermawan[1], and Resti Marlina[2*]

[1] Department of Forest Products, IPB University, Bogor, Indonesia
[2] Research Center for Biomass and Bioproducts, National Research and Innovation Agency (BRIN), Cibinong, Indonesia
[3] Department of Chemical Engineering, University of Indonesia, Depok, Indonesia
[4] Departement of Agricultural Engineering, Universitas Brawijaya, Malang, Indonesia

*E-mail: resti.marlina@brin.go.id

Abstract. Fire retardants are materials designed to inhibit or slow down the primary phases of fire development. Lignin, a naturally fire-resistant material, can be combined with other substances to achieve specific properties without compromising its inherent fire-retardant qualities. Nanocellulose, particularly cellulose nanofibrils (CNF), offers excellent reinforcement, gas barrier properties, and thermal stability, making it a promising material to enhance the performance of lignin layers. This study focuses on improving the fire resistance of composite products, such as roof tiles, by utilizing bentonite-cellulose-lignin composites applied through three methods: lamination, mixing, and coating. The resulting composite demonstrates higher thermal degradation temperatures compared to commercial lignin. Furthermore, roof tiles treated with the laminated and coated methods exhibit enhanced mechanical, physical, and fire-resistant properties.

1. Introduction

The quality of composite boards has been extensively developed by researchers, focusing on physical, mechanical, and flexibility properties. Composite technology enables the creation of alternative materials made from two or more constituent materials that are more economical. One aspect of composites that is still being researched to date is the fire resistance aspect, both from the intrinsic properties of the composites made and the fire-retardant materials. Fire retardants inhibit or slow down the primary phases of fire development [1]. Moreover, studies related to the fire resistance of composite boards are still limited. Research on fire-retardant materials has been widely explored, with some fire retardants showing potential for combination with other materials to create fire-resistant composites that maintain good mechanical quality without reducing the thermal conductivity of the resulting composite. In previous research, a composite roofing board with good physical and mechanical strength was developed, showing medium-scale fire resistance potential [2]. Given this potential, enhancing fire resistance by applying fire-retardant composites to these boards is deemed necessary. One commonly found fire-retardant material in Indonesia with the potential for combination with other additives is lignin [3], [4].

Burhani and Nugroho (2017) explain that lignin purified with 10% PAC meets the criteria as a fire-retardant additive. In their study, the char produced from purified lignin reached

50.78%wt (Burhani and Nugroho, 2017), indicating higher fire resistance, as increased char content reduces flammability [6]. Lignin content in plants varies significantly, such as sugarcane bagasse with 28.4% lignin content [7], sorghum with 20-32% lignin [8], and oil palm shells (OPS), which have a high lignin content of 49% [9]. In addition to lignin, oil palm plants also contain other useful components, such as cellulose at 41.3%-46.5% [10]. Zhao et al. (2021) explain that cellulose, when reduced to nano size, offers good reinforcement, gas barrier properties, and thermal stability [11]. This aligns with the research of Jancic et al. (2020), which found that nanoscale cellulose exhibits high mechanical strength, excellent optical transparency, high specific surface area, low weight, and biodegradability [12]. These advantages make cellulose nanofibrils (CNF) a promising reinforcement for lignin layers, adding thermal stability for layers to be applied to composite roofing boards.

Zhao et al. (2021) state that lignin combined with cellulose nanofibrils (CNF) in a starch-based solution can be used as a fire-retardant composite [11]. However, in this study, bentonite is used to replace starch due to its similar properties and greater cost-effectiveness. According to Syuhada (2019), Indonesia has abundant bentonite reserves, approximately 380 million tons, which is a valuable resource that should be optimally utilized [13]. Bentonite's high fire resistance makes it a suitable material to combine with lignin and CNF for enhanced fire-retardant performance.

2. Materials and Methods

2.1 Materials and Equipment

The equipment used in this study includes disc mill, 4-20 mesh sieve, oven, digital scale, rotary drum mixer, spray gun, square mold (35 cm × 35 cm x 0.6cm), teflon sheet, hot press, calliper, and desiccator while the materials used in this work are sorghum fiber, molasses as adhesive, oil palm empty fruit bunches, oil palm kernel shell, bentonite, Poly Aluminium Chloride (PAC) technical grade, Sulfuric Acid (H_2SO_4), and aquades. Besides that, Fourier-Transform InfraRed (FTIR), X-Ray Diffraction (XRD), and Thermogravimetric Analysis (TGA) were used to characterize fire-retardant materials while Universal Testing Machine (UTM) was used to measure mechanical properties.

2.2 Methods

The raw materials used in this study include cleaned Oil Palm Empty Fruit Bunches and Oil Palm Kernel Shell, which will be dried in an oven at 105°C for 24 hours to achieve a moisture content of 5%. Sorghum and molasses were also used as materials in the production of roof tiles composite according to [2] using sorghum bagasse-based boards with a target density of 0.8 g/cm^3 and a molasses adhesive content of 20%. Fire retardant composites material was produces by mixing lignin, cellulose nano fibrils (CNFs), and bentonite with percentage of each component are CNF 25%, bentonite 40%, dan lignin 35%.

Lignin isolation was carried out following the procedures conducted by by Burhani and Nugroho (2017) using variation of PAC 10% and H_2SO_4 5% with target yield of 98.6% [5]. The lignin used in this study is derived from black liquor, a by-product of CKS processing. Moreover, Cellulose Nano-fibril (CNF) was obtained from TKKS fiber using a bleaching method based on the study by Athanasia et al. (2018) [14]. The TKKS fiber undergoes a double bleaching process, resulting in a cellulose content of 84.1±0.0 (%w/w). The bleached OPEFB is then processed through mechanical process using ultrafine super grinder machine at 2% (w/v) concentration in

Journal of Physics: Conference Series **2980** (2025) 012013 doi:10.1088/1742-6596/2980/1/012013

DI water, producing Cellulose Nano-fibril (CNF). Then, fire-retardant composite layer was produced using effective formulation that consist of 25% of CNF, 40% of bentonite, and 35% of lignin. Click or tap here to enter text.The final step is the application of the fire-retardant composite layer onto the composite roofing board. In this study, three techniques are used for applying the fire-retardant composite layer to the composite roofing board: laminated, mixed, and coated, followed by testing mechanical and fire resistant properties.

3. Results and Discussion

3.1 Material's characterization

3.1.1 X-Ray Diffraction (XRD)

X-Ray diffraction is a non-destructive method used to characterize materials, determine the crystallographic structure of crystals, and analyze the chemical composition of both natural and synthetic materials [15]. The purpose of the characterization test is to observe the diffraction pattern of the composite membrane and compare it with the diffraction patterns of its constituent elements: bentonite, CNF, and lignin.

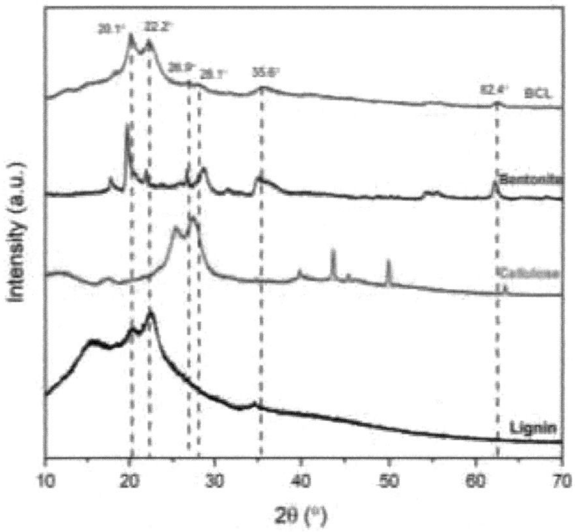

Figure 1. XRD Results of BCL, Lignin, Cellulose Nano-Fibril (CNF), and Bentonite.

Based on Figure 1, the phases appearing at specific 2θ angles indicate the identity of the composite components. In the XRD graph of the BCL35 fire-retardant composite sample, three dominant peaks are observed at 2θ angles of 20°, 22°, and 35°. In the lignin sample, dominant peaks are observed at 2θ angles of 20° and 22°, matching the peaks in the fire-retardant composite sample. In the composite sample, a third peak appears at 2θ angles of 35° and 62°, which aligns with the peaks observed in the bentonite sample. Additionally, a shoulder is present in the fire-retardant composite sample at 2θ of 26°, corresponding with peaks in both the bentonite and 2% CNF samples at the same angle. This indicates that the shoulder observed at 2θ of 26° originates from the bentonite and 2% CNF components.

Journal of Physics: Conference Series **2980** (2025) 012013 doi:10.1088/1742-6596/2980/1/012013

3.1.2 Fourier-Transform InfraRed (FTIR)

FTIR testing was conducted to analyze the functional groups of the material and the reactions formed in the fire-retardant layer based on IR spectra frequencies ranging from 400 to 4000 cm^{-1}. The absorbance value is influenced by the concentration of substances within the sample; the higher the concentration of a substance, the more molecules will absorb light at specific wavelengths, resulting in greater absorbance values [16].

Table 1. FTIR Functional Groups and Wavenumber Spectrum

Wave Number	Functional Groups	Lignin	Cellulose	Bentonite	BCL
2990-3680	O-H	3302	3302	3619	3328
1650-1600	C=C Stretching (Medium)	1600	1645	1626	1608
995-985	C=C bending	-	989	989	996

Based on Table 1, the data indicate that during the formation process of the fire-retardant composite layer (BCL), there is a decomposition of the chemical structure. Several changes in functional groups occur within the fire-retardant composite layer (BCL).

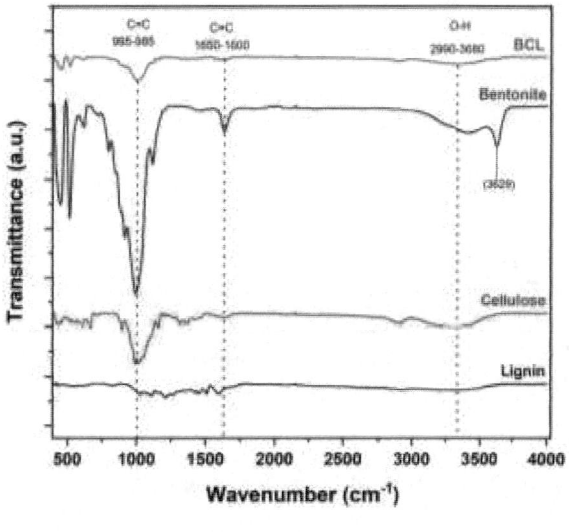

Figure 2. FTIR Results of BCL, Lignin, Cellulose Nano-Fibril (CNF), and Bentonite

The FTIR spectral measurement data show satisfactory results for the bentonite, 2% CNF, and lignin samples, with spectra characteristics closely matching those of BCL (Figure 2). Dominant peaks appear at wavenumbers 995-985 cm^{-1} and 1600-1650 cm^{-1}, indicating C=C stretching associated with C=C bond stretching and deformation. Additionally, the OH stretching bond observed around 3619 cm^{-1} in bentonite exhibits a shift in wavenumber during the formation of the fire-retardant composite layer (BCL). This shift in the OH group wavenumber suggests the occurrence of hydrogen bonding interactions between bentonite, 2% CNF, and lignin during the composite formation process.

Journal of Physics: Conference Series **2980** (2025) 012013 doi:10.1088/1742-6596/2980/1/012013

3.1.3 Thermogravimetric analysis (TGA)

Thermogravimetric analysis (TGA) is used to determine the mass loss of a sample during the heating process. The rate and extent of mass loss during the analysis indicate the reactivity of the sample material. The thermogravimetric (TG) and derivative thermogravimetric (DTG) curves of BCL (the fire-retardant layer) are shown in Figure 3.

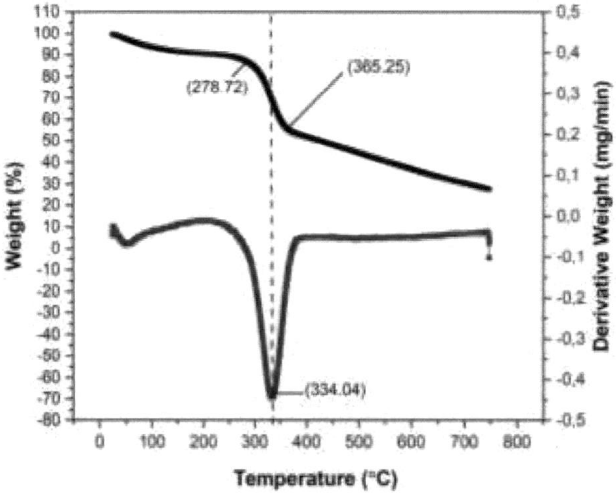

Figure 3. TGA/DTG Thermogram of the Fire-Retardant Layer (BCL)

Based on the results (Figure 3), the BCL fire-retardant layer experiences weight loss starting at 278°C and continuing to 365°C, with a degradation temperature of 334°C. This indicates an increase in degradation temperature compared to lignin before being composited with CNF and bentonite. According to Burhani and Nugroho (2017), alkali lignin from Sigma Aldrich begins to lose weight at temperatures ranging from 204°C to 343°C. These results show that lignin composited with CNF and bentonite has an increased degradation temperature, making it suitable for coating composite boards to enhance fire resistance.

3.2 Mechanical and Physical Properties of Roof Tile Composite

3.2.1 Density

Density represents the ratio between the weight of the particle board and the volume of the resulting board. Statistical analysis shows that the application of the fire-retardant layer does not significantly affect the density of the composite roofing board. The density of the composite roofing boards produced ranges from 0.75 to 0.78 g/cm^3 (Figure 4). Overall, the density values of the particle boards meet the JIS A 5908-2003 standard (0.4 g/cm^3 - 0.9 g/cm^3). The sorghum bagasse particles used in this study are a raw material with a low density of 0.125 g/cm^3 [17]. This low density allows the particles to be more easily compacted during the pressing process, resulting in better particle-to-particle contact [18].

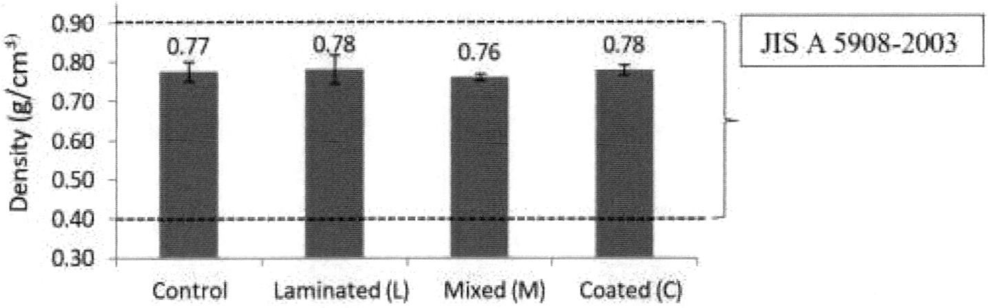

Figure 4. Density Test Results on Control Board, Laminated (L), Mixed (M), Coated (C)

3.2.2 Water Absorption Test

Water absorption capacity is a physical property that indicates the particle board's ability to absorb water during immersion [19]. According to JIS A 5908-2003, there is no specified requirement for water absorption capacity. The water absorption values for the composite roofing tiles produced range from 45% to 53% (Figure 5).

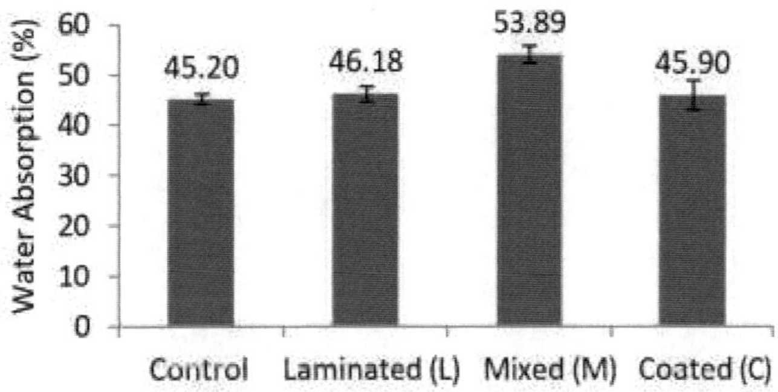

Figure 5. Water Absorption Test Results on Control Board, Laminated (L), Mixed (M), Coated (M)

Statistical analysis shows that the application of the fire-retardant layer does not significantly affect the water absorption capacity of the composite roofing board. Overall, the water absorption test results for boards treated with the fire-retardant composite layer indicate high water absorption values, ranging from 46% to 53%. These results differ from those reported by Malau et al. (2015), who explained that as the density of particle boards increases, the particle bonds become more compact, reducing pore size and making it more difficult for water to fill the spaces, thus lowering water absorption [20]. The high-water absorption observed here may be due to the additional compounds in the treated boards, which increase their water-absorbing properties. One material with a high hygroscopic or water-absorbing characteristic is bentonite [21].

3.2.3 Modulus of Elasticity (MOE)

The modulus of elasticity (MOE) is a mechanical property that indicates a composite product's ability to maintain its shape and stiffness. Detecting the mechanical properties of composite products is a crucial parameter in determining the suitability of their intended applications [22].

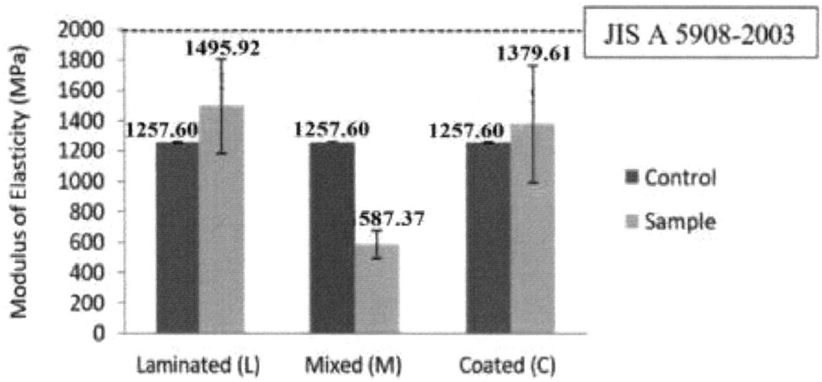

Figure 6. MOE Test Results on Control Board, Laminated (L), Mixed (M), Coated (C)

Statistical analysis indicates that the application of the fire-retardant layer significantly affects the MOE of the composite roofing board. The resulting MOE values range from 587.37 to 1495.92 MPa (Figure 6). These results show that the laminated and coated boards have higher MOE values than the control board, while the mixed board has a lower MOE value compared to the control board.

3.2.4 Modulus of Rupture (MOR)

The modulus of rupture (MOR) is a mechanical property that indicates the composite product's ability to withstand maximum load. The MOR values obtained range from 4.10 to 10.36 MPa (Figure 7).

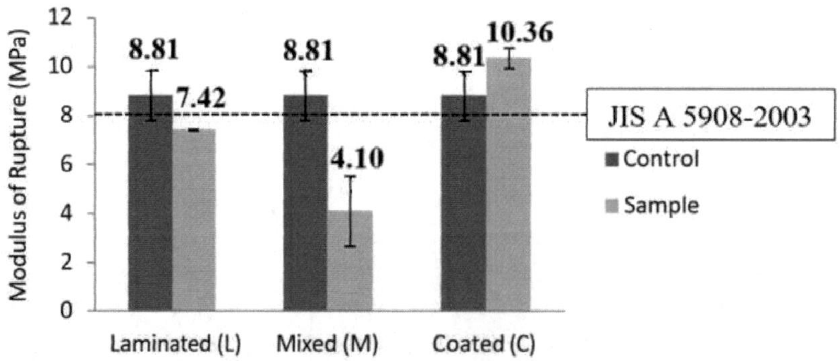

Figure 7. MOR Test Results on Control Board, Laminated (L), Mixed (M), Coated (C)

Statistical analysis shows that the application of the fire-retardant layer does not significantly affect the MOR of the composite roofing board. The coated and control boards meet the JIS A 5908-2003 standard, with values exceeding 8 MPa. The coated board exhibits a higher MOR compared to the other boards.

Journal of Physics: Conference Series **2980** (2025) 012013 doi:10.1088/1742-6596/2980/1/012013

3.3 Fire Burn Test

3.3.1 Burn-Through Test

The burn-through test aims to determine the fire resistance of a material by measuring the time it takes for fire to penetrate a composite board vertically. The longer it takes for the board to be penetrated by fire, the better its fire resistance. The time required for fire to penetrate the composite roofing board ranges from 158 to 170 seconds (Figure 8).

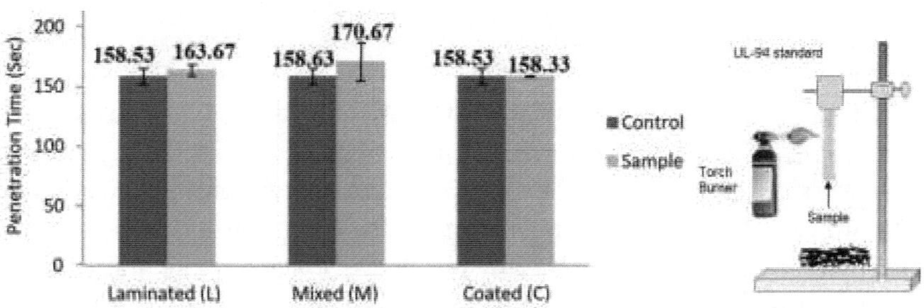

Figure 8. Burn-Through Test Results on Control Board, Laminated (L), Mixed (M), Coated (C)

Statistical analysis shows that the application of the fire-retardant layer does not significantly affect the burn-through test results for the composite roofing board. As shown in Figure 8, the board treated with the mixed (M) method demonstrated the longest fire resistance time, likely due to the even distribution of the fire-retardant solution throughout the board's thickness. The laminated (L) board achieved a better burn-through time than the control and coated boards, reaching 163.67 seconds. This improvement is attributed to the high-temperature pressure applied after the fire-retardant solution was applied, enhancing penetration depth and allowing the solution to permeate more deeply into the composite board. The coated-treated board had a burn-through time comparable to the control board, around 159 seconds. This similarity is likely because the coating treatment did not undergo high temperature pressing, leaving the fire-retardant solution on the surface and making the board more prone to combustion.

3.3.2 Vertical Burn Test

The vertical burn test aims to assess a material's fire resistance by measuring the duration of flame on a board ignited vertically for 20 seconds. This test differs from the burn-through test, where a longer flame duration indicates poorer fire resistance. In this test, the flame duration on the composite roofing board ranges from 1.07 to 2.07 seconds (Figure 9).

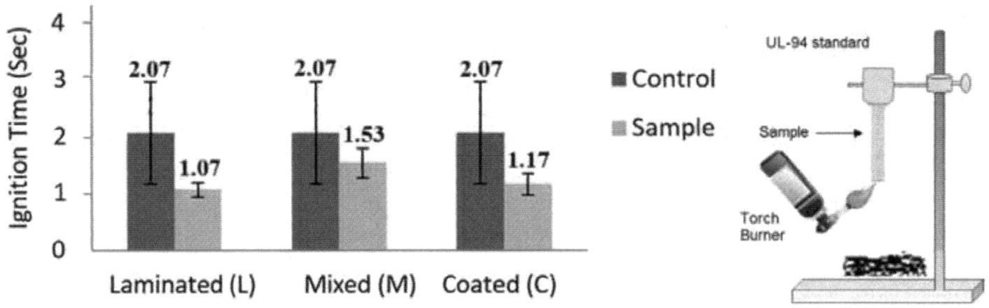

Figure 9. Vertical Burn Test Results on Control Board (C), Lamination (H), Mix (X), Coating (M)

Statistical analysis indicates that applying the fire-retardant layer does not significantly affect the vertical burn test results for the composite roofing board. The results show that the laminated (L) board had the shortest flame duration, attributed to the effective penetration depth of the fire-retardant solution, providing better internal protection compared to the coated (C) board. The vertical burn test for the mixed (M) board displayed the longest flame duration among all boards. This result differs from the burn-through test due to the mixing process's inefficiency, which may leave some fire-retardant particles unevenly distributed across the board or residual in the mixer, thus reducing its effectiveness in resisting flames.

3.3.3 Horizontal Burn Test

The horizontal burn test aims to assess a material's fire resistance by measuring the flame duration on a board ignited horizontally for 2 x 10 seconds. This test follows the same principle as the vertical burn test. The flame duration on the composite roofing board ranges from 3.02 to 6.00 seconds (Figure 10).

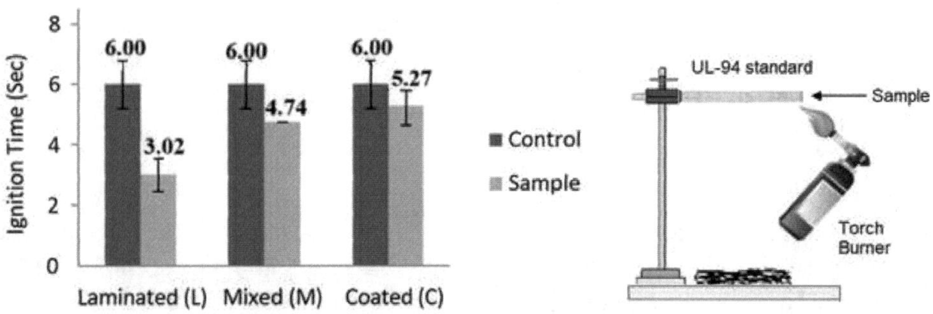

Figure 10. Horizontal Burn Test Results on Control Board, Laminated (L), Mixed (M), Coated (C)

Statistical analysis indicates that the application of the fire-retardant layer does not significantly affect the horizontal burn test results for the composite roofing board. The results show that the laminated (L) board had the shortest flame duration, consistent with the previous vertical burn test. This is due to the effective penetration depth achieved by the fire-retardant solution applied to the board. In contrast, the coated (C) board exhibited the longest flame duration among all boards in the horizontal burn test. This difference, compared to previous burn tests, can be attributed to the horizontal burning process, which causes the flame to reach the center of the coated (C) board more quickly, thereby depleting the fire-retardant layer more rapidly.

4. Conclusion

The fire-retardant composite layer (lignin-CNF-Bentonite) demonstrates an increased degradation temperature compared to commercial lignin. Although the method of applying the fire-retardant layer does not have a significant effect across the three burn tests, the fire-retardant layer has proven effective in enhancing the fire resistance of the composite roofing board. Composite roofing boards coated with the laminated and coated methods tend to exhibit improved mechanical and physical properties.

References

[1] Z. A. Ayesha Kausar Irum Rafique and B. Muhammad, "Recent Developments in Different Types of Flame Retardants and Effect on Fire Retardancy of Epoxy Composite," *Polym Plast Technol Eng*, vol. 55, no. 14, pp. 1512–1535, 2016, doi: 10.1080/03602559.2016.1163607.

[2] A. Syahfitri *et al.*, "Conversion of agro-industrial wastes of sorghum bagasse and molasses into lightweight roof tile composite," *Biomass Convers Biorefin*, vol. 14, no. 1, pp. 1001–1015, Jan. 2024, doi: 10.1007/s13399-022-02435-y.

[3] L. Liu *et al.*, "A polyphosphoramide-grafted lignin enabled thermostable and fire-retardant polylactide with preserved mechanical properties," *Compos Part A Appl Sci Manuf*, vol. 160, Sep. 2022, doi: 10.1016/j.compositesa.2022.107028.

[4] N. Surtiyeni, R. Rahmadani, N. Kurniasih, Khairurrijal, and M. Abdullah, "A Fire-Retardant Composite Made from Domestic Waste and PVA," *Advances in Materials Science and Engineering*, vol. 2016, 2016, doi: 10.1155/2016/7516278.

[5] D. Burhani and S. Nugroho, "The Potential of Lignin from Black Liquor Wastewater of Bioethanol Production as Flame Retardant Additive," in *4th International Symposiumon Innovative Bioproduction Indonesia*, Bogor, Sep. 2017, pp. 48–57. [Online]. Available: https://www.researchgate.net/publication/326477065

[6] E. M. Pearce, Y. P. Khanna, and D. Raucher, "CHAPTER 8 - Thermal Analysis in Polymer Flammability," in *Thermal Characterization of Polymeric Materials*, E. A. TURI, Ed., Academic Press, 1981, pp. 793–843. doi: https://doi.org/10.1016/B978-0-12-703780-6.50013-4.

[7] S. You *et al.*, "Effect of high antioxidant activity on bacteriostasis of lignin from sugarcane bagasse," *Biochem Eng J*, vol. 180, Mar. 2022, doi: 10.1016/j.bej.2022.108335.

[8] S. Srinivasan and S. Venkatachalam, "One pot green process for facile fractionation of sorghum biomass to lignin, cellulose and hemicellulose nanoparticles using deep eutectic solvent," *Int J Biol Macromol*, vol. 277, Oct. 2024, doi: 10.1016/j.ijbiomac.2024.134295.

[9] M. Ameen *et al.*, "Effect of acid catalysts on hydrothermal carbonization of Malaysian oil palm residues (leaves, fronds, and shells) for hydrochar production," *Biomass Convers Biorefin*, vol. 12, pp. 103–114, 2022, doi: 10.1007/s13399-020-01201-2/Published.

[10] M. Han, Y. Kim, S. W. Kim, and G.-W. Choi, "High efficiency bioethanol production from OPEFB using pilot pretreatment reactor," *Journal of Chemical Technology & Biotechnology*, vol. 86, no. 12, pp. 1527–1534, 2011, doi: https://doi.org/10.1002/jctb.2668.

[11] Y. Zhao, C. Troedsson, J. M. Bouquet, E. M. Thompson, B. Zheng, and M. Wang, "Mechanically reinforced, flexible, hydrophobic and uv impermeable starch-cellulose nanofibers (Cnf)-lignin composites with good barrier and thermal properties," *Polymers (Basel)*, vol. 13, no. 24, Dec. 2021, doi: 10.3390/polym13244346.

[12] U. Jančič, M. Bračič, A. Ojstršek, M. Božič, T. Mohan, and S. Gorgieva, "Consolidation of cellulose nanofibrils with lignosulphonate bio-waste into excellent flame retardant and UV blocking membranes," *Carbohydr Polym*, vol. 251, p. 117126, 2021, doi: https://doi.org/10.1016/j.carbpol.2020.117126.

[13] H. R. S. Warsono, W. Kurniawan, and H. Hinode, "Utilization of Modified Indonesia Natural Bentonite for Dye Removal," *ASEAN Journal of Chemical Engineering*, vol. 18, no. 2, pp. 13–21, 2018.

[14] A. A. Septevani, D. Burhani, and S. Sudiyarmanto, "Pengaruh Proses Pemutihan Multi Tahap Serat Selulosa Dari Limbah Tandan Kosong Kelapa Sawit," *Jurnal Kimia dan Kemasan*, vol. 40, no. 2, p. 71, Oct. 2018, doi: 10.24817/jkk.v40i2.3508.

[15] W. Li, H. Wang, D. Ren, Y. S. Yu, and Y. Yu, "Wood modification with furfuryl alcohol catalysed by a new composite acidic catalyst," *Wood Sci Technol*, vol. 49, no. 4, pp. 845–856, Jul. 2015, doi: 10.1007/s00226-015-0721-0.

[16] G. L. Liu and S. G. Kazarian, "Recent advances and applications to cultural heritage using ATR-FTIR spectroscopy and ATR-FTIR spectroscopic imaging," Mar. 30, 2022, *Royal Society of Chemistry*. doi: 10.1039/d2an00005a.

[17] S. S. Kusumah, K. Umemura, K. Yoshioka, H. Miyafuji, and K. Kanayama, "Utilization of sweet sorghum bagasse and citric acid for manufacturing of particleboard I: Effects of pre-drying treatment and citric acid content on the board properties," *Ind Crops Prod*, vol. 84, pp. 34–42, 2016, doi: https://doi.org/10.1016/j.indcrop.2016.01.042.

[18] J. Sutiawan, D. Hermawan, S. Surya Kusumah, B. A. Widyaningrum, and E. Sukara, "Pemanfaatan Maltodextrin Singkong untuk Perekat Ramah Lingkungan dalam Pembuatan Papan Partikel dari Bagas Sorgum Utilization of Cassava Maltodextrin for Eco-friendly Adhesive in the Manufacturing of Sorghum Bagasse Particleboard," *Jurnal Sylva Lestari ISSN*, vol. 8, no. 2, pp. 144–154, 2020.

[19] Y. Li and S. Ren, Eds., "2 - Basic Properties of Building Decorative Materials," in *Building Decorative Materials*, in Woodhead Publishing Series in Civil and Structural Engineering. , Woodhead Publishing, 2011, pp. 10–24. doi: https://doi.org/10.1533/9780857092588.10.

[20] J. C. Malau, T. Sucipto, and A. H. Iswanto, "Kualitas Papan Partikel Batang Pisang Barangan Berdasarkan Variasi Kadar Perekat Phenol Formaldehida (Particle Board Quality from Barangan Banana Stem Variation Based On Phenol Formaldehyde Resin Levels)," *Peronema Forestry Science Journal*, vol. 5, no. 1, pp. 1–9, 2016.

[21] J. Ahmad, K. J. Kontoleon, M. Z. Al-Mulali, S. Shaik, M. H. El Ouni, and M. A. El-Shorbagy, "Partial Substitution of Binding Material by Bentonite Clay (BC) in Concrete: A Review," *Buildings*, vol. 12, no. 5, May 2022, doi: 10.3390/buildings12050634.

[22] S. Zhang *et al.*, "Phosphorus containing group and lignin toward intrinsically flame retardant cellulose nanofibril-based film with enhanced mechanical properties," *Compos B Eng*, vol. 212, p. 108699, 2021, doi: https://doi.org/10.1016/j.compositesb.2021.108699.

Journal of Physics: Conference Series **2980** (2025) 012014 doi:10.1088/1742-6596/2980/1/012014

Green synthesis of silver nanoparticles using extract of unbaked Cilembu sweet potato with the presence of microwave irradiation in alkaline condition

Bryan S Aliwarga[1], A'rifaturrohmah[1], Lia A T W Asri[2], Akfiny H Aimon[3] and Arie Wibowo[2,4*]

[1]Undergraduate student, Department of Materials Engineering, Faculty of Mechanical and Aerospace Engineering, Institut Teknologi Bandung, Bandung 40132, West Java, Indonesia
[2]Materials Science and Engineering Research Group, Faculty of Mechanical and Aerospace Engineering, Institut Teknologi Bandung, Bandung 40132, West Java, Indonesia,
[3]Department of Physics, Faculty of Mathematical and Natural Sciences, Institut Teknologi Bandung, Bandung 40132, West Java, Indonesia
[4]Research Center for Nanoscience and Nanotechnology, Institut Teknologi Bandung, Bandung 40132, West Java, Indonesia

*Corresponding author: ariewibowo@itb.ac.id

Abstract. Green synthesis approach has garnered a lot of attention for fabrication of nanomaterials such as silver nanoparticles (AgNPs) because this technique offers a simple method with non-toxic product. This study reported about preparation of AgNPs through green synthesis approach using extract of unbaked Cilembu (CSP) sweet potato with microwave-assisted heating in alkaline condition (pH=9). In this research, microwave irradiation time was varied (0; 30; 60 and 90 s) to identify the effects of microwave irradiation time on the particle size, uniformity and morphology of the synthesized AgNPs. Characterization such as UV-visible spectroscopy, dynamic light scattering and transmission electron microscopy was conducted on the obtained samples of synthesized AgNPs to identify the surface plasmon resonance characteristic, average particles size and morphologies of the obtained AgNPs respectively.

1. Introduction

Nanomaterials, defined as a material that has at least one of their dimensions in the range of 1 to 100 nm, have become an interesting new class of materials in high demand with a large range of practical applications. One of the most popular nanomaterials is silver nanoparticles (AgNPs) due to its wide range of applications. Eswaran et al explored the use of AgNPs as photocatalyst for toxic dye degradation [1], while Maghimaa et al found that AgNPs coating on cotton fabrics proved to be effective as an antimicrobial element and promotes wound healing [2].

Various AgNPs synthesis methods has been explored, including both top-down and bottom-up approach. Synthesis using top-down approach reduced the bulk material into smaller size. One of the successful AgNPs synthesis for biomedical applications using this approach is through laser liquid ablation [3]. The downside of this approach is the lack of particle size control and requires complex equipment [4]. Bottom-up approach, which assembles smaller atomic and molecular component into larger particles is the more common approach for AgNPs synthesis due to better particle size control and simpler process, this is usually achieved through chemical synthesis.

However, reducing agents that commonly used for synthesis AgNPs through chemical synthesis are sodium borohydrate ($NaBH_4$) and hydrazine (N_2H_4) which are toxic and harmful subtances towards human body which limits the biomedical applications [5]. Hence, a safer and enviromentaly friendlier synthesis approach is required.

Green synthesis represents an easy, straightforward and sustainable method for the synthesis of nanoparticles through the utilization of bacteria, fungi, plants, and plant extracts, which is preferred alternative to other synthetic approaches [6]. However, plant extract was preferred above other biological methods due to its uncomplicated use, accessibility, cost-effectiveness and compatibility with biomedical applications [7]. Our previous study examined the green synthesis of AgNPs using extract of Cilembu sweet potato/CSP (*Ipomoea batatas L var. Rancing*) [8, 9]. The findings of this research indicated that the presence of AgNPs was successfully promote osteogenesis of human wharton's jelly mesenchymal stem cells [9] and a baking procedure was required to degrade starch into reducing sugars such as glucose and maltose as a sign that AgNPs was formed [8]. While the using of unbaked CSP resulted in the formation of AgCl instead of AgNPs [8]. However, it was discovered that the baking procedure which took 60 minutes at 120 °C and required a significant amount of time and energy. Previously, we introduced microwave heating as a fast and energy-efficient heating source for fabrication of AgNPs using extract of seminyak leaf [10]. The results showed that the microwave irradiation causes the reduction of Ag^+ ions to Ag^0 which indicated the formation of AgNPs, which was not found on sample without microwave irradiation. Therefore, microwave irradiation was used in this research to replace baking procedure to enhance maltose and other reducing sugar content in the extract of unbaked CSP.

During synthesis, pH is one of crucial parameter in controlling particle size of AgNPs and an increase in pH will cause the nucleation centres to rise, leading to a faster reduction of the nanoparticles [11]. Previously, unbaked CSP extract was used to prepare AgNPs in acidic condition [8]. However, it was reported that highly uniform and smaller particle size can be obtained in alkaline condition [12]. For example, Philip varied the pH of the solution from 6.5 to 8.5, and the findings indicated that pH 8.5 resulted in the smallest AgNPs size [13]. Since particle size have a significant impact on controlling the physical, chemical, optical, and electronic properties of AgNPs that synthesized using green method [14] and smaller particles size lead to higher antibacterial properties [8], this research performed fabrication of AgNPs using extract of unbaked CSP and microwave-heating in an alkaline condition (pH=9).

2. Materials and methods

2.1 Materials

A pro analytical (p.a.) grade of silver nitrate ($AgNO_3$) and 1 M sodium hydroxide (NaOH) were purchased from Sigma Aldrich. Demineralized water was obtained from Materials Processing Laboratory at Department of Materials Engineering, Bandung Institute of Technology. CSP were purchased in Dago, Bandung, from a local traditional market.

2.2 Preparation of extract of unbaked CSP and AgNPs

CSP extract was prepared based on previous report [8]. Briefly, CSPs were peeled and chopped to get the freshly meat. 100 mL of demineralized water was added into a 500 mL beaker glass that contains 100 g of finely chopped CSP. The solution was heated to 100 °C and stirred using magnetic stirrer for 10 minutes at 350 rpm. The large solid precipitate was removed from the extract solution of CSP by filtering it using metal sieve. Second filtering was performed to separate the smaller solid precipitate from the CSP extract solution using a Whatman filter paper (No. 42).

AgNPs were prepared by mixing 10 mL of 20 mM $AgNO_3$, 5 mL of demineralized water and 5 mL of CSP extract solution. Then, NaOH was added until the pH solution is equals to 9. The mixed solution was microwave irradiated for 60 seconds based on study by Ananda et al. [10]. To evaluate the effects of irradiation time, the mixed solution was given microwave irradiation treatment for shorter and longer periods of time, specifically at 0, 30, 60 and 90 seconds.

2.3 Characterization
UV-Visible spectra were conducted using Agilent Technologies 8453 spectrophotometer to investigate the surface plasmon resonance of AgNPs. Particle average size and distribution were determined using Horiba SZ-100 dynamic light scattering. The Hitachi HT 7700 transmission electron microscopy (TEM) was used to examine the size and morphology of the AgNPs particles.

3. Results and discussion

Visual observation of samples in Figure 1 presents that the colour of sample become darker as microwave irradiation increased. This visual observation could be an early indication that AgNPs might be formed in sample with microwave irradiation. Further characterization using UV-Visible spectroscopy, dynamic light scattering, and transmission electron microscopy are needed to confirm this result.

Figure 1. Comparison of visual appearance of samples in alkaline condition a) 0 seconds (b) 30 seconds (c) 60 seconds (d) 90 seconds

Compared to UV-Vis result of sample without microwave irradiation (0 s; red line), all samples with microwave irradiation (30; 60 and 90 s) showed a peak at 400-500 nm, which is the peak of surface plasmon resonance characteristics of AgNPs (Figure 2). Generally, smaller particles size of plasmonic particles such as AgNPs exhibit a sharper curve with more prominent peaks and the location of the peaks shifted towards to lower wavelengths [15].

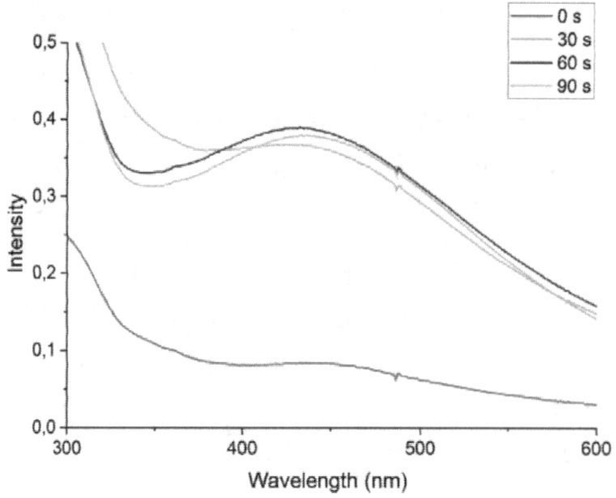

Figure 2. UV-Visible absorption spectra chart of samples as a function of microwave irradiation time.

The low-intensity peak observed in the sample without microwave irradiation is similar with unbaked CSP extract that was previously reported, suggesting the absences of AgNPs in sample [8]. This can be validated further through the DLS data and TEM image later. Since sample with 60 s of microwave irradiation showed the highest absorbance, it could be expected that this sample might have the smallest particle size, which needs to be confirmed further.

The average particle size of samples and their distribution were evaluated using dynamic light scattering, and the average particles size results were presented in Table 1 while their distribution were presented in Figure 3 respectively. As can be seen in Table 1, the average particles size of sample without microwave irradiation is 3132.5 nm, suggesting agglomeration was formed in this sample. It is noting that polydispersity index (PDI) of sample without microwave irradiation is 1.0 (Table 1) and dominated by micrometer size of particles (Figure 3), which showed a huge disparity of particle size formed in the sample.

Table 1. The average value of diameter and polydispersity index.

Microwave irradiation time (second)	Average particle diameter (nm)	Polydispersity index
0	3132.5	1.0
30	221.9	0.4
60	182.2	0.3
90	150.5	0.4

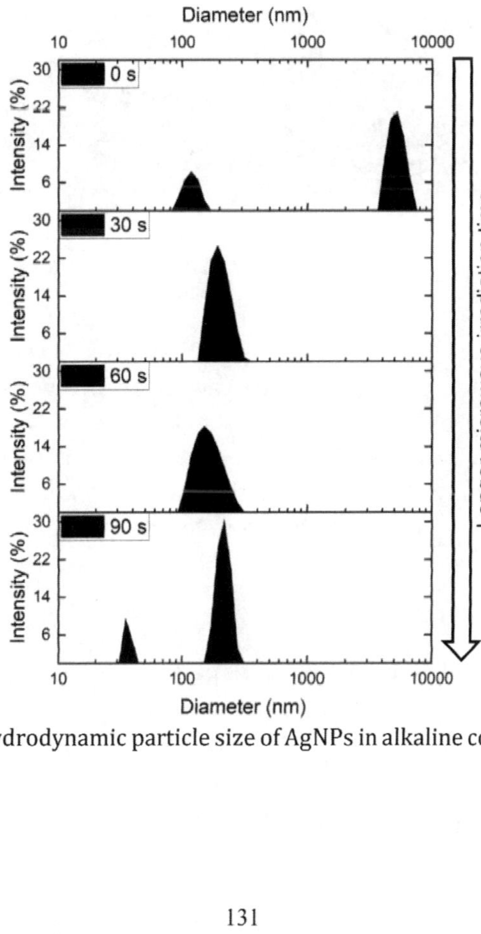

Figure 3. Distribution of hydrodynamic particle size of AgNPs in alkaline condition at various microwave irradiation time.

Sample with no irradiation (0 s) has the largest particle size, this occurs because the starch in CSP has not degrade yet to form a reducing sugar. Microwave irradiation facilitates the rapid initial heating required for the formation of AgNPs and resulting in an increase of reaction rate [16]. The microwave irradiation degrades starch into reducing sugar, leading to the AgNPs formation. Study [17] was successful in forming AgNPs through green synthesis and utilizing the microwave irradiation, resulting in higher crystallinity and monodispersed nanoparticles with a short reaction time [17]. Sample that irradiated for 90 s exhibited particles bimodal distribution and agglomeration, this phenomenon can be explained by the evaporation of a significant volume of water resulting in an increase of the solution concentration, as shown in Figure 1d.

By applying microwave irradiation, smaller particle size was observed, and the particles size of sample tend to decrease as increasing microwave irradiation time (Table 1). It is noteworthy that the sample with 60 s microwave irradiation time is still considered as homogeneous particles because PDI of this sample is 0.3. Meanwhile, samples with 30 s and 90 s of microwave irradiation time have PDI value 0.4, which is slightly higher than 0.3 as the limit for particle distribution to be considered as homogeneous particles [18].

Figure 4 shows TEM images of the sample in alkaline condition with variety of microwave irradiation time. As shown in Figure 4a, sample without microwave irradiation is predominantly composed of micrometer-sized Ag particles. By applying microwave irradiation, nanosized particles were observed and the smallest particles was observed at sample with 60s of microwave irradiation. This finding is corelating with the UV-Vis data (Figure 2) and DLS data (Figure 3), which indicate that AgNPs only begin to form when the sample is irradiated with microwaves [19].

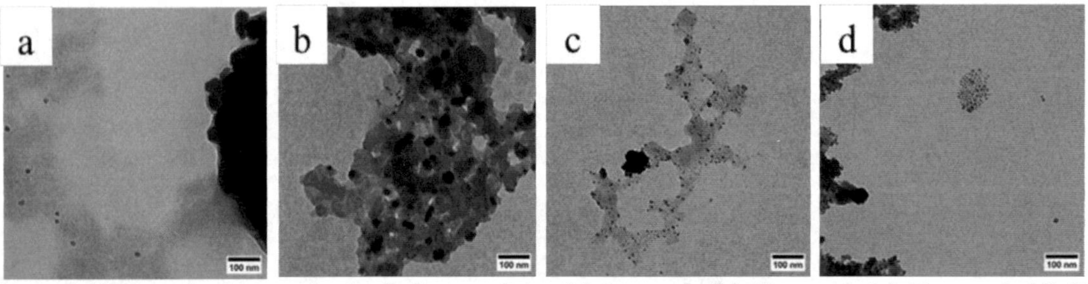

Figure 4. TEM results for samples in alkaline condition (a) 0 seconds (b) 30 seconds (c) 60 seconds (d) 90 seconds

4. Conclusion

This study successfully shows the result of green synthesis of AgNPs in alkaline condition with microwave irradiation using unbaked CSP extract. It is shown that AgNPs are formed when microwave irradiation was used, proven by the appearance of surface plasmon resonance of AgNPs (400–500 nm). The study also shows that the optimal condition to obtain smaller and more uniform AgNPs is achieved on sample with microwave irradiation for 60 seconds.

References

[1] S. G. Eswaran, H. Narayan and N. Vasimalai, "Reductive photocatalytic degradation of toxic aniline blue dye using green synthesized banyan aerial root extract derived silver nanoparticles," *Biocatalysis and Agricurtural Biotechnology,* vol. 36, p. 102140, 2021.

[2] M. Maghimaa and S. A. Alharbi, "Green synthesis of silver nanoparticles from Curcuma longa L. and coating on the cotton fabrics for antimicrobial applications and wound healing activity," *Journal of Photochemistry and Photobiology B: Biology,* vol. 204, p. 111806, 2020.

[3] N. M. Mwenze, M. Juma, M. Maaza, Z. Birech and M. S. Dhlamini, "Laser liquid ablation for silver nanoparticles synthesis and conjugation with hydroxychloroquine for Covid-19 treatment," *Materials Today: Proceedings,* 2023.

[4] N. P. U. Nguyen, N. T. Dang, L. Doan and T. T. H. Nguyen, "Synthesis of Silver Nanoparticles: From Conventional to 'Modern' Methods—A Review," *Processes,* vol. 11, no. 9, p. 2617, 2023.

[5] R. Vishwanath and B. Negi, "Conventional and green methods of synthesis of silver nanoparticles and their antimicrobial properties," *Current Research in Green and Sustainable Chemistry,* vol. 4, p. 100205, 2021.

[6] M. Rafique, I. Sadaf, M. S. Rafique and M. B. Tahir, "A review on green synthesis of silver nanoparticles and their applications," *Artificial Cells, Nanomedicine, and Biotechnology,* vol. 45, no. 7, p. 1272–1291, 2017.

[7] E. T. Assefa, G. Shumi, K. M. Gendo, G. Kenasa and N. Roba, "Review on green synthesis, characterization, and antibacterial activity of CuO nanoparticles using biomolecules of plant extract," *Results in Chemistry,* vol. 8, p. 101606, 2024.

[8] A. Wibowo, G. U. N. Tajalla, M. A. Marsudi, G. Cooper, L. A. T. W. Asri, F. Liu, H. Ardy and P. J. D. S. Bartolo, "Green Synthesis of Silver Nanoparticles Using Extract of Cilembu Sweet Potatoes (Ipomoea batatas L var. Rancing) as Potential Filler for 3D Printed Electroactive and Anti-Infection Scaffolds," *Molecules,* vol. 26, no. 7, p. 2042, 2021.

[9] M. Mira, A. Wibowo, G. U. N. Tajalla, G. Cooper, P. J. D. S. Bartolo and A. Barlian, "Osteogenic Potential of 3D printed Silver Nanoparticles-based Electroactive Scaffold for Bone Tissue Engineering Using Human Wharton's Jelly Mesenchymal Stem Cells", *Materials Advances*, vol 4, pp. 6407-6418, 2023.

[10] M. B. Ananda, F. A. S. Sanjaya, T. Bachrurozy, H. M. Ar Rasyid, A. Barlian, A. H. Aimon, F. Qulub, P. Widiyanti and A. Wibowo, "Fast Microwave-Assisted Green Synthesis of Silver Nanoparticles Using Low Concentration of Seminyak (Champeria sp.) Leaf Extract," *Indonesian Journal of Chemistry,* vol. 23, pp. 395-404, 2023.

[11] K. Vijayaraghavan and T. Ashokkumar, "Plant-mediated biosynthesis of metallic nanoparticles: A review of literature, factors affecting synthesis, characterization techniques and applications," *Journal of Environmental Chemical Engineering,* vol. 5, no. 5, pp. 4866-4883, 2017.

[12] C. D. Gutierrez-Sanchez, L. T´ellez-Jurado and H. J. Dorantes-Rosales, "Synthesis of zirconia nanoparticles by sol-gel. Influence of acidity-basicity," *Ceramics International,* vol. 50, no. 11, pp. 20547-20560, 2024.

[13] D. Philip, "Honey mediated green synthesis of silver nanoparticles," *Spectrochimica Acta Part A: Molecular and Biomolecular Spectroscopy,* vol. 75, p. 1078–1081, 2010.

[14] M. Bordoloi, R. K. Sahoo, K. J. Tamuli, S. Saikia and P. P. Dutta, "Plant Extracts Promoted Preparation of Silver and Gold Nanoparticles: A Systematic Review," *Nano,* vol. 15, no. 2, pp. 2030001-1 - 2030001-24.

[15] A. Amirjani, F. Firouzi and D. F. Haghshenas, "Predicting the Size of Silver Nanoparticles from Their Optical Properties," *Plasmonics,* vol. 15, p. 1077–1082, January 2020.

[16] N. Kaur, A. Singh and W. Ahmad, "Microwave Assisted Green Synthesis of Silver Nanoparticles and Its Application: A Review," *Journal of Inorganic and Organometallic Polymers and Materials,* vol. 33, pp. 883-672, 2023.

Journal of Physics: Conference Series **2980** (2025) 012014 doi:10.1088/1742-6596/2980/1/012014

[17] S. Francis, S. Joseph, E. P. Koshy and B. Mathew, "Microwave assisted green synthesis of silver nanoparticles using leaf extract of elephantopus scaber and its environmental and biological applications," *Artificial Cells, Nanomedicine, and Biotechnology,* vol. 46, pp. 795-804, 2018.

[18] F. J. Heiligtag and M. Niederberger, "The fascinating world of nanoparticle research," *Materials Today,* vol. 16, no. 7-8, pp. 262-271, 7 2013.

[19] A. P. Reverberi, M. Vocciante, M. Salerno, M. Ferretti and B. Fabiano, "Green Synthesis of Silver Nanoparticles by Low-Energy Wet Bead Milling of Metal Spheres," *Materials,* vol. 13, no. 1, p. 63, 2019.

Journal of Physics: Conference Series **2980** (2025) 012015
doi:10.1088/1742-6596/2980/1/012015

Fabrication and Characterization of Low-Cost Dye-sensitized Solar Cells (DSSCs) with Natural Light Absorbers of Henna-Based Natural Dye and N719 Mixtures

Waode Sukmawati Arsyad[1], Muh. Oheo Bijaksawan[1], La Ode Muhamad Nurrakhmad Arsyad[2], La Aba[1], La Ode Ahmad Nur Ramadhan[3], Ida Usman[1], Rahmat Hidayat[4]

[1]Physics Department, Halu Oleo University, Kendari, Indonesia
[2]Civil Engineering, Halu Oleo University, kendari, Indonesia
[3]Chemistry Department, Halu Oleo University, kendari, Indonesia
[4]Physics Department, Bandung Institute of Technology, Bandung, Indonesia

*E-mail: wdsukmawati@uho.ac.id

Abstract. To reduce the production cost of DSSC, we fabricated DSSC with natural dye extracted from Henna (*Lawsonia inermis L.)* using the maceration method and a mixture of Henna dye and N719 Ruthenium dye. For Henna dye (with and without evaporation), the TiO_2 electrodes were immersed for a time variation of (24, 48, and 72) hours. For the mixture, the TiO_2 electrodes were immersed for 24 hours with N719 concentration variation. The photovoltaic characteristics were measured using a standard solar simulator with a 100 mW/cm^2 power input. For Henna dye, the current density (J_{sc}) increases from (0.65 to 0.67) mA/cm^2, (0.68 to 0.75) mA/cm^2, and (0.74 to 0.75) mA/cm^2 for immersion time of (24, 48 and 72) hours respectively, when the dye solution was evaporated. The cells using unevaporated Henna for the mixture show better performance whereas the J_{sc} increased as much as 39.27%. The rise of N719 concentration from (0.1 to 0.2) mM and (0.2 to 0.3) mM increased the J_{sc} by 23.01% and 11.49% respectively. In contrast, the open circuit voltage (V_{oc}) decreased by as much as (2.04, 1.5, and 3.85) %. The highest efficiency of 0.276% was obtained for the DSSC with dye from the mixture of unevaporated Henna and 0.3 mM N719.

1. Introduction

Silicon-based solar cells have now reached an efficiency of 36.1% [1], however, its complexity in the fabrication process, including the materials purity, and high-temperature processing, has limited its development, especially in low-middle income countries. The transition to the thin film solar cells which also seems promising with the efficiency has reached 23.6% for CIGS [2] Additionally, there are drawbacks related to the use of toxins and the scarcity of materials. This has led to the development of dye-sensitized solar cells (DSSC), which offer ease in the fabrication process and allow the use of various materials without requiring high-purity materials.

Journal of Physics: Conference Series **2980** (2025) 012015 doi:10.1088/1742-6596/2980/1/012015

Over the past 20 years, DSSC has reached an efficiency of 15.2% [3]. however, most reported DSSCs still use inorganic and synthetic dyes as the light absorber which are expensive and chemically difficult to synthesize. Therefore, many scientists are trying to find cheaper alternatives for the light absorber material. Among others, natural dye extracted from plants is the most interesting choice due to its environmental friendliness, simple preparation techniques, and abundance [4], besides being renewable and biodegradable [5]. The dye can be extracted from different parts of plants such as leaves [6], bark [7], roots [8], flowers [9], seeds [10], and fruits [11]. Extraction is the separation of the desired color component by the cell using physical or chemical techniques from a plant into a solvent medium under employment conditions [12, 7, 5]. Methods for extracting dye from its natural sources include the aqueous extraction method [13], solvent extraction method [14], acid and alkali extraction [15], supercritical fluid extraction [16], and ultrasonic extraction techniques [17].

One of the dyes that can be extracted from leaves is Henna (*Lawsonia inermis L.*) which contains Lawsone [18] that results in a yellow and orange color [19]. The Henna dye is often used as a nail and hair color [20], and in healthcare, usually utilized as an anti-inflammation [21], anti-fungal [22], and anti-microbial [23], anti-oxidant [24], and anti-cancer [25]. Moreover, Henna's extract is also used to reduce surfactant adsorption on Kaolinite [26], and as a natural corrosion inhibitor [27]. Recently, many attempts have been made to employ Henna dye as a light absorber or light harvester material in DSSCs [28]. Lawsone molecules form polymeric structures through O-H••O hydrogen bonding. Since Lawsone is a redox-active molecule, it can undergo reversible conversion to naphthosemiquinone and catechol by accepting or donating electrons. Moreover, Lawsone absorbs visible light in the 400–600 nm range [29]. Theoretical studies have shown that electron density percolates through the lawsone molecules via intermolecular hydrogen bonding. These properties suggest that Lawsone could be a promising candidate for use in light-harvesting materials [29]. Jinchu et al., [30], reported using a lawsone metal complex as an effective sensitizer for TiO_2-based DSSCs, achieving an efficiency of 2.11% for the lawsone Cr^{3+} complex, compared to 3.04% for the reference cell with N719 dye. Similar work was reported by Lakshmi et al, using a Lawsone-Al metal complex with 1.36% efficiency [31]. Said et al., [32] calculated the HOMO and LUMO of Lawsone in different mediums and solvents, and found the HOMO and LUMO position of -7.341 eV and -3.584 eV with an energy gap of 3.847 eV for Lawsone in ethanol. This result shows that Lawsone can be used as a sensitizer in DSSC since its LUMO position is higher than the conduction band of TiO_2 (-4.1 eV) [33] which ensures the electron injection from Lawsone to the TiO_2.

In this paper, we reported the utilization of the Henna dye extracted with the maceration method as the light absorber material in DSSC. Furthermore, we investigated the effect of the evaporation vacuum process on the Henna dye solution and the mix of Henna dye with N719 Ruthenium dye, the immersion time variation on the TiO_2 electrode, and the N719 dye concentration variation on the performance of the DSSC

2. Methodology

2.1 Materials Preparation

2.1.1 Dye Extraction from Henna Leaves

The henna leaves were blended in a blender. 100 ml of ethanol was poured into 75 gr of smashed Henna leaves inside a closed-wrapped jar and macerated for 24 hours. The extract was filtered using Whatman filter paper. The solution was evaporated using a rotary evaporator at

Journal of Physics: Conference Series **2980** (2025) 012015 doi:10.1088/1742-6596/2980/1/012015

70°C until the solution became more concentrated. The solution was kept inside a dark bottle and stored in the refrigerator before use.

2.1.2 Mixed Henna Dye and N719 Ruthenium Dye

0.1, 0.2, and 0.3 mM of N719 Ruthenium dye (Solaronix) dissolved in unevaporated and evaporated Henna Dye, and stirred using a stirrer bar for 20 minutes until homogeneous. The solutions were kept inside a dark bottle and stored at room temperature before use.

2.1.3 DSSC Fabrication

DSSC's fabrication follows the same method as reported in our previous research [34, 35, 36, 37]. FTO substrate (TCO30-8, 5x5 cm from Solaronix) cut into six parts (1.6x2.5) cm in size. It was cleaned with soap and water and subjected to an ultrasonic bath (12 minutes) inside a solution of isopropanol p.a (Merck) and ethanol p.a (Merck) with a 1:1 volume ratio. The conductive substrate was then dried with a hot gun and ready to use. A (5x5) mm of TiO_2 transparent layer (Ti-Nanoxide T/SP, Solaronix) was deposited on top of a conductive substrate and heated on a hot plate at 120°C for six minutes. The second TiO_2 layer was deposited again on top of it. The same step applies until the third transparent layer. The TiO_2 layer was annealed at a hotplate with a gradually increasing temperature from (120 to 475)°C with an increase of 30°C every minute. Once it reached 475°C, the temperature was held for 30 minutes, before slowly cooling until it reached room temperature. We repeated the same step for one TiO_2 scattering layer (Ti-Nanoxide R/SP, Solaronix). The total TiO_2 thickness was around 12 μm.

The photo-anode was immersed in a solution of unevaporated and evaporated Henna Dye, with time variations of (24, 48, and 72) hours, and in the mix of Henna dye (unevaporated and evaporated) and N719 Ruthenium dye with variations in the concentration of (0.1, 0.2, and 0.3) mM. The photo-anode and the counter-electrode (Platinum coated FTO with drilled hole, Solaronix) were connected using a 50 nm thickness of Surlyn® (Solaronix) as a spacer. Using a vacuum pump, the space was filled with liquid electrolytes (Mosalyte TDE-250, Solaronix). A similar method was employed to make the reference cell using standard N719 Ruthenium (0.7 mM) as its light absorber material.

2.2 Dye and DSSC Characterization

The molecular structure, crystalline structure, and optical properties of the Henna dye (unevaporated and evaporated) were characterized using FTIR (Shimadzu), XRD (Smartlab, Rigaku), and UV-Vis spectrometer (OCEAN OPTICS 2000). The photovoltaic characteristics of DSSC were measured using a standard solar simulator AM 1.5 with a power input (P_{in}) of 100 mA/cm^2. The current density (J_{sc}), open circuit voltage (V_{oc}), fill factor (FF), and efficiency (η) values are obtained from the datasheet provided during measurement. However, the J_{sc} and V_{oc} values can also be extracted from the current density-voltage (J-V) data/graphs, while the fill factor and efficiency are calculated using equations 1–3.

$$P = j_{sc}\ x\ FF\ x\ V_{oc} \tag{1}$$

$$FF = \frac{V_{opt} \cdot I_{opt}}{V_{oc} \cdot I_{sc}} \tag{2}$$

$$\eta = \frac{V_{opt} \cdot I_{opt}}{P_{in}} = \frac{V_{oc} \cdot I_{sc} \cdot FF}{P_{in}} \tag{3}$$

3. Results and Discussions

3.1 Chemical and Physical properties of unevaporated and evaporated Henna Dye

Unevaporated and evaporated Henna dye shows yellow brown as shown in Figure 1. There was no color change in the evaporation process.

Figure 1. Unevaporated (a), and evaporated (b) Henna dye solution.

The FTIR spectra of unevaporated and evaporated Henna dye are shown in Figure 2. The spectrum shows high-intensity absorption at a frequency of 3274 cm^{-1} associated with stretch vibration from O-H (phenol) originating from the Lawsone aromatic ring [38]. Absorption at a frequency of 2926 cm^{-1} originating from C-H aliphatic stretch [26]. The absorption band at a frequency of 2374 cm^{-1} is associated with the presence of the C-N functional group [34]. There are two absorptions at a frequency of 1722 and 1624 cm^{-1}, indicating two carbonyl bands (C=O) that conjugate with α, β-unsaturated carbonyl band which supports the possibility of having a 1,4-naphthoquinone [27]. The existence of an O-H broad absorption band and C=O at a frequency of 1722 cm^{-1} indicates a carboxyl acid functional group. The absorption band at frequencies of 1436 and 1381 cm^{-1} originates from C=C vibration in the aromatic benzene ring [40]. The absorption band at frequencies (1215, 1076, and 1033) cm^{-1} might originate from the C-O vibration of the alkyl-substituted ether functional group [41]. The presence of aromatic rings in the chemical structure of Lawsone supported by the sharp peak at a frequency of 769 cm^{-1} originated from bend vibration (out of plane) of aromatic C-H [42]. Based on this FTIR analysis, it can be said that Lawsone was the main constituent in Henna leaves identified from its three main functional groups: O-H in phenol, C=O, and C=C.

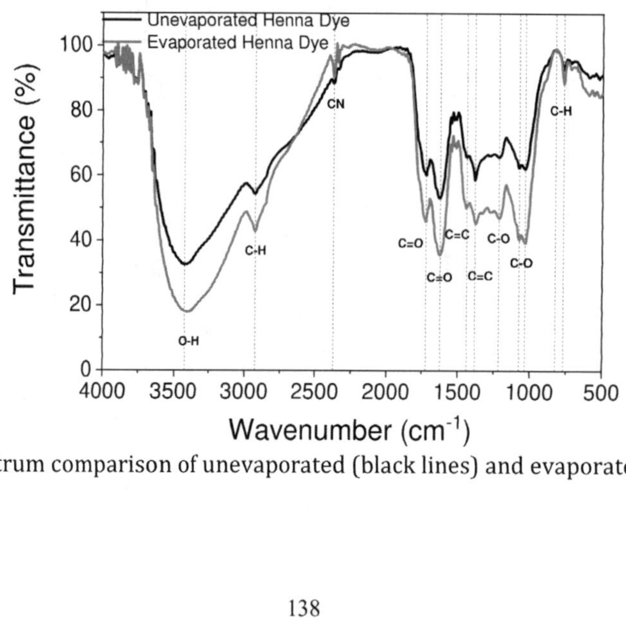

Figure 2. FTIR spectrum comparison of unevaporated (black lines) and evaporated (red line) Henna dye

The shift and intensity increase in the absorption band peak of the evaporated Henna indicates the augmentation of compound number in Henna extract. This might be caused by the ethanol solvent separation not attached to the compounds in the Henna dye during evaporation. The solvent evaporation can decrease the solvent effect and clearly shows the content of active compounds in Henna dye [43, 44].

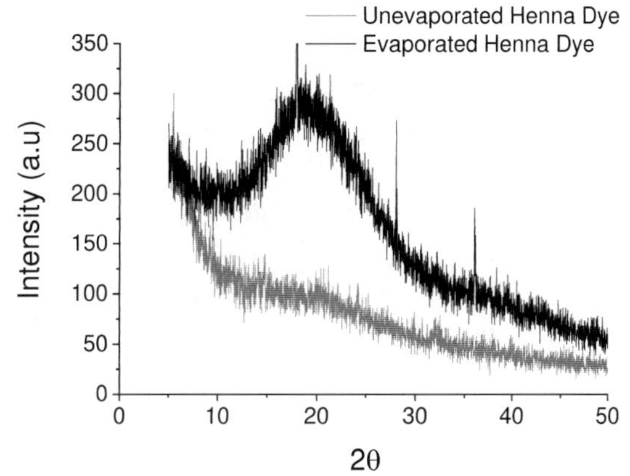

Figure 3. XRD spectrum comparison of unevaporated (red lines) and evaporated (black lines) Henna dye

From the XRD spectrum in Figure 3, the evaporated Henna shows a more prominent amorphous structure with a broad non-Bragg peak at $2\boldsymbol{\theta}$ position range from (15-30)°. Figure 4 shows the absorbance spectra of both unevaporated and evaporated Henna dye, along with the absorbance spectrum of N719 Ruthenium dye for comparison.

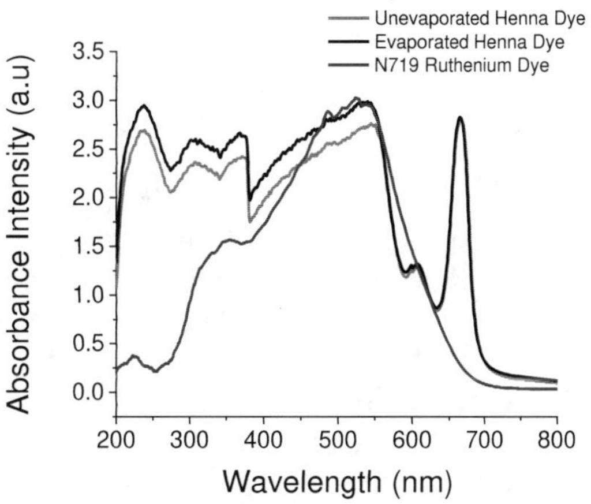

Figure 4. Absorbance spectrum comparison of unevaporated (red line) and evaporated (black line) Henna dye, and N719 Ruthenium dye (blue line)

The absorbance peak at 238 nm and 370 nm originated from Lawsone molecules. Theoretically [45], the absorbance peak at 238 nm and 370 nm comes from electronic transition configuration from the excited condition of $\boldsymbol{H-3 \rightarrow L}$, and $\boldsymbol{H \rightarrow L}$, respectively. The absorption spectrum range of the Henna dye is wider than the reference dye.

3.2 Photovoltaic Characteristics of DSSC

3.2.1 Henna Dye as light absorber material

To ensure that we made good quality DSSC, we made a standard reference cell using N719 Ruthenium dye. The current density – open voltage (J-V) characteristic is presented in Figure 5. The short circuit current density (J_{sc}), open circuit voltage (V_{oc}), and overall conversion efficiency are 21.92 mA/cm^2, 0.68 Volt, and 6.8%, respectively. This indicates that the fabrication method used to produce the DSSC was correct, as the results closely align with those of another researcher who used the same N719 dye [46]. To analyze the effect of vacuum evaporation on the performance of the DSSC, we made eight cells using unevaporated and evaporated Henna dye as the light absorber, where the immersion time variated for (24, 48, and 72) hours. The results are shown in Figure 6 (a, b, and c) and Table 1.

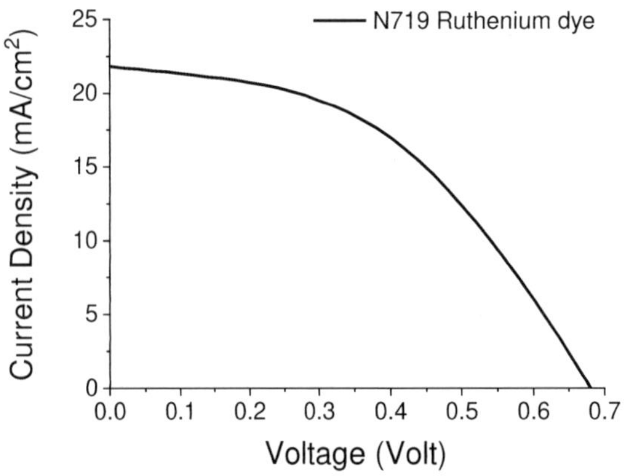

Figure 5. J-V curve of DSSC based on standard N719 Ruthenium dye

For Henna dye, the evaporation process affected the J_{sc} and V_{oc} of the cells. The J_{sc} increases as the dye is evaporated. The evaporation process drowns out the excess solvent and water in the extract since water can easily bond with Ti in TiO_2. Therefore, the evaporation process enhances the bonding probability of the carboxylate functional group in natural dye with Ti in TiO_2, escalating the charge transfer process from dye molecules to the TiO_2 conduction band and eventually decreasing electrons recombination rate to produce more current. Furthermore, the rise in the J_{sc} might be related to the increase in the absorption intensities of Henna dye after evaporating, particularly in the wavelength area from (200 – 540) nm (Figure 4). Immersion time affects the J_{sc} value, with J_{sc} increasing as immersion time increases (Table 1). This was observed in both unevaporated and evaporated henna dye-based DSSCs. This effect is likely since longer immersion times allow more dye to be absorbed onto the TiO_2 surface, enabling more photons to be absorbed and converted into current. However, a different trend was observed for the V_{oc} value, where the V_{oc} reached its maximum at an immersion time of 48 hours and then decreased with immersion times up to 72 hours.

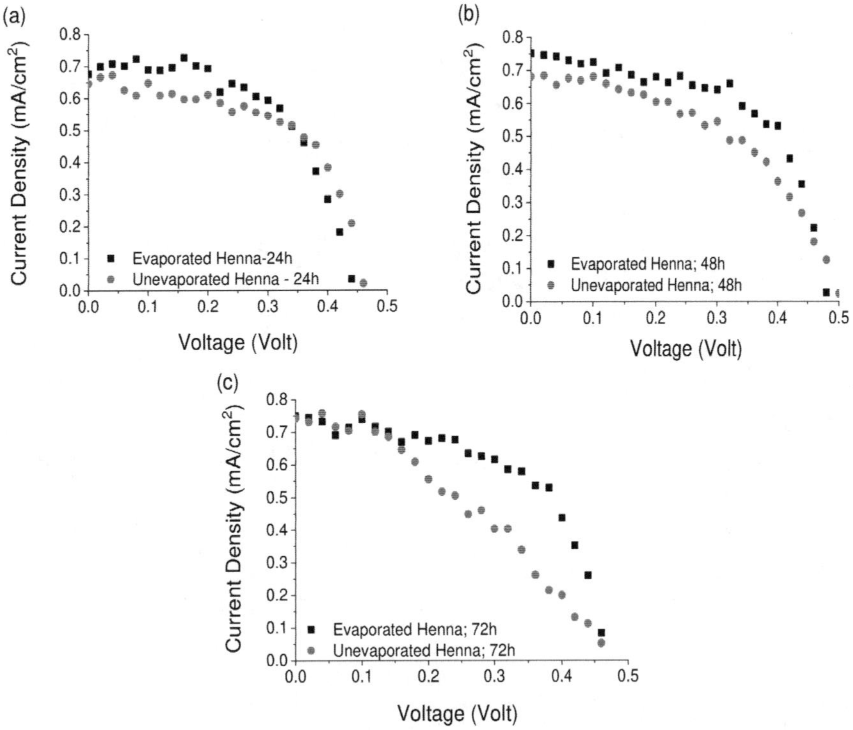

Figure 6. J-V curve of DSSC with light absorber using unevaporated and evaporated Henna dye with immersion time variation of (24, 48, and 72) hours.

Table 1. Photovoltaic parameters of the DSSC based on natural dye from unevaporated and evaporated Henna with immersion time variation

Henna Dye		V_{oc} (Volt)	J_{sc} (mA/cm^2)	Fill Factor (%)	Efficiency (%)
Evaporated	24h	0.44	0.67	60.7	0.18
	48h	0.48	0.75	58.6	0.21
	72h	0.47	0.75	57.3	0.20
Unevaporated	24h	0.46	0.64	58.8	0.18
	48h	0.50	0.68	48	0.17
	72h	0.47	0.74	36.9	0.13

Ideally, soaking for up to 72 hours increases the dye uptake, which raises the number of electrons at the LUMO level of the dye, allowing more electrons to be injected into the TiO_2 conduction band. This leads to an upward shift in the Fermi level, resulting in an increased energy gap between the TiO_2 Fermi and electrolyte energy level, corresponding to the V_{oc} [47]

value in the DSSC cell. However, the widening of this energy gap due to the shift in the Fermi level is also accompanied by an increased likelihood of electron recombination from the TiO_2 conduction band to the electrolyte, as the electron concentration increases [37]. These two mechanisms compete with each other. If electron recombination becomes dominant, it will negatively affect the V_{oc} value, which is why, as observed in Table 1, the V_{oc} value decreases with increasing soaking time up to 72 hours. Further measurements, such as impedance measurements, are needed to confirm the statement above.

3.2.2 Mixture of Henna Dye and N719 Ruthenium Dye as light absorber material

The J-V curves of DSSC with light absorber from the mixture of Henna dye (unevaporated and evaporated) and N719 Ruthenium dye are shown in Figure 7. The DSSC cells that utilize a mixture of unevaporated Henna dye and N719 dye as a light harvester material showed better performance, which can be seen in its higher J_{sc}. This contradictory the results obtained for DSSC cells using Henna dye without N719, where the evaporation process increases the J_{sc} when the Henna dye is evaporated.

The higher J_{sc} for DSSC based on the mixture of unevaporated Henna and N719 Ruthenium dye might be related to the ability of the N719 dye to dissolve more easily due to the solvent's (ethanol) presence. Meanwhile, the solvent was completely diminished in the evaporated Henna and N719 Ruthenium dye mixture.

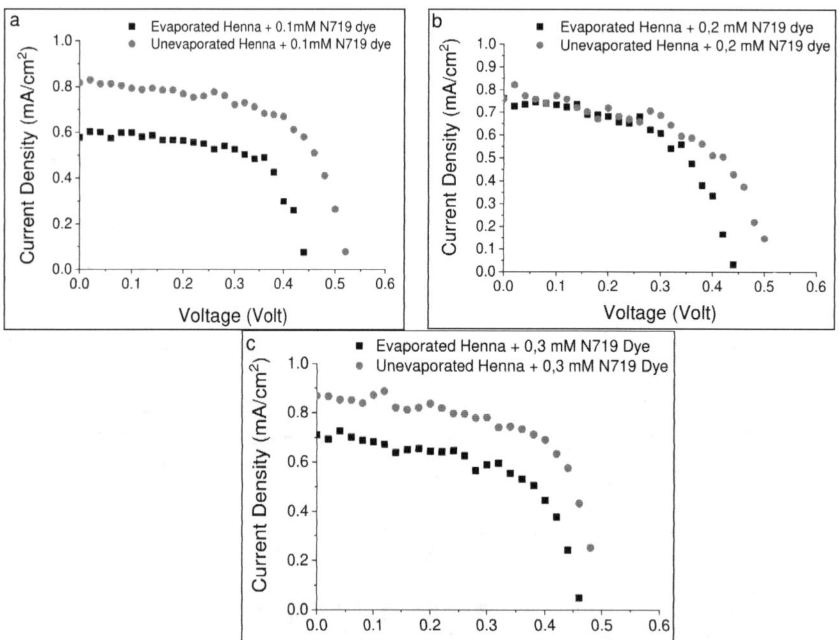

Figure 7. J-V curve of DSSC with light absorber using a mixture of Henna dye (unevaporated and evaporated) and N719 Ruthenium dye, with 48 hours of immersion time.

Table 2 shows the effect of N719 concentration in the mixture. Adding N719 dye into unevaporated Henna increased J_{sc} by as much as 39.27%. The J_{sc} increased as much as 23.01% and 11.49% for N719 dye concentration from (0.1 to 0.2) mM and (0.2 to 0.3) mM, respectively, however, it lowers the V_{oc} equivalent to 2.04%, 1.51%, and 3.85% for N719 concentrations of (0.1, 0.2, and 0.3) mM, respectively. Different patterns for J_{sc} and V_{oc} were observed in the DSSC based on a mixture of N719 dye and evaporated Henna. There was no linear correlation between N719 concentration with J_{sc} and V_{oc}. Instead, there was optimum J_{sc} and V_{oc} for specific concentrations (Table 2).

Table 2. Photovoltaic parameters of the DSSC based on light absorber from a mixture of unevaporated and evaporated Henna with inorganic N719 Ruthenium dye (variation concentration of N719 dye)

Henna Dye	N719 Concentration (mM)	Jsc (mA/cm^2)	Voc (Volt)	FF (%)	η (%)
Unevaporated	0	0.443	0.538	57.3	0.137
	0.1	0.617	0.527	62.1	0.267
	0.2	0.759	0.519	54.2	0.213
	0.3	0.869	0.499	63.7	0.276
Evaporated	0	0.594	0.465	58.0	0.160
	0.1	0.579	0.447	68.2	0.176
	0.2	0.763	0.444	56.2	0.190
	0.3	0.710	0.464	58.3	0.192

The highest overall conversion efficiency (η) was obtained for 0.3 mM N719 dye in unevaporated Henna dye. Initially, we attempted to match the concentration of N719 dye in the mixture with the concentration used to prepare the reference cells, which contained only N719 dye at 0.7 mM in ethanol. However, upon testing, we found that 0.3 mM was the maximum N719 dye that could completely dissolve in the Henna dye solution. Beyond this concentration, N719 dye became difficult to dissolve. Therefore, we chose 0.3 mM as the maximum concentration for this study.

Conclusion

The henna dye can be used as a light absorber material in DSSC. We found that the evaporation process of the Henna dye and immersion time affect the performance of DSSC. The highest overall efficiency was achieved for the DSSC using Henna dye that had been evaporated and then immersed for 48 hours. Once the duration exceeds 48 hours, the efficiency of the cell decreases. As for the mixture of Henna dye and inorganic N719 dye, the unevaporated Henna with 0.3 Mm N719 dye gives the highest overall efficiency.

It can be said that when Henna dye is mixed with other inorganic or organic synthetic dyes, a solvent is still required to dissolve it. Future research will focus on the purification and isolation of Lawsone from Henna leaves and their respective performance when used as light absorber material in DSSC.

Acknowledgment

This research was funded by DRTPM DIKTI through the Regular Fundamental Research Scheme, under Contract Nos. 049/E5/PG.02.00.PL/2024 and 24/UN29.20/PG/2024.

References

[1] Schygulla P., Müller R., Höhn O., Schachtner M., Chojniak D., Cordaro A., Tabernig S., Bläsi B., Polman A., Siefer G., Lackner D. and Dimroth F., 2023, *Proc. EUPVSEC Conference*, Lisbon.

[2] Buonomenna M. G., 2023, Inorganic Thin-Film Solar Cells: Challenges at the Terawatt-Scale, *Symmetri*, 15 (9), 1718.

[3] Ren Y., Zhang D., Suo J., Cao Y., Eickemeyer F. T., Vlachopoulos N., Zakeeruddin S. M., Hagfeldt A. and Grätzel M., 2022, Hydroxamic acid pre-adsorption raises the efficiency of cosensitized solar cells, *Nature*, 613, 60-65.

[4] Narayan MR, 2012, Review: Dye sensitized solar cells based on natural photosensitizers, *Renew. Sust. Energ. Rev*, 16, 2018-215.

[5] Ragab M M, Hassabo A A and Othman H A, 2022, An Overview of Natural Dyes Extraction Techniques for Valuable Utilization on Textile Fabrics, *J. Text. Color. Polym. Sci*, 19 (2), 137-153.

[6] Chang H and Lo Y.J, 2010, Pomegranate leaves and mulberry fruit as natural sensitizers for dye-sensitized solar cells, *Sol. Energy*, 84, 1833-1837.

[7] Habib N, Akram W, Adeel S, Amin N, Hosseinnezhad M and Haq E U, 2022, Environmental friendly extraction of peepal (ficus religiosa) bark-based reddish brown tannin natural dye for silk coloration, *Environ, Sci. Pollut. Res. Int.*, 29 (23), 35048-35060.

[8] Yamazaki E, Murayama M, Nishikawa N, Hashimoto N, Shoyama M and Kurita O, 2007, Utilization of natural carotenoids as photosensitizers for dye-sensitized solar cells, *Sol. Energ.*, 81, 512–516.

[9] Chandra Maurya I, Singh S, Srivastava P, Maiti B and Bahadur L, 2019, Natural dye extract from Cassia fistula and its application in dye-sensitized solar cell: Experimental and density functional theory studies, *Opt. Mater. (Amst).*, 90, 273-280.

[10] Gomez Ortiz NM, Vasquez Maldonado I A , Perez Espadas AR, Mena Rejon GJ, Azamar Barios JA and Oskam G, 2010, Dye-sensitized solar cells with natural dyes extacted from achiote seeds, *Sol. Energ. Mat. Sol. Cells*, 94, 40-44.

[11] Abebe RW, Delele WA, Nigus GH and Yeshitila AT, 2020, Anthocyanin Component For Dye-Sensitized Solar Cells Extracted From Teclea Shimperi Fruit As Light-Harvesting Materials, *Mater Sci Energy Technol*, 2, 889-895.

[12] Kumbasar E.P.A, 2011, Natural Dyes, *InTech*, Croatia.

[13] R. Mia, M.-M. Islam, T. Ahmed, M.-A. Waqar, N.-J. Khanam, S. Sultana, M.-S.-K. Bhuiyan and M.-N. Uddin, 2022, Natural dye extracted from Triadica Sebifera in aqueous medium for sustainable dyeing and functionalizing of viscose fabric, *Clean. Eng. Technol.*, 8, 100471.

[14] Borges M., Tejera R., Díaz L., Esparza P. and Ibáñez E., 2012, Natural dyes extraction from cochineal (Dactylopiuscoccus). New extraction methods, *Food Chemistry*, 132 (4), 1855-60.

[15] Samanta A. K., Awwad N. S. and Algarni H. M., 2020, Chemistry and technology of natural and synthetic dyes and pigments, *IntechOpen*, London, United Kingdom.

[16] Elmaaty T. A., . Sayed-Ahmed K., Magdi M. and Elsisi H., 2023, An eco-friendly method of extracting alizarin from Rubia tinctorum roots under supercritical carbon dioxide and its application to wool dyeing, *Sci. Rep.*, 13 (30).

[17] Muruganandham M., Sivasubramanian K., Velmurugan P., Suresh-Kumar S., Arumugam N., Almansour A. I., Suresh-Kumar R., Manickam S., Pang C. H. and Sivakumar S., 2023, An eco-friendly ultrasound approach to extracting yellow dye from Cassia alata flower petals: Characterization, dyeing, and antibacterial properties," *Ultrason Sonochem*, 98, 106519.

[18] Semwal R. B., Semwal D. K., Combrinck S., Cartwright-Jones C., and Viljoen A., 2014, Lawsonia inermis L. (henna): Ethnobotanical, phytochemical and pharmacological aspects, *J. Ethnopharmacol.*, 155 (1), 80-103.

[19] Qadariyah L, Azizah N, Syafa'atullah A Q, Bhuana D S and Mahfud M, 2018, The Extraction of Natural Dyes from Henna Leaves (Lawsonia Inermis L.) by Ultrasound-assisted Method," in *IOP Conf. Series: Materials Science and Engineering*, Padang.

[20] Cartwright-Jones C., 2006, Developing Guidelines on Henna: A Geographical Approach, *Kent State University*, Kent, Ohio, USA.

[21] Imam H., Uddin Mahbub N., Khan F., Kabir Hana H. and Rahman Sarker M., 2013, Alpha amylase enzyme inhibitory and anti-inflammatory effect of Lawsonia inermis, *PJBS*, 16, 1796–1800.

[22] Rahmoun N., Boucherit-Otman Z., Boucherit K., Benabdallah M. and Choukchou-Braham N., 2013, Antifungal activity of the Algerian Lawsonia inermis (henna), *Pharm. Biol.*, 51, 131–135.
[23] Volpato G., Kourkova P., and Zeleny V., 2012, Healing war wounds and perfuming exile: the use of vegetal, animal, and mineral products for perfumes, cosmetics, and skin healing among Sahrawi refugees of Western Sahara, *J. Ethnobiol. Ethnomed.*, 8, 49-68.
[24] Adetutu A., Owoade O. A., and Oyekunle O. S., 2013, Comparative effects of some medicinal plants on sodium arsenite-induced clastogenicity," *IJPBS*, 4, B777–B783.
[25] Kumar M., Kumar S., and Kaur S., 2014, Identification of polyphenols in leaf extracts of Lawsonia inermis L. with antioxidant, antigenotoxic and antiproliferative potential, *Int. J. Green Pharm.*, 8, 23–36.
[26] Musa M., Sulaiman W., Majid Z., Idris A., and Rajaei K., 2020, Henna extract as a potential sacrificial agent in reducing surfactant adsorption on kaolinite: The role of salinity, *JKSUES*, 32 (8), 543-547.
[27] Nik W., Zulkifli F., Sulaiman O., Samo K., and Rosliza R., 2012, Study of Henna (Lawsonia inermis) as Natural Corrosion Inhibitor for Aluminum Alloy in Seawater, *IOP Conference Series: Materials Science and Engineering*, 1 (2), 723-728.
[28] Selva E, Kamatchi D, Sheeba and Meenakhi S, 2022, Natural dye sensitized solar cells using henna leaves with ZnO nanoparticle at various pH values, *J. Adv. Sci. Res.*, 60-69.
[29] Khadtare S. S., Ware A. P., Salunke-Gawali S., Jadkar S. R., Pingale S. S. and Pathan H. M., 2015. Dye Sensitized Solar Cell with Lawsone Dye Using ZnO Photoanode: Experimental and TD-DFT Study, *RSC Adv.,* 5, 17647-17652.
[30] Jinchu I., Sreekola C. O., and Sreelatha K. S., 2020. Lawsone metal complex as an effective sensitizer for dye sensitized solar cells, *Mater. Today Proc.,* 38(2) 1356-1360.
[31] Lakshmi R., Krishnakumar G., Joseph L. J., Sreelatha K. S., and Jinchu I., 2016. Lawsone dye complex: An efficient sensitizer for dye sensitized solar cell, in *International Conference on Electrical, Electronics, and Optimization Techniques (ICEEOT)*, Chennai, India.
[32] Said M., Datti Y., Ahmad A., Buhari T. R., and Ahmed A. B., 2021. Computational Study of Lawsonia Inermis as Potential and Promising Candidate for Production of Solar Cell, *EJ-CHEM,* 2(1), 21-27.
[33] Chen L. C., Hsu C. H., Chan P. S., Zhang X., and Huang C. J., 2014. Improving the performance of dye-sensitized solar cells with TiO_2/graphene/TiO_2 sandwich structure, *Nanoscale Res. Lett,* 9, 380.
[34] Arsyad W. S, Herman, Prijamboedi B and Hidayat R, 2018, Revealing the limiting factors that are responsible for the working performance of quasi-solid state DSSCs using an ionic liquid and organosiloxane-based polymer gel electrolyte, *Ionics*, 24 (3).
[35] Arsyad W. S, Januari A, Jumalia, Ilmawati WS, Usman I and Hidayat R, 2018, Non-ohmic resistance effects of the AZO and TiZO as a blocking layer in dye-sensitized solar cells (DSSCs), *JNOPM*, 27 (2), 1850.
[36] Arsyad W. S., R. Suhardiman, I. Usman , L. Aba, S. Suryani, L. O. A. N. Ramadhan, M. Nurdin and R. Hidayat, 2024, Improvement of Dye-Sensitized Solar Cells (DSSCs) performance using Crude Brazilein Extract from Sappanwood (Caesalpinia Sappan L.) with the incorporation of ZnO nanoparticles, *J. Mol. Struct*, 1303, 137548.
[37] Usman I., Arsyad W. S., Herman H., and Hidayat R., 2023, Low carrier recombination in polysiloxane gel electrolyte for high-performance DSSC, *Mat. Res. Express*, 10 (4), 046201.
[38] Saadaoui S., Mohamed A., Moufida B., Rached G., Emanueles S., Salvo M., Alessandra A., and Rosaria A., 2017, Performance of natural dye sensitized solar cells by ZnO nanorod and nanowall enhaced photoelectrodes, *Belstein J. Nanotechnol*, 8, 287-295.
[39] William D., and Fleming I., 1980, Spectroscopic methods in organic chemistry, *McGraw-Hill Book*, United Kingdom.
[40] Hamdy A., and El-Gendy N., 2013, Thermodynamic, adsorption and electrochemical studies for corrosion inhibition of carbon steel by henna extract in acid medium, *Egypt. J. Pet*, 22, 17-25.
[41] Wagini N., Amira S., Mohamed S., Yaser A., and El-Saady M., 2014, Phytochemichal Analysis of Nigerian and Egyptian Henna (Lawsonia Inermis L.) Leaves Using TLC, FTIR and GCMS, *Plants*, 2 (3), 2732.
[42] Wahyuningsih S., Wulandari L., Wartono M., Munawaroh H., and Ramelan A., 2017, The Effect of pH and Color Stability of Anthocyanin on Food Colorant, *IOP Conference Series: Materials Science and Engineering*, 193, 012047.
[43] Stuart B., 2004, Infrared Spectroscopy: Fundamentals and Applications, *John Wiley & Sons*, UK.
]44] Smith B., 2011, Fundamentals of Fourier Transform Infrared Spectroscopy, *CRC Press*, Boca Raton.
[45] Madili N., Pogrebnoi A., and Pogrebnaya T., 2018, Theoretical Design of Complex Molecule via Combination of Natural Lawsone and Synthetic Indoline D131 Dyes for Dye Sensitized Solar Cells Application," *Comput. Chem.*, 6 (4).
[46] Qin Y., and Peng Q., 2012, Ruthenium sensitizers and their applications in dye-sensitized solar cells, *Int. J. Photoenergy*, 12, 291579.
[47] Narayan-Chanda D. N., Ho-Joon L., Won-Youl C., and Jae-Joon L., 2013. Electrochemical approach to enhance the open-circuit voltage (Voc) of dye-sensitized solar cells (DSSCs), Electrochim. Acta, 109, 39-45.

Production of HHO Gas by Water Electrolysis with High Electrical Currents: New Technique to Measure Gas Released

Messiah C. Sangian[1], Hanny F. Sangian[2], Fatimah A. Noor[1*]

[1] Physics and Technology of Advanced Materials Research Division, Department of Physics, Faculty of Mathematics and Natural Sciences, Institut Teknologi Bandung, Bandung 40132, Indonesia
[2] Department of Physics, Faculty of Mathematics and Natural Science, Sam Ratulangi University, 95115 Manado, Sulawesi Utara, Indonesia

*E-mail: fatimah@fi.itb.ac.id

Abstract. *This work aimed to produce HHO gas through water electrolysis, investigating the effect of varying the electric current from* 10 A *to* 100 A *as well as changing the quantity of catalyst. The apparatus was set up in an acrylic container, with nine pairs of electrode plates, each of area* 30 cm × 30 cm, *separated by a diaphragm. The container was filled with* 25 L *of electrolyte, and each electrode was connected to a power supply that allowed the current and voltage to be adjusted. The initial experiment involved producing HHO gas using pure water without a catalyst, adjusting the electric current as required. Additional experiments, used up to* 400 g *of potassium hydroxide catalyst. The gas production was measured using a novel water displacement technique. It was found that the highest volume of HHO gas,* 1,300 mL *in* 113 s, *was produced with a current of* 70 A *and a catalyst mass of* 400 g, *whereas the volume at a current of* 10 A *without a catalyst was* 200 mL *in* 139 s.

Keywords: Water electrolysis, HHO gas, electric current, catalyst.

1. Introduction

The global energy landscape is undergoing a significant transformation driven by the pressing need to reduce fossil fuel usage. Fossil fuels, which have been the backbone of global energy supply, are finite resources, and their combustion is a major source of greenhouse gas emissions, contributing substantially to climate change [1]. Indonesia, like many other countries, relies heavily on fossil fuel imports, which burdens its economy and exposes it to the volatility of global oil prices [2]. In response, Indonesia has made notable progress in developing and commercializing renewable energy sources, particularly biodiesel and bioethanol, as part of its strategy to achieve energy security and sustainability [3].

Hydrogen gas has emerged as a promising alternative energy carrier due to its high energy density, harmless combustion products, and potential for renewable production. Hydrogen can be produced from various raw materials, including biomass, natural gas, oil, and water, through different technological pathways [4]. Among these, water electrolysis is particularly attractive because it offers a direct and efficient method to produce hydrogen using electrical energy [5]. This process involves splitting water molecules into hydrogen and oxygen gas using an electric current in the presence of an electrolyte. The efficiency of water electrolysis can be significantly enhanced by using catalysts that lower the reaction's activation energy [6].

Three primary technologies dominate the field of water electrolysis: alkaline water electrolysis (AWE), proton exchange membrane (PEM) electrolysis, and solid-oxide electrolysis (SOE). Each of these has its own set of advantages and challenges. AWE is well-established and cost-effective but is less efficient than PEM electrolysis and SOE [7]. PEM electrolysis offers a higher efficiency and quicker start-up times, making it suitable for variable renewable energy sources, though it requires costly materials such as platinum for catalysts [8]. SOE operates at high temperatures, which can enhance efficiency but also poses challenges related to material durability and system complexity [9].

Recent advancements in water electrolysis have focused on improving the efficiency and reducing the cost of hydrogen production. This has included the development of non-precious metal catalysts, hybrid electrolysis processes, and novel reactor designs [10], [11]. These innovations aim to make hydrogen a viable alternative to fossil fuels for a range of applications, including transportation, industry, and energy storage, supporting the transition to sustainable energy source [12].

However, most previous studies have primarily focused on low-current electrolysis methods and traditional diaphragm techniques, which often limit the gas production rate and overall efficiency. These limitations hinder the feasibility of scaling up hydrogen production for industrial applications. This study aims to explore the production of oxyhydrogen gas (commonly referred to as HHO gas) through water electrolysis using high electric currents ranging from 10 A to 100 A. The term "HHO" is widely use in certain technical and industrial context to denote a specific mixture of hydrogen (H_2) and oxygen (O_2) gases in a 2:1 molar ratio, produced via water electrolysis [13]. The research employs a novel electrolysis apparatus with a transparent plastic diaphragm and investigates the effect of varying the amount of potassium hydroxide (KOH) catalyst on gas production. By using higher currents and larger quantities of catalyst, this study seeks to address the limitations of previous studies, demonstrating how these factors can significantly enhance gas output and improve the efficiency of electrolysis process[14].

The study also introduces a new displacement method for measuring the gas produced. This technique calculates the volume of water displaced from the container due to the pressure buildup of HHO gas, which has not been widely reported in prior research. The experimental setup involves an acrylic container with stainless steel electrodes and a power supply capable of delivering high currents. The use of KOH as a catalyst is analyzed for its effectiveness in enhancing gas production under different current settings [15].

The results of this study should provide valuable insights into optimizing the electrolysis process for large-scale hydrogen production. This research contributes to the broader effort to develop sustainable energy technologies by demonstrating the potential of high-current electrolysis with advanced catalyst configurations. The findings could help pave the way for more efficient and cost-effective hydrogen production methods, supporting global efforts to transition to a hydrogen-based energy economy [16].

2. Materials and Methods

2.1 Construction of the electrolysis apparatus

The electrolysis apparatus developed for this study was designed and constructed to maximize the production rate of HHO gas. The setup comprised a custom-made acrylic container with dimensions of 1,000 mm × 2,000 mm × 4 mm, chosen for its chemical resistance and transparency, which facilitated observation of the electrolysis process.

The electrode assembly consisted of nine pairs of stainless steel (SS-316) plates, each with dimensions of 30 cm × 30 cm and a thickness of 1.2 mm. Stainless steel was selected for its excellent conductivity and corrosion resistance, which are critical for sustained electrolysis operations. The plates were configured alternately as anodes and cathodes, separated by a specialized plastic diaphragm. This diaphragm, designed for high porosity and durability, prevented short circuits while allowing ionic exchange between the electrodes.

To power the electrodes, a high-current DC power supply (a Rhino TIG-HT 160 welding machine) was employed. This power supply could deliver adjustable currents ranging from 10 A to 100 A, providing the flexibility needed to explore the effects of varying electric current on HHO gas production. High-quality, heat-resistant cables were used to connect the electrodes to the power supply, ensuring they could withstand the thermal stresses and high currents without degradation.

The electrolyte used in the apparatus was a solution of potassium hydroxide (KOH) in tap water. KOH was chosen as the catalyst for its ability to enhance the conductivity of the electrolyte and accelerate the electrolysis process. The electrolyte solution was prepared by dissolving specified amounts of KOH in 25 L of water, ensuring consistent and repeatable results. The KOH concentration was varied in different experimental setups to study its impact on gas production efficiency.

Monitoring and measurement equipment were integral to the apparatus design. A multimeter (400 A, 600 V) was employed to measure the current and voltage across the electrodes accurately, ensuring precise control and monitoring of the electrical parameters during the electrolysis process. Additionally, a 2,000 mL measuring flask was used to capture and measure the displaced water. The volume of displaced water served as a direct measure of the volume of gas produced, providing a reliable and straightforward method of quantification.

Temperature control and monitoring were also crucial, given the exothermic nature of the electrolysis reaction and the possibly that high temperature could damage or reduce the efficiency of the apparatus [17]. A manual mercury thermometer (0–100 °C) was used to track the temperature of the electrolyte solution, ensuring it remained within optimal operational limits.

This comprehensive approach to constructing the electrolysis apparatus provided robust data on the production of HHO gas under various conditions. The innovative use of high direct currents, combined with the optimized catalyst concentrations and precise measurement techniques, enhances our understanding of the electrolysis process and offers practical insights for scaling up hydrogen production technologies.

2.2 Experimental Methods

The procedure began with the production of HHO gas without the use of a KOH catalyst, establishing a baseline for subsequent experiments in which concentration of the catalyst was varied. First, 25 L of deionized water was introduced into the electrolysis container, ensuring the electrodes were completely submerged. Connections were made to the positive and negative electrode plates and to the DC power supply using the high-quality cables. An ammeter was placed in series with one of the electrodes, and a voltmeter was placed in parallel across the electrodes.

The power supply was activated, initially set to a current of 10 A. As the current began to flow through the electrodes, the generation of gas bubbles was observed. The electric current flowing between the electrodes, along with the voltage across the plates, was continuously recorded.

Additionally, any increase in temperature due to energy dissipation was monitored and documented.

Intermittently, hydrogen and oxygen gas produced during electrolysis became trapped at the water surface, increasing the internal air pressure. This pressure displaced water through a connected pipe into a measuring flask. The volume of displaced water was considered equivalent to the volume of the generated gas. To verify this assumption, the ideal gas law $pV = nRT$ was employed, using the parameters $p = 1$ atm, $T = 303$ K (room temperature), and $R = 8.314$ J/mol·K, with n representing the number of moles of gas.

The baseline experiment, conducted without a catalyst, was repeated four times, incrementing the current from 10 A to 40 A, 70 A, and 100 A. To observe the effect of KOH concentration, subsequent trials introduced incremental amounts of KOH, with concentrations set at 200 g, 250 g, 300 g, 350 g, and 400 g per 25 L for deionized water. Each concentration was tested under the current settings of 10 A, 40 A, 70 A, and 100 A to measure its influence on gas production rates.

In each trial, steps were taken to ensure the reliability and reproducibility of the results. The electrolyte solution was carefully prepared, and the apparatus was thoroughly inspected before each experiment to avoid any contamination or inconsistencies. The measurements of current, voltage, and temperature were taken at regular intervals to monitor the process dynamically. Additionally, the volume of gas produced was measured multiple times to account for any fluctuations and ensure precise data collection.

The current was incrementally increased during each experiment as described, allowing the monitoring of gas generation at different current levels. Any variations in gas production or temperature were carefully documented for each current increment, providing detailed insight into the effect of electrical input on the electrolysis efficiency.

The data analysis involved comparing the experimental volumes of HHO gas to the predicted volumes calculated using the ideal gas law. This comparison validated the experimental setup and ensured the accuracy of the results. Observations were made regarding the impact of increased catalyst mass on overall efficiency, particularly concerning potential heat dissipation issues and their effect on gas production rates.

This comprehensive approach provided significant insights into optimizing the parameters for efficient HHO gas production via water electrolysis. It highlighted the influence of electrical current and catalyst concentration on the gas generation process, offering valuable contributions to the field of renewable energy research.

3. Results and Discussion

3.1 Production of HHO gas without a catalyst

Table 1 presents the experimental and predicted volumes of HHO gas produced at different time intervals when using a current of 10 A without a catalyst. The experimental setup was validated by comparing these values. The experimental results consistently showed that the measured volumes of HHO gas were closely aligned with the predicted volumes, typically within a 5% margin of error. This high level of accuracy indicates that the method of measuring the displaced water volume as an indirect measure of HHO gas production is valid and reliable. For instance, at 77 seconds, the experimental volume was 100.00 mL and the predicted volume was 97.15 mL. At 380 seconds, the experimental volume was 540.00 mL and the predicted volume was 524.61 mL.

Figure 1 illustrates the volumes of HHO gas generated over time using electric currents of 10 A, 40 A, 70 A, and 100 A without a catalyst. The data show a strong positive correlation between

Table 1. Comparison between experimental and predicted volumes of HHO gas

Time (s)	Experimental volume (mL)	Predicted volume (mL)
77	100.00	97.15
139	200.00	194.30
158	260.00	252.59
197	280.00	272.02
205	300.00	291.45
227	320.00	310.88
241	340.00	330.31
250	360.00	349.74
260	380.00	369.17
275	400.00	388.60
294	420.00	408.03
306	440.00	427.46
317	460.00	446.889
330	480.00	466.32
347	500.00	485.75
367	520.00	505.18
380	540.00	524.61

the electric current supplied and the volume of gas produced. At the lowest current of 10 A, the gas volume reached 200 mL at 139 seconds.

As the current was increased to 40 A, the rate of gas production accelerated, achieving 260 mL at 127 seconds. Further increases in current to 70 A and 100 A resulted in even more pronounced increases in gas volume, with 280 mL produced at 121 seconds and 340 mL at 124 seconds, respectively. These results demonstrate that higher electric currents enhance the electrolysis process, leading to faster and more efficient HHO gas production. The increased energy input at higher currents accelerates the dissociation of water molecules into hydrogen and oxygen gas.

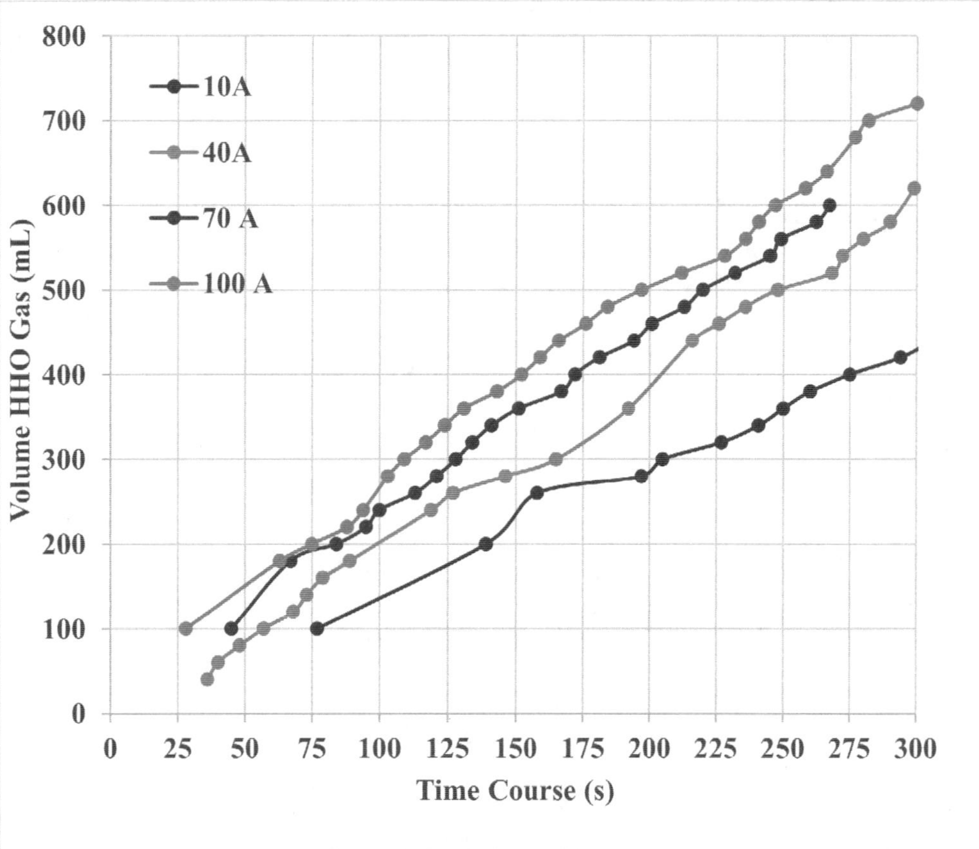

Figure 1. Volume of HHO gas generated over time without Catalyst

When comparing these results with previous research, it is evident that the use of higher electric currents improves the efficiency of HHO gas production. While De Souza et al. [14] achieved high efficiency (94.5%) in hydrogen production using ionic liquids as electrolytes with platinum electrodes ubder low current densities (20 mA/cm^2), this study demonstrates substantial gas production rates using deionized water and SS-316 stainless steel electrodes under high current conditions. This set up, leveraging the high conductivity and corrosion resistance of stainless steel, achieved production volumes up to 340 mL at 100 A, showcasing the scalability of high-current electrolysis.

The findings align with previous research that indicates a direct relationship between the magnitude of electric current and the rate of gas production. Higher currents provide more energy, which speeds up the electrolysis process and increases the overall gas yield. This study confirms the effectiveness of using high electric currents to optimize the electrolysis process for HHO gas production. The clear correlation between the current supplied and the volume of gas produced underscores the importance of electrical current strength in determining the efficiency of the electrolysis process.

Journal of Physics: Conference Series **2980** (2025) 012016 doi:10.1088/1742-6596/2980/1/012016

3.2 Production of HHO gas with 200 g of catalyst

The influence of a 200 g KOH catalyst on the production of HHO gas through water electrolysis was investigated under electric currents of 10 A, 40 A, 70 A, and 100 A. The experimental setup was the same as the previous experiments, except for the addition of the KOH catalyst to enhance the electrolysis process.

Figure 2 illustrates the volumes of HHO gas generated over time using electrical currents of 10 A, 40 A, 70 A, and 100 A with a catalyst.

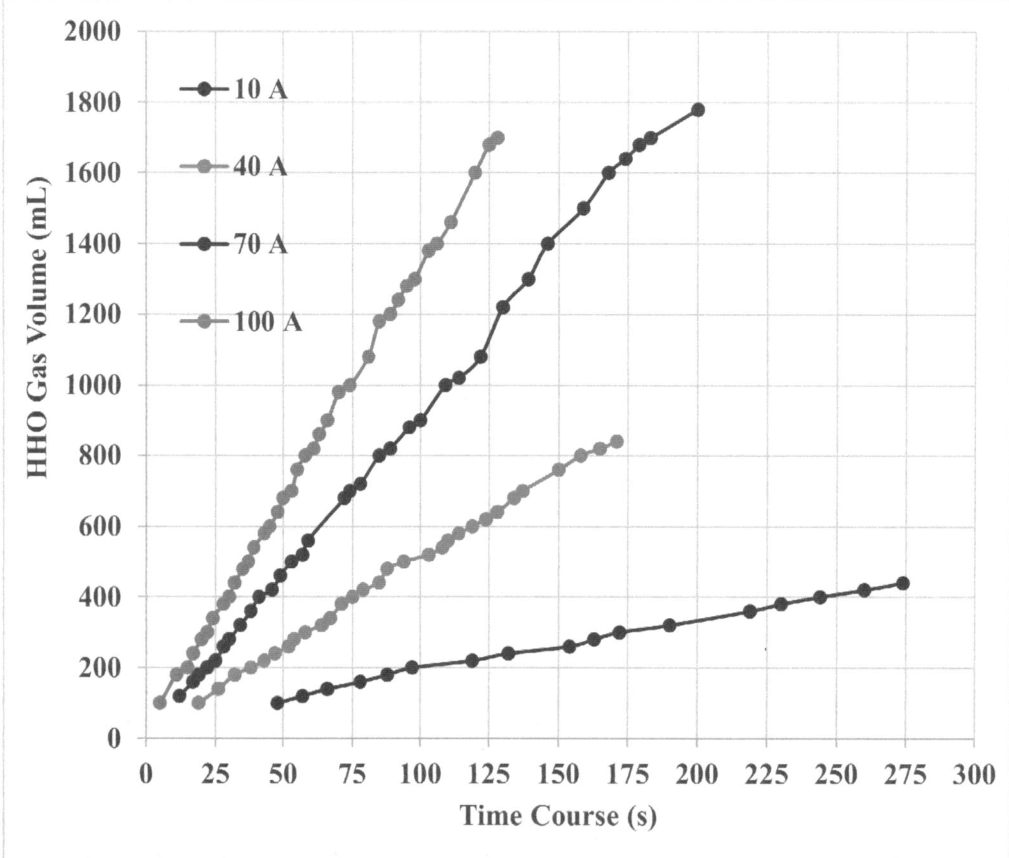

Figure 2. Volume of HHO gas generated over time with a 200 g of KOH catalyst at different currents

The results show a marked increase in the volume of HHO gas produced at each current level when the catalyst was used, compared with the experiments without the catalyst. At 10 A, the gas volume reached 240 mL at 132 seconds, compared with 200 mL at 139 seconds without the catalyst. This represents a 20% increase in gas production efficiency. At 40 A, the gas volume rose to 640 mL at 128 seconds, significantly higher than the 260 mL produced at 127 seconds without the catalyst, reflecting a 146% increase in efficiency. At 70 A, the production reached 1,080 mL at 122 seconds, compared with 280 mL at 121 seconds without the catalyst, indicating a 286% improvement. At 100 A, the volume of HHO gas produced was 1,700 mL at 128 seconds, a substantial increase from 340 mL at 124 seconds without the catalyst, representing a 400% enhancement.

The significant improvement in HHO gas production can be attributed to the catalyst's role in lowering the energy barrier for the dissociation of water molecules, thus accelerating the electrolysis process. The KOH catalyst enhances the conductivity of the electrolyte, facilitating ion exchange and electron flow between the electrodes.

When comparing these results with previous research, it is evident that increasing the electric currents improved the efficiency of HHO gas production. Studies conducted by Masjuki et al. [18] and El Soly et al. [19] have shown that the addition of catalysts significantly improves the efficiency of hydrogen production through water electrolysis. In particular, Masjuki et al. [18] reported an HHO production volume of approximately 2,150 mL using 1% KOH concentration and a lower current setting (7-10 A) over a duration of 15 minutes, which differs from this study's use of 200 g KOH concentration at currents up to 100 A in only 128 seconds, demonstatring a significant faster production rate and highlighting the efficiency of high-current electrolysis combined with higher catalyst concectrations.

Similarly, El Soly et al. [19] achieved production rates of 866 mL/min for dry cells and 1,375 mL/min for wet cells using 5-20 g/L NaOH concentrations and currents up to 27 A. In comparison, this study utilized a higher KOH concentrations (200 g) and currents up to 100 A, resulting in much higher production rates of up to 1,700 mL at 100 A. While El Soly et al. required 30-90 minutes to reach stable production rates, this study achieved its maximum production in just 128 seconds, demonstrating a significantly faster production process. However, the production rates achieved in this study surpass those reported in earlier research.

The findings of this study were compared with previous research to evaluate the relative efficiency of the HHO gas production process. The addition of a 200 g KOH catalyst to the water electrolysis process significantly increased the rate of production of the HHO gas. Higher electric currents combined with the catalyst resulted in a substantial increase in gas production efficiency. Compared with previous studies, the results of this research demonstrate a notable improvement, reinforcing the potential of using high-current electrolysis with advanced catalyst configurations for efficient hydrogen generation.

3.3 Production of HHO gas with 250 g of catalyst

The study further explored the production of HHO gas using a 250 g KOH catalyst under electric currents of 10 A, 40 A, and 70 A. Unfortunately, measurements at 100 A were not performed due to the overheating of the cables, highlighting a significant difficulty when operating at such high currents. This subsection discusses the results obtained at the three tested current levels, comparing them with those achieved using 200 g of the catalyst as well as findings from previous research.

Figure 3 illustrates the volumes of HHO gas generated over time using electric currents of 10 A, 40 A, and 70 A with 250 g of KOH catalyst. The results show a substantial increase in gas production compared with the lower mass of catalyst.

Journal of Physics: Conference Series **2980** (2025) 012016 doi:10.1088/1742-6596/2980/1/012016

At 10 A, the volume of HHO gas reached 400 mL at 129 seconds, significantly higher than the 240 mL produced at 132 seconds using 200 g of the catalyst. This represents a 66.7% increase in efficiency. At 40 A, the gas volume rose to 700 mL at 125 seconds, compared with 640 mL at 128 seconds for 200 g of the catalyst, showing a 9.4% improvement. At 70 A, the production reached 1,080 mL at 118 seconds, matching the volume produced with 200 g of the catalyst, indicating no further efficiency gains at this current level.

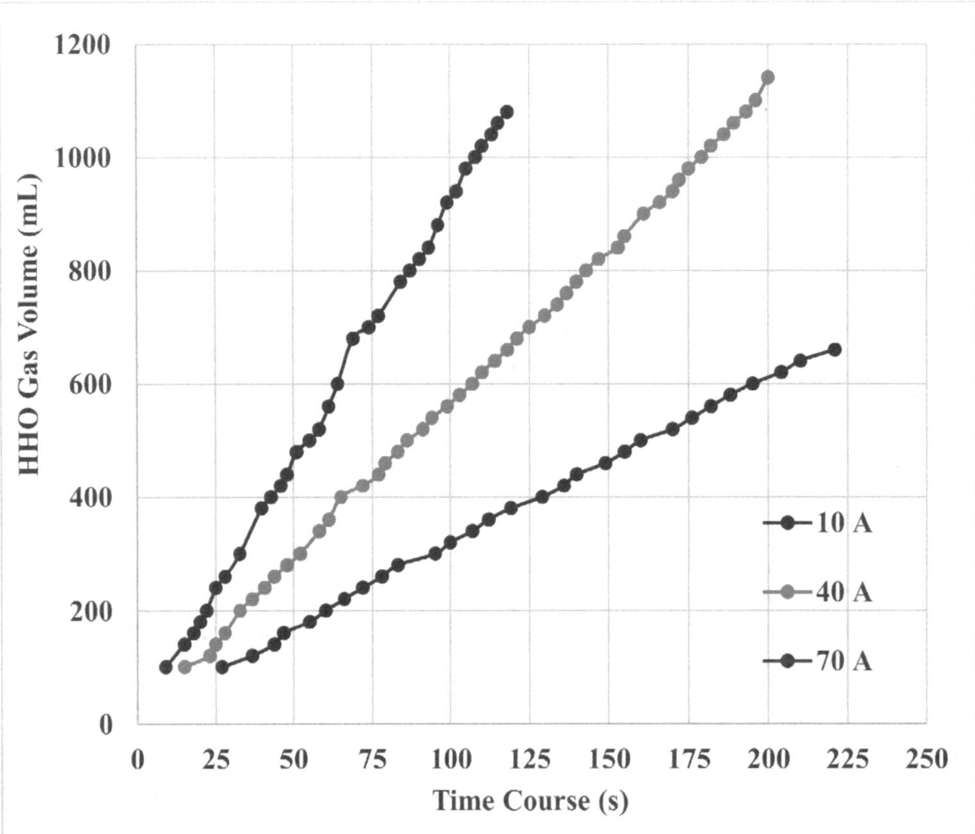

Figure 3. Volume of HHO gas generated over time with 250 g of KOH catalyst, at different currents

The overheating problem that occurred with a current of 100 A underscores a critical challenge in scalling up the current for HHO gas production. High currents generate significant heat, which can degrade the system's components and reduce its overall efficiency. To mitigate this issue, incorporating a condenser or cooling system would be effective solution.

A water cooled condenser could be integrated into the setup to absorp and dissipate the heat generated during the electrolysis process. This would involve circulating cool water around the reaction chamber and electrodes, helping to maintain a stable temperature and preventing overheating. The condenser could be connected to water reservoir or a recirculating cooling unit to ensure continuous cooling during high-current operation.

Alternatively, an air-cooled system with fans directing airflow over the electrodes and reacton chamber could also provide effective heat dissipation. This approach would be simpler to implement and maintain, especially if a water cooling system is not feasible.

3.4 Production of HHO gas with 300 g of catalyst

The study further examined the production of HHO gas using 300 g of KOH catalyst under electric currents of 10 A, 40 A, and 70 A. Measurements at 100 A were not conducted due to the overheating problem discussed in Section 3.3. This subsection discusses the results obtained at the three tested current levels, comparing them with those achieved with 250 g of the catalyst and findings from previous research.

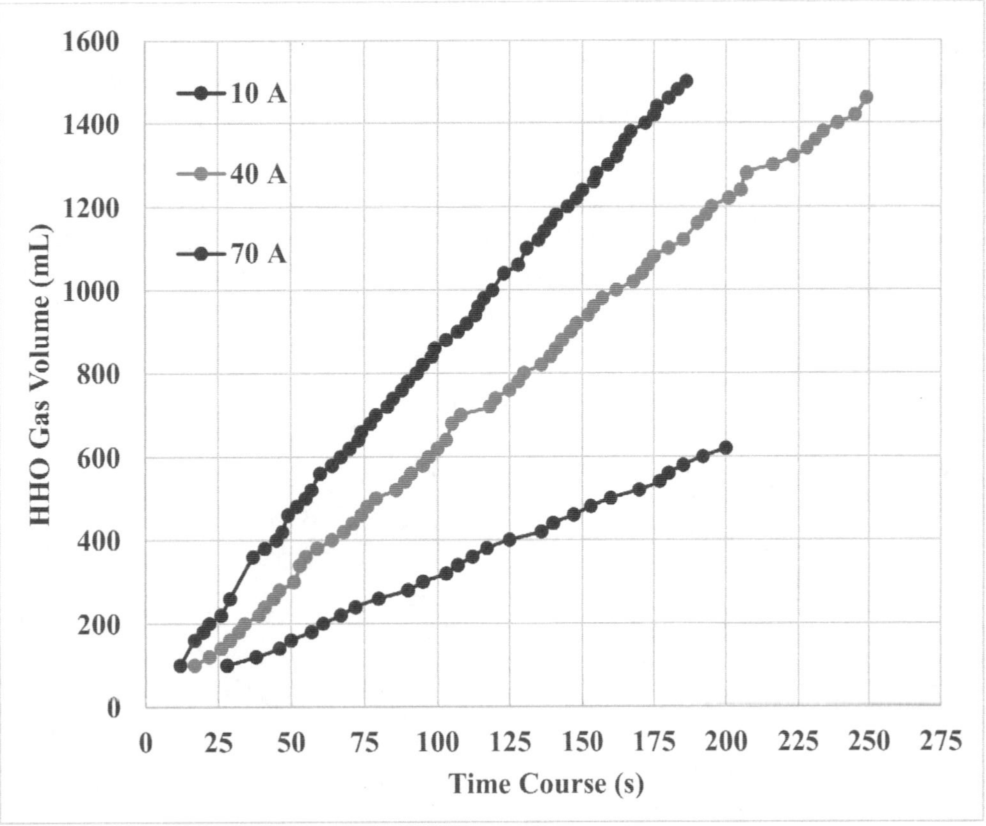

Figure 4. Volume of HHO gas generated over time with a 300 g of KOH catalyst, at different currents

Figure 4 presents the volumes of HHO gas generated over time using electric currents of 10 A, 40 A, and 70 A with a 300 g KOH catalyst. The results reveal a notable increase in gas production compared with using 250 g of catalyst.

At 10 A, the volume of HHO gas reached 400 mL at 125 seconds, which is consistent with the volume produced using 250 g of the catalyst. This suggests that increasing the catalyst mass beyond 250 g does not significantly enhance the production efficiency at lower currents. Specifically, when the catalyst mass was increased to 300 g, the volume of HHO gas produced at 10 A remained the same at 400 mL, but the production time was reduced by only 1.5% (from 129 seconds to 125 seconds). This minimal change in production time indicates that the additional catalyst does not provide a substantial gain in efficiency at lower currents, highlighting that the catalyst concentration has diminishing returns at lower current levels. At 40 A, the gas volume was 760 mL at 125 seconds, compared with 700 mL at 125 seconds when using 250 g of the catalyst, an 8.6% improvement. At 70 A, the production reached 1,000 mL at 119 seconds, compared with 1,080 mL at 118 seconds using 250 g of the catalyst, indicating a slight decrease in efficiency.

When comparing these results with previous research, it is evident that increasing the electric currents improved the efficiency of HHO gas production. Essuman et al. [20] reported optimal HHO gas yield rates with an electrolyte concentration of 0.025 M KOH, achieving a yield rate of 2.27 cm^3/s (or 136.2 mL/min) under voltage settings between 9 to 13 V. The maximum yield improvements noted by Essuman reached up to 71.96%, significantly influenced by factors such as electrode configuration and temperature. In contrast, this study achieved much higher gas production volumes, such as 1,700 mL at 100 A with 200 g KOH in 128 seconds, equating to a production rate of 796.88 mL/min. This demonstrates that increased KOH concentrations and higher current levels have substantial impact on production efficiency, exceeding those reported in Essuman's low voltage, low current setup.

3.5 Production of HHO gas with 350 g of catalyst

The study expanded the exploration of HHO gas production using a 350 g KOH catalyst under electric currents of 10 A, 40 A, and 70 A. Like the previous experiments, measurements at 100 A were not performed due to the overheating of the cables. This subsection discusses the results obtained at the three tested current levels, comparing them with those achieved using 300 g of the catalyst as well as findings from previous research.

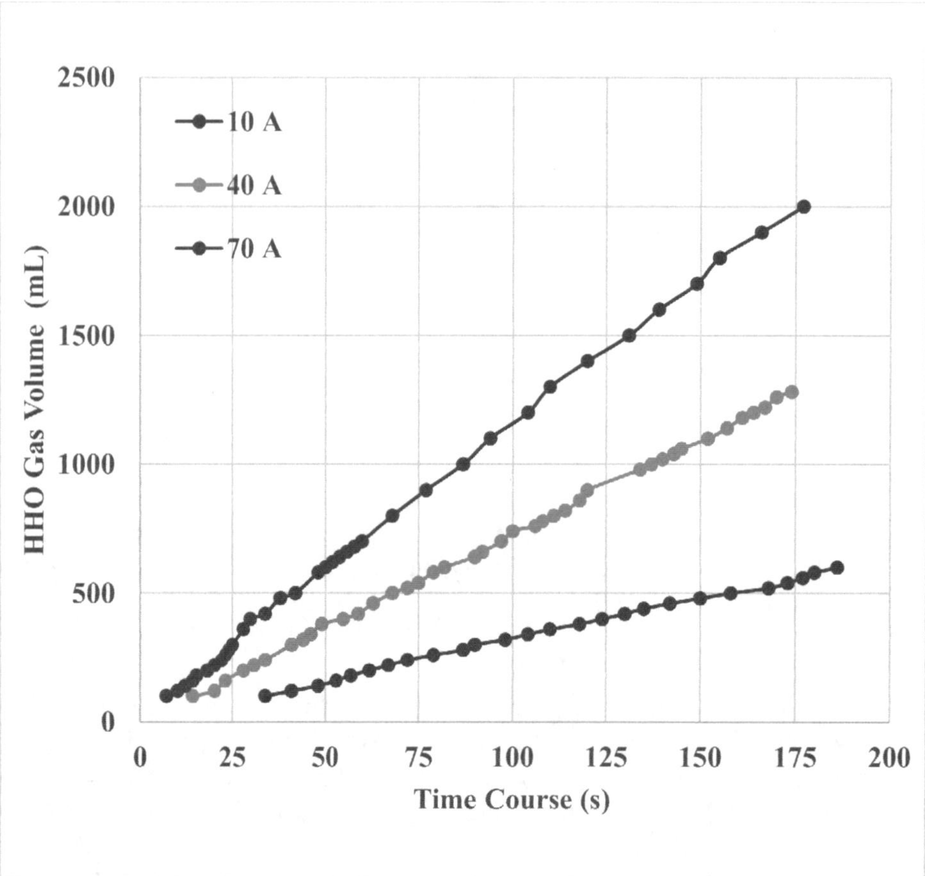

Figure 5. Volume of HHO gas generated over time with 350 g KOH of KOH catalyst, at different currents

Figure 5 illustrates the volumes of HHO gas generated over time using electric currents of 10 A, 40 A, and 70 A with 350 g of KOH catalyst. The results reveal a significant increase in gas production compared with using 300 g of catalyst.

However at 10 A, the volume of HHO gas reached 400 mL at 124 seconds, which is consistent with the volume produced using 300 g of the catalyst. This suggests that increasing the catalyst mass beyond 300 g does not significantly enhance the production efficiency at lower currents. At 40 A, the gas volume rose to 900 mL at 120 seconds, compared with 760 mL at 125 seconds when using 300 g of the catalyst, an 18.4% improvement. At 70 A, the production reached 1,300 mL at 110 seconds, compared with 1,000 mL at 119 seconds for 300 g, indicating a 30% improvement.

3.6 Production of HHO gas with 400 g of catalyst

The study further explored HHO gas production using 400 g of KOH catalyst under electric currents of 10 A, 40 A, and 70 A. Measurements at 100 A were not performed due to the overheating problem. This section discusses the results obtained at the three tested current levels,

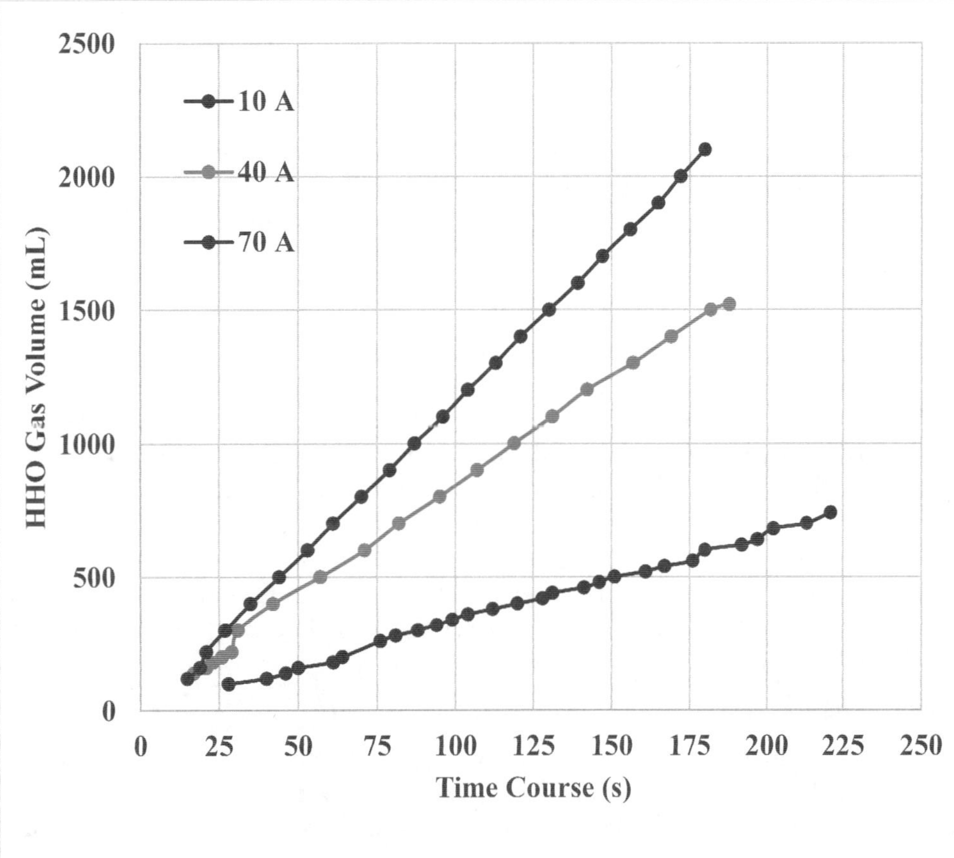

Figure 6. Volume of HHO Gas Generated Over Time with a 400 g KOH Catalyst at Currents of 10 A, 40 A, and 70 A

compares them with those achieved when using 350 g of the catalyst, and analyzes findings from previous research.

Figure 6 illustrates the volumes of HHO gas generated over time using electric currents of 10 A, 40 A, and 70 A with 400 g of KOH catalyst. The results reveal a significant increase in gas production compared with using 350 g of the catalyst.

At 10 A, the volume of HHO gas reached 400 mL at 120 seconds, consistent with the volume produced using 350 g of the catalyst, suggesting that increasing the catalyst mass beyond 350 g does not significantly enhance the production efficiency at lower currents. However, at 40 A, the gas volume rose to 1,000 mL at 119 seconds, compared with 900 mL at 120 seconds for 350 g catalyst, an 11.1% improvement. At 70 A, the production reached 1300 mL at 113 seconds, the same as with the 350 g catalyst, indicating no further efficiency gains at this higher current.

The enhanced gas production at higher currents with the 400 g catalyst can be attributed to the increased KOH concentration, which improves the electrolyte's conductivity, leading to more efficient ion exchange and electron flow between the electrodes. However, at very high currents, the efficiency gains plateau due to the system's inability to manage the increased heat and potential saturation effects of the electrolyte.

The catalyst's ability to lower the activation energy of the electrolysis reaction is key to these efficiency improvements. By reducing the energy barrier for the dissociation of water molecule, the KOH catalyst accelerates the electrolysis process, leading to higher gas production rates. This catalytic effect is particularly significant at higher currents, for which the energy input is substantial enough to overcome the activation energy with the help of the catalyst.

However, at a current of 10 A, the addition of more catalyst does not significantly increase the production rate. This can be attributed to the fact that at lower currents, the limiting factor is not the availability of ions for the reaction but the quantity of electrical energy provided. Thus, adding more catalyst does not substantially change the production dynamics as the system is already operating at an optimal ion concentration for the given electrical input.

4. Conclusions

This study demonstrated that increasing the mass of KOH catalyst and electric current significantly increases the production rate of HHO gas through water electrolysis. Using 400 g of KOH with currents of 10 A, 40 A, and 70 A resulted in the largest gas volumes, notably achieving 1300 mL at 70 A in 113 seconds. Comparison with smaller catalyst masses (200 g, 250 g, 300 g, and 350 g) and previous studies confirmed the superior efficiency of the catalyst configuration and high-current electrolysis. However, limitations at 100 A due to cable overheating highlight the need for better thermal management. These findings demonstrate the importance of balancing catalyst mass and electric current for efficient hydrogen generation, underscoring the potential of high-current electrolysis with advanced catalysts for sustainable energy production.

Acknowledgments

The authors would like to acknowledge the financial support provided by the University of Sam Ratulangi Manado and the Indonesian Higher Education Department, which made this study possible. We extend our heartfelt thanks to our peers and colleagues who reviewed this study and provided valuable feedback. We owe special gratitude to the dean of FMIPA University of Sam Ratulangi, Dr. Gerald H. Tamuntuan, for granting permission and providing the laboratory and electrolysis apparatus, ensuring the successful execution of our experiments.

References

[1] N. Abas, A. Kalair, and N. Khan, "Review of fossil fuels and future energy technologies," *Futures*, vol. 69, pp. 31–49, 2015.

[2] A. Rahman, P. Dargusch, and D. Wadley, "The political economy of oil supply in Indonesia and the implications for renewable energy development," *Renewable and Sustainable Energy Reviews*, vol. 144, p. 111027, 2021.

[3] A. Halimatussadiah, D. Nainggolan, S. Yui, F. R. Moeis, and A. A. Siregar, "Progressive biodiesel policy in Indonesia: Does the Government's economic proposition hold?," *Renewable and Sustainable Energy Reviews*, vol. 150, p. 111431, 2021.

[4] N. L. Panwar, S. C. Kaushik, and S. Kothari, "Role of renewable energy sources in environmental protection: A review," *Renewable and sustainable energy reviews*, vol. 15, no. 3, pp. 1513–1524, 2011.

[5] I. K. Kapdan and F. Kargi, "Bio-hydrogen production from waste materials," *Enzyme Microb Technol*, vol. 38, no. 5, pp. 569–582, 2006.

[6] K. Chandrasekhar, Y.-J. Lee, and D.-W. Lee, "Biohydrogen production: strategies to improve process efficiency through microbial routes," *Int J Mol Sci*, vol. 16, no. 4, pp. 8266–8293, 2015.

[7] S. Sebbahi, N. Nabil, A. Alaoui-Belghiti, S. Laasri, S. Rachidi, and A. Hajjaji, "Assessment of the three most developed water electrolysis technologies: alkaline water electrolysis, proton exchange membrane and solid-oxide electrolysis," *Mater Today Proc*, vol. 66, pp. 140–145, 2022.

[8] S. Anwar, F. Khan, Y. Zhang, and A. Djire, "Recent development in electrocatalysts for hydrogen production through water electrolysis," *Int J Hydrogen Energy*, vol. 46, no. 63, pp. 32284–32317, 2021.

[9] F. Liu *et al.*, "Elevated-temperature bio-ethanol-assisted water electrolysis for efficient hydrogen production," *Chemical Engineering Journal*, vol. 434, p. 134699, 2022.

[10] X. Li *et al.*, "Bifunctional iron doped CuS catalysts towards highly efficient overall water electrolysis in the alkaline electrolyte," *Int J Hydrogen Energy*, vol. 47, no. 38, pp. 16719–16728, 2022.

[11] Z. Chen, W. Wei, L. Song, and B.-J. Ni, "Hybrid water electrolysis: a new sustainable avenue for energy-saving hydrogen production," *Sustainable Horizons*, vol. 1, p. 100002, 2022.

[12] M. Ji and J. Wang, "Review and comparison of various hydrogen production methods based on costs and life cycle impact assessment indicators," *Int J Hydrogen Energy*, vol. 46, no. 78, pp. 38612–38635, 2021.

[13] T. Nabil and M. M. K. Dawood, "Enabling efficient use of oxy-hydrogen gas (HHO) in selected engineering applications; transportation and sustainable power generation," *J Clean Prod*, vol. 237, p. 117798, 2019.

[14] R. F. De Souza, J. C. Padilha, R. S. Gonçalves, and J. Rault-Berthelot, "Dialkylimidazolium ionic liquids as electrolytes for hydrogen production from water electrolysis," *Electrochem commun*, vol. 8, no. 2, pp. 211–216, 2006.

[15] A. Sa'ed and A. A. Al-Rousan, "Effect of HHO gas on combustion emissions in gasoline engines," *Fuel*, vol. 90, no. 10, pp. 3066–3070, 2011.

Journal of Physics: Conference Series **2980** (2025) 012016 doi:10.1088/1742-6596/2980/1/012016

[16] O. H. Ghazal, "Performance and combustion characteristic of CI engine fueled with hydrogen enriched diesel," *Int J Hydrogen Energy*, vol. 38, no. 35, pp. 15469–15476, 2013.

[17] S. Rashidi, N. Karimi, B. Sunden, K. C. Kim, A. G. Olabi, and O. Mahian, "Progress and challenges on the thermal management of electrochemical energy conversion and storage technologies: Fuel cells, electrolysers, and supercapacitors," *Prog Energy Combust Sci*, vol. 88, p. 100966, 2022.

[18] H. H. Masjuki, A. M. Ruhul, N. N. Mustafi, M. A. Kalam, M. I. Arbab, and I. M. R. Fattah, "Study of production optimization and effect of hydroxyl gas on a CI engine performance and emission fueled with biodiesel blends," *Int J Hydrogen Energy*, vol. 41, no. 33, pp. 14519–14528, 2016.

[19] A. K. El Soly, M. A. El Kady, A. E. F. Farrag, and M. S. Gad, "Comparative experimental investigation of oxyhydrogen (HHO) production rate using dry and wet cells," *Int J Hydrogen Energy*, vol. 46, no. 24, pp. 12639–12653, 2021.

[20] S. P. K. Essuman, A. Nyamful, V. Agbodemegbe, and S. K. Debrah, "Experimental studies of the effect of electrolyte strength, voltage and time on the production of brown's (HHO) gas using oxyhydrogen generator," *Open Journal of Energy Efficiency*, vol. 8, no. 2, pp. 64–80, 2019.

Journal of Physics: Conference Series **2980** (2025) 012017 doi:10.1088/1742-6596/2980/1/012017

Effect Of Manganese Alloy Concentration on the Intercalation Voltage Variation in $Na_3V_{2-x}Mn_x(PO_4)_3$ Cathode Material

Himawan Widiyanto[1,2*], Ganes Shukri[1,3], Mohammad Kemal Agusta[1,3], Qory Cipta Scientika[1], Afthon Ilman Huda Isyfi[1,2], and Dhias Pratama Lazuarfy[1,2]**

[1] Nano and Quantum Technology Research Group, Faculty of Industrial Technology, Bandung Institute of Technology, Jalan Ganesa No. 10, Bandung, West Java, Indonesia
[2] PT Perusahan Listrik Negara (Persero), Jakarta, Indonesia
[3] Research Centre for Nanoscience and Nanotechnology (RCNN), Bandung Institute of Technology, Jalan Ganesa No. 10, Bandung, West Java, Indonesia

*E-mail: 23322324@mahasiswa.itb.ac.id
**E-mail: ganes_tf07@itb.ac.id

Abstract. Vanadium as one of the components in the $Na_3V_2(PO_4)_3$ (NVP) cathode Na-ion batteries material is a toxic element and has limited natural abundance. Partial or even complete replacement of vanadium without significantly sacrificing the NVP overall electrochemical performance is desirable. Herein, by using comprehensive first-principles density functional theory (DFT)-based calculations, we investigate the effect of manganese (Mn) alloying on the intercalation voltage of NVP cathode. Our study substantiates that variations in manganese concentration up to 100% can potentially sustain structural stability. Based on stability, we justify that a 50% manganese alloy is the safest in terms of minimizing the risk of self-discharge. In terms of battery performance metrics, specifically intercalation voltage, manganese alloys demonstrate a reduction in the range of intercalation voltages but an enhancement in the voltage levels. In general, our findings show that controlling the alloy concentration is important to obtain the desired range of intercalation voltage values depending on the intended application the Mn-alloy NVP-based cathode. The addition of 25% Mn alloy changes the intercalation voltage range of pristine NVP *i.e.* 1.38 V ~2.38 V to 1.72~2.56 V. Whereas, the complete replacement of V with Mn *i.e.* 100% Mn changes the intercalation voltage of NMnP to 2.37 V ~ 3.01 V. These enhancements make manganese-alloyed NVP a promising choice for high-performance sodium-ion battery cathodes.

1. Introduction

In 1976, a material with the formula $Na_{1+x}Zr_2P_{3-x}Si_xO_{12}$ ($0 \leqslant x \leqslant 3$) was named NASICON, which stands for Sodium (Na) Super Ionic Conductor [1]. Materials of the NASICON type typically refer to materials with a chemical composition of the formula AMM'(PO4)3, where the 'A' site can be occupied by alkali ions (Li^+, Na^+, K^+, Rb^+, and Cs^+), alkaline earth ions (Mg^{2+}, Ca^{2+}, Sr^{2+}, and Ba^{2+}), transition metals (Cu^{2+}, Ag^+, Pb^{2+}, Cd^{2+}, Mn^{2+}, Co^{2+}, Ni^{2+}, Zn^{2+}, Al^{3+}, Ge^{4+}, Zr^{4+}, and Hf^{4+}), and ion-molecules (H_3O^+ and NH^{4+}) [2]. The cathode material in a battery cell plays a crucial role in determining the cell's capacity and battery life. As illustrated in **Figure 1**, the sodium-based cathode material $Na_3V_2(PO_4)_3$ features a three-dimensional lattice framework with a NASICON (Sodium Super Ionic Conductor) structure. This structure provides rapid Na^+ diffusion characteristics, excellent structural stability, and commendable electrochemical performance

[3,4]. However, vanadium, a key component of $Na_3V_2(PO_4)_3$, is toxic and naturally scarce, leading to high production costs [5]. These factors present significant challenges in the development of $Na_3V_2(PO_4)_3$ materials. Researchers have extensively explored substituting vanadium with more abundant and cost-effective elements in various experiments.

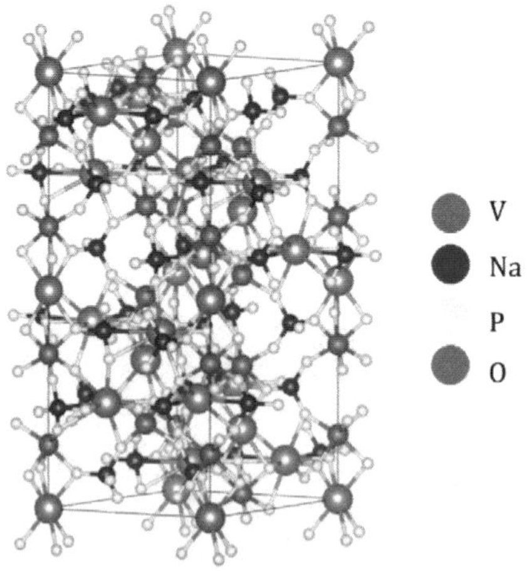

Figure 1. Visualization of $Na_3V_2(PO_4)_3$

Improving the performance of sodium-ion battery cells is crucial and one key enabler for that is by tunning the Na intercalation potential in $Na_3V_2(PO_4)_3$. One effective method to achieve this is through element alloying. Previous studies have focused on enhancing the performance of sodium-ion batteries by incorporating various transition metals, such as iron, magnesium, nickel, molybdenum, and aluminium, into the sodium and/or vanadium structure of the battery. The primary goal has been to achieve higher energy capacity and faster charging times [6–8]. Experimental studies have indicated significant improvements in the electrochemical characteristics of NVP cathodes when doped with manganese. Specifically, the introduction of manganese as an alloy enhances the discharge capacity to 107 $mAhg^{-1}$ with a 90% capacity retention after 100 cycles [9]. The use of manganese alloy has also shown stability successful measurement of a discharge capacity of 80 mAh/g for the manganese-doped material in $Na_3MnTi(PO_4)_3$. Furthermore, the addition of manganese alloy activates the redox couple reactions $Mn^{2+/3+}$ and $Mn^{3+/4+}$ within the voltage range of 2.5 – 4.2 V. Thermodynamic equilibrium is obtained at approximately 3.6 V and 4.0 V, respectively [10].

Studies on experimental fronts have shown promising results in enhancing the electrochemical properties of $Na_3V_2(PO_4)_3$ cathodes through doping or alloying with various elements. However, earlier research has fallen short in providing a comprehensive understanding of how each dopant influences the physical characteristics of these modified cathodes. To address this gap, we leverage computational approaches, such as density functional theory (DFT), to systematically understand and predict the effects of alloying on both the physical properties and electrochemical performance of cathodes [11]. This research investigated the effects of manganese alloying at various concentrations to substitute vanadium. The primary objectives of this research are to quantify the effects of alloy addition on the electrochemical properties, particularly focusing on the intercalation voltage, and to understand the redox mechanism of the cathode resulting from the addition of alloys. By achieving these objectives, we aim to provide a deeper understanding of how alloying with different elements can optimize the performance of NVP cathodes.

2. Computational Details

In this study, an analysis of NVP stability across various concentrations of sodium ions was conducted, using a range from the lowest sodium content (Na_1VP) to the highest (Na_4VP) [10].This approach thus represents the charging and discharging conditions of a battery. The models for the Na_1VP, Na_2VP, Na_3VP, and Na_4VP structures were developed based on the insertion/extraction references from previous study by Wang et al [12]. Theoretically, the NaSICON structure can accommodate up to four sodium ions, forming N_4VP within a rhombohedral crystal framework [12–15]. The insertion and extraction of Na^+ ions are generally limited to three ions per two vanadium atoms, covering a composition range from N_1VP to N_4VP. On the other hand, the extraction of the fourth sodium ion from N_1VP has proven to be challenging [16–19]. This research investigated the effects of manganese alloying at various concentrations, including the pristine condition, 25% manganese, 50% manganese, and 100% manganese substitution for vanadium (as depicted in **Figure 2**).

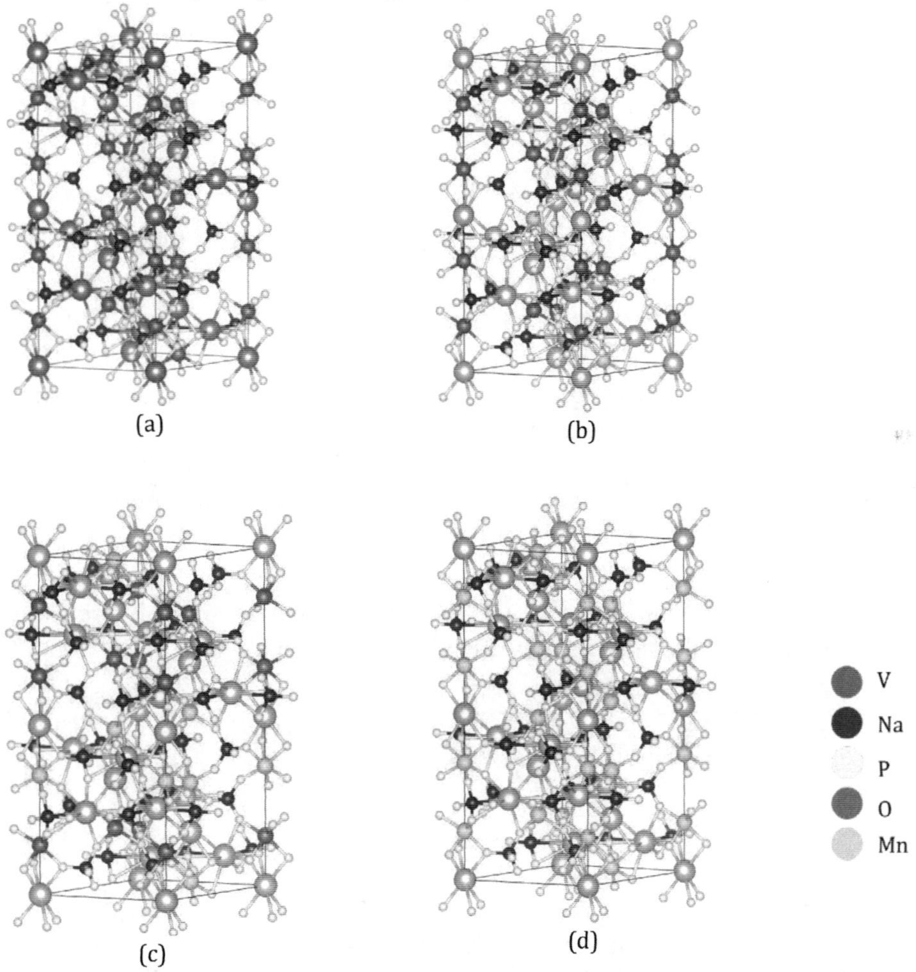

Figure 2. Visualization of $Na_3V_2(PO_4)_3$ at (a) pristine condition (b) 25% Mn (c) 50% Mn (d)100% Mn.

The intercalation voltage and formation energy were calculated using the spin-polarized Density Functional Theory (DFT) as implemented in the Vienna *Ab-initio* Simulation Package (VASP). Core electron interactions are modelled using pseudopotentials of the Projector-

Augmented Wave (PAW) type. The exchange-correlation functional utilised was the Generalized Gradient Approximation (GGA) of Perdew-Burke-Ernzerhof (PBE). The convergence threshold for electronic relaxation (EDIFF parameter in VASP) was set at 1×10^{-4} eV, while the ionic relaxation threshold (EDIFFG in VASP) was set on -0.05 eV/Å. The cutoff energy for calculations was determined to be 400 eV. Brillouin zone sampling was performed using a 3 x 3 x 3 *k*-points grid with the Monkhorst-Pack scheme.

To establish the general phase behaviour of sodium into the structure, we analysed the stability of NVP with varying sodium ion concentrations. We investigated the range between the lowest (Na_1VP) and highest (Na_4VP) sodium contents using **Equation 1**. This equation compares the energy of NVP at a specific sodium content (Na_xVP), x ($1 \le x \le 4$) called as mixing energy [20]. Mixing energy indicates the thermodynamic stability of cathode materials.

$$\Delta E_{mixing}(x) = E[Na_xVP] - \left(\frac{4-x}{3}\right)E[Na_1VP] - \left(\frac{x-1}{3}\right)E[Na_4VP] \tag{1}$$

The reversible reaction equation for the insertion/extraction of Na^+ into/from the N_xVP structure is shown in the **Equation 2,** with *x* representing the initial concentration of Sodium and *y* representing the number of insertions/extractions of Na^+ and ΔG is the Gibbs free energy of the reaction [20].

$$Na_xV_2(PO_4)_3 + yNa^+ + ye^- \overset{\Delta G^o}{\longleftrightarrow} Na_{x+y}V_2(PO_4)_3 \tag{2}$$

The total energy calculated by DFT is an approximation of the Gibbs free energy value, as shown in **Equation 3**.

$$\frac{\Delta G}{-yf} = \Delta\phi \tag{3}$$

To evaluate the electrochemical performance of a battery material, we analyse it by calculating the intercalation voltage. The intercalation voltage can be calculated using the formula presented in **Equation 3**, with *y* representing the number of Na^+ insertions/removals and *f* representing the Faraday constant, and $\Delta\phi$ representing the potential difference. The predicted value of the intercalation voltage (V_{int}) is calculated using the total energy data from DFT calculations of the $Na_3V_{2-x}M_x(PO_4)_3$ cathode system, employing a thermodynamic approach, which is the potential difference value in **Equation 4**. Thus, the potential difference, which is the value of the intercalation voltage, is indicated in $Na_3V_{2-x}M_x(PO_4)_3$ [20].

$$V_{int} = -\frac{\Delta G}{yf} = -\frac{E(N_{x+y}VP) - [E_x(NVP) + y\mu_{Na}]}{yf} \tag{4}$$

3. Results and Discussion

3.1 Structural Properties

This research explores the effects of varying manganese alloy compositions, specifically at 25%, 50%, and 100%. As illustrated in **Table 1**, During fully charging-fully discharging conditions (lowest sodium content (Na_1VP) to the highest (Na_4VP)), the addition of manganese alloy enhances the range of percentage change in structural volume. The addition of Mn alloy causes changes in the lattice volume of $Na_3V_2(PO_4)_3$ ranging from 6.87% (Mn 25%) to 8.09% (Mn 100%)

Journal of Physics: Conference Series **2980** (2025) 012017 doi:10.1088/1742-6596/2980/1/012017

The variation in lattice volume is associated with structural stability, with the observed range remaining below 8.26% [21,22]. Thus, varying manganese alloys can still maintain considerable stability, as evidenced by the minimal change in lattice volume over a full charge-discharge cycle.

Table 1. Lattice parameter variations in manganese alloying

Structure based on Concentration		**Lattice Parameter**						
Alloy	Na	a [Å]	b [Å]	c [Å]	α [°]	β [°]	γ [°]	V [Å³]
Pristine	Na1	8.49	8.49	21.35	90.35	89.65	120.06	1331.38
	Na2	8.63	8.63	21.34	90.13	89.87	120.47	1368.79 *Δ =5.30%*
	Na3	8.75	8.75	21.27	90.25	89.76	120.19	1405.93
	Na4	8.87	8.87	21.11	90.00	90.00	120.00	1438.49
Mn 25%	Na1	8.43	8.42	21.35	90.04	89.99	120.00	1312.74
	Na2	8.63	8.62	21.37	90.61	89.33	120.66	1368.09 *Δ =6.87%*
	Na3	8.75	8.75	21.29	90.34	89.76	120.19	1409.52
	Na4	8.89	8.89	21.10	90.00	90.00	120.00	1443.11
Mn 50%	Na1	8.45	8.45	21.20	90.19	89.83	119.96	1310.38
	Na2	8.64	8.59	21.39	90.79	89.19	120.73	1364.37 *Δ =6.99%*
	Na3	8.74	8.75	21.32	90.62	89.32	120.31	1408.87
	Na4	8.89	8.89	21.14	90.00	90.00	120.00	1447.22
Mn 100%	Na1	8.36	8.36	21.29	90.12	89.88	119.95	1289.56
	Na2	8.56	8.56	21.33	90.78	89.22	120.56	1346.85 *Δ =8.09%*
	Na3	8.71	8.71	21.38	90.37	89.63	120.07	1403.08
	Na4	8.88	8.88	21.13	90.00	90.00	120.00	1444.46

3.2 Mixing Energy

Energy mixing is correlated with stability. A negative mixing energy suggests that the material will naturally tend to form a homogeneous mixture. If the mixing energy is positive, there is a tendency for the components to separate into distinct phases, which can affect the material's performance. Low mixing energy indicates the stability of the formed phase, thus it is relatively safe from the potential for self-discharge. As illustrated in **Figure 3**, the lowest mixing energy profile is observed for Na_3VP in its pristine state and at desodiation levels of 25%, 50%, and 100%. A global minimum in the mixing energy curve is consistent with previous study [23]. During the desodiation process, the mixing energy profiles for these states are calculated under identical conditions. When examining the desodiation process with varying manganese alloy additions of 25%, 50%, and 100%, it becomes evident that the lowest mixing energy is achieved with a 50% manganese alloy addition.

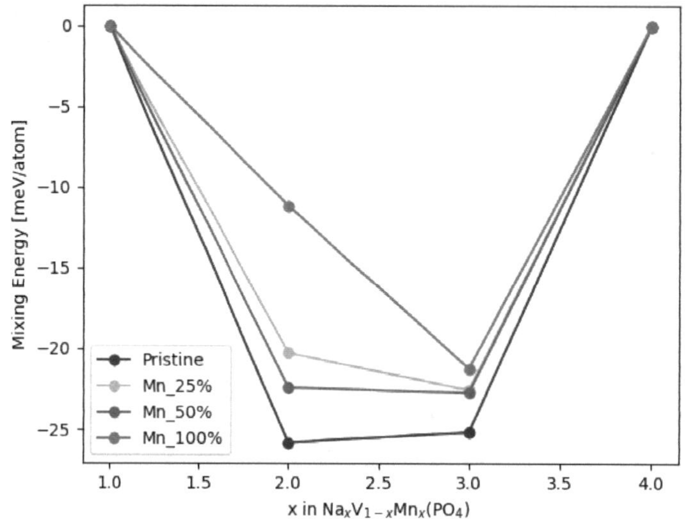

Figure 3. Mixing energies of pristine and Mn-alloyed NVP materials

3.3 Intercalation Voltage

The intercalation voltage is an important parameter for evaluating the electrochemical performance of a battery material. As illustrated in **Figure 4**, the intercalation voltage ranges for the pristine material and for manganese alloy additions of 25%, 50%, and 100% are 1.00, 0.84, 0.97, and 0.64, respectively.

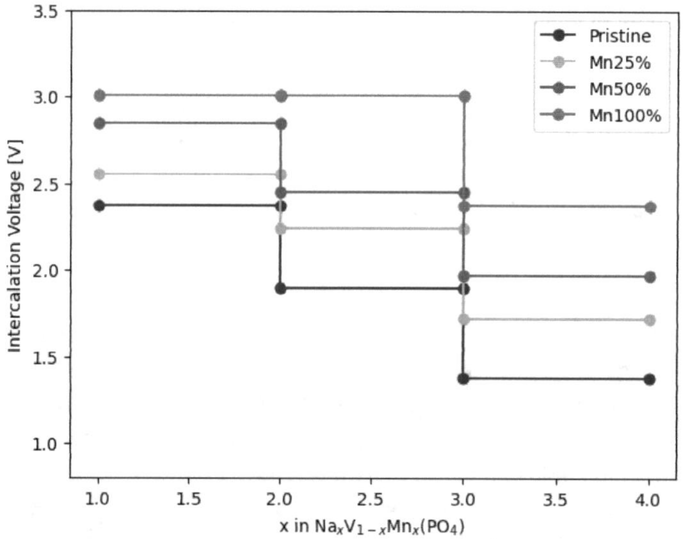

Figure 4. Intercalation voltage of pristine and Mn-alloyed NVP materials

Our findings indicate that increasing the amount of manganese alloy results in a decrease in the voltage range. The results of the intercalation voltage calculations indicate that increasing the addition of manganese alloy leads to a reduction in the voltage range. However, as shown in **Table 2**, the addition of manganese alloy in Na_1VP into Na_4VP demonstrates an increase in the intercalation voltage level.

Journal of Physics: Conference Series **2980** (2025) 012017 doi:10.1088/1742-6596/2980/1/012017

Table 2. Intercalation voltage during fully charging into fully discharging condition

Structure based on Na Concentration		Intercalation Voltage (V)	
Pristine	Na_1VP - Na_2VP	2.38	Δ = 1,00
	Na_2VP - Na_3VP	1.90	
	Na_3VP - Na_4VP	1.38	
Mn 25%	Na_1VP - Na_2VP	2.56	Δ = 0,84
	Na_2VP - Na_3VP	2,24	
	Na_3VP - Na_4VP	1.72	
Mn 50%	Na_1VP - Na_2VP	2,85	Δ = 0,88
	Na_2VP - Na_3VP	2,45	
	Na_3VP - Na_4VP	1,97	
Mn 100%	Na_1VP - Na_2VP	3,01	Δ = 0,64
	Na_2VP - Na_3VP	3,01	
	Na_3VP - Na_4VP	2,37	

3.4 Electrochemical Redox Behaviour

The intercalation voltage value is a result of the redox (reduction-oxidation) reaction mechanism. To rationalize the variation in predicted intercalation voltage values, an analysis of the redox mechanism and magnetic moment is also conducted, which is derived from the output of structural optimization calculations using the DFT method.

The electron configuration of an atom, such as vanadium, adheres to specific principles. Electrons occupy orbitals starting from the lowest energy levels, with each orbital accommodating a maximum of two electrons with opposite spins. As a transition metal, vanadium has a distinctive electron configuration that enables it to exhibit multiple oxidation states. The electron configuration of vanadium is denoted as $1s^2\ 2s^2\ 2p^6\ 3d^3\ 4s^2$.

During the processes of (de)intercalation or the formation of sodium ion (Na) vacancies, vanadium atoms (V) experience changes in their electron configuration to balance the charge of the phosphate anion (PO_4^{3-}). This altered electron configuration in vanadium can result in either a high-spin or low-spin state, contingent upon the number of paired electrons in its sub-orbitals. The high-spin state is marked by a greater number of unpaired electrons, rendering the material paramagnetic. Conversely, the low-spin state features more paired electrons, leading to a diamagnetic material.

These adjustments in electron configuration are crucial for understanding the magnetic properties and reactivity of vanadium-containing compounds, particularly in applications involving energy storage and conversion. By analysing the changes of the electronic state, it is expected that one can better tailor materials for specific technological applications, enhancing both efficiency and performance.

Journal of Physics: Conference Series **2980** (2025) 012017 doi:10.1088/1742-6596/2980/1/012017

Table 3. Magnetic moment and redox mechanism during fully charging into fully discharging condition

Structure based on Na Concentration		Magnetic Moment [μB]		Compound
		V	Mn	
Pristine	Na_1VP	0.99 – 1.03	-	$Na_1V^{4+}V^{4+}(PO_4)_3$
	Na_2VP	1.40 – 1.47	-	$Na_2V^{3+}V^{4+}(PO_4)_3$
	Na_3VP	1.83 – 1.85	-	$Na_3V^{3+}V^{3+}(PO_4)_3$
	Na_4VP	2.23 – 2.23	-	$Na_4V^{2+}V^{3+}(PO_4)_3$
Mn_25%	Na_1VP	0.81 – 1.04	2.95 – 3.11	$Na_1V^{4+}Mn^{4+}(PO_4)_3$
	Na_2VP	-1.02 – 1.34	3.69 – 3.82	$Na_2V^{3+}Mn^{4+}(PO_4)_3$ $Na_2V^{4+}Mn^{3+}(PO4)_3$
	Na_3VP	1.58 – 1.77	4.17 – 4.28	$Na_3V^{3+}Mn^{3+}(PO_4)_3$
	Na_4VP	2.05 – 2.21	4.46 – 4.46	$Na_4V^{2+}Mn^{3+}(PO_4)_3$ $Na_4V^{3+}Mn^{2+}(PO4)_3$
Mn_50%	Na_1VP	0.81 – 0.93	2.77 – 2.87	$Na_1V^{4+}Mn^{4+}(PO_4)_3$
	Na_2VP	1.01 – 1.27	3.29 – 3.66	$Na_2V^{3+}Mn^{4+}(PO_4)_3$ $Na_2V^{4+}Mn^{3+}(PO4)_3$
	Na_3VP	1.32 – 1.63	3.79 – 4.23	$Na_3V^{3+}Mn^{3+}(PO_4)_3$
	Na_4VP	1.35 – 2.09	4.34 – 4.46	$Na_2V^{2+}Mn^{3+}(PO_4)_3$ $Na_2V^{3+}Mn^{2+}(PO4)_3$
Mn_100%	Na_1VP	-	2.66 – 2.66	$Na_1Mn^{4+}Mn^{4+}(PO_4)_3$
	Na_2VP	-	3.19 – 3.19	$Na_2Mn^{3+}Mn^{4+}(PO_4)_3$
	Na_3VP	-	1.87 – 3.69	$Na_3Mn^{3+}Mn^{3+}(PO_4)_3$
	Na_4VP	-	4.08 – 4.08	$Na_4Mn^{2+}Mn^{3+}(PO_4)_3$

Magnetic moment correlates with the charge value approximation of transition metals (vanadium and manganese), thus resulting in the generation of a magnetic force source. Changes in the magnetic moment value indicate a change in charge. As shown in **Table 3,** within the range Na_1VP-Na_4VP, vanadium and manganese exhibit an increase in magnetic moment values, indicating an increase in the number of unpaired electrons.

3.5 Density of State (DOS)

The intercalation voltage value is the result of the redox reaction mechanism, thus to rationalize the variation in predicted intercalation voltage values, an analysis of the electronic density of states (DOS) data is also conducted, which is derived from the output of structural optimization calculations using the DFT method. The DOS represents the number of quantum states that can be occupied by electrons per unit energy interval. From **Figure 5**, the redox mechanism is predominantly driven by vanadium and manganese, as indicated by the abundance of these atoms around the Fermi energy level and their shifts. During fully charging-fully discharging conditions (lowest sodium content (Na_1VP) to the highest (Na_4VP)), there is a shift toward the left side of the fermi level, and significant changes occur in the states of Transition Metals (V and Mn). The shift in the graph at the Fermi level to the left corresponds with the decreased atomic charges of vanadium and manganese.

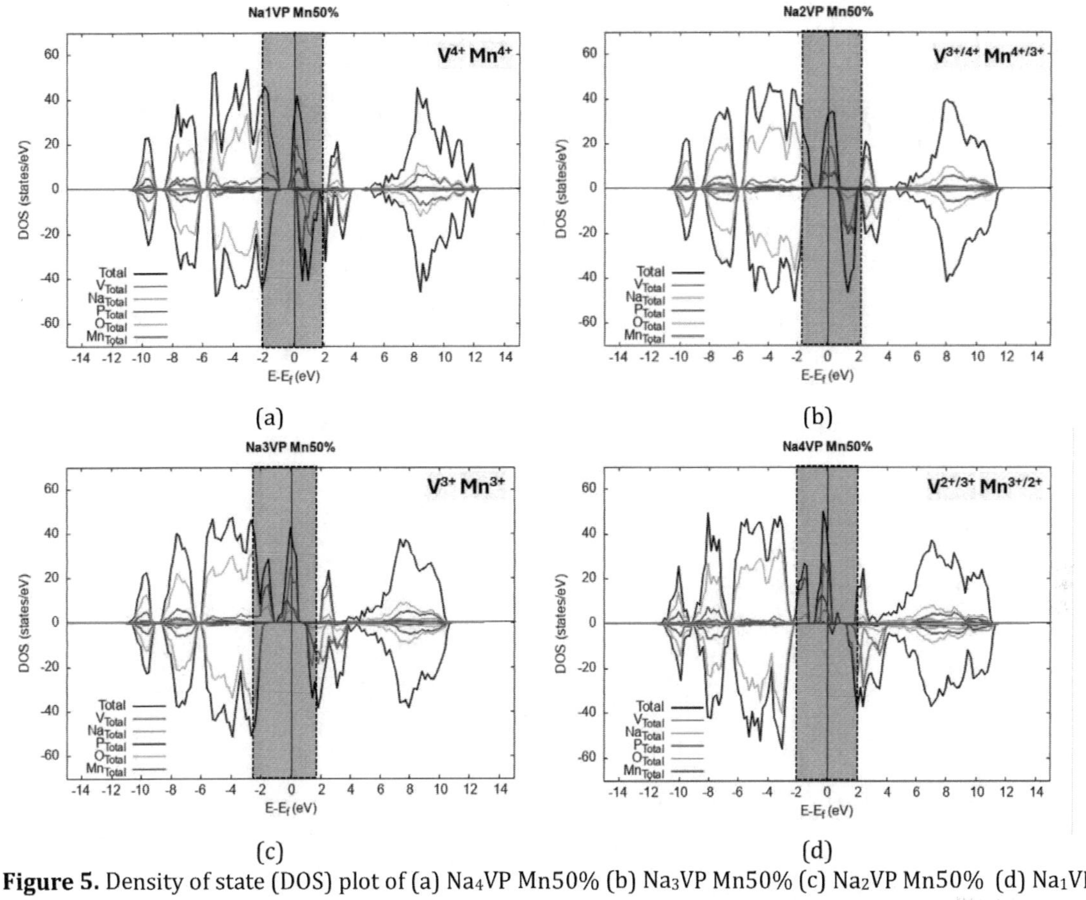

Figure 5. Density of state (DOS) plot of (a) Na_4VP Mn50% (b) Na_3VP Mn50% (c) Na_2VP Mn50% (d) Na_1VP 50% Mn

In our investigation of manganese variation effects, we conducted Density of States (DOS) analyses for Na_3VP with varying manganese concentrations. As illustrated in **Figure 6**, transition metals, specifically vanadium and manganese, remain predominant. Notably, vanadium continues to be the more dominant redox-active cation compared to manganese. We observe that even with the addition of manganese at 25% and 50% concentrations, vanadium in 3*d* orbital is still the dominant states near the Fermi level. This suggests that during the charging-discharging condition process, vanadium may still the more dominantly redox cation than the Mn in particular at the beginning of desodiation process (i.e. $Na_{3.0}$ to $Na_{2.0}$). However, it can also be noticed that there are some additions of occupied 3*d* states of Mn at around -2.0 eV - -1.0 eV energy window in the 25% and 50% Mn-alloyed NVP. These occupied Mn 3*d* could be states responsible for the increase in the intercalation voltage level of Mn-alloyed NVP as they give more electrons to the modified NVP. Manganese plays a role in reducing the surrounding vanadium and in donating electrons. Thus, it can be stated that manganese linearly increases the working voltage. Our results are somewhat consistent with previous study reported that the addition of Mn increases the overall charge/discharge capacity [9]. In general, we have shown that alloying NVP with Mn can be one viable route in tuning the intercalation voltage window of NVP given the intended specific application.

Journal of Physics: Conference Series **2980** (2025) 012017 doi:10.1088/1742-6596/2980/1/012017

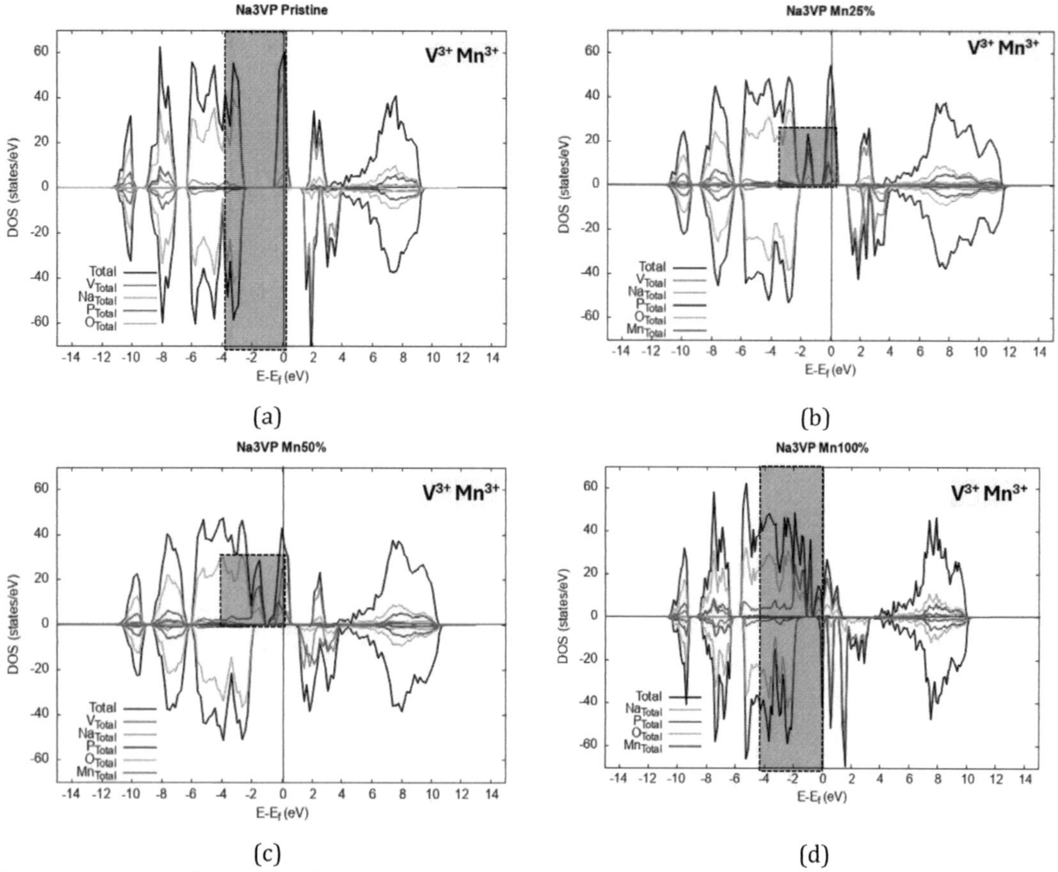

Figure 6. Density of state (DOS) plot of (a) Na_3VP Pristine (b) Na_3VP Mn25% (c) Na_3VP Mn50% (d) Na_4VP Mn100%

4. Conclusion

Our study substantiates that variations in manganese concentration up to 100% can potentially sustain structural stability. We provide evidence that a 50% manganese alloy content exhibits the greatest stability, thereby minimizing the risk of self-discharge. In terms of battery performance metrics, specifically intercalation voltage, manganese alloys demonstrate a reduction in the range of intercalation voltages but an enhancement in the voltage levels. These findings support the selection of voltage parameters tailored to specific system requirements. Our findings highlight the importance of controlling the alloy concentration to achieve the desired range of intercalation voltage values, depending on the intended application of the Mn-alloyed NVP-based cathode. These enhancements make manganese-alloyed NVP a promising choice for high-performance sodium-ion battery cathodes.

Acknowledgements

G.S. would like to acknowledge research fund endowed by the Indonesian Ministry of Research, Technology and Higher Education (Ristekdikti) through the "Penelitian Fundamental 2024" scheme.

Journal of Physics: Conference Series **2980** (2025) 012017 doi:10.1088/1742-6596/2980/1/012017

References

[1] Goodenough J B, Hong H Y-P and Kafalas J A, 1976, Mater Res Bull, 11, 203–20
[2] Liu Y, Rong X, Xie F, Lu Y, Zhao J, Chen L and Hu Y S, 2023, Materials Futures, 2
[3] Hwang J Y, Myung S T and Sun Y K, 2017, Chem Soc Rev, 46, 3529–614
[4] Lyu Y, Liu Y, Yu Z E, Su N, Liu Y, Li W, Li Q, Guo B and Liu B, 2019, Sustainable Materials and Technologies, 21
[5] Chen M, Liu Q, Wang S, Wang E, Guo X and Chou S, 2019, Adv Energy Mater, 9
[6] Li P, Gao M, Wang D, Li Z, Liu Y, Liu X, Li H, Sun Y, Liu Y, Niu X, Zhong B, Wu Z-G and Guo X, 2023, ACS Appl Mater Interfaces, 15, 9475–85
[7] Wu C, Tong J, Gao J, Li J, Li X, Zhu J, Gu M and Zhou W, 2021, ACS Appl Energy Mater, 4, 1120–9
[8] Bag S, Murarka H, Zhou C, Bhattacharya A, Jokhakar D, Pol V G and Thangadurai V, 2020, ACS Appl Energy Mater, 3, 8475–86
[9] Bhaskara Rao Y, Achary K R and Patro L N, 2022, ACS Omega 7, 48192–201
[10] Gao H, Li Y, Park K and Goodenough J B, 2016, Chemistry of Materials, 28, 6553–9
[11] Nenni, Saputro A G, Shukri G, Mobarak N N, Aprilyanti F D, Nuruddin A and Dipojono H K, 2023, New Journal of Chemistry, 47, 3773–82
[12] Wang Z, Park S, Deng Z, Carlier D, Chotard J N, Croguennec L, Gautam G S, Cheetham A K, Masquelier C and Canepa P, 2022, J Mater Chem A Mater, 10, 209–17
[13] Kabbour H, Coillot D, Colmont M, Masquelier C and Mentré O, 2011, J Am Chem Soc, 133, 11900–3
[14] Lim S Y, Kim H, Shakoor R A, Jung Y and Choi J W, 2012, J Electrochem Soc, 159, A1393–7
[15] Yabuuchi N, Kubota K, Dahbi M and Komaba S, 2014, Chem Rev, 114, 11636–82
[16] Ishado Y, Inoishi A and Okada S, 2020, Electrochemistry, 88, 457–62
[17] Lalère F, Seznec V, Courty M, David R, Chotard J N and Masquelier C, 2015, J Mater Chem A Mater, 3, 16198–205
[18] Noguchi Y, Kobayashi E, Plashnitsa L S, Okada S and Yamaki J, 2013, Electrochim Acta, 101, 59–65
[19] Wang Z, Park S, Deng Z, Carlier D, Chotard J N, Croguennec L, Gautam G S, Cheetham A K, Masquelier C and Canepa P, 2022, J. Mater. Chem. A, 10, 209–17
[20] Singh B, Wang Z, Park S, Gautam G S, Chotard J N, Croguennec L, Carlier D, Cheetham A K, Masquelier C and Canepa P, 2021, J Mater Chem A Mater, 9, 281–92
[21] Jian Z, Yuan C, Han W, Lu X, Gu L, Xi X, Hu Y S, Li H, Chen W, Chen D, Ikuhara Y and Chen L, 2014, A Adv Funct Mater, 24, 4265–72
[22] Hu J, Li X, Liang Q, Xu L, Ding C, Liu Y and Gao Y, 2024, Nanomicro Lett, 17
[23] Deng Z, Sai Gautam G, Kolli S K, Chotard J N, Cheetham A K, Masquelier C and Canepa P, 2020, Chemistry of Materials, 32, 7908–20

Journal of Physics: Conference Series **2980** (2025) 012018 doi:10.1088/1742-6596/2980/1/012018

The Influence of Nickel and Cobalt Alloy Modifications on Intercalation Voltage in Sodium-Ion Battery Cathodes $Na_3V_2(PO_4)_3$

A.Hafizh Rifa'i[1,2]*, Qory Cipta Scientika[1], Fadjar Fathurahman[1,3]. Ganes Shukri[1,3], Dhias Pratama Lazuarfy[1,2] and Afthon Ilman Huda Isyfi[1,2]**

[1]Nano and Quantum Technology Research Group, Faculty of Industrial Technology, Bandung Institute of Technology, Jalan Ganesa No. 10, Bandung, West Java, Indonesia
[2]PT Perusahan Listrik Negara (Persero), Jakarta, Indonesia
[3]Research Centre for Nanoscience and Nanotechnology (RCNN), Bandung Institute of Technology, Jalan Ganesa No. 10, Bandung, West Java, Indonesia

*E-mail: 23322323@mahasiswa.itb.ac.id
**E-mail: ganes_tf07@itb.ac.id

Abstract. The Sodium Super Ionic CONductor (NaSICON) $Na_3V_2(PO4)_3$ / NVP is a promising and extensively researched cathode for Na-ion battery due to its rapid Na^+ diffusion, good structural stability and high energy density. On the other hand, Vanadium is a toxic element, limited in nature, and expensive. Partial replacement of V with other element is thus desirable without degrading the inherent good electrochemical property of the NVP itself. Here we use Density Functional Theory (DFT) calculations to study the effects of alloying NVP with Nickel and Cobalt in $Na_xV_{1.5}M_{0.5}(PO_4)_3$ [M= Co, Ni] cathode. We find that the intercalation voltage range for $Na_xV_2(PO4)_3$ (x=1 to 4) in pristine condition were 1.38 to 2.38 V ($\triangle$V=1V). The range of intercalation voltage for $Na_xV_{1.5}Ni_{0.5}(PO4)_3$ (x=1 to 4) were 1,81 to 2.72 V ($\triangle$V=0.91V). The range of intercalation voltage for $Na_xV_{1.5}Co_{0.5}(PO4)_3$ (x=1 to 4) were 1.73 to 2.65 V ($\triangle$V=0.92V). In general, our results indicate that the NVP cathode with Ni/Co alloys in 25% concentrations possibly increase the intercalation voltage level value but decrease the intercalation voltage range. Our results thus reveal that partially alloying the NVP with Ni and/or Co is a promising alternative method to tune the intercalation voltage of the NVP.

1. Introduction

The use of traditional electrochemical batteries such as lead-acid and nickel-cadmium batteries is prone to heavy metal pollution, limiting their widespread application [1-2]. Lithium-ion batteries (LIBs) have been successfully used in small energy storage devices, such as mobile phones and notebooks [3]. However, lithium content in the Earth's crust is only about 0.0065%, and more than 76% of the global lithium reserves are concentrated in South America [4]. The limited and uneven distribution of lithium resources has driven some researchers to seek alternative battery types. Sodium, which belongs to the same group as lithium, has an abundance of about 2.8% and shares similar physical and chemical properties with lithium [5-6]. Therefore, sodium-ion batteries (NIBs) have the potential to become large-scale energy storage systems to address the challenges posed by conventional and expensive lithium-ion batteries [7-8].

Journal of Physics: Conference Series **2980** (2025) 012018 doi:10.1088/1742-6596/2980/1/012018

Various layered metal oxides, polyanion compounds, and organic materials have been widely investigated as potential cathode materials for Sodium-ion Batteries (NIBs). Metal oxides undergo irreversible structural changes during charge/discharge cycles, while organic compounds often exhibit low reaction kinetics. On the other hand, polyanion compounds offer fast Na-ion diffusion, high structural stability, high energy density, and excellent electrochemical performance [9-10]. The theoretical capacity of the NVP electrode at a 3.4 V plateau is 117.6 mAh/g, which is higher than many other electrode materials [11-12]. Among the polyanion compounds, $Na_3V_2(PO4)_3$ (NVP) exhibits a 3D NASICON (Na Super Ionic Conductor) structure which provides larger interstitial spaces to accommodate Na^+ ions during the battery cycling process [13].

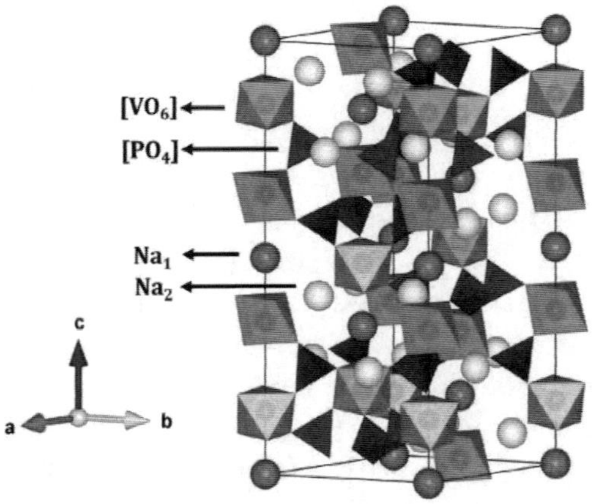

Figure 1. Crystal structure of $Na_3V_2(PO4)_3$

The $Na_3V_2(PO4)_3$ which commonly crystalize in the rhombohedral (R$\bar{3}$c) symmetry is among the most promising cathode material for NIB. It is formed by the corner-shared VO_6 octahedra and PO_4 tetrahedra connected by oxygen atoms as shown in **Figure 1**. One of the challenges in developing NVP materials is the need for a complete or partial replacement of the toxic and expensive V element. Vanadium is a relatively rare metal, with a content of about 0.015% in the Earth's crust. As a result, vanadium is more expensive than most other metals [14]. Additionally, the large distance between VO_6 octahedral units results in the lower intrinsic electronic conductivity of NVP and is considered a major drawback of NVP-based cathodes. To address these issues, various studies on NVP electrode modifications have become a major focus, including carbon material modification, doping/alloy, and nanocrystallization of the material. Carbon layer modification is typically used to improve the electronic conductivity of composite materials, but it cannot enhance the intrinsic electronic and ionic conductivity of NVP or the charge/discharge voltage platform of the material [15]. Element doping/alloying is considered an effective modification strategy. Common doping/alloying elements in NVP include transition metal atoms to replace Na(1) and V sites, which can significantly increase lattice spacing, enhance intercalation voltage, improve the electronic conductivity of the material, and modulate anion sites to raise the working voltage of the material, thus increasing its energy density [15].

Various metals have been studied by researchers for use as doping/alloying materials in NVP; however, in this study, the authors will focus on Nickel (Ni) and Cobalt (Co) alloy, which are relatively more abundant in nature, particularly in Indonesia. According to the 2024 Mineral Commodity Summaries report published by the U.S. Geological Survey in January 2024, Indonesia is the world's largest nickel producer in 2023, with a total production of 1.8 million tons (50% of the world's total production in 2023). Indonesia's nickel reserves are estimated at 55 million tons (42.3% of the world's total reserves). According to the same source, Indonesia is the second-largest cobalt producer in the world, following Congo. Of the 230,000 tons of global cobalt production in 2023, Indonesia contributed 17,000 tons, accounting for approximately 7% of the total global production [16]. In this study, we employ density functional theory (DFT) to comprehensively investigate the effects of nickel and cobalt alloying on the intercalation voltage profile of Na3V1.5Ni0.5(PO4)3 (M = Ni, Co) and to elucidate the associated redox mechanisms occurring during the (de)sodiation process (i.e., charge-discharge cycle).

2. Computational Details

In this study, the composition of the $Na_3V_2(PO_4)_3$ cathode was modified by replacing 25% of the vanadium (V) atoms with Nickel (Ni) and Cobalt (Co) alloy. For each alloy variation, structural optimization was carried out by manually varying the positions of Nickel and Cobalt relative to the Vanadium atoms in different configurations. The study aimed to determine the most stable parameters and structures of $Na_xV_2(PO_4)_3$, $Na_xV_{1.5}Ni_{0.5}(PO_4)_3$ and $Na_xV_{1.5}Co_{0.5}(PO_4)_3$ using density functional theory (DFT) calculations, implemented in the Vienna *Ab-Initio* Simulation Package (VASP)[16]. The Kohn-Sham equation was solved using the projector augmented-wave (PAW) method, with specific PAW potentials for sodium, vanadium, Nickel, and Cobalt. The exchange-correlation energy was calculated using the generalized gradient approximation of Perdew-Burke-Ernzerhof (GGA-PBE) functional, with a plane-wave cutoff energy of 520 eV and a Monkhorst-Pack *k*-point grid of 3 x 3 x 1 for all Brillouin zone integrations. Convergence criteria for forces and energy were set at 0.05 eV/Å and 10^{-4} eV, respectively. The NVP structure was based on the standard crystal database (CCSD), consisting of 126 atoms ($Na_4V_2(PO_4)_3$) in the $R\bar{3}c$ rhombohedral space group, which served as the parent structure for various $Na_xV_2(PO_4)_3$ concentrations, where $1 \le x \le 4$, as shown in **Figure 2**.

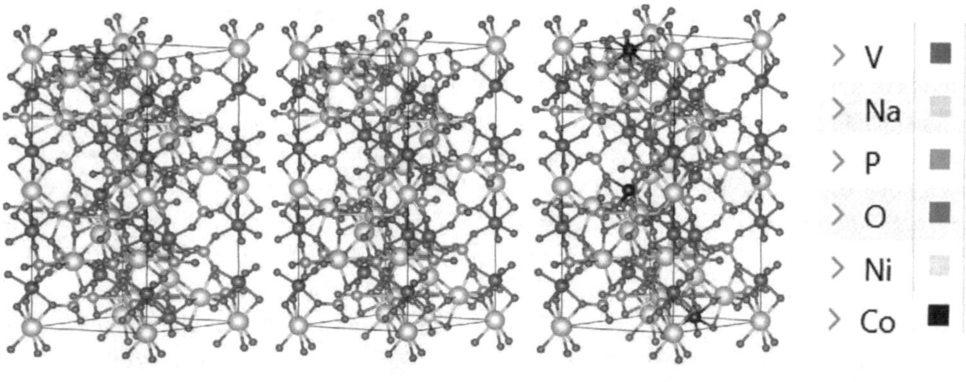

(a) **(b)** **(c)**

Figure 2. Visualization of $Na_3V_2(PO_4)_3$ at [a] pristine condition [b] With Ni alloy, $Na_3V_{1.5}$ $Ni_{0.5}$ $(PO_4)_3$ [c] With Cobalt alloy $Na_3V_{1.5}$ $Co_{0.5}$ $(PO_4)_3$.

The structures Na_1VP and Na_4VP represent the NVP structure with the lowest and highest Na ion concentrations thus serving as the lower and upper limits in the present work. Based on the (de)sodiation equation in NVP crystals, as well as the upper and lower limits of the NVP structure, the spontaneity of the intercalation reaction can be expressed as a change in Gibbs free energy, which is formulated as follows :

$$\Delta G_f(Na_xVP) = G[Na_xVP] - \left(\frac{4-x}{3}\right) G[Na_1VP] - \left(\frac{x-1}{3}\right) G[Na_4VP] \quad (2.1)$$

Using the approximation ΔG ≈ ΔU = ΔE_DFT, the formation energy (ΔE_f) for vacancy formation or the process of intercalation or deintercalation of Na ions in bulk NVP becomes:

$$\Delta E_f(Na_xVP) = E[Na_xVP] - \left(\frac{4-x}{3}\right) E[Na_1VP] - \left(\frac{x-1}{3}\right) E[Na_4VP] \quad (2.2)$$

$[Na_xVP]$ represents the energy of the bulk NVP under consideration, while $E[Na_1VP]$ and $E[Na_4VP]$ represent the energies of the bulk NVP with Na_1VP and Na_4VP ion concentrations, respectively [17]. The calculation of intercalation voltage in the NVP cathode follows the formula derived as follows:

$$\Delta G^\circ = -nFV \rightarrow n = (x - y) \quad (2.3)$$

$$V = -\frac{\Delta G^\circ}{(x-y)F} \quad (2.4)$$

Using the approximation of the equation $\Delta G \approx \Delta U = \Delta E_{DFT}$, for the change in Gibbs free energy, the intercalation voltage becomes:

$$V = -\frac{E(Na_xV_2(PO_4)_3) - [E(Na_yV_2(PO_4)_3 + (x-y)\mu_{Na}]}{(x-y)F} \quad (2.5)$$

Where $\mu_{Na}$is the chemical potential of the Na crystal, and F is the Faraday constant, valued at 96,845 C/mol or 26,801 mAh/mol.

3. Result

3.1 Structral Properties

The lattice parameters of the selected $Na_xV_2(PO_4)_3$ (pristine Condition), $Na_xV_{1.5}Ni_{0.5}(PO_4)_3$ and $Na_xV_{1.5}Co_{0.5}(PO_4)_3$ structures in each configuration, calculated using the PBE method, are presented in **Table 1**.

From the data of **Tabel 1**, several key analyses can be derived: (i) the presence of Ni/Co decreases the cell volume of pure NVP at each concentration. (ii) In the pristine NVP, generally, the insertion of Na^+ into the NASICON structure increases the *a* and *b* parameter but decreases the *c* parameter (along the hexagonal axis). The reduction in the *c* parameter was caused by the increased van der Waals forces between V-O bonds, which subsequently intensified the electrostatic repulsion between the three parallel oxygen ions and Na1, as the desodiation makes the screening effect becomes lesser [18]. In contrast, the incorporation of Ni and Co alloys reduced the lattice parameters (*a*, *b*, and *c*) of the NVP material. This is due to the smaller atomic sizes of Nickel and Cobalt compared to Vanadium, which help preserve shorter V-O bond lengths. Based on **Table 1**, it is also observed that alloying NVP with nickel results in a lattice volume change of 6.52%, while alloying with cobalt induces a lattice volume change of 8.5%. Lattice volume changes in NVP below 8.8% due to nickel and cobalt alloying are considered acceptable, as they do not significantly compromise the overall structure of NVP [19].

Table 1. Lattice constants (in Å and °), volume (in Å³) and space group (Spg.) computed with PBE.

Structure based on Na concentration		*a [Å]*	*b [Å]*	*c [Å]*	α [°]	β [°]	γ [°]	V [Å^3]	
NVP Pristine	$Na_1V_2(PO_4)_3$	8.49	8.49	21.34	90.35	89.65	120.06	1331.38	ΔV = 5.3%
	$Na_2V_2(PO_4)_3$	8.63	8.63	21.40	90.13	89.87	120.47	1368.79	
	$Na_3V_2(PO_4)_3$	8.74	8.74	21.27	90.25	89.76	120.19	1405.93	
	$Na_4V_2(PO_4)_3$	8.87	8.87	21.11	90.00	90.00	120.00	1438.49	
NVPNi	$Na_1V_{1.5}Ni_{0.5}(PO_4)_3$	8.42	8.42	21.28	90.01	89.99	120.03	1306.29	ΔV = 6.5%
	$Na_2V_{1.5}Ni_{0.5}(PO_4)_3$	8.61	8.61	21.32	90.46	89.78	120.47	1362.55	
	$Na_3V_{1.5}Ni_{0.5}(PO_4)_3$	8.70	8.70	21.35	90.44	89.70	120.20	1397.49	
	$Na_4V_{1.5}Ni_{0.5}(PO_4)_3$	8.86	8.86	22.04	90.00	90.00	120.00	1499.08	
NVPCo	$Na_1V_{1.5}Co_{0.5}(PO_4)_3$	8.36	8.36	21.20	90.00	89.99	120.03	1283.34	ΔV = 8.5%
	$Na_2V_{1.5}Co_{0.5}(PO_4)_3$	8.62	8.63	21.35	90.31	89.78	120.47	1367.96	
	$Na_3V_{1.5}Co_{0.5}(PO_4)_3$	8.74	8.73	21.26	90.29	89.82	120.16	1402.45	
	$Na_4V_{1.5}Co_{0.5}(PO_4)_3$	8.85	8.85	21.11	90.00	90.00	120.00	1433.54	

[a] Na intercalation in fully charge is Na_1 and the fully discharged phase is Na_4.

3.2 Electrochemical Stability and Intercalation Voltage

The results from DFT calculations reveal the formation energy plot for pristine NVP, as well as NVP with the addition of Nickel or Cobalt alloys, as shown in **Figure 3**. The minimum points of the formation energy represent the most stable Na / vacancy configurations (at 0 K) at each Na concentration, x=1,2,3,4 [17].

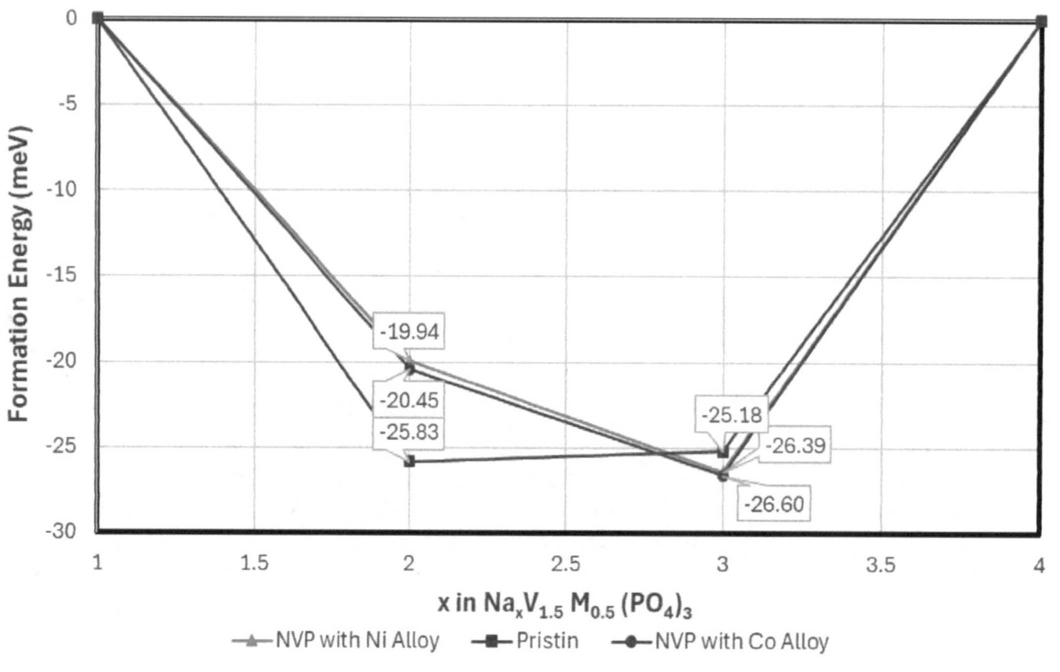

Figure 3. The Formation energy profiles of pristine and Nickel Cobalt Alloy.

As depicted in **Figure 3**, we found that the formation energy of Na_3 with a 25% concentration of Nickel or Cobalt alloy has a lower value compared to the formation energy of pristine (pure Vanadium without any alloy) Na_3 (NVP with Nickel alloy=-26.60 meV, NVP with Cobalt Alloy =-26,39 meV and NVP Pristine = -25.18 meV). It can be concluded that the addition of both nickel and cobalt, tends to decrease the formation energy of the material. The effect of nickel alloy is more significant in reducing formation energy compared to Cobalt.

After obtaining the formation energy values for pristine NVP and NVP with the addition of Nickel or Cobalt alloys, intercalation voltage calculations were subsequently performed using DFT. The results are shown in **Figure 4**.

Based on the processed intercalation voltage data from **Figure 4**, it is found that the addition of Nickel and Cobalt alloys increases the intercalation voltage compared to the intercalation voltage of pristine NVP. The intercalation voltage for $Na_xV_{1.5}M_{0.5}(PO_4)_3$ with the addition of a Nickel alloy is found to be 2.72 V for Na_1-Na_2, 2.49 V for Na_2-Na_3 and 1.81 V for Na_3-Na_4. The intercalation voltage for $Na_xV_{1.5}M_{0.5}(PO_4)_3$ with the addition of a Cobalt alloy is found to be 2.65 V for Na_1-Na_2, 2.40 V for Na_2-Na_3 and 1.73 V for Na_3-Na_4. Meanwhile. the intercalation voltage for pristine $Na_{x1}V_2(PO_4)_3$ is 2.38 V for Na_1-Na_2, 1.90 V for Na_2-Na_3 and 1.38 V for Na_3-Na_4. From the calculated intercalation voltage data, it is also known that the intercalation voltages range for $Na_xV_2(PO_4)_3$ (x=1 to 4) were 1.38 to 2.38 V ($\triangle$V=1V), under pristine condition. The range of intercalation voltage for $Na_xV_{1.5}Ni_{0.5}(PO_4)_3$ (x=1 to 4) were 1.81 to 2.72 V ($\triangle$V=0.91V). The range of intercalation voltage for $Na_xV_{1.5}Co_{0.5}(PO_4)_3$ (x=1 to 4) were 1.73 to 2.65 V ($\triangle$V=0.92V).

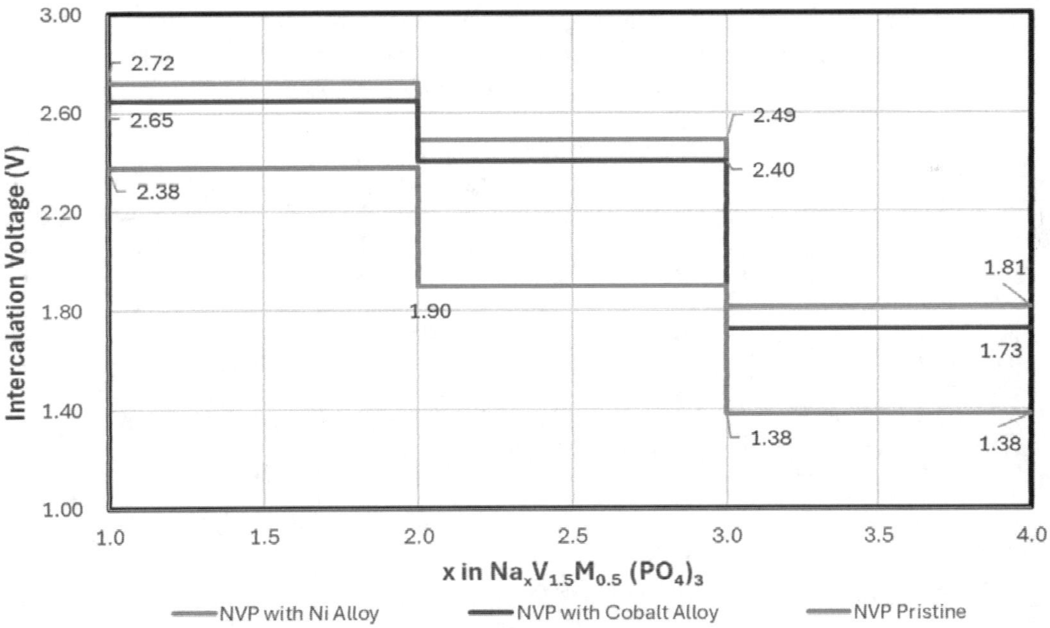

Figure 4. The intercalation Voltage profiles of pristine and Nickel Cobalt Alloy.

3.3 Ectrochemical Redox Behavior

The electron configuration of a material adheres to the Aufbau principle. Hund's rule. and the Pauli exclusion principle. The Aufbau principle states that electrons fill orbitals starting from the lowest energy levels. while Hund's rule indicates that electrons are added to orbitals or suborbitals until they are half-filled before pairing occurs. The Pauli exclusion principle asserts

Journal of Physics: Conference Series **2980** (2025) 012018 doi:10.1088/1742-6596/2980/1/012018

that electrons. being fermions. cannot occupy the same quantum state; thus. a single suborbital or quantum number *l* can be occupied by only two electrons with opposite spins (spin up and spin down).

Vanadium. a transition metal. contains 23 electrons and has the electron configuration [Ar] $3d^3$ $4s^2$. In the process of ion formation. the electrons that are removed first come from the outermost orbital (4s). even though the 3*d* orbital has a higher energy level. Vanadium can display several oxidation states. with the electron configurations for its ions being: (I) V^{2+} is produced when both electrons in the $4s^2$ shell are removed. (II) V^{3+} is formed when one electron from the 3d orbital is lost. and (III) V^{4+} is created when two electrons from the 3d orbital are removed.

The magnetic moment changes of Vanadium ions in the PBE calculations show a fairly linear trend with respect to Na concentration.

The values shown in **Figure 5.** exhibit consistency across all V atoms (1-12) with minimal variation. indicating stable magnetic configurations in pristine NVP. This stability implies that pristine NVP maintains balanced spin interactions. leading to uniform magnetic properties throughout its crystal structure.

As shown in **figure 6**. the magnetization moments for Vanadium atoms (positions 1-9) remain relatively stable across all sodium concentrations. with values consistently around 1.5 to 2 μB. Notably. the magnetization moment for Nickel atoms (positions 10-12) also shows minimal variation. indicating that the presence of Nickel does not significantly alter the overall magnetic characteristics of the material. Overall. the data suggest that both Vanadium and Nickel contribute to stable magnetic configurations within the $Na_xV_{1.5}Ni_{0.5}(PO_4)_3$ structure. regardless of the sodium concentration. This stability in magnetization moments indicates a robust interaction between the atoms. supporting the uniform magnetic properties of the compound

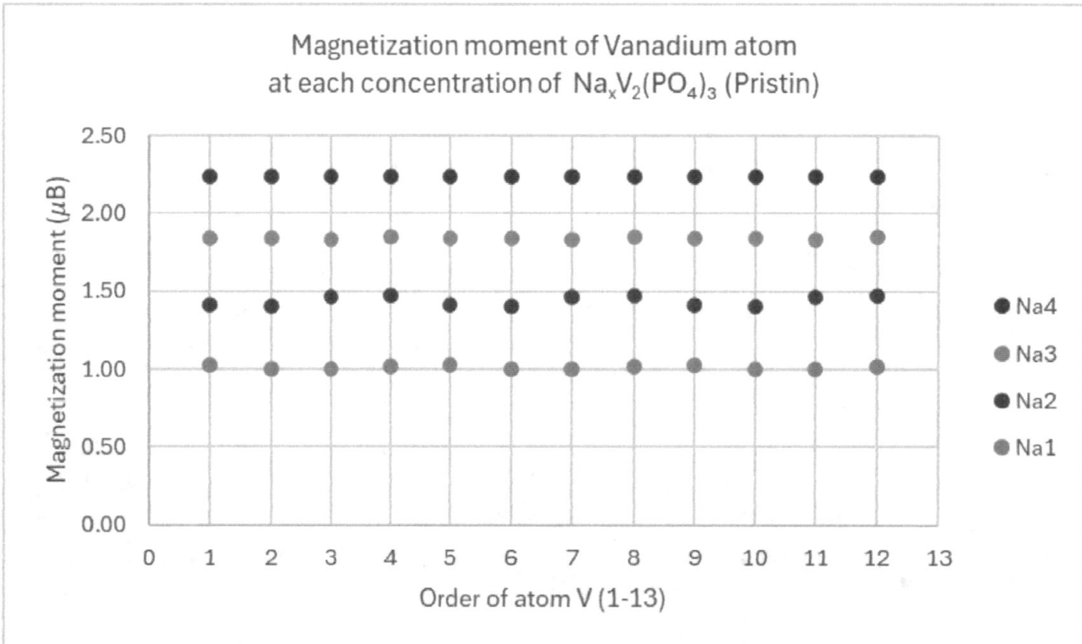

Figure 5. Magnetization moment profiles of NVP in pristine.

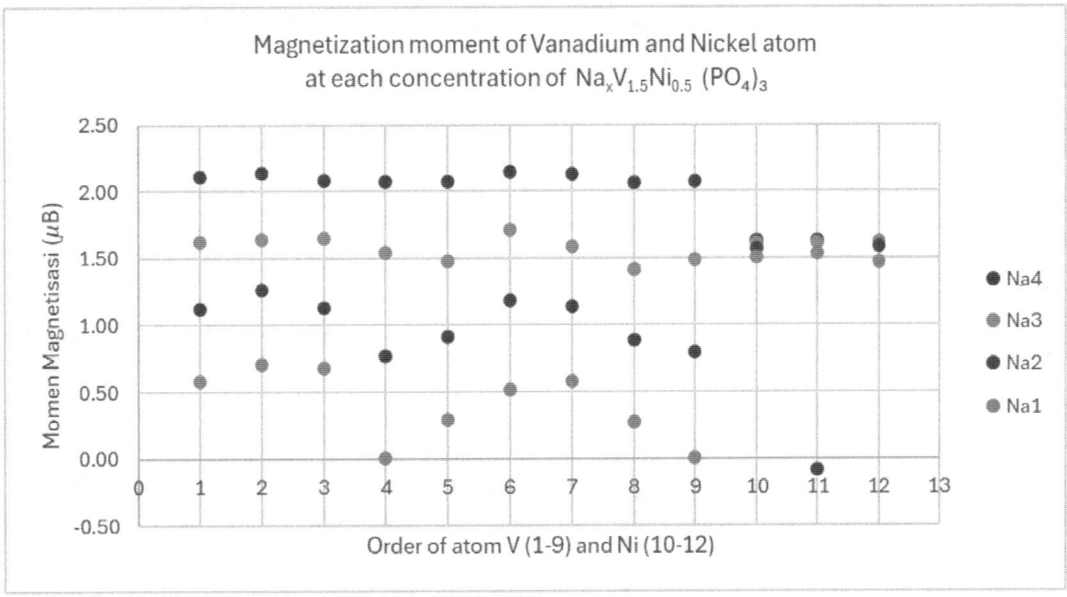

Figure 6. Magnetization moment profiles of NVP with Nickel Alloy.

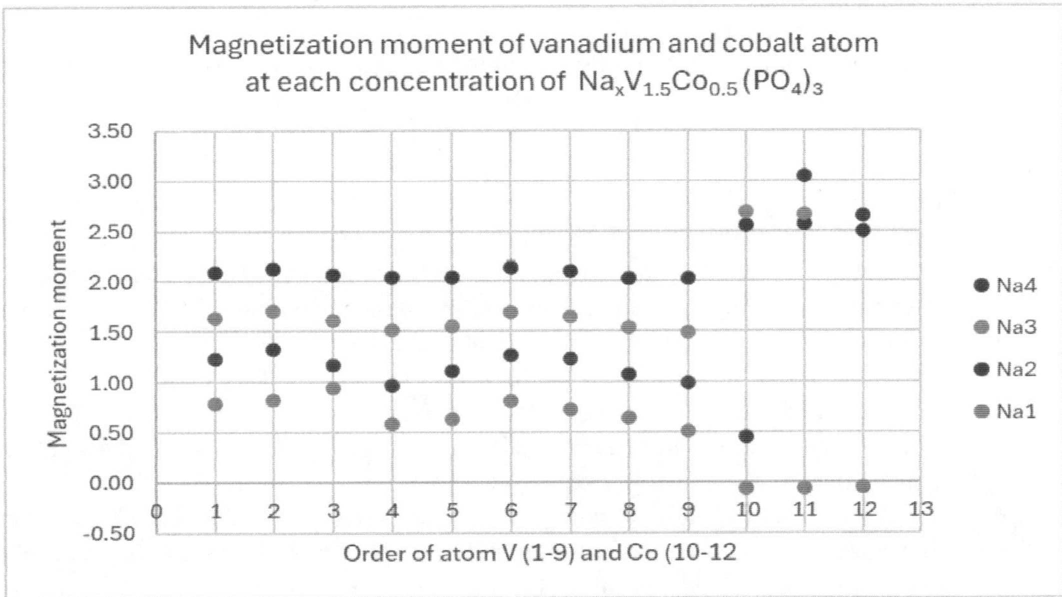

Figure 7. Magnetization moment profiles of NVP with Cobalt Alloy.

The magnetization moments for Vanadium atoms (positions 1-9) demonstrate a consistent range. typically between 1.80 and 2.20 μB. across all sodium concentrations. This indicates that the magnetic properties of Vanadium remain stable despite variations in sodium content. In contrast. the magnetization moments for Cobalt atoms (positions 10-12) exhibit a slight increase. particularly at higher sodium concentrations. reaching values around 3.0 μB. This suggests that the presence of Cobalt enhances the overall magnetic response of the material. especially as the sodium concentration increases. Meanwhile. the magnetization. the magnetization moment of Na_1 is consistently close to zero. as shown in **Figure 7**.

Based on the DFT calculations, the magnetization moment data is subsequently used to approximate the charge states of the transition metals. as shown in **Table 2** [20].

Table 2. Electronic structure. magnetic moment and bond length of vanadium ion in NVP and NVPNi/Co. Calculated with PBE method.

Na concentration. x =	Magnetic Moment [μB]			Compound		
	NVP	$Na_xV_{1.5}Ni_{0.5}(PO_4)_3$	$Na_xV_{1.5}Co_{0.5}(PO_4)_3$	NVP	$Na_xV_{1.5}Ni_{0.5}(PO_4)_3$	$Na_xV_{1.5}Co_{0.5}(PO_4)_3$
Na =1	0.997–1.026	0.004 – 1.533	-0.066–0.935	$Na_1V^{4+}V^{4+}(PO_4)_3$	$Na_1V^{5+}/^{4+}Ni^{2+}(PO_4)_3$	$Na_1V^{5+/4+}Co^{2+}(PO_4)_3$
Na = 2	1.400 – 1.469	-0.087 – 1.591	0.454 – 3.048	$NaV^{4+}V^{3+}(PO_4)_3$	$Na_2V^{4+/3+}Ni^{4+/2+}(PO_4)_3$	$Na_2V^{4+/3+}Co^{3+/2+}(PO_4)_3$
Na = 3	1.832 – 1.849	1.415 – 1.716	1.489 – 2.681	$Na_3V^{3+}V^{3+}(PO_4)_3$	$Na_3V^{3+/2+}Ni^{2+}(PO_4)_3$	$Na_3V^{3+}Co^{3+/2+}(PO_4)_3$
Na = 4	2.234	1.593–1.632	2.026 – 2.559	$Na_4V^{3+}V^{2+}(PO_4)_3$	$Na_2V^{3+/2+}Ni^{2+}(PO_4)_3$	$Na_2V^{3+/2+}Co^{3+}(PO_4)_3$

For sodium concentration x=1. the magnetic moments for $Na_1V_{1.5}$ range from 0.997 to 1.026 μB. while the magnetic moments for $Na_1V_{1.5}Ni_{0.5}$ and $Na_1V_{1.5}Co_{0.5}$ exhibit values of 0.004 to 1.533 μB and -0.066 to 0.935 μB. respectively. At x=2. the magnetic moments for $Na_2V_{1.5}$ increase to a range of 1.400 to 1.469 μB. The magnetic moments for $Na_2V_{1.5}Ni_{0.5}$ and $Na_2V_{1.5}Co_{0.5}$ show significant increases. with values of -0.087 to 3.048 μB. indicating that the presence of Ni and Co enhances the magnetic response. For x=3. the magnetic moments for $Na_3V_{1.5}$ further increase. ranging from 1.832 to 1.849 μB. The magnetic moments for the nickel and cobalt compounds also increase. reaching values of 1.415 to 2.681 μB. At x=4. the magnetic moment for $Na_4V_{1.5}$ reaches 2.234 μB. while the magnetic moments for the nickel and cobalt variants range from 1.593 to 2.559 μB. Overall, the data as shown in **Table 2** demonstrate that an increase in the number of Na atoms in NVP (Pristine) leads to an enhancement of the magnetization moment. Furthermore. the data indicate that as the Na atom content increases in $NaxV_{1.5}Ni_{0.5}(PO_4)_3$ and $NaxV_{1.5}Co_{0.5}(PO_4)_3$. the range (Δ) of the magnetization moment also increases.

To analyze the effects of nickel and cobalt alloying on NVP, this study also examines the reduction and oxidation processes of transition metals using Density of States (DOS) data. The electronic Density of States (DOS) was obtained for the transition metal atoms (nickel, cobalt, and vanadium) in each condition, providing a representation of their electronic states.

From the DOS plot as shown in **Figure 8**, it can be observed that the addition of nickel and cobalt alloying at 25% concentration introduces additional electronic states to the left of the Fermi energy, indicating these states are occupied by electrons. This phenomenon confirms that nickel (atomic number 28) and cobalt (atomic number 27), positioned to the right of vanadium (atomic number 23) in the periodic table, act as electron donors to the system. The presence of these electron donors aligns with the observed increase in intercalation voltage values. In summary, this study demonstrates that alloying NVP with nickel and cobalt is a feasible approach to tuning the intercalation voltage of NVP, making it adaptable for specific desired applications.

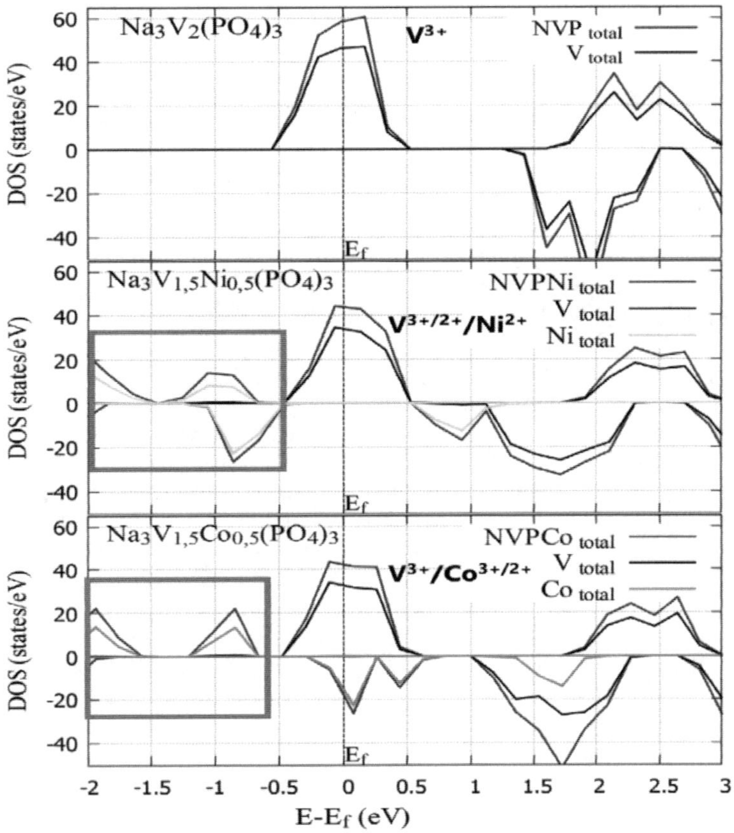

Figure 8. DOS Profile of $Na_3V_2(PO_4)_3$, $Na_3V_{1,5}Ni_{0,5}(PO_4)_3$ dan $Na_3V_{1,5}Co_{0,5}(PO_4)_3$.

4. Conclusion

The intercalation voltages range for $Na_xV_2(PO_4)_3$ (x=1 to 4) were 1.38 to 2.38 V ($\triangle$V=1 V). under pristine condition. The range of intercalation voltage for $Na_xV_{1.5}Ni_{0.5}(PO_4)_3$ (x=1 to 4) were 1.81 to 2.72 V ($\triangle$V=0.91V). The range of intercalation voltage for $Na_xV_{1.5}Co_{0.5}(PO_4)_3$ (x=1 to 4) were 1.73 to 2.85 V ($\triangle$V=0.92V). In general. our results indicate that the $Na_3V_2(PO_4)_3$ cathode with nickel and cobalt alloy in 25% concentrations possibly increase the intercalation voltage but decrease the intercalation voltage range which can be beneficial for further utilization of $Na_3V_2(PO_4)_3$ based cathode materials.

Acknowledgements

G.S. would like to acknowledge financial research support granted by the Ministry of Research, Technology and Higher Education (MENRISTEKDIKTI) of the Republic of Indonesia under the 'Riset Fundamental 2024' funding scheme no. 036/E5/PG.02.00.PL/2024. Financial research support granted by the Bandung Institute of Technology under the "Riset Unggulan 2023" scheme no. 308/IT1.B07.1/TA.00/2023 is alsoacknowledged.

Journal of Physics: Conference Series **2980** (2025) 012018 doi:10.1088/1742-6596/2980/1/012018

References

[1] G.B, Xu. Z, Chen. X,Liu. Y, Zhang. X,L. Wei.L,W. Yang, P.K. Chu, 2019, Electrochim. Acta, 309, 177–186.
[2] L. Li, X. Liu, L. Tang. H, Liu Y, G. Wang, 2019, J. Alloy. Compd, 790, 203–211.
[3] Z. He, Y. Jiang, W. Meng, J. Zhu, Y Liu, L. Dai, L. Wang, 2016, Electrochim. Acta, 222, 1491–1500.
[4] Z. He, Y. Jiang, J. Zhu, Y. Li, Z. Jiang, H. Zhou, W. Meng, L. Wang, L. Dai, 2018, J. Alloy. Compd, 731, 32–38.
[5] X. Cao, Q. Sun, L. Zhu, L. Xie, 2019.,J. Alloy. Compd, 791, 296–306.
[6] H. Yi, L. Lin, M. Ling, Z. Lv, R. Li, Q. Fu, H. Zhang, Q. Zheng, X. Li, 2019, ACS Energy Letters, 4, 1565–1571
[7] N. Yabuuchi, K. Kubota , M. Dahbi, S. Komaba,2014. Chem. Rev.. 114. 11636–11682.
[8] Y. B. Rao, K. K. Bharathi, L. N. Patro,2021, Solid State Ionics, 366-367,115671.
[9] J. Y. Hwang, S. T. Myung, Y. K. Sun., 2017, Chemical. Soc. Rev, 46, 3529–3614.
[10] Y. Lyu, Z. E.Yu, N. Su, 2019. Sustainable Materials and Technologies. e00098
[11] S. Wang, J. Zhao, L. Wang, X. Liu, Y. Wu, J. Xu, 2015, Ionics, 21,2633-2638.
[12] H. Wang, D. Jiang, Y. Zhang, G. Li, X. Lan, H. Zhong, Z. Zhang, Y. Jiang, 2015, Electrochim. Acta, 155, 23-28.
[13] X. Zhang, X. Rui, D. Chen, H. Tan, D.Yang, S.Huang, Y.Yu, 2019, Nanoscale,11, 2556–2576.
[14] M. Sanchayan, D. Sayan, G.Debakshi, 2024, Journal of Electroanalytical Chemistry, 955, 118046.
[15] Yuanqiang Z, Hui X, Ji M, Pengdong C, Yong Chen, 2023, Journal of Solid State Chemistry, 317, Part A,123669.
[16] US Geological Survey. 2021.Mineral Commodity Summaries. U.S. Department of the Interior.
[17] Singh. B. et al., 2021, Journal of Materials Chemistry A, 9,281–292.
[18] Park S,Wang Z, Deng Z, Moog I, Canepa P, Fauth F, 2022. *Chem. Mater,* 34, 451–62.
[19] Shen, F., Huang, Q., Tietz, F., Hao, J., Huang, C. T., Hernandha, F. H., Rahman, 2023., Advanced Science, 01490
[20] L. Chaoping, K. Fantai, L. Roberto, The Journal of Physical Chemistry,2016,120,6832-6393.

Journal of Physics: Conference Series **2980** (2025) 012019 doi:10.1088/1742-6596/2980/1/012019

On the Variation of Intercalation Voltage in Si-doped ($Na_{3+y}V_2(PO_4)_{3-y}(SiO_4)_y$) Cathode Material: A First-Principles Study

Qory Cipta Scientika[1*], Sasfan Arman Wella[2], Fadjar Fathurrahman[1,3], Ahmad Nuruddin[1,3], Isa Anshori[3,4], Aditya Wibawa Sakti[5], Ganes Shukri[1,2]**

[1] Nano and Quantum Technology Research Group, Faculty of Industrial Technology, Bandung Institute of Technology, Bandung, Indonesia
[2] Research Centre for Quantum Physics, National Research and Innovation Agency (BRIN), South Tangerang, Indonesia
[3] Research Centre for Nanoscience and Nanotechnology (RCNN), Bandung Institute of Technology, Bandung, Indonesia
[4] Lab-on-Chip Laboratory, Biomedical Engineering Department, Bandung Institute of Technology, Indonesia
[5] Global Center for Science and Engineering, Waseda University, Tokyo, Japan

*E-mail: 23322301@mahasiswa.itb.ac.id
**E-mail: ganes_tf07@itb.ac.id

Abstract. The $Na_3V_2(PO_4)_3$ (i.e. NVP) is a Natrium Super-ionic Conductor (NaSICON) type compound which provides a 3D open framework that has strong covalent bonds, making this material a promising cathode material with high-power density. Increasing its ability to deliver Sodium-ion at a faster rate and at a higher electronic conductivity have been becoming target researches for its further utilization. Here, by utilizing the first-principles calculation, we investigated the effect of partial substitution of P with Si on the intercalation voltage of silicon-doped NVP. We determine the thermodynamically stable phase of the pristine NVP and Si-doped NVP or $Na_{3.1}V_2(PO_4)_{2.9}(SiO_4)_{0.1}$ (i.e. NVPS0.1) by considering various sodium configurations. Based on the mixing energy results, we then calculated the intercalation voltage and found that the addition of Si dopant slightly increases the range of intercalation voltage of pure NVP from 1.43-2.36 V, $\Delta V = 0.93$ V and 1.24-2.44 V for NVPS0.1 with $\Delta V = 1.20$ V, increases 0.27 V. In general, our finding shows the potential of Si to modulate the overall intercalation voltage window of a NVP-based cathode material for Na-ion battery.

1. Introduction

Traditional commercial batteries such as lead-acid batteries, nickel-cadmium batteries, and other electrochemical batteries are prone to heavy metal pollution and have limited applications. Therefore, over the past three decades, lithium-ion batteries (LIBs), with their advantages of high working voltage, low self-discharge rate, long life cycle, and being more environmentally friendly, have dominated products ranging from portable electronic devices to electric vehicles [1][2]. However, the production costs of lithium-ion batteries are very high because raw materials like lithium, nickel, and cobalt are becoming increasingly scarce [3]. It is predicted that by 2030, there will be a deficit of these materials, driving prices up exponentially. Sodium (Na), the sixth most abundant element on Earth (about 2.74%, or 284 times more abundant than lithium), can be directly extracted from seawater, making it a potential alternative to replace lithium [4][5]. In addition, sodium shares many physical and chemical properties with lithium [6] and has a relatively high energy density (200 W h/kg). Although the use of sodium has solved some of the issues associated with lithium mentioned earlier, the performance and life cycle of Sodium-Ion Batteries (SIBs) still do not match that of LIBs [7]. This challenge has led many researchers to investigate promising electrode materials for SIBs [2][8]. Sodium-Ion Batteries (SIBs) generally operate on the same principle as LIBs. During the charging process, Na ions leave the cathode and migrate through the electrolyte towards the anode, while electrons flow from the cathode to the anode. This ion and electron transport process from the cathode to the anode or vice versa continues during both the charging and discharging processes[2][9].

The positive electrode or the cathode in a rechargeable battery is a limiting factor that plays a crucial role in ion movement and determines a battery specific capacity, cycling ability, and specific energy [10]. To date, layered metal oxides (NaM_xO_y, M = transition metals), transition metal fluorides (Na_xF_3, X = transition metals), polyanionic compounds, and organic materials have been identified as potential cathode materials for SIBs [11]. Metal oxides experience irreversible structural changes during charge/discharge cycles, while organic compounds often exhibit low reaction kinetics. On the other hand, polyanionic compounds offer fast Na-ion diffusion, high structural stability, high energy density, and excellent electrochemical performance. Among polyanionic compounds, $Na_3V_2(PO_4)_3$ (NVP) presents a 3D NASICON (Na Super Ion CONductor) framework. NVP offers larger interstitial space (~8.9%) to accommodate Na-ions, an essential aspect to achieve high cycle stability, which is in line with battery lifespan [12].

Experimentally, the NASICON-type NVP structure has a rhombohedral crystal structure with space group *R*3*c*, where each primitive cell consists of six $Na_3V_2(PO_4)_3$ formula units (meaning there are 18 Na atoms, 12 V atoms, 18 P atoms, and 72 O atoms in one primitive cell). It has been identified that there are two distinct sodium (Na) sites—Na(1) and Na(2)—each with different coordination numbers to oxygen atoms. The Na(1) site coordinates with six oxygen atoms and is located between two adjacent [$V_2(PO_4)_3$] units, while the Na(2) site has eight coordination bonds, with two of these bonds sharing coordination with [$V_2(PO_4)_3$] to accommodate two Na-ions [13]. Jian and colleagues found that during the extraction of Na ions from $Na_3V_2(PO_4)_3$ to form $NaV_2(PO_4)_3$, only ions occupying the Na(2) site could be eliminated, while the sodium atoms in the Na(1) site remained immobile. From this finding, it was concluded that the conduction mechanism in the NVP structure is more readily facilitated along the Na(2) → Na(2) pathway [12]. However, Hu et al. disputed this claim. In their recent study, Hu et al. demonstrated that the Na(1) site could also participate in the conduction mechanism by activating the Na(1) site. This activation could occur if the pure NVP structure was intercalated with other cations such as Zn^{2+}. Hu et al. also cautioned about the spatial arrangement of atoms, noting that the M–O–M bond cannot be directly linked due to the resulting electron delocalization, which would lead to very low electronic conductivity [14]. These studies provide guidelines for determining the NVP structure and serve as a starting point for reconstructing other NVP structures, especially for doped NVP materials.

An important factor in improving the performance of sodium-ion battery cells is by increasing the ionic conductivity of NVP. One method to achieve this is through element doping. Previous efforts have included substituting parts of Na and/or V with various transition metals such as Fe^{3+}, Mg^{2+}, Ni^{3+}, Mo^{6+}, Al^{3+}, and others to achieve high capacity and enhance the rate performance of sodium-ion battery cells [10][15][16][17]. In an experiment by Rao et al., partially replacing V with Fe-Mn was found to increase the discharge capacity to 107 mA h g^{-1}. However, Pal et al., discovered a more effective way to improve the ionic conductivity of NVP by expanding the rhombohedral crystal lattice of NVP through Si^{4+} substitution, which could increase the number of Na-ions per unit formula of NVP. The expanded open lattice structure is expected to provide more space for sodium ion transfer, thereby enhancing the cathode's rate capacity, especially at high current densities. This is due to the higher reversibility of the redox reaction when Si ions are substituted at the P site in NVP. This scheme also results in a good specific capacity value of 112 mA h/g at C/10 and 85 mA h/g at 1C, making a $Na_{3.1}V_2(PO_4)_{2.9}(SiO_4)_{0.1}$ a promising cathode material for sodium-ion batteries [18].

In this study, the effect of Si substitution at the P site on the redox reaction and its impact on intercalation voltage in the cathode of sodium-ion batteries will be investigated in more details using by means of first-principles density functional theory (DFT) calculations. Two systems will be studied: pure $Na_xV_2(PO_4)_3$ and $Na_xV_2(PO_4)_{2.9}(SiO_4)_{0.1}$ (or NVPS0.1) with aggregate variations of x = 4, 3, 2, 1 for pristine and x = 4, 3.1, 2.1, 1 for NVPS0.1. This study will focus on identifying the energetics, geometry, and redox properties of the pristine NVP and Si-doped NVP systems which can provide more insight into the detailed physicochemical mechanisms occur in NVP-based material during a battery operation.

2. Method

We used the Vienna Ab Initio Simulation Package (VASP) based on density functional theory (DFT) to determine the most stable parameters and structures of NVP and NVPS0.1. Those calculation solved the Kohn-Sham equation for many-particle system using the projector augmented-wave (PAW) method [19]. The PAW potentials specifically used for describing the core electrons are: Na 08Apr2002 $3s^1$, V_pv 07Sep2000 $3p^6$ $3d^4$ $4s^1$, P 06Sep2000 $2s^2$ $3p^3$, and O 08Apr2002 $2s^2$ $2p^4$. For the exchange-correlation energy were calculated by generalized gradient approximation (GGA) with the Perdew-Burke-Ernzerhof (PBE) functional [20]. We set the plane-wave cutoff energy at 520 eV to expand the Kohn-Sham wavefunctions for the valence electrons, and performed Brillouin zone integration using the Monkhorst-Pack method with a 3 × 3 × 3 k-point grid. The convergence criteria for forces and energy were set at 0.05 eV/Å and 10^{-4} eV per atom, respectively.

In this study, we derived the NVP structure from the standard crystal database (CCSD), where fully sodiated NVP with $R\bar{3}c$ rhombohedral space group consists of 126 atoms (24 Na, 12 V, 18 P, and 72 O atoms) or equivalent to $Na_4V_2(PO_4)_3$, shown on **Figure 1.** This structure served as the parent structure for N_xVP concentrations, where $1 \le x \le 4$. Based on the parent structure described above, many possible de-dosiated structures of NVP could be generated. However, due to computational cost limitations resulting from the demanding nature of DFT calculations and because of metastable structure of NVP, we only generated a few structural models for Na_3 and Na_2. Then, we optimized the structure using the previous parameter described above and calculated the lattice parameter at differences concentration. In total, we carried out DFT structural relaxation (i.e. relax the cell volume, ionic position, and cell shape) for total of 16 structures:1 structure for N_4VP, 7 for N_3VP, 4 for N_2VP, and 4 structure for N_1VP. Meanwhile, for the

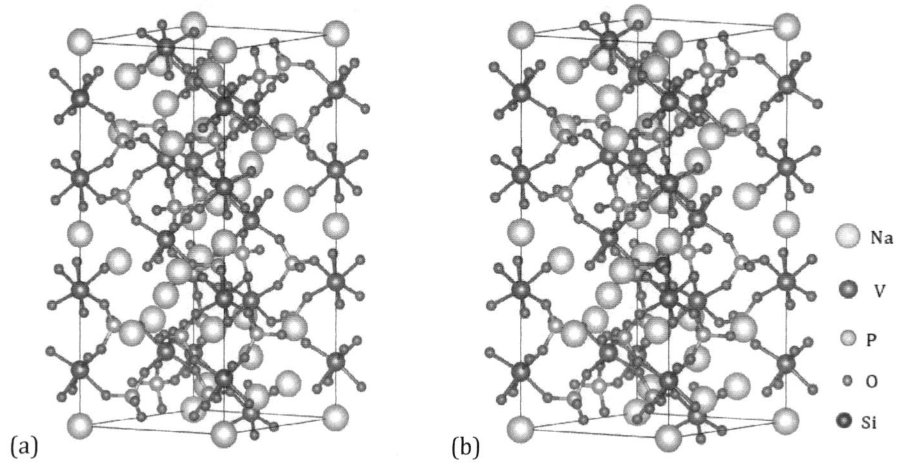

Figure 1. Optimize structure of Na_4VP (a) and $Na_4VPS0.1$ (b).

optimized structures of NVPS0.1, we only used the NVP structures with the lowest energy value in a convex hull of mixing energy for each concentration.

The stability of desodiation state was evaluated by calculating mixing energy (E_m) at different Na composition to construct the 0 K phase diagram can be defined as eq 1.

$$\Delta E_m(Na_xVP(S)) = E[Na_xVP(S)] - \left(\frac{4-x}{3}\right)E[Na_1VP(S)] - \left(\frac{x-1}{3}\right)E[Na_4VP(S)] \quad (1)$$

where $E[Na_xVP(S)]$ represents the energy of the bulk NVP or NVPS0.1 partially (de)sodiated, while $E[Na_1VP(S)]$ and $E[Na_4VP(S)]$ correspond to the energy of fully sodiated and fully (de)sodiated. $E[Na_xVP(S)]$ was calculated, for given x, by finding the structure with the lowest energy among various arrangements of vacancies of Na^+. The intercalation voltage of Na in the NVP and NVPS0.1 was studied by condisering the following reaction

$$Na_xVP(S) + (y-x)Na \rightleftharpoons Na_yVP(S) \quad (2)$$

with $(y-x)$ representating the concentration of intercalated Na. The intercalation voltage (V) was given by

$$V = -\frac{\Delta G^o}{(y-x).e} \quad (3)$$

for NVP, $V = -\frac{E(Na_yV_2(PO_4)_3)-[E(Na_xV_2(PO_4)_3+(y-x)\mu_{Na}]}{(y-x).e}$ (4)

and for NVPS0.1, $V = -\frac{E(Na_yV_2(PO_4)_{2.9}(SiO_4)_{0.1})-[E(Na_xV_2(PO_4)_{2.9}(SiO_4)_{0.1})+(y-x)\mu_{Na}]}{(y-x).\ e}$ (5)

where μ_{Na} is the chemical potential of BCC Na (2 atoms), valued at -2.60454101 eV.

Journal of Physics: Conference Series **2980** (2025) 012019 doi:10.1088/1742-6596/2980/1/012019

3. Result

3.1 Structural Properties

Table 1. Lattice constants (in Å and °), volume (in Å^3) and space group (Spg.) computed with PBE.

Structure based on Na concentration		*a*	*b*	*c*	α	β	γ	Vol.	Spg.
Na_1	Exp Ref. (Sunkyu Park, 2022)	8.43	8.43	21.48	90.00	90.00	120.00		$R\bar{3}c$
	NVP	8.54	8.54	21.77	90.04	89.96	120.01	1373.99	$R\bar{3}c$
	NVPS0.1	8.55	8.55	21.67	89.98	90.00	120.00	1372.38	$R\bar{3}c$
Na_2	Exp Ref. (Sunkyu Park, 2022)	15.24	8.61	8.74		126.28			$P2_1/c$
	NVP	8.71	8.71	21.65	90.17	89.83	120.48	1414.03	$R\bar{3}c$
	NVPS0.1	8.70	8.71	21.64	90.13	89.78	120.45	1414.32	$R\bar{3}c$
$Na_{2.1}$	NVPS0.1	8.72	8.73	21.63	90.14	89.72	120.46	1420.09	
Na_3	Exp Ref. (Sunkyu Park, 2022)	15.41	8.73	8.82		126.11			$C2/c$
	NVP	8.85	8.85	21.63	90.32	89.68	120.13	1464.50	$R\bar{3}c$
	NVPS0.1	8.84	8.85	21.60	90.22	89.71	120.15	1460.73	
$Na_{3.1}$	Exp Ref. (Santosh Kumar, 2020)	8.73	8.73	21.87	90.00	90.00	120.00	1444.50	$R\bar{3}c$
	NVPS0.1	8.87	8.86	21.57	90.25	89.78	120.11	1467.91	$R\bar{3}c$
Na_4	Exp Ref. (Sunkyu Park, 2022)	8.94	8.94	21.36	90.00	90.00	120.00		$R\bar{3}c$
	NVP	9.02	9.02	21.35	90.00	90.00	120.00	1503.57	$R\bar{3}c$
	NVPS0.1	9.01	9.01	21.36	89.98	90.00	119.98	1502.56	$R\bar{3}c$

* Na intercalation in fully charge is $N_1VP/N_1VPS0.1$ and the fully discharged phase is $N_4VP/N_4VPS0.1$.

On the basis of the mixing energy, we selected the most stable 16 structures for the DFT calculation. The structure with the lowest energy from the DFT calculations was considered for further calculation. The lattice parameter of the selected structures NVP and NVPS0.1 material in each configuration, calculated by using the PBE method, are shown in **Table 1.** Based on the table, several analyses can be drawn: (i) the addition of Si doping at the phosphate sites does not significantly affect the changes in lattice parameters, such as the lengths of lattice *a, b,* and *c*, or even the volume. But Si-doping tends to increase the cell volume of pure NVP if the number of ions per formula unit (f.u.) is adjusted with the addition of Si aggregates, as reported by Pal et.al. [18]. (ii) In both compounds (pure NVP and Si-doped NVP), generally, the insertion of Na^+ into the NaSICON structure increases the *a* parameter and decreases the *c* parameter (along the hexagonal axis). We argue that the reduction in the *c* parameter is due to an increase in the interaction between V-O during a (de)sodiation process, which in turn increases the electrostatic repulsion between the parallel three oxygen ion and Na(1) as it becomes less screened by the Na^+ ions [21].

3.2 Electrochemical Stability and Intercalation Voltage

The mixing energy (E_f) is derived from the implementation and expansion of the Gibbs free energy equation to assess the spontaneity of a reaction at 0 K. **Figure 2** shows the mixing energy profile of pristine and Si-doped NVP structure. Based on these figures, it is evident that the minimum points (indicated by the solid line) represent the most stable Na/vacancy configurations (at 0 K) at each Na concentration, *x*=1,2,3,4. Meanwhile, the points above the minimum correspond to metastable (or unstable) Na/vacancy configurations. Our DFT calculations reveal a previously unexplored, thermodynamically stable Na-vacancy configuration at $Na_{3.1}V_2(PO_4)_{2.9}(SiO_4)_{0.1}$ with a minimum mixing energy of -32.61 eV/atom. Thus, it can be stated that this configuration represents the most stable composition for the NVPS0.1 compound and also become a global minima in the convex hull for both NVP and NVPS0.1. Several other stable configurations are also observed in the convex hull such as $Na_{2.1}V_2(PO_4)_{2.9}(SiO_4)_{0.1}$ which is even more stable than the $Na_3V_2(PO_4)_{2.9}(SiO_4)_{0.1}$ configuration. In general, the E_f of the Si-doped materials were marginally more negative compared to the pristine NVP material, indicating that Si doping contributed to stabilizing the partially deintercalated states.

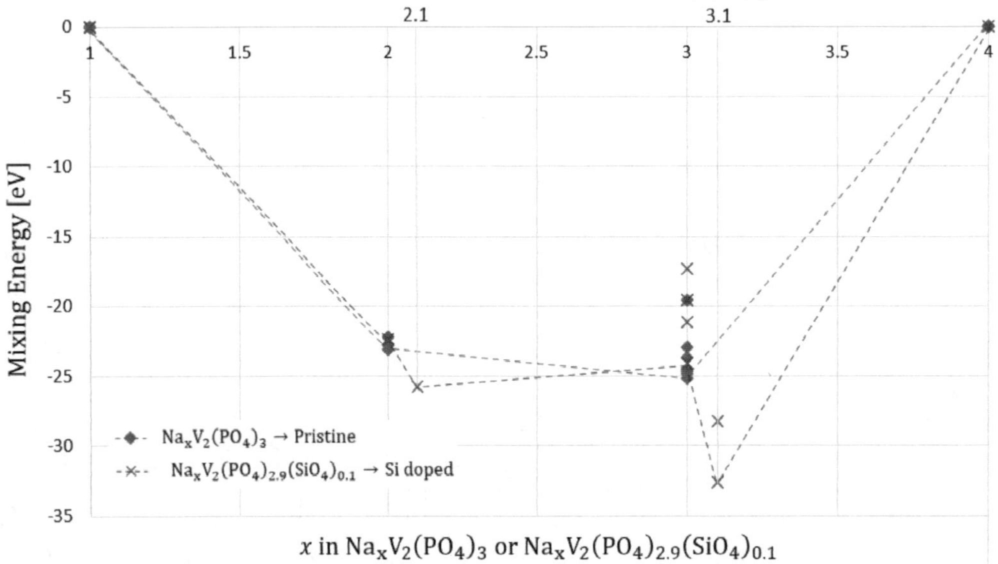

Figure 2. Formation energies of pristine and Si-doped NVP materials for all Na/vacancy ordering vs. Na content (x) in N_xVP or $N_xVPS0.1$. The dashed line, show the convex hull envelope constructed by the most stable phase. The PBE functional was used in the calculation.

One of the key parameters for evaluating the electrochemical performance of a battery material is its intercalation voltage. For some applications of an rechargable battery, cathode materials with a high intercalation voltage is desirable, but it must remain lower than the available potential window of the electrolyte.

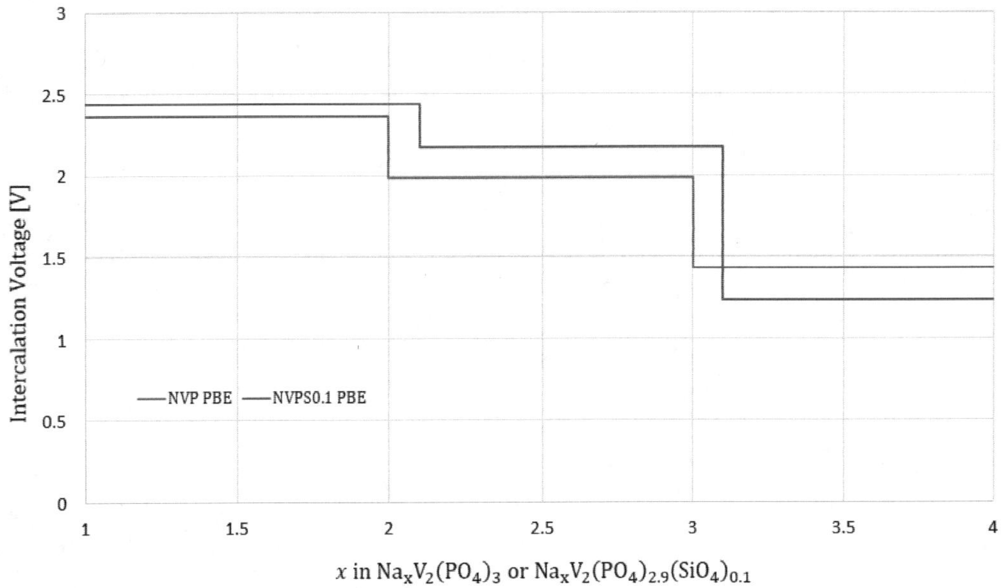

Figure 3. Voltage profiles of the undoped and Si-doped NVP cathode materials.

Figure 3. shows that the addition of Si doping at the phosphate site broadens the intercalation voltage range, as observed by the PBE methods. The intercalation voltage range of pure NVP is 1.43–2.36 V ($\Delta V = 0.93\ V$), while for the NVPS0.1 compound, the intercalation voltage range is 1.24–2.44 V ($\Delta V = 1.20\ V$). This means that the addition of Si doping modulates the NVP intercalation voltage by up to 0.27 V. It is important to note that our present work utilizes results obtained from the PBE method, which generally underestimated the voltage by 1 V [22]. Therefore, to achieve more accurate results, a correction method, such as dispersion correction or Hubbard U correction, can be applied. However, the trend shown by our present prediction offers an initial insight on how the addition of Si can potentially modulate the NVP voltage window.

3.3 Electrochemical Redox Behavior

The electron configuration of a material adheres to the Aufbau principle, Hund's rule, and the Pauli exclusion principle. The Aufbau principle states that electrons fill orbitals starting from the lowest energy levels, while Hund's rule indicates that in general electrons are added to orbitals or suborbitals until they are half-filled before pairing occurs. The Pauli exclusion principle asserts that electrons, being fermions, cannot occupy the same quantum state; thus, a single suborbital or quantum number l can be occupied by only two electrons with opposite spins (i.e. spin up and spin down).

(a)
(b)
(c)

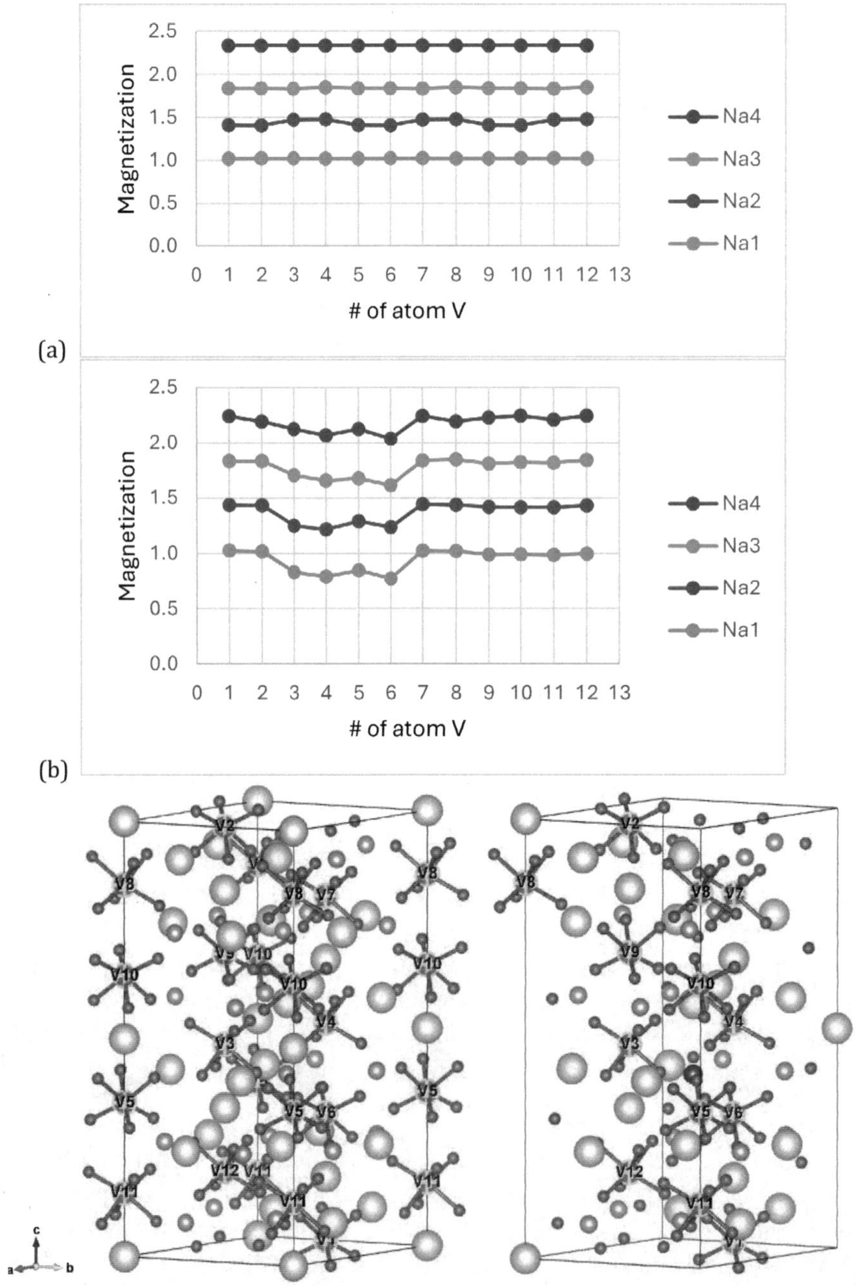

Figure 4. Magnetic moment of Vanadium atom in all concentration of N_xVP, (a), and $N_xVPS0.1$, (b). The position of each Vanadium atom in a bulk cell, (c). The PBE functional was used in the calculation.

Journal of Physics: Conference Series **2980** (2025) 012019 doi:10.1088/1742-6596/2980/1/012019

Vanadium, a transition metal, contains 23 electrons and has the electron configuration [Ar] $3d^3$ $4s^2$. In the process of ion formation, the electrons that are removed first come from the outermost orbital (4s), despite the fact that the 3d orbital has a higher energy level. Vanadium can display several oxidation states, with the electron configurations for its ions being: (I) V^{2+} is produced when both electrons in the $4s^2$ shell are removed, (II) V^{3+} is formed when one electron from the 3d orbital is lost, and (III) V^{4+} is created when two electrons from the 3d orbital are removed. When NVP and NVPS0.1 undergo (de)intercalation or the formation of Na vacancies, the V atoms will compensate for the charge from the anion groups $(PO_4)^{3-}$ (for NVP) and $(PO_{3.9})^{3-}$ + $(SiO_{0.1})^{4-}$ (for NVPS0.1) by altering their electron configuration. The more unpaired electrons present, the more paramagnetic the material will be, while a higher number of paired electrons will lead to increased diamagnetism. The magnetic moment changes of V ions in the PBE calculations show a fairly linear trend with respect to Na concentration. The magnetic moment calculations for NVPS0.1 also exhibit similar results with the PBE method, where the magnetic moment changes linearly with the Na concentration. According to studies conducted by Singh, B. (2021)[23] and Wang, Z. (2022) [24], the magnetic moment of the V atom shows values of 2.7, 1.9, and 1.0 μB for the oxidation states V^{2+}, V^{3+}, and V^{4+}, respectively. Based on the simulation results obtained, PBE method successfully indicate the presence of V atoms in the V^{2+}, V^{3+} and V^{4+} oxidation states, shown in **Figure 4**.

Table 2. presents the orbital characteristics, magnetic moments, and V – O bond radius for each oxidation state of V. The magnetic moment values indicate the number of spin-up electrons around the Fermi level. A magnetic moment close to zero suggests a minimal presence of spins in that state. It can also be observed that V^{4+} ions have fewer t_{2g} orbitals occupied compared to other oxidation states (leaving only one spin electron in the $d_{xy}/d_{xz}/d_{yz}$ states), resulting in a near-zero magnetization value. The vacancy in these orbitals is filled by electrons from the O(2p) atoms, which leads to the shortening of the bond between V^{4+} and O (as indicated by the shortest orbital radius compared to other oxidation states), and this also influences the Coulomb interaction. In contrast, the V^{2+} and V^{3+} oxidation states are in a high-spin configuration (with only spin-up electrons occupying each orbital in the d-shell), as evidenced by the high magnetization values of these ions (in line with Hund's rule).

The presence of transition metal oxidation states for each composition of NVP or NVPS0.1 is shown in **Table 3** for all V-atom and **Table 4** shown the local geometries for V-atom that close to the doping site. In general, Si doping reduces the V-O bond radius for every Na concentration. This result contrasts with the increasing cell volume of NVP as more Si is added, suggesting that Si provides more space for Na insertion, which positively impacts sodium diffusion mechanisms. In the N_4VP compound, we were unable to distinguish between two V sites. According to research by Wang et al., the N_4VP composition exhibits metallic behaviour (with a very small band gap of approximately 0.3 eV), indicating a low likelihood of electron pairing [24].

Journal of Physics: Conference Series **2980** (2025) 012019 doi:10.1088/1742-6596/2980/1/012019

Table 2. Electronic structure, magnetic moment and bond length of vanadium ion in NVP and NVPS0.1. Calculated with PBE method.

V-ion	Orbitals	Magnetic Moment [μB]		V – O [Å]	
		NVP	NVPS0.1	NVP	NVPS0.1
V^{2+}	$t_{2g}^3e_g^0$	2.336	2.034 – 2.243	1.98 – 2.05	1.87 – 1.93
V^{3+}	$t_{2g}^2\ e_g^0$	1.477 – 1.847	1.416 – 1.847	1.92 – 2.00	1.82 – 1.91
V^{4+}	$t_{2g}^1\ e_g^0$	1.021 – 1.407	0.772 – 1.219	1.77 – 1.90	1.74 – 1.79

Table 3. The compounds formed based on the magnetic moment values and V-O bond lengths in NVP and NVPS0.1. Calculated with PBE method.

Na concentration, x =	Magnetic Moment [μB]		V – O [Å]		Compound	
	NVP	NVPS0.1	NVP	NVPS0.1	NVP	NVPS0.1
Na =1	1.021 – 1.024	0.772 – 1.025	1.86 – 1.90	1.74 – 1.79	$Na_1V^{4+}V^{4+}(PO_4)_3$	$Na_1V^{4+}V^{4+}(PO_4)_{2.9}(SiO_4)_{0.1}$
Na = 2	1.407 – 1.477	1.217 – 1.436	1.77 – 1.92	1.75 – 1.82	$Na_2V^{4+}V^{3+}(PO_4)_3$	$Na_2V^{4+}V^{3+}(PO_4)_{2.9}(SiO_4)_{0.1}$
Na = 3	1.832 – 1.847	1.613 – 1.847	1.98 – 2.00	1.87 – 1.91	$Na_3V^{3+}V^{3+}(PO_4)_3$	$Na_3V^{3+}V^{3+}(PO_4)_{2.9}(SiO_4)_{0.1}$
Na = 4	2.336	2.034 – 2.243	1.98 – 2.05	1.87 – 1.93	$Na_4V^{2+}V^{2+}(PO_4)_3$	$Na_4V^{3+}V^{2+}(PO_4)_{2.9}(SiO_4)_{0.1}$

Journal of Physics: Conference Series **2980** (2025) 012019 doi:10.1088/1742-6596/2980/1/012019

Table 4. Local geometries difference NVP pristine and NVPS0.1 of Vanadium near the doping site.

Na concentration, x =	V - O [Å]		Na(1) - O [Å]	
	NVP	NVPS0.1	NVP	NVPS0.1
Na = 2	1.96	1.91	2.49	2.52
Na = 3	2.04	1.98	2.41	2.49

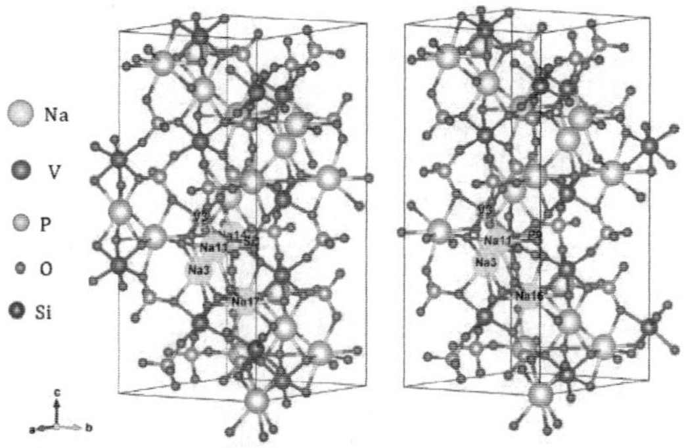

Based on the electronic density of states (DOS) result, as shown in **Figure 5** and **Figure 6**, it is demonstrated that in both the NVP and NVP-Si systems, the redox mechanism is dominated by vanadium atoms (indicated by the shift of V 3*d* near the Fermi level from Na4 → Na1). Therefore, in both compounds, the cation (particularly the transition metal, V) plays a crucial role in the oxidation-reduction mechanism compared to the anions. It can also be observed that as the (de)sodiation process increases, the overlap shifts further to the right (positive values) of the Fermi level. This indicates that the occupancy of V decreases, leading to an increase in the redox number of V, which further shows that under charging-discharging condition an oxidation process occurs. This is further supported by the change of the calculated magnetic moment shown in **Table 3**. Additionally, based on the DOS, the introduction of Si does not bring additional electrons in metal transition V, which is also mentioned by Li, et.al., [16]. We note that assigning an exact oxidation number of the V in each Na concentration directly from the computed magnetic moments is in some cases difficult. Therefore, we simultaneously observe the occupancy of the DOS and consider the basic concept of charge neutrality of an ionic compound to further determine the oxidation number of the cation. The result of **Table 3** also comparable with previous study by Wang, et.al., 2022 [24]. From the DOS plot, we can also observe that the introduction of Si in NVP bulk leads to reduced band gap, indicating increased conductivity, albeit minimal.

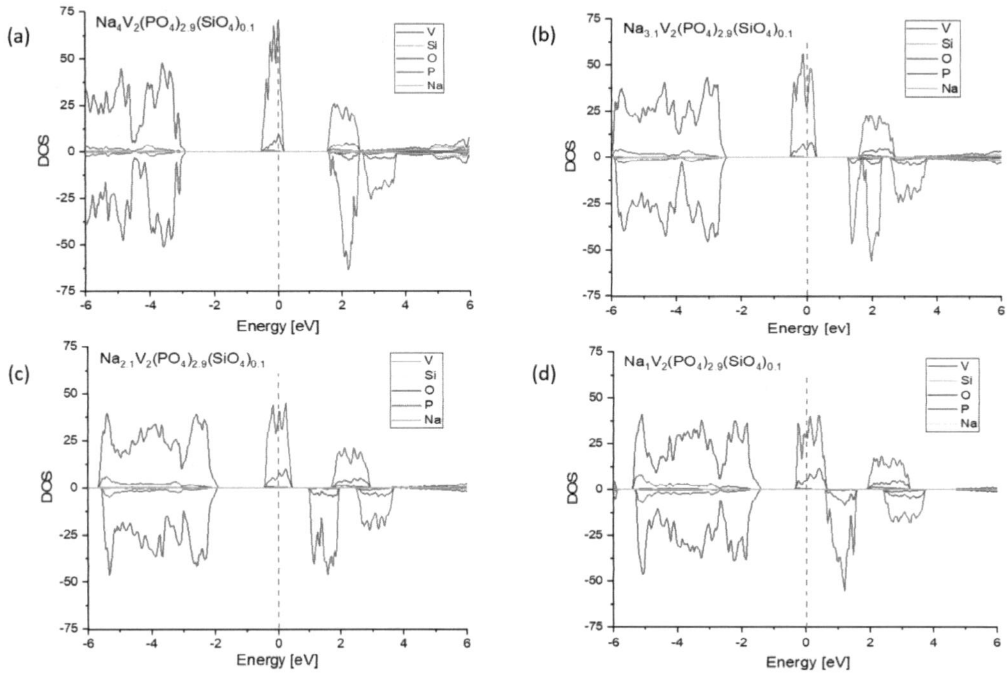

Figure 5. Density of state (DOS) plot of NVP for each stable concentration, (a)–(d).

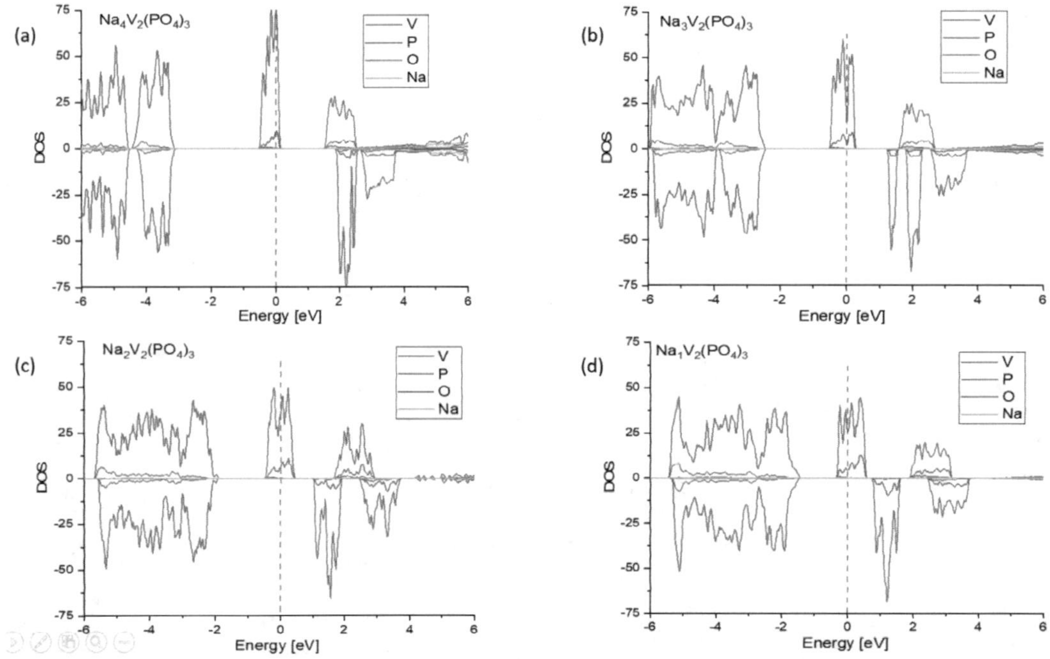

Figure 6. Density of state (DOS) plot of NVPS0.1 for each stable concentration, (a)–(d).

4. Conclusion

The partial replacement of phosphorus (P) with silicon (Si) in the NVP cathode has been shown to slightly increases the intercalation voltage range of sodium-ion battery cathode. We showed how substituting pentavalent P^{5+} ($r^{5+}_{IV,P} = 0.17$ Å) by larger tetravalent Si^{4+} ($r^{4+}_{IV,Si} = 0.26$ Å) : (i) enlarge the NVP lattice, (ii) increase the number of Na-ions per formula unit (f.u.), and (iii) increase structure stability for pristine structure by strengthen V – O bond and by reducing the Coulombs repulsion. Our result, also discovered other NVPS0.1 stable configuration, $Na_{2.1}V_2(PO_4)_{2.9}(SiO_4)_{0.1}$, which may be insightful in further research.

Acknowledgements

G.S. would like to acknowledge research fund endowed by the Indonesian Ministry of Research, Technology and Higher Education (ristekdikti) through the "Penelitian Fundamental 2024" scheme.

References

[1] Zhang B, Ma K, Lv X, Shi K, Wang Y, Nian Z, Li Y, Wang L, Dai L dan He Z 2021 Recent advances of NASICON-Na3V2(PO4)3 as cathode for sodium-ion batteries: Synthesis, modifications, and perspectives *J. Alloys Compd.* **867** 159060

[2] Abraham K M 2020 How Comparable Are Sodium-Ion Batteries to Lithium-Ion Counterparts? *ACS Energy Lett.* **5** 3544–7

[3] Olivetti E A, Ceder G, Gaustad G G dan Fu X 2017 Lithium-Ion Battery Supply Chain Considerations: Analysis of Potential Bottlenecks in Critical Metals *Joule* **1** 229–43

[4] Zhao L, Zhang T, Li W, Li T, Zhang L, Zhang X dan Wang Z 2023 Engineering of Sodium-Ion Batteries: Opportunities and Challenges *Engineering* **24** 172–83

[5] Slater M D, Kim D, Lee E dan Johnson C S 2013 Sodium-ion batteries *Adv. Funct. Mater.* **23** 947–58

[6] Ma X, Cao X, Zhou Y, Guo S, Shi X, Fang G, Pan A, Lu B, Zhou J dan Liang S 2020 Tuning crystal structure and redox potential of NASICON-type cathodes for sodium-ion batteries *Nano Res.* **13** 3330–7

[7] Thirupathi R, Kumari V, Chakrabarty S dan Omar S 2023 Recent progress and prospects of NASICON framework electrodes for Na-ion batteries *Prog. Mater. Sci.* **137** 101128

[8] Pan H, Hu Y S dan Chen L 2013 Room-temperature stationary sodium-ion batteries for large-scale electric energy storage *Energy Environ. Sci.* **6** 2338–60

[9] Xie F, Zhang L, Ye C, Jaroniec M dan Qiao S Z 2019 The Application of Hollow Structured Anodes for Sodium-Ion Batteries: From Simple to Complex Systems *Adv. Mater.* **31** 1–8

[10] Bhaskara Rao Y, Achary K R dan Patro L N 2022 Enhanced Electrochemical Performance of the Na3V2(PO4)3/C Cathode Material upon Doping with Mn/Fe for Na-Ion Batteries *ACS Omega* **7** 48192–201

[11] Zhang X, Rui X, Chen D, Tan H, Yang D, Huang S dan Yu Y 2019 Na 3 V 2 (PO 4) 3 : An advanced cathode for sodium-ion batteries *Nanoscale* **11** 2556–76

[12] Jian Z, Yuan C, Han W, Lu X, Gu L, Xi X, Hu Y S, Li H, Chen W, Chen D, Ikuhara Y dan Chen L 2014 Atomic structure and kinetics of NASICON NaxV 2(PO4)3 cathode for sodium-ion batteries *Adv. Funct. Mater.* **24** 4265–72

[13] Zheng Q, Yi H, Li X dan Zhang H 2018 Progress and prospect for NASICON-type Na3V2(PO4)3 for electrochemical energy storage *J. Energy Chem.* **27** 1597–617

[14] Hu P, Zou Z, Sun X, Wang D, Ma J, Kong Q, Xiao D, Gu L, Zhou X, Zhao J, Dong S, He B, Avdeev M, Shi S, Cui G dan Chen L 2020 Uncovering the Potential of M1-Site-Activated NASICON Cathodes for Zn-Ion Batteries *Adv. Mater.* **32** 1–9

[15] Wu C, Tong J, Gao J, Li J, Li X, Zhu J, Gu M dan Zhou W 2021 Studies on the Sodium Storage Performances of Na3AlxV2- x(PO4)3@C Composites from Calculations and Experimental Analysis *ACS Appl. Energy Mater.* **4** 1120–9

[16] Li P, Gao M, Wang D, Li Z, Liu Y, Liu X, Li H, Sun Y, Liu Y, Niu X, Zhong B, Wu Z G dan Guo X 2022 Optimizing Vanadium Redox Reaction in Na3V2(PO4)3Cathodes for Sodium-Ion Batteries by the Synergistic Effect of Additional Electrons from Heteroatoms *ACS Appl. Mater. Interfaces* **2**

[17] Bag S, Murarka H, Zhou C, Bhattacharya A, Jokhakar D, Pol V G dan Thangadurai V 2020 Understanding the Na-Ion Storage Mechanism in Na3+ xV2- xMx(PO4)3(M = Ni2+, Co2+, Mg2+x = 0.1-0.5) Cathodes *ACS Appl. Energy Mater.* **3** 8475–86

[18] Pal S K, Thirupathi R, Chakrabarty S dan Omar S 2020 Improving the Electrochemical Performance of Na3V2(PO4)3 Cathode in Na-Ion Batteries by Si-Doping *ACS Appl. Energy Mater.* **3** 12054–65

[19] Joubert D 1999 From ultrasoft pseudopotentials to the projector augmented-wave method *Phys. Rev. B - Condens. Matter Mater. Phys.* **59** 1758–75

[20] Perdew J P, Burke K dan Ernzerhof M 1996 Generalized gradient approximation made simple *Phys. Rev. Lett.* **77** 3865–8

[21] Park S, Wang Z, Deng Z, Moog I, Canepa P, Fauth F, Carlier D, Croguennec L, Masquelier C dan Chotard J N 2022 Crystal Structure of Na2V2(PO4)3, an Intriguing Phase Spotted in the Na3V2(PO4)3-Na1V2(PO4)3System *Chem. Mater.* **34** 451–62

[22] Breuer O, Chakraborty A, Liu J, Kravchuk T, Burstein L, Grinblat J, Kauffman Y, Gladkih A, Nayak P, Tsubery M, Frenkel A I, Talianker M, Major D T, Markovsky B dan Aurbach D 2018 Understanding the Role of Minor Molybdenum Doping in LiNi0.5Co0.2Mn0.3O2 Electrodes: From Structural and Surface Analyses and Theoretical Modeling to Practical Electrochemical Cells *ACS Appl. Mater. Interfaces* **10** 29608–21

[23] Singh B, Wang Z, Park S, Gautam G S, Chotard J N, Croguennec L, Carlier D, Cheetham A K, Masquelier C dan Canepa P 2021 A chemical map of NaSICON electrode materials for sodium-ion batteries *J. Mater. Chem. A* **9** 281–92

[24] Wang Z, Park S, Deng Z, Carlier D, Chotard J N, Croguennec L, Gautam G S, Cheetham A K, Masquelier C dan Canepa P 2022 Phase stability and sodium-vacancy orderings in a NaSICON electrode *J. Mater. Chem. A* **10** 209–17

Journal of Physics: Conference Series **2980** (2025) 012020 doi:10.1088/1742-6596/2980/1/012020

Optical Characterization of Intrinsic Silicon Thin Films Growth via Hot Wire in Plasma-Very High-Frequency PECVD

A W Nuayi[1,2], R Ruslan[1], F A Noor[1]*, T Winata[1], and Irzaman[3]**

[1]Physics of Electronic Materials Research Division, Department of Physics, Faculty of Mathematics and Natural Sciences, Institut Teknologi Bandung, Jln. Ganesha No 10, Bandung,40132 Indonesia

[2] Physics of Materials Group, Department of Physics, Faculty of Mathematics and Natural Sciences, Universitas Negeri Gorontalo, Jln. Jend. Sudirman No. 06 Kota Gorontalo, Gorontalo, 96128 Indonesia

[3]Department of Physics Faculty of Mathematics and Natural Sciences, IPB University, Jln. Raya Dramaga, Kecamatan Dramaga, Kab. Bogor,16680 Indonesia

E-mail: fatimah@itb.ac.id*, wahidin.awn2023@ung.ac.id**

Abstract. This study investigated the growth of a thin intrinsic silicon layer on a 1.5 x 2 cm glass substrate. The growth utilized a Hot Wire in Plasma-Very High-Frequency Plasma Enhanced Chemical Vapor Deposition (HW IP-VHF PECVD) system at varied substrate temperatures (200, 225, 250, and 275 °C). The VHF PECVD method served as a comparative benchmark. Silane gas (SiH_4) was introduced at 70 sccm to enhance growth. This research seeks to preliminarily optimize the optical properties of type-i polycrystalline silicon thin films for future application in p-i-n silicon-based graphene solar cells. Optical characterization was conducted using a UV-Vis and NIR spectrometer across the 300 – 1100 nm spectrum. The findings illustrate the intrinsic silicon film's optical response at wavelengths exceeding 600 nm. Comparative analysis indicates lower transmittance for films grown via HW IP-VHF PECVD than VHF PECVD. Notably, at 275 °C substrate temperature, the intrinsic silicon film exhibited a 54.65% increase in average absorption compared to VHF PECVD. Furthermore, average absorbance increases of 1.61, 10.51, and 14.24 were recorded at 200, 225, and 250 °C substrate temperatures, respectively. The HW IP-VHF PECVD method yielded reduced E_{gap} values of 7.81, 12.90, 2.82, and 4.82% at 200, 225, 250, and 275 °C substrate temperatures. This study underscores the potential for optimizing optical properties through controlled growth conditions using HW IP-VHF PECVD.

1. Introduction

In facing global challenges in meeting sustainable energy needs, the development of solar cell technology has become the main focus in efforts to find alternative energy sources. One of the fundamental components in photovoltaic cell advancement is the optical characterization of the substances employed, encompassing intrinsic silicon thin films. Intrinsic silicon thin films are imperative in transmuting solar energy into electrical energy. Advancing methodologies for cultivating these films have become indispensable in guaranteeing the optimal quality and efficacy of solar cells.

Journal of Physics: Conference Series **2980** (2025) 012020 doi:10.1088/1742-6596/2980/1/012020

The challenge for the photovoltaic industry, including the fabrication of thin film-based solar cells, is a cost-effective process that uses fewer raw materials without reducing the quality obtained. In solar cell fabrication technology, c-Si material is generally used, but the price is relatively more expensive, so various efforts have been made to look for alternative materials. One of the primary materials of this alternative is p-Si, which is not only thermally efficient but has a higher thermal conductivity than c-Si [1,2]. Therefore, p-Si thin films are a cost-saving alternative for solar cell production. Si-based thin films, including p-Si, are usually made by decomposing osilane gas (SiH_4) in a plasma generated by a radio frequency amplifier (glow discharge) at a relatively low temperature (250 °C) [3].

Various deposition techniques for growing thin films and p-Si-based devices from the gas phase have been discovered and developed. Hammond [4] stated that in the early stages of development (1963-1970), the physical vapor deposition (PVD) method was used. Still, the step coverage was not uniform due to damage caused by X-rays and unwanted impurity ions. Low productivity was not comparable to costs incurred, and this method is not widely used. Then, in the early stages of manufacture (1970-1976), the chemical vapor deposition (CVD) method was used in a horizontal atmospheric pressure reactor for bipolar integrated circuits, the results of which were better than the PVD method. Since then, various CVD modification methods have been developed to produce p-Si layers, such as electron cyclotron resonance CVD (ECR CVD) [5], closed chamber cyclic CVD (CCC-CVD) [6], rapid thermal CVD (RTCVD) [7], plasma enhanced CVD (PECVD) [8], dan hot wire CVD (HW-CVD) [9,10].

The hot wire in plasma-very high frequency-(HWIP-VHF-PECVD) method is another form of development or modification of the conventional PECVD method. Physics of Electronic Materials Research Division, Department of Physics ITB first developed this method in the 2000s by adopting several advantages of the HW-PECVD and VHF-PECVD methods and aimed at overcoming the limitations of these two methods [11]. The difference between this method and the HWC-PECVD method lies in using a different resource frequency, whereas the HW IP-VHF-PECVD method uses RF 70 MHz. Assuming that the molecules entering this electrode, area are already unstable species, the electric field generated from the 70 MHz RF power generator can more efficiently help break down these species into simpler radicals.

Based on the previous descriptions, this study is directed at the optical characterization of thin intrinsic silicon films obtained through the PECVD process with very high frequencies using hot wires in plasma. The main objective of this study is to analyze the optical characteristics of intrinsic silicon thin films produced by varying the substrate temperature between 200-275 °C with an increase every 25 °C on the optical properties of p-Si thin films, which are grown using the HWIP-VHF-PECVD method. As a comparison, growth was also carried out using the VHF PECVD method. Optical characterization analysis is vital because the thin film's transmission, reflection, and absorption properties will affect the solar cell's overall performance. This study will also investigate the influence of growth process parameters on the optical characteristics of the resulting thin films.

It is hoped that the results of this study will provide a deeper understanding of the potential application of the HWIP-VHF- PECVD method in the growth of intrinsic silicon thin films for multi-layer graphene (MLG) based solar cell applications, which will be carried out in the following study. With a better understanding and control of the optical characteristics of these thin films, this study has the potential to significantly contribute to developing more efficient and sustainable MLG-based solar cell technologies.

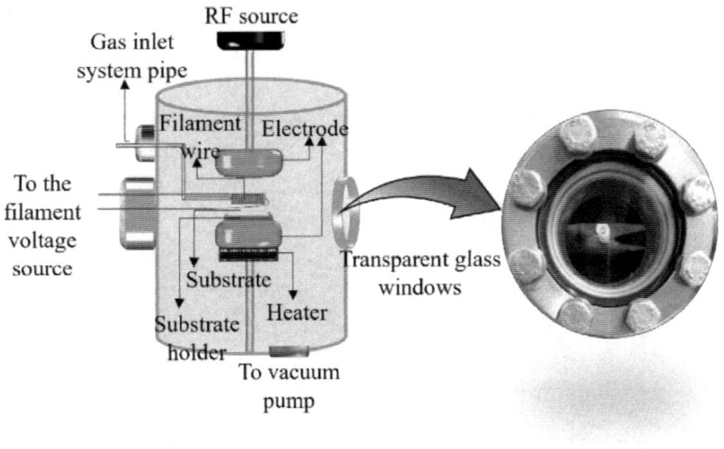

Figure 1. HW IP-VHF PECVD chamber schematic

2. Experimental methods

2.1 Substrate preparation

Before the deposition process, the initial step is to prepare the glass substrate, which is cut to a size of 1.5 x 2 cm, followed by the substrate cleaning process using ultrasonication, which is carried out by vibrating it in an ultrasonic bath for 15 minutes successively in a solution of distilled water, acetone (CH_3COCH_3) pro analysis, methanol (CH_3OH) pro analysis, and ending with distilled water solution again. After the process, it is dried by spraying a nitrogen gas gun.

2.2 Silicon intrinsic layer growth process

In this study, a thin intrinsic silicon layer was grown on a glass substrate in the HWIP-VHF PECVD reactor system. The gas source used in this study is SiH_4 gas with a concentration of 10% in hydrogen (H_2), passed through the middle of the filament coil (tungsten) and continues to flow between two electrodes, as seen in Figure 1.

The cleaned substrate is positioned on the holder and introduced into the access door. Following the establishment of a vacuum in the access door, the gate valve between the access door and reactor one opens, allowing the substrate's transfer to the reactor via a load lock. Once the substrate is situated in the reactor chamber above the heater, the load lock returns to its original state, and the gate valve is subsequently closed. The substrate is then enclosed by a shutter

Table 1. Intrinsic silicon layer growth parameters

Parameters	Method	
	HW IP-VHF PECVD	VHF PECVD
Substrate	Glass 15 x 20 mm	Glass 15 x 20 mm
Substrate temperature	200, 225, 250, dan 275 °C	200, 225, 250, dan 275 °C
RF frequency	70 MHz	70 MHz
RF power	30 watts	30 watts
Deposition pressure	300 mTorr	300 mTorr
SiH_4 flow rate	70 sccm	70 sccm
Voltage	3 Volt	3 Volt
Deposition time	30 minutes	30 minutes
Distance between the substrate and filament	13.31 mm	13.,31 mm

between the top electrode and subjected to heating. Upon reaching the designated growth temperature, the HW system is activated with a direct electric power supply, while gas is introduced through controlled mass flow (MFC). A radio frequency of 70 MHz is employed to generate plasma. The total reactor pressure is manually adjusted using a butterfly valve until the desired deposition pressure is attained. The growth duration is measured from when the shutter opens until it is closed again. Following the growth phase, the substrate undergoes cooling via N_2 gas flow (purging to 80 °C) and remains in the chamber until achieving an ambient temperature of 27 °C.

The VHF PECVD technique employs a similar procedure without activating the filament system. The deposition process for HWIP-VHF PECVD and VHF PECVD techniques is optimized by adjusting substrate temperature parameters between 200–275 °C, increasing in increments of 25 °C. Deposition parameters encompass SiH_4 gas flow rate, chamber pressure, RF power, voltage, deposition time, and the spatial relationship between the filament wire and the substrate, as detailed in Table 1.

2.3 Characterization

The p-Si thin layer was then characterized using X-ray diffraction (XRD) to see its crystal structure (carried out on samples grown at a substrate temperature of 275 °C). To determine the optical properties of this layer, measurements were carried out using Ultraviolet-Visible-Near-Infrared (UV-Vis-NIR) spectroscopy. From this UV-Vis-NIR spectrum data, the absorption coefficient value can be calculated. The ability of a material to absorb/absorb radiation is called absorptivity, expressed in the form of an absorbance coefficient (α), and the absorbance value for semiconductor materials or materials is located in a specific wavelength range. The absorbance coefficient and gap energy of the intrinsic silicon layer that is grown are calculated using equations (1) and (2)

$$\alpha = \frac{Ln\,[1/T]}{d} \tag{1}$$

$$\alpha \propto (hv - E_g)^n \tag{2}$$

with α represents the absorbance coefficient, T is the transmittance value, d is the film layer thickness (cm), hv is the photon energy (eV), λ is the wavelength (nm), and n is a constant that states the electron transition in the semiconductor (1/2 for direct transitions and 2 for indirect transitions [12,13]. Using the absorbance coefficient (α) and the Tauc plot method, the E_{gap} value of the p-Si thin film in this study can be determined, namely by extrapolating to the part of the linear plot that cuts the basis of the curve $(hv - E_g)^{1/n}$ to hv [12,14].

3. Result and discussion

3.1 X-Ray Diffraction (XRD) Characterization

Characterization via XRD seeks to ascertain the amorphous crystallinity of thin silicon layers produced by varying methods. Analysis was performed on a layer synthesized at 275 °C within the 2θ = 10°–80° range. The XRD results for the intrinsic silicon layer, illustrated in Figure 2 (a), reveal a broad spectrum at 2θ = 10°–36°, indicative of the amorphous phase. Typically, the peak intensity for layers formed via HW IP-VHF PECVD is diminished compared to those from VHF PECVD. This observation suggests that the amorphous characteristics in films from HW IP-VHF PECVD are inferior to those from VHF PECVD.

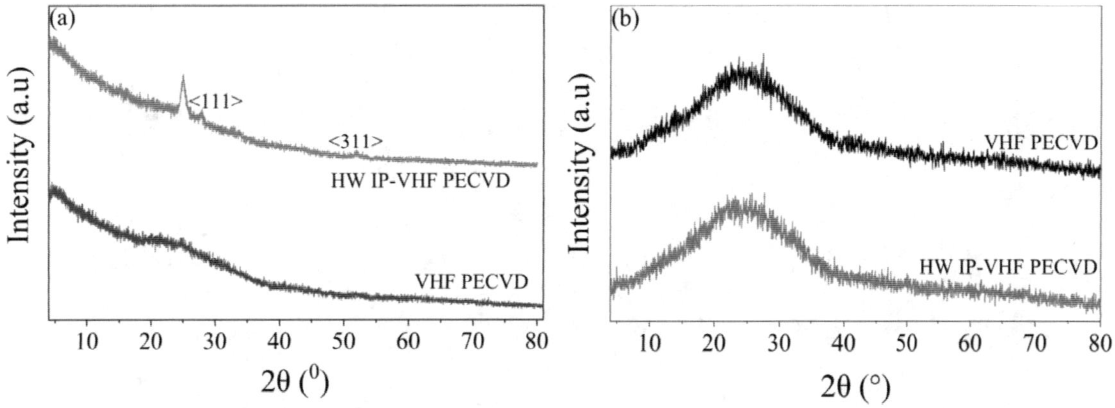

Figure 2. XRD spectrum of intrinsic silicon thin films grown using the VHF PECVD and HW IP-VHF PECVD methods at substrate temperatures (a) 275 °C, (b) 200 °C

The reduced peak intensity observed in the HW IP-VHF PECVD layers at 275 °C implies a decline in the amorphous phase, favoring crystalline phase formation, as evidenced by peaks at 2θ = 28.04° and 52.07°. Consequently, it is inferred that the intrinsic silicon layer from VHF PECVD exhibits an amorphous structure, corroborated by the diffraction peak between 2θ = 10°–36°, consistent with Schropp [15] findings on amorphous silicon. In contrast, silicon layers from HW IP-VHF PECVD at the same temperature exhibit microcrystalline structures, evidenced by low-intensity peaks at 2θ = 52.07° corresponding to <311> orientation. According to Ichikawa et al. [16] XRD peaks for silicon crystals with <311> orientation are found within 2θ = 51-56°. Therefore, as per the material classification by Schropp et al. [15,17,18], the intrinsic silicon layer produced via HW IP-VHF PECVD is classified as a dual phase with a microcrystalline structure due to the presence of peaks at 2θ = 28.04° and 52.07°.

One of the factors contributing to the reduced intensity in the XRD results produced by the HW IP-VHF PECVD method is the use of filament/hot wire in plasma, which is not activated in the VHF PECVD method. Using hot wire in plasma causes higher energy in the SiH_4 gas decomposition process, which produces more Si radicals that increase the crystal structure of the silicon film. In addition, the additional energy through the use of hot wire in plasma helps improve the mobility of Si atoms from decomposition. It allows for more regular film growth than the VHF PECVD method. This can be seen in the SEM results of a thin layer of silicon grown at a substrate temperature of 275 °C (Figure 3). The SEM results show that the particle sizes of both growth methods look homogeneous. Still, the intrinsic silicon thin layer grown using the HW IP-VHF-PECVD method shows significant surface morphology and cross-section with crystal grains spread throughout the surface of the thin layer. The grain size and grain boundaries appear to have been formed well. This differs from the surface morphology and cross-section of thin films grown by the VHF-PECVD method at the same substrate temperature, where the crystalline grains have not formed properly, so the layer is still in the amorphous phase. Using the ImageJ application, the average diameter size for samples grown by the HW IP-VHF PECVD method was obtained at 133.41 ± 5.78 nm, and 87.14 ± 4.50 nm for the VHF-PECVD method. Meanwhile, the average thicknesses are 436.50 nm and 154.50 nm, respectively.

Although the diffraction peaks formed from the XRD measurement results have not shown enough intensity, the formation of crystalline grains in the SEM results suggests that the HWIP-VHF-PECVD method can grow thin films in the microcrystalline phase.

Journal of Physics: Conference Series **2980** (2025) 012020 doi:10.1088/1742-6596/2980/1/012020

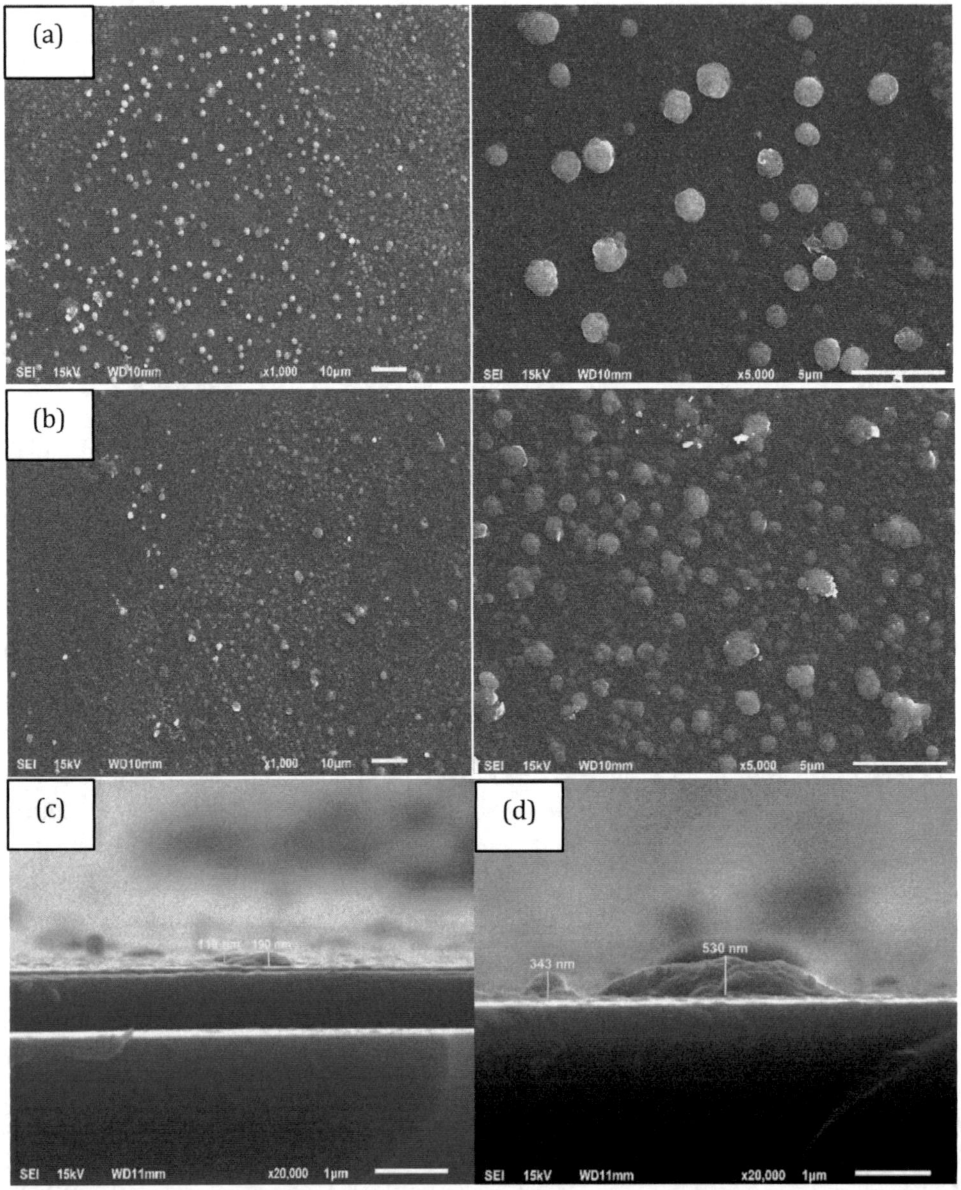

Figure 3. SEM image of the surface of the intrinsic silicon thin film grown at a substrate temperature of 275 0C at 1000x and 5000x magnifications using the (a) VHF-PECVD method, (b) HW IP-VHF-PECVD, and SEM cross-sectional photos of intrinsic silicon thin films grown at a substrate temperature of 275 °C using the (c) VHF-PECVD and (d) HW IP-VHF- PECVD methods taken at 20,000x magnifications

The XRD characterization of intrinsic silicon thin films reveals a broad-spectrum indicative of the amorphous phase for both growth methods. The peak intensity of the XRD spectrum for films from the HW IP-VHF PECVD method is notably lower and narrower than that of the VHF PECVD method. These findings suggest that the amorphous quality of films produced by HW IP-VHF PECVD is lower than that of VHF PECVD, with both methods yielding films with an amorphous structure at a substrate temperature of 200 °C.

Journal of Physics: Conference Series **2980** (2025) 012020 doi:10.1088/1742-6596/2980/1/012020

3.2 Optical characterization

Characterization of intrinsic silicon thin films' optical properties was conducted using UV-VIS NIR spectroscopy at wavelengths from 200 to 1100 nm, varying substrate temperatures from 200 to 275 °C.

Figure 4 and Table 2 show the transmittance of the UV-VIS NIR thin layer of intrinsic silicon produced via the VHF PECVD technique at various substrate temperatures, both with and without post-growth annealing. The optical transparency for unannealed and annealed samples exhibits transmittance values ranging from 1–86% and 1–87.58% at wavelengths λ > 461 nm and λ 522 nm, respectively. Additionally, the transmission spectrum is observed to shift towards longer wavelengths with increased substrate temperatures, indicating that both substrate temperature and annealing influence the transmittance of the intrinsic silicon thin layer. This data evidences that elevated substrate temperatures and post-growth annealing treatments significantly alter the transmittance characteristics of the intrinsic silicon thin layer.

Furthermore, Figure 5 (a) illustrates that the intrinsic silicon thin layer is synthesized using the HW IP-VHF PECVD technique at diverse substrate temperatures. This evidence suggests that increasing substrate temperature during the growth of intrinsic silicon thin films via the HW IP-VHF PECVD method leads to a notable reduction in transmittance, paralleling the findings from the VHF PECVD method.

The HW IP-VHF PECVD method treats intrinsic silicon thin layers similarly to the VHF-PECVD process. An annealing process is performed post-growth at various temperatures. Figure 5 (b) demonstrates that annealing decreases the average transmittance of the intrinsic silicon thin layer with temperature variations.

This data shows that increasing the substrate temperature and providing post-growth annealing treatment affects changes in the intrinsic silicon thin layer's transmittance. This indicates that, with increasing substrate temperature, there are more radical molecules from SiH_4 gas, which are precursors that react with the substrate, so the more significant the layer formed, the ability of the resulting thin layer to transmit UV and NIR waves also decreases.

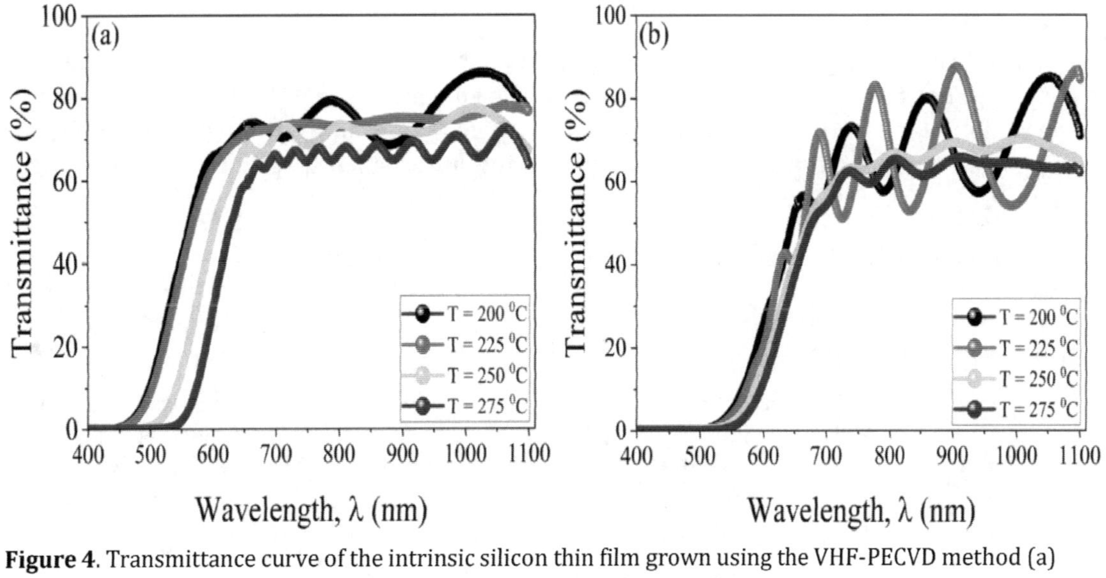

Figure 4. Transmittance curve of the intrinsic silicon thin film grown using the VHF-PECVD method (a) without annealing, (b) annealed for 3 hours at 450 °C after the growth process.

Journal of Physics: Conference Series **2980** (2025) 012020 doi:10.1088/1742-6596/2980/1/012020

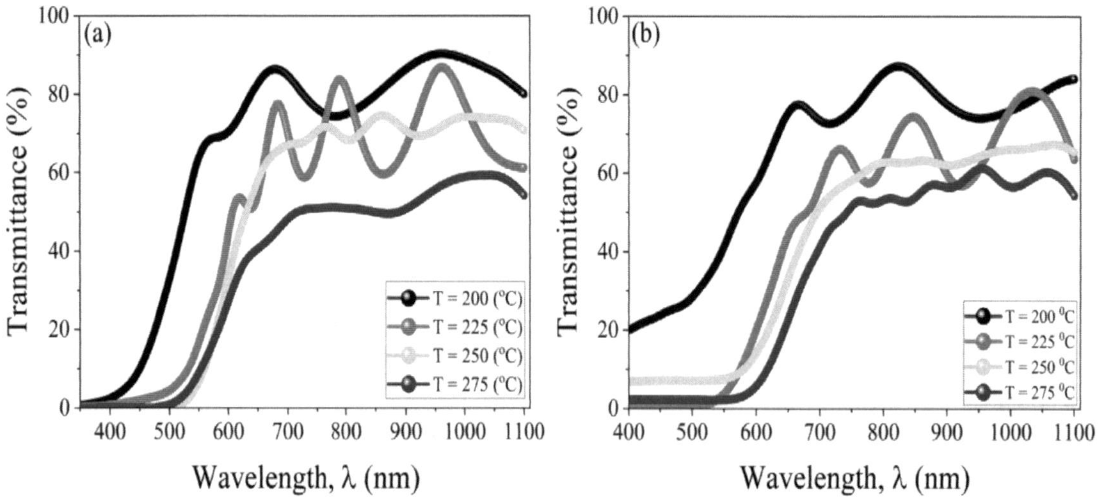

Figure 5. Transmittance curve of the intrinsic silicon thin film grown using the HW IP-VHF PECVD method (a) without annealing, (b) annealed for 3 hours at a 450 °C after the growth process.

Table 2 Maximum transmittance data and average transmittance of intrinsic silicon thin films grown using the HWC PECVD and HW IP-VHF PECV methods

T (°C)	VHF PECVD				HW IP-VHF PECVD			
	Non-annealing		With annealing		Non-annealing		With annealing	
	Max	Avg	Max	Avg	Max	Avg	Max	Avg
200	86.15	60.23	87.59	55.97	90.31	69.11	87.29	64.88
225	78.35	58.32	85.19	54.07	86.84	58.89	80.97	45.41
250	77.47	53.08	70.30	51.49	74.65	50.23	67.34	42.85
275	72.75	47.11	65.79	48.40	59.36	37.85	61.25	35.65

The transmittance spectrum analysis of intrinsic silicon thin films produced via VHF PECVD and HW IP-VHF PECVD methods, as shown in Figures 4 and 5, reveals an optical response in the λ region exceeding 600 nm. This finding indicates that the intrinsic silicon thin layer has a favorable response in the infrared region, shifting towards longer wavelengths or lower photon energy, accompanied by oscillations in the transmittance and absorbance spectrum due to thin layer interference.

3.3 Determination of thickness and refractive index of Intrinsic Silicon Thin Film

Determining layer thickness is crucial for calculating the deposition rate and optical gap of intrinsic silicon thin layers using the Swanepoel equation. The Swanepoel equation effectively determines thin layer thickness through UV-Vis or NIR spectrometry transmittance data. The process begins with graphing transmittance values and then identifying maximum (T_M) and minimum (T_m) transmittance values at each wavelength, leading to peaks or valleys[19]. The T_M and T_m values yield the refractive index of the layer materials at corresponding wavelengths. Consequently, the thin layer's thickness is ascertainable via Equation 3, based on the refractive index.

$$d = \frac{\lambda_1 \lambda_2}{2\,(\lambda_1 n_2 - \lambda_2 n_1)} \tag{3}$$

Wavelengths λ_1 and λ_2 correspond to the peaks or troughs of adjacent waves, with n_1 and n_2 representing the thin layer's refractive indices at these wavelengths, as reported in references [19–21]. Swanepoel posited that the refractive index equation applies to regions where $\alpha \neq 0$ and $T < 1$, indicating weak to medium absorption. In this context, equation (4) employs variable N to ascertain the refractive index, calculated via equation (5).

$$n = [N + (N^2 - s^2)^{\frac{1}{2}}]^{\frac{1}{2}} \tag{4}$$

$$N = 2s\left(\frac{T_M - T_m}{T_M T_m}\right) + \frac{s^2 + 1}{2} \tag{5}$$

with n indicating the refractive index of the thin film, s reflecting the substrate's refractive index (1.5), T_M signifying transmittance at peak points, and T_m showing the transmittance at valley points [19].

In this study, the envelope method within the Origin application enhanced the precision of T_M and T_m value determination. The thickness of the intrinsic silicon thin layer was calculated by averaging T_M and T_m values of adjacent peaks (or valleys) λ_1 and λ_2. The average thickness value derived from equation (3) facilitated determining the deposition rate for the intrinsic silicon layer created by VHF PECVD and HW IP-VHF PECVD methods, as shown in Table 2. Additionally, Figure 5 illustrates the relationship between substrate temperature variations, average thickness, and deposition rate of the intrinsic silicon thin layer.

The data from Table 3 and Figure 6 indicates that the intrinsic silicon thin film thickness increases with rising substrate temperature during thin film growth. This is attributed to the enhanced decomposition of SiH_4 into simpler radicals [22,23]. Increasing the substrate temperature from 200 to 275 °C elevates the diffusion rate and the number of interacting radicals, consequently increasing the thin film deposition rate and thickness. This observation corroborates prior transmittance and absorbance results, where higher deposition rates were noted with elevated substrate temperatures during thin layer growth. Thus, more SiH_4-derived radical molecules interacting with the substrate result in increased layer thickness. This finding is crucial for the resultant thin layer's capability to transmit UV and NIR waves, which diminishes, as illustrated in Figures 4 (a) and 5 (a). Nonetheless, post-growth annealing promotes a more substantial thin layer formation. This reveals that the annealing method enhances the quality of the resulting layer, proven by the diminished transmittance of the thin layer cultivated through both strategies against the non-annealed intrinsic silicon thin layer, as represented in Figures 4 (b) and 5 (b).

Compared to the VHF PECVD method, the HW IP-VHF PECVD method results in a lower silicon thin layer deposition rate, yielding a thinner intrinsic silicon layer. This is evident when

Table 3 Results of the calculation of the average thickness value and the deposition rate of the thin layer of intrinsic silicon

T (°C)	Average layer thickness, $\bar{d}$ (Å)				Deposition rate (Å/s)	
	A_1	B_1	A_2	B_2	A	B
200	9219.18	1167.50	11664.48	8985.46	4.66	0.589
225	12198.32	2117.03	14685.43	15955.68	6.16	1.069
250	15273.96	12293.58	20750.61	24004.22	7.71	6.209
275	17983.40	16718.15	24096.16	39192.78	9.08	8.444

Remarks: A=VHF PECVD method; B= HW IP-VHF PECVD method; 1= without annealing; 2= with annealing

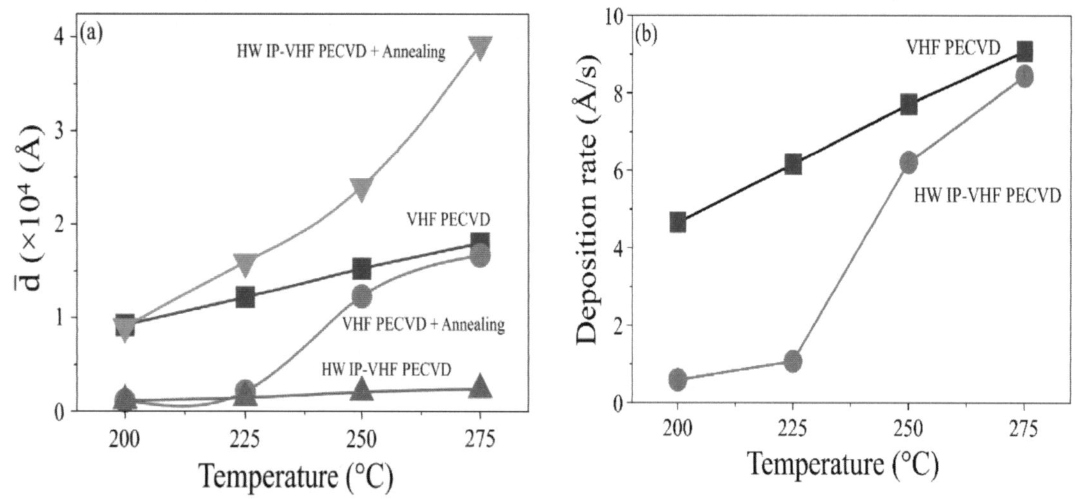

Figure 6. The effect of variations in substrate temperature in the growth of intrinsic silicon thin films using the VHF PECVD and HWC-VHF PECVD methods on (a) the average thickness of the thin film, (b) the rate of layer deposition

contrasting with intrinsic silicon thin films produced using the VHF PECVD method. The use of wires in the VHF-PECVD system causes the SiH_4 gas molecules to decompose into silicon radical atoms before deposition on the substrate surface. Consequently, fewer radical atoms are available for silicon layer formation on the substrate.

A more pronounced effect of substrate temperature on the thickness and deposition rate of silicon thin films grown via the HW IP-VHF PECVD method is noted compared to the VHF PECVD method. This observation reflects the notable gradient of thickness variation and thin layer deposition rate in the HW IP-VHF PECVD method. The steep gradient changes in growth and deposition rates result from an increase in H atoms in the plasma due to hot filament decomposition, which leads to a reduction in the mean free path and an increase in particle collisions in the plasma during the SiH_4 decomposition and conversion process [23,24].

The intrinsic silicon layer's thickness, as shown in Table 3, aligns with the definition of a thin layer, defined by small dimensions from atomic condensation from 10^{-6}–10^{-9} meters [25] study's intrinsic silicon layer qualifies as a thin film.

3.4 Determination of the Optical Gap of Intrinsic Silicon Thin Films

The optical value of the intrinsic silicon thin layer was determined via optical transmission measurements utilizing a UV-Vis spectrometer. The Tauc equation (equation (2)) is used to determine the value which, when reformulated as equation (6):

$$(\alpha h\nu)^{1/\gamma} = \mathrm{B}(h\nu - E_{gap}) \tag{6}$$

where B is a constant, and γ is a constant that states whether the electron transition between valence and conduction bands occurs directly or indirectly (1/2 for direct and 2 for indirect transitions). Assuming that the electron transition within the intrinsic silicon thin layer, as produced in this study, occurs directly ($\gamma = ½$).

From equation 6, the value of E_{gap} will coincide with hν when $(\alpha h\nu)^{1/\gamma}$Is equal to zero. The optical band gap is determined via the Tauc plot by extrapolating the linear region to intersect the energy axis [26]. This intersection signifies the E_{gap} size, applicable when α exceeds 10^3 cm^{-1}.

Journal of Physics: Conference Series **2980** (2025) 012020 doi:10.1088/1742-6596/2980/1/012020

Figure 7 and Table 4 show E_{gap} values for intrinsic silicon layers using the VHF PECVD and HW IP- VHF PECVD method at 200, 225, 250, and 275 °C substrate temperatures. These results indicate enhanced intrinsic silicon layer structure is achieved through increased substrate temperatures and 450 °C annealing for 3 hours, evidenced by lower E_{gap} values. The E_{gap} obtained aligns closely with Liu et al. [27], the reported value is 1.66 – 1.92 eV for polycrystalline silicon via the rf-PECVD method. When compared with other studies, as reported by Ichikawa et al. [16], reported polycrystalline silicon structures obtained under growth conditions using a silane precursor gas without dilution of dissolved hydrogen at a growth pressure of 0.1 Torr with a deposited film growth rate of 0.9–0.4 nm/s, and a substrate temperature and hot wire 400 and 1700 °C respectively. Compared with the results obtained on silicon layer growth using the HWIP-VHF PECVD method, the hot wire in plasma can reduce the temperature of the substrate and filament by 31.25 and 70.59%. Even though the resulting structure is still in the form of microcrystals (the crystal structure is still low). However, by optimizing the pressure and flow rate of the precursor gas, it is suspected that polycrystalline silicon layers can be produced using the HW IP-VHF PECVD method.

From Table 4, it can be seen that there is a decrease in E_{gap} after the annealing process for 3 hours at a temperature of 450. The decrease in the Egap value of the intrinsic silicon layer after the post-deposition annealing process is because the annealing process helps in the recombination process. This is because many defects and impurities can form in the silicon crystal structure during the deposition process. The annealing process helps in recombining and eliminating these defects, reducing the energy level of the defects in the energy gap. This also allows the crystal structure to return to a more stable and ordered state. As defects and impurities decrease, the band energy levels in the material also change. This can cause a shift in the valence and conduction bands, resulting in a decrease in E_{gap}. In addition, annealing can improve the crystallinity of the material by increasing the grain size and reducing amorphism. With increasing crystallinity, the energy band becomes wider and reduces the energy level shift in the energy gap, thereby reducing the E_{gap}. Additionally, the annealing process can improve the microscopic

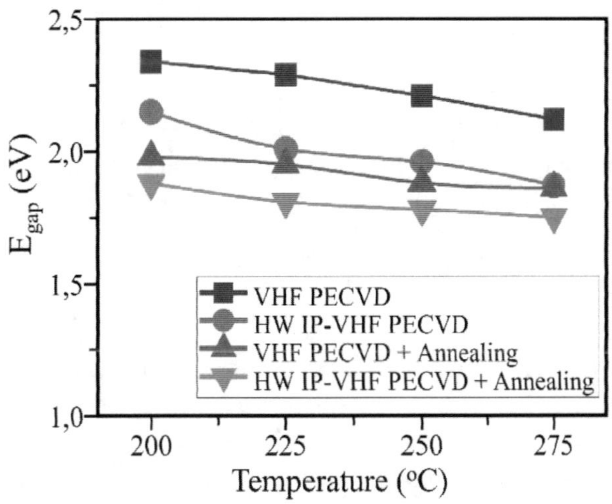

Figure 7. The effect of variations in substrate temperature in the growth of intrinsic silicon thin films using the VHF PECVD and HWC-VHF PECVD methods on E_{gap} value

Table 4 Results of the calculation of the E_{gap} value of the thin layer of intrinsic silicon

T (°C)	E_{gap} (eV)			
	A_1	B_1	A_2	B_2
200	2.34	2.15	1.98	1.88
225	2.29	2.01	1.95	1.81
250	2.21	1.96	1.88	1.78
275	2.12	1.87	1.87	1.75

Remarks: A=VHF PECVD method; B= HW IP-VHF PECVD method;
1= without annealing; 2= with annealing

structure of thin films by promoting atomic reorientation and better bonding. This can reduce mechanical stress and improve coating quality, thereby positively impacting electronic properties, including E_{gap}.

Additionally, the reduction in E_{gap} is attributed to substrate temperature effects on hydrogen content in SiH_4 gas, which acts as a precursor for the silicon layer growth. This decrease in hydrogen is likely due to the increased dissociation of SiH_4 into simpler radicals at higher temperatures, which enhances radical diffusion on the substrate and reduces hydrogen coverage, leading to a smaller E_{gap}.

The HW IP-VHF PECVD method produces intrinsic silicon thin films with lower E_{gap} values than the VHF PECVD method at varying substrate temperatures. This method resulted in E_{gap} reductions of 7.81, 12.90, 2.82, and 4.82% for substrate temperatures of 200, 225, 250, and 275 °C, respectively, compared to VHF PECVD-grown films. Figure 7 illustrates the relationship between substrate temperature and E_{gap} value for the silicon thin films produced by both techniques.

The reduction in E_{gap} value in HW IP-VHF PECVD-grown intrinsic silicon layers is linked to the filament's presence in the hot wire system. The filament, aligned with the gas outlet, facilitates gas molecule decomposition in plasma through thermal effects. This process generates ionic radicals, influenced by electrons accelerated by the RF power electric field. These ionic radicals serve as chemical etching agents, reducing the deposition rate of the intrinsic silicon layer. Moreover, the lower E_{gap} value indicates an enhancement in the crystal structure, transitioning from amorphous to microcrystalline when the intrinsic silicon layer is developed at 275 °C, corroborated by prior XRD analyses.

4. Conclusion

The study reveals that the HW IP-VHF PECVD technique produces intrinsic silicon thin films with reduced E_{gap} values that decline with increased substrate temperature. Additionally, these films are thinner than those produced by the VHF PECVD method. Annealing after growth increased film thickness while decreasing E_{gap} values for both methods, with reductions between 15.12% and 6.34%. The HW IP-VHF PECVD method achieved lower E_{gap} values (approximately 7.81 to 12.90%) under the same substrate temperature, indicating enhanced suitability for optical applications. Furthermore, this method exhibited a lower deposition rate, resulting in thinner films than the VHF PECVD technique. XRD analysis showed that the HW IP-VHF PECVD method facilitated a microcrystalline structure at a designated substrate temperature. Post-growth annealing further augmented film thickness and diminished E_{gap} values across both growth methods. The annealing process significantly lowered E_{gap} values for each method, ranging from 15.12% to 6.34%. In summary, this research offers crucial insights into the fabrication and utilization of intrinsic silicon thin films in electronic and optical applications.

The HW IP-VHF PECVD method exhibits superiority in generating intrinsic silicon thin films with reduced E_{gap} values, enhancing optical efficiency. This method also yields thinner layers, which benefits devices with stringent thickness constraints. Furthermore, post-growth annealing can augment the thin film structure, broadening its applicability across diverse technological fields.

Acknowledgment

This research was supported by "Penelitian, Pengabdian kepada Masyarakat, dan Inovasi (PPMI)" in the fiscal year 2024. AWN gratefully acknowledges the Centre for Higher Education Fund (Balai Pembiayaan Pendidikan Tinggi, BPPT), the Education Financing Service Centre (Puslapdik), the Ministry of Education, Culture, Research and Technology of the Republic of Indonesia and Indonesian Endowment Fun for Education (LPDP) for funding support through the Indonesia Education Scholarship program (BPI) in 2021 with decree number 1775/J5. 2.3/BPI.06/10/2021

References

[1] Gordon I, van Gestel D, van Nieuwenhuysen K, Carnel L, Beaucarne G and Poortmans J 2005 Thin-film polycrystalline silicon solar cells on ceramic substrates by aluminium-induced crystallization *Thin Solid Films* **487** 113–7

[2] Tüzün Ö, Qiu Y, Slaoui A, Gordon I, Maurice C, Venkatachalam S, Chatterjee S, Beaucarne G and Poortmans J 2010 Properties of n-type polycrystalline silicon solar cells formed by aluminium induced crystallization and CVD thickening *Solar Energy Materials and Solar Cells* **94** 1869–74

[3] Beaucarne G 2007 Silicon Thin-Film Solar Cells *Adv Optoelectron* **2007** 1–12

[4] Hammond M L 1990 Poly-Si - A most important material *MRS Proceedings* **182** 3

[5] Hsiao H L, Hwang H L, Yang A B, Chen L W and Yew T R 1999 Study on low temperature facetting growth of polycrystalline silicon thin films by ECR downstream plasma CVD with different hydrogen dilution *Appl Surf Sci* **142** 316–21

[6] Koynov S, Grebner S, Radojkovic P, Hartmann E, Schwarz R, Vasilev L, Krankenhagen R, Sieber I, Henrion W and Schmidt M 1996 Initial stages of microcrystalline silicon film growth *J Non Cryst Solids* **198–200** 1012–6

[7] Monna R, Angermeier D, Slaoui A and Muller J C 1996 Polycrystalline silicon thin films by rapid thermal chemical vapor deposition (RTCVD) on graphite substrates *Conference Record of the Twenty Fifth IEEE Photovoltaic Specialists Conference - 1996* (Washington, DC USA: IEEE) pp 701–4

[8] Chen W, Stuckelberger J, Wang W, Phang S P, Macdonald D, Wan Y and Yan D 2021 N-type polysilicon passivating contacts using ultra-thin PECVD silicon oxynitrides as the interfacial layer *Solar Energy Materials and Solar Cells* **232** 111356

[9] Khorani E, Scheul T E, Tarazona A, Nutter J, Rahman T and Boden S A 2020 p+ polycrystalline silicon growth via hot wire chemical vapour deposition for silicon solar cells *Thin Solid Films* **705** 137978

[10] Rath J K, Barbon A and Schropp R E I 1998 Limited influence of grain boundary defects in hot-wire CVD polysilicon films on solar cell performance *J Non Cryst Solids* **227–230** 1277–81

[11] Usman I and Winata T 2008 Pengaruh Ketebalan Lapisan Aktif terhadap Karakteristik Sel Surya Berbasis a-Si:H yang Ditumbuhkan dengan Teknik HWC-VHF-PECVD *Jurnal Matematika dan Sains* **13** 109–13

[12] Tauc J and Menth A 1972 States in the gap *J Non Cryst Solids* **8–10** 569–85

[13] Leng W J, Yang C R, Zhang J H, Chen H W, Ji H, Fu C L and Liao J X 2006 Structural and optical properties of $Ba_x Sr_{1-x}TiO_3$ thin films on indium tin oxide/quartz substrates prepared by radio-frequency magnetron sputtering *J Appl Phys* **99**

[14] Umar, Faanzir, Nuayi A W, Siskandar R, Syafutra H and Alatas dan Irzaman H 2013 Uji Sifat Optik Pada Film Tipis $Ba_{0,55}Sr_{0,45}TiO_3$ *Simposium Nasional Inovasi dan Pembelajaran Sains 2013* ed Acep Purqon, Fourier D.E. Latief,

Journal of Physics: Conference Series **2980** (2025) 012020 doi:10.1088/1742-6596/2980/1/012020

Sparisoma Viridi, Syeilendra Pramuditya and Dede Enan (Bandung: Program Studi Magister Pengajaran Fisika Fakultas Matematika dan Ilmu Pengetahuan Alam Institut Teknologi Bandung) pp 280–4
[15] Schropp R E I 2006 Amorphous (Protocrystalline) and Microcrystalline Thin Film Silicon Solar Cells *Nanostructured Materials for Solar Energy Conversion* ed T Soga (Elsevier) pp 131–66
[16] Ichikawa M, Tsushima T, Yamada A and Konagai M 2000 Amorphous-to-polycrystalline silicon transition in hot wire cell method *Jpn J Appl Phys* **39** 4712–5
[17] van Sark W G J H M, Korte L and Roca F 2012 Physics and Technology of Amorphous-Crystalline Heterostructure Silicon Solar Cells *Engineering Materials* Engineering Materials vol 0, ed R E I Schropp (Berlin, Heidelberg: Springer Berlin Heidelberg) pp 1–12
[18] Schropp R E I, Carius R and Beaucarne G 2007 Amorphous silicon, microcrystalline silicon, and thin-film polycrystalline silicon solar cells *MRS Bull* **32** 219–24
[19] Swanepoel R 1983 Determination of the thickness and optical constants of amorphous silicon *J Phys E* **16** 1214–22
[20] Shaaban E R, Almohammedi A, Yousef E S, Ali G A M, Chong K F, Adel A and Ashour A 2018 Structural, optical and electrical characteristics of sulfur incorporated ZnSe thin films *Optik (Stuttg)* **164** 527–37
[21] Minkov D, Angelov G, Nestorov R, Nezhdanov A, Usanov D, Kudryashov M and Mashin A 2020 Optical characterization of As_xTe_{100-x} films grown by plasma deposition based on the advanced optimizing envelope method *Materials* **13** 1–18
[22] Chittick R C, Alexander J H and Sterling H F 1969 The Preparation and Properties of Amorphous Silicon *J Electrochem Soc* **116** 77
[23] Kumbhar A A and Kshirsagar S T 1996 Comparative study of properties of a-Si:H films produced by hot-filament CVD, glow discharge CVD and their hybrid version *Thin Solid Films* **283** 49–56
[24] Mursal, Supu A, Usman I, Winata T, Sukirno and Barmawi M 2004 Growth of Amorphous Silicon Germanium (a-SiGe:H) Alloys Thin Film by PECVD *Indonesian Journal of Physics* **15** 13–6
[25] Nuayi Abd W, Alatas H, Husein I S and Rahmat M 2014 Enhancement of Photon Absorption on $Ba_xSr_{1-x}TiO_3$ Thin-Film Semiconductor Using Photonic Crystal Thin-Film Semiconductor Using Photonic Crystal *Int J Opt* **2014** 1–8
[26] Sze S M and Lee M K 2012 *Semiconductor Devices Physics and Technology* ed D Sayre (New York: John Wiley & Sons, Inc.)
[27] Liu J, Zhang Y, Fan Z, Sun H and Shan F 2020 Low-temperature deposition of large-grain polycrystalline Si thin films on polyethylene terephthalate *Thin Solid Films* **707** 138065

Journal of Physics: Conference Series **2980** (2025) 012021 doi:10.1088/1742-6596/2980/1/012021

Effect of microwave roasting in the leaching of the transition metals Ni, Fe, and Co from the low-grade nickel laterite

La Agusu[1,4*], W. O. S. Arsyad[1], I. N. Sudiana[1], A. Alimin[2], M. Z. Firihu[1], L. O. Ahmad[2], L. M. Darusman[2], A. J. Muhammad[3], S. Mitsudo[3], T. Asano[3], K. Nakagawa[4], T. Nishiumi[3], Y. Ishikawa[4], Y. Fujii[4], and Y. Tatematsu[4]

[1] Department of Physics, Faculty of Mathematics and Natural Sciences, Halu Oleo University, Kendari, Indonesia
[2] Department of Chemistry, Faculty of Mathematics and Natural Sciences, Halu Oleo University, Kendari Indonesia
[3] Department of Applied Physics, Faculty of Engineering, University of Fukui, Fukui, Japan
[4] Research Center for Development of Far Infrared Region, University of Fukui, Fukui, Japan

*E-mail: laagusu@uho.ac.id

Abstract. In many cases, low-grade nickel laterite ores with nickel (Ni) content of 1.2% or less have been left as land disposal at the mining areas. We are looking for an alternative, more environmentally friendly method to process this waste in a hydrometallurgical scheme. This study investigates the effectiveness of microwave-assisted recovery of transition metals—specifically nickel (Ni), iron (Fe), and cobalt (Co)—from low-grade nickel laterite (Ni<1.5%). The methodology involves roasting the sulfuric acid-nickel laterite mixture and leaching it at room temperature. The results reveal that microwave roasting significantly enhances metal recovery, achieving remarkable rates of 98.6% for nickel, 96.9% for iron, and 99.0% for cobalt, even under low microwave powers of 100 W, 180 W, and 300 W, with only 5 minutes of heating time. A solid-to-solution ratio of 1.2 g/mL was maintained throughout the process. While higher microwave power did lead to a slight decrease in recovery efficiency due to temperatures exceeding the boiling point of sulfuric acid, the overall process remained highly efficient. The mineralogical analysis indicated that the nickel laterite is primarily composed of dolomite, calcite, goethite, nepouite, vermiculite, and cristobalite, highlighting the complexity of the material. Additionally, extending the heating duration at lower powers can reduce acid consumption, contributing to the environmental sustainability of the extraction method compared to conventional techniques. These findings underscore not only the potential of microwave technology in mineral processing but also promote more sustainable practices in the recovery of valuable metals from low-grade ores.

Journal of Physics: Conference Series **2980** (2025) 012021 doi:10.1088/1742-6596/2980/1/012021

1. Introduction

The increasing demand for transition metals, particularly nickel, iron, and cobalt, has necessitated innovative extraction methods to enhance recovery from low-grade ores [1]. Nickel laterite, often characterized by low nickel concentrations (Ni < 1.5%), poses significant challenges for traditional extraction processes, which can be inefficient and environmentally detrimental [2]. As the global market for these metals expands, there is a pressing need to develop more efficient and sustainable recovery techniques [3]. Recent advancements in microwave technology present a promising alternative that can potentially optimize the extraction processes for these valuable metals from complex mineral matrices [4].

Microwave-assisted extraction (MAE) has emerged as a promising technique across various fields, particularly in mineral processing, owing to its capacity for selective heating and its effectiveness in enhancing chemical reactions. The underlying mechanism of microwave heating involves the rapid agitation of polar molecules, which facilitates efficient energy transfer and localized heating. This distinctive characteristic enables shorter processing times and reduced energy consumption compared to traditional heating methods. Research indicates that MAE can significantly enhance the dissolution of metal-bearing minerals, thereby improving the recovery rates of transition metals during leaching processes [5].

In this context, the combination of microwave roasting with sulfuric acid treatment offers a novel approach to extract nickel, iron, and cobalt from low-grade nickel laterite. The roasting process facilitates the conversion of metal oxides and hydroxides into soluble metal sulfates, enhancing their subsequent recovery during leaching [6]. Preliminary investigations indicate that this method can yield high recovery rates, even under relatively mild conditions, which is particularly advantageous when working with low-grade ores that would otherwise be deemed unprofitable [1,7].

The mineralogical composition of the nickel laterite significantly influences the recovery outcomes [2]. Common minerals such as dolomite, calcite, goethite, nepouite, vermiculite, and cristobalite can impact the efficiency of the acid leaching process. Understanding the interactions between these minerals and the acid is essential for optimizing the extraction process. Furthermore, maintaining an appropriate solid-to-solution ratio during leaching can maximize metal recovery while minimizing reagent consumption, thus aligning with sustainability goals.

This study aims to investigate the effectiveness of microwave-assisted recovery of transition metals from low-grade nickel laterite. The findings will provide insights into optimal microwave power settings, heating times, and the influence of mineral composition on recovery rates. By highlighting the potential of this innovative approach, we hope to contribute to the development of more sustainable practices in the mining and metallurgical industries, ultimately reducing the environmental impact associated with metal extraction.

2. Materials and Method

Nickel laterite samples were collected from the North Konawe area in Southeast Sulawesi Province and blended to achieve a nickel content of approximately 1.2%. The samples underwent washing, drying, and grinding to pass through a #200 mesh sieve. Characterization of the raw materials was performed using X-ray diffraction (XRD), X-ray fluorescence (XRF), and scanning electron microscopy with energy dispersive X-ray spectroscopy (SEM-EDS). Technical-grade sulfuric acid with a concentration of 96% was utilized as the solvent. Nickel laterite powder was combined with sulfuric acid at a solid-to-liquid ratio of 1.2 g/mL and placed in a high-temperature

Journal of Physics: Conference Series **2980** (2025) 012021 doi:10.1088/1742-6596/2980/1/012021

glass tube fitted with a K-type thermocouple and an exhaust gas outlet. The sample-containing beaker was positioned in a modified microwave oven operating as a chemical reactor, miximum power set to 800 watts. The samples were subjected to heating for varying durations of 5, 10, and 20 minutes at microwave powers ranging from 100 W to 800 W. After heating, the samples were cooled and dissolved in 16 mL of distilled water. The concentrations of dissolved nickel, iron, and cobalt were determined using atomic absorption spectroscopy (AAS), while insoluble solids were analyzed using SEM-EDS and XRF.

The metal recovery was determined using equation

$$\text{Recovery} = \frac{Vc}{(Rr + Vc)} \text{ x } 100 \quad (1)$$

where V denotes the volume (L) of the pregnant leach solution, c represents the metal concentration (mg/L) in the solution, R signifies the mass (g) of the solid leaching residue, and r indicates the grade of the metal within that residue.

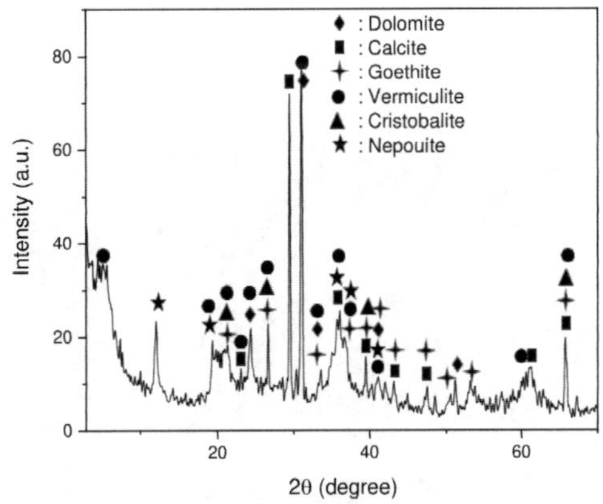

Figure 1. XRD pattern of raw material. It contains dolomite (♦), calcite (■), goethite (✦), nepouite (★), vermiculite (●), and cristobalite (▲)

3. Result and Discussion

3.1 Raw materials

The raw material has the main compounds such as Ni, Fe_2O_3, MgO, Co, Al_2O_3, CaO, Cu, Cr_2O_3, MnO, and SiO_2 with a total content of 1.2%, 19.23%, 16.11%, 0.03%, 3.10%, 1.59%, 0.01%, 1.02%, 0.32%, and 46.39%, respectively. The loss of ignition (LOI) is 10.85%. The Si and Mg ratio is 2.88, a fairly high value, close to the upper limit permitted in the nickel processing using the pyrometallurgy technique. The detailed content of raw material measured using the XRF method is shown in Table 1.

Figure 1 shows the diffraction peaks of XRD data of raw material. It consists of dolomite ($Ca_{0.5}Mg_{0.5}CO_3$), calcite ($Ca_{0.936}Mg_{0.064}CO_3$), goethite (FeOOH), nepouite ($Mg_{0.24}Ni_{2.76}Si_2O_9H_4$), vermiculite ($C_{12}H_2Al_{1.43}Fe_{0.48}Mg_{2.37}N_2O_{16}Si_{2.72}$), and cristobalite ($SiO_2$). The respective crystalline degrees of dolomite, calcite, goethite, zeolite, silica, and magnetite are 51.8%, 17.2%, 12.2%, 7.7%, 6.9%, and 4.2%. It was calculated from the area under the corresponding peaks of each mineral. Nickel can exist in nickel-bearing dolomite and/or goethite [8,9]. In the case of nickel-bearing dolomite, some Mg ions are replaced by Ni ions resulting in a composition that can be represented as $Ca(Mg_{1-x}Ni_x)CO_3$. Nickel-bearing goethite is a variation of goethite, an iron oxide mineral (α-

Journal of Physics: Conference Series **2980** (2025) 012021 doi:10.1088/1742-6596/2980/1/012021

Table 1 XRF data of raw material

Compound	Content (%)	Compound	Content (%)
Ni	1.20	SiO_2	46.39
Fe_2O_3	19.23	P_2O_5	0.01
MgO	16.11	TiO_2	0.08
Co	0.03	Zn	0.02
Al_2O_3	3.10	K_2O	0.05
CaO	1.59	SO_3	0.00
Cu	0.01	Na_2O	0.00
Cr_2O_3	1.02	LOI	10.85
MnO	0.32	**Total**	100.00

FeO(OH)), that contains significant amounts of nickel forming $Fe_{1-x}Ni_xO(OH)$. Here, *x* indicates the proportion of Ni substituting iron or magnesium in the dolomite or goethite mineral lattice.

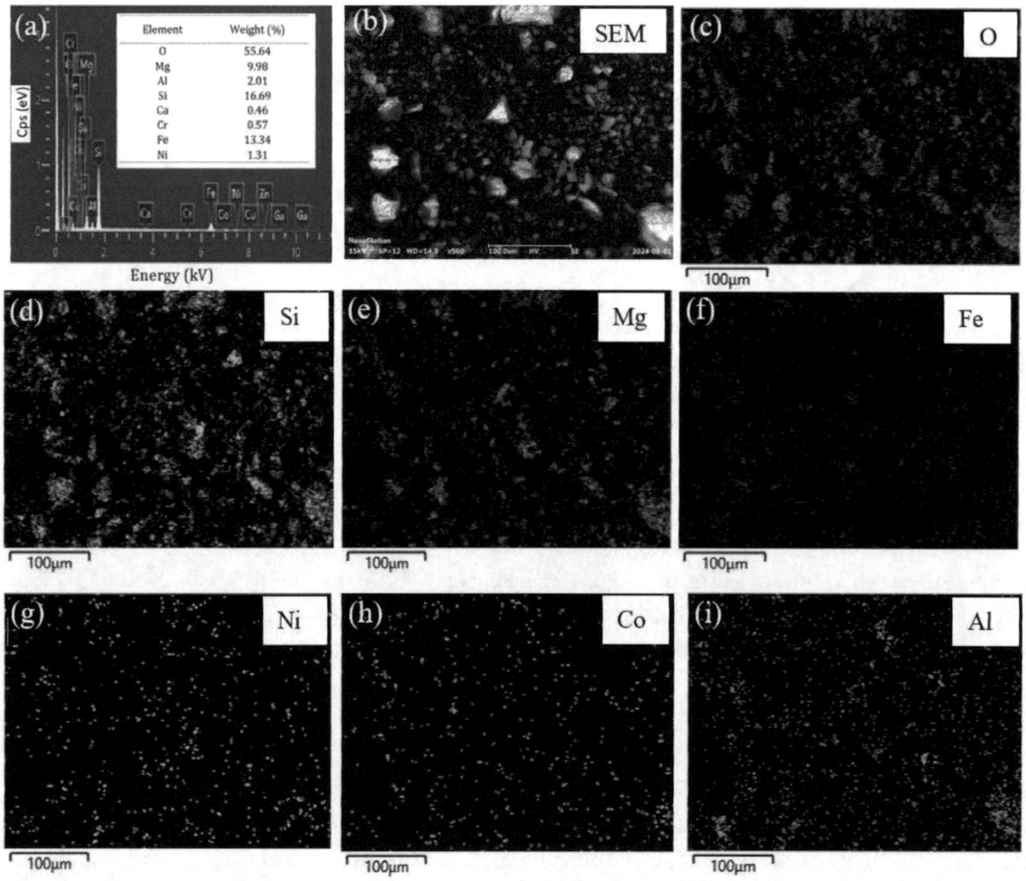

Figure 2. Results of SEM-EDS analysis of raw material, EDS spectrum and content list (a), SEM image (b), and the maps of elements (c-i)

Figure 2(a) presents a map of the energy spectrum alongside the corresponding elemental composition. The weight percentages of oxygen (O), magnesium (Mg), aluminum (Al), silicon (Si), calcium (Ca), chromium (Cr), iron (Fe), and nickel (Ni) are 55.64%, 9.98%, 2.01%, 16.69%, 0.46%, 0.57%, 13.34%, and 1.31%, respectively. Notably, magnesium (Mg), silicon (Si), and oxygen (O) consistently co-localize, as observed in Figures 2(c-e), which supports the XRD data identifying dolomite minerals. Additionally, the presence of a Ca atom further confirms the dolomite signature, as corroborated by the XRD analysis in Figure 1. The mapping also reveals a strong association between iron (Fe) and oxygen, providing substantial evidence of goethite and magnetite within the sample.

Nickel-bearing dolomite exhibits limited solubility in water; however, it readily dissolves under acidic conditions, especially when exposed to strong acids (such as sulfuric acid). Similarly, while goethite is generally insoluble in water, nickel-bearing goethite can dissolve in acidic environments, facilitating the release of nickel into solution. The reactivity of nickel-bearing

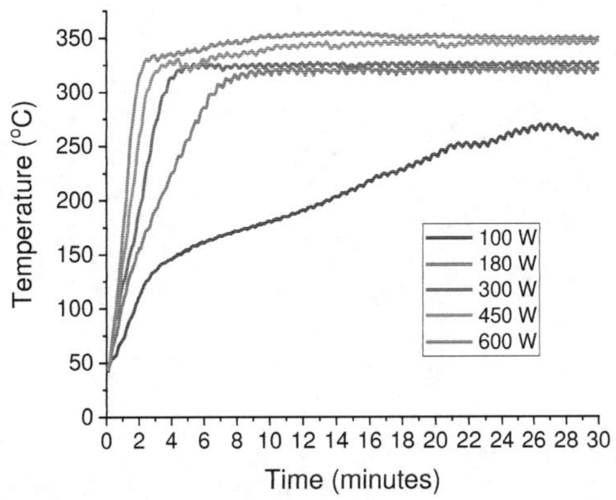

Figure 3. Temperature profile of microwave oven for roasting process of the mixture of nickel laterite ore and H_2SO_4. The microwave power is adjusted across a range of levels: 100 W, 180 W, 300 W, 450 W, and 600 W, with the maximum set at 800 W.

goethite during leaching processes is directly affected by its nickel content, with higher concentrations enhancing the dissolution efficiency during acid leaching. This characteristic makes nickel-bearing goethite a valuable target for extraction in hydrometallurgical operations.

3.2 Microwave roasting

A mixture of nickel laterite and H_2SO_4 was subjected to microwave roasting at varying power levels (100 W, 180 W, 300 W, 450 W, 600 W, and 800 W) and various heating durations (5, 10, and 20 minutes). Figure 3 shows the temperature profile of samples heated by microwave. The temperature rise rate per minute for power 100 W, 180 W, 300 W, 450 W, and 600 W is 8.9, 36.8, 60.6, 91.2, and 116.4, respectively. The maximum achievable temperature depends on the usage power level; higher microwave energy leads to a greater maximum temperature. Notably, even when heated for up to 30 minutes, the maximum temperature remains below the boiling point of H_2SO_4, which is 337 °C.

Figure 4 presents the roasting results, including a photograph of the powder product (a), a SEM image of the product (b), elemental mapping on SEM image (c), the EDS spectrum (d), and the elemental composition derived from the SEM-EDS analysis. Cobalt is more readily detectable

Journal of Physics: Conference Series **2980** (2025) 012021
doi:10.1088/1742-6596/2980/1/012021

with SEM-EDS following the roasting process, as compared to the SEM-EDS data for the raw materials presented in Figure 2. Sulfur content is 20.74%, indicating the sulfate form of metallic part of nickel laterite ore.

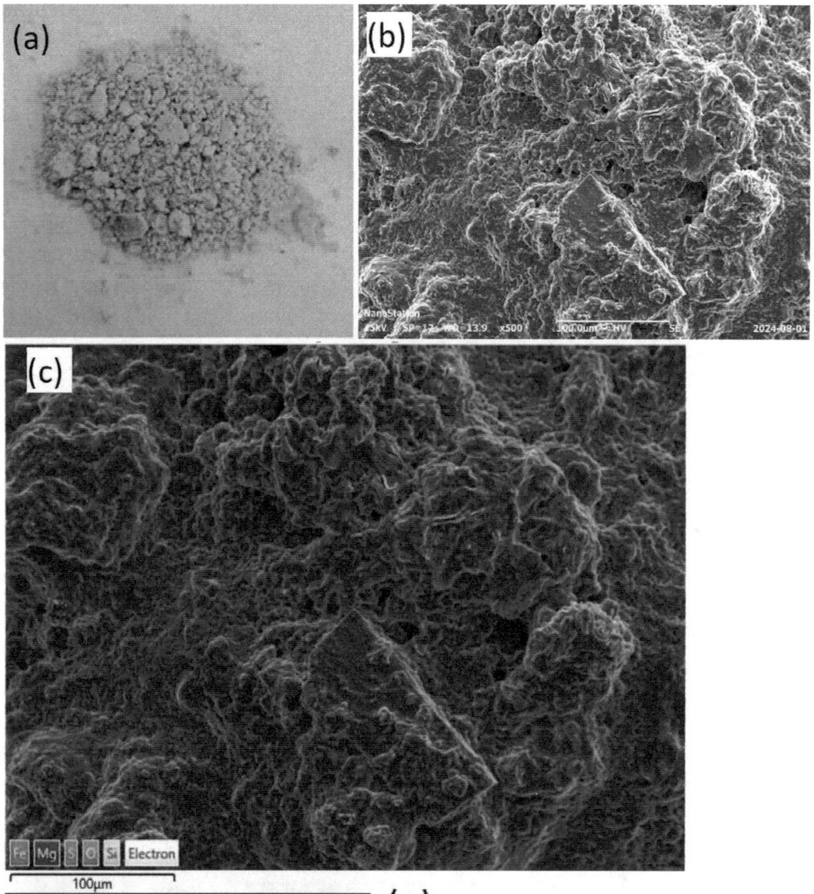

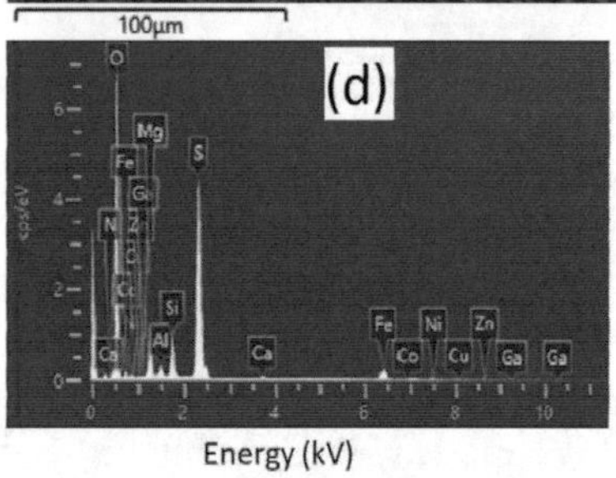

(e)

Element	Weight (%)	Element	Weight (%)
O	60.67	Fe	7.15
Mg	4.97	Co	0.03
Al	0.95	Ni	0.70
Si	3.61	Zn	0.14
S	20.74	Ga	0.04
Ca	0.48		

Figure 4. SEM-EDS data of roasted sample at P = 450 W and t = 5 minutes. Here, (a) a photograph of the sample after roasting, (b) SEM image at magnitude of 500x, (c) EDS mapping, (d) EDS spectra, and (e) elemental content.

3.3 Ni, Fe, and Co recoveries

Figure 5(a) illustrates the results of metallic ion extraction using distilled water following the roasting process. Here, the baking time is set at 5 minutes, with microwave power levels varying from 100 W to 800 W. The resulting solution predominantly exhibits a green color, indicating high nickel concentrations and a relatively low iron content. In contrast, a higher iron content would result in a reddish-green coloration of the solution. The SEM-EDS data for the residue at the operating conditions of 5 minutes and 450 W is shown in Figure 5(b-e). The residue

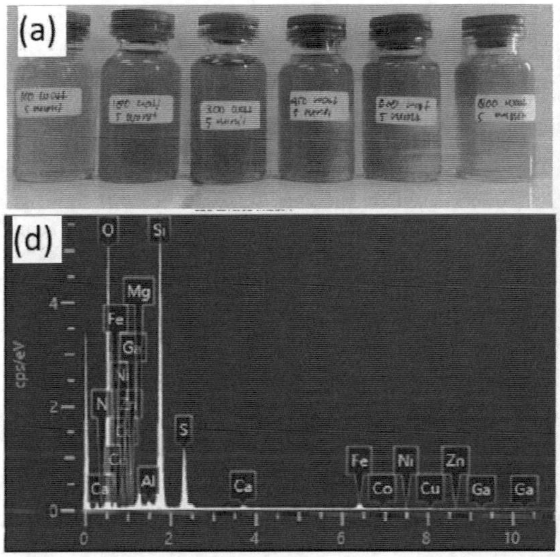

(e)

Element	Weight (%)	Atomic (%)
O	57.79	72.08
Mg	2.00	1.64
Al	0.40	0.3
Si	27.22	19.34
S	7.89	4.91
Ca	0.42	0.21
Fe	3.76	1.34
Co	0.03	0.01
Ni	0.49	0.17

Figure 5. SEM-EDS of residue and leachate solution. Here, (a) a photograph of leachate solution at t = 5 minutes and microwave power of 100 W, 180 W, 300 W, 450 W, 600 W, and 800 W, (b) an SEM image of solid residue at a magnitude of 500x, (c) EDS elemental mapping, (d) map of the spectrum, and (e) elemental contents.

predominantly consists of silicon (Si), sulfur (S), magnesium (Mg), iron (Fe), aluminum (Al), and oxygen (O) atoms. Notably, cobalt (0.03%) and nickel (0.49%) are still detectable, suggesting that not all of these metals can be effectively dissolved in distilled water during the extraction process.

The residue was found to still contain a relatively high Mg concentration, approximately 2.0%. The Fe atoms are around 3.76%, indicating that the Fe remaining in the residue is still quite large. This incomplete solubility aligns with findings from similar studies, which indicate that the chemical form and interactions of metal ions in solution can significantly influence their extraction efficiency [10,11]. However, S atom content in the residue has significantly decreased to approximately 7.89%, down from 20.74% immediately following the roasting process (see Figure 4(e)), suggesting that a portion of the sulfur has dissolved together with the metallic ions.

Figure 6 illustrates the recovery results of nickel (Ni), iron (Fe), and cobalt (Co) during the leaching process with distilled water at room temperature, following a microwave roasting treatment involving a mixture of sulfuric acid and nickel laterite. The microwave roasting process demonstrates remarkable efficiency, achieving minimum recoveries of 98.6% for Ni, 96.9% for Fe, and 99.0% for Co, even with a brief heating duration of just 5 minutes at a low power setting of 100 W. This efficiency underscores the ability of microwave roasting to facilitate the formation of soluble metal sulfates, which readily dissolve in water. However, when higher power levels (450 W to 800 W) are employed, recovery rates decline. At these higher power settings, the sample temperatures reach 342 °C and 352 °C, respectively, exceeding the boiling point of sulfuric acid. Under these conditions, sulfuric acid can evaporate, and prolonged heating may lead to the formation of metal oxides, which exhibit significantly lower solubility in water compared to their sulfate counterparts.

High recovery of Ni and Fe can be explained as the reaction between nickel-bearing goethite and sulfuric acid [9,12]

$$Fe_{1-x}Ni_xO(OH) + 2H_2SO_4 \rightarrow Fe_2(SO_4)_3 + NiSO_4 + 4H_2O + O_2 \quad (2)$$

The presence of sulfuric acid facilitates the breakdown of the goethite structure, releasing iron and nickel ions into the solution. The acid reacts with the hydroxyl groups and iron oxides in the goethite. The iron and nickel ions released combine with sulfate ions (SO_4^{2-}) from the sulfuric acid to form soluble iron(III) sulfate ($Fe_2(SO_4)_3$) and nickel sulfate ($NiSO_4$). This solubilization is critical for the extraction of these metals. Water and oxygen gas act as the by-products.

The reaction between nickel-bearing dolomite, represented as $Ca(Mg_{1-x}Ni_x)CO_3$ and H_2SO_4 involves the dissolution of the carbonate mineral and the formation of soluble metal sulfates including Ni and Mg. The general reaction can be expressed as follows [8]:

$$Ca(Mg_{1-x}Ni_x)CO_3 + H_2SO_4 \rightarrow CaSO_4 + MgSO_4 + NiSO_4 + H_2O + CO_2 \quad (3)$$

In this reaction, sulfuric acid reacts with the carbonate component of nickel-bearing dolomite. The acid protonates the carbonate, releasing carbon dioxide (CO_2) gas, which can escape into the atmosphere. The calcium, magnesium, and nickel ions released from the dolomite react with sulfate ions (SO_4^{2-}) from sulfuric acid to form calcium sulfate ($CaSO_4$), magnesium sulfate ($MgSO_4$), and nickel sulfate ($NiSO_4$). These sulfates are soluble in water, allowing nickel and other metals to be recovered from the solution.

Cobalt can be recovered from cobalt-goethite bearing $Fe_{1-x}Co_xO(OH)$ and cobalt-dolomite bearing $Ca(Mg_{1-x}Co_x)CO_3$ where the reaction with sulfuric acid is as follows [13,14]

$$Fe_{1-x}Co_xO(OH) + 2H_2SO_4 \rightarrow Fe_2(SO_4)_3 + CoSO_4 + 4H_2O + O_2 \quad (4)$$

$$Ca(Mg_{1-x}Co_x)CO_3 + H_2SO_4 \rightarrow CaSO_4 + MgSO_4 + CoSO_4 + H_2O + CO_2 \quad (5)$$

Figure 6 shows that in microwave roasting of nickel laterite, high nickel recovery was observed during the roasting process at lower power levels (100 W, 180 W, and 300 W). In contrast, recovery was comparatively lower at higher power settings. This phenomenon can be attributed to the maximum temperatures achieved at different power levels. For instance, at operational powers of 450 W and 600 W, the maximum temperatures reach 342 °C and 352 °C, respectively, exceeding sulfuric acid's boiling point. As a result, some sulfate may evaporate, forming metal oxides instead of metal sulfates.

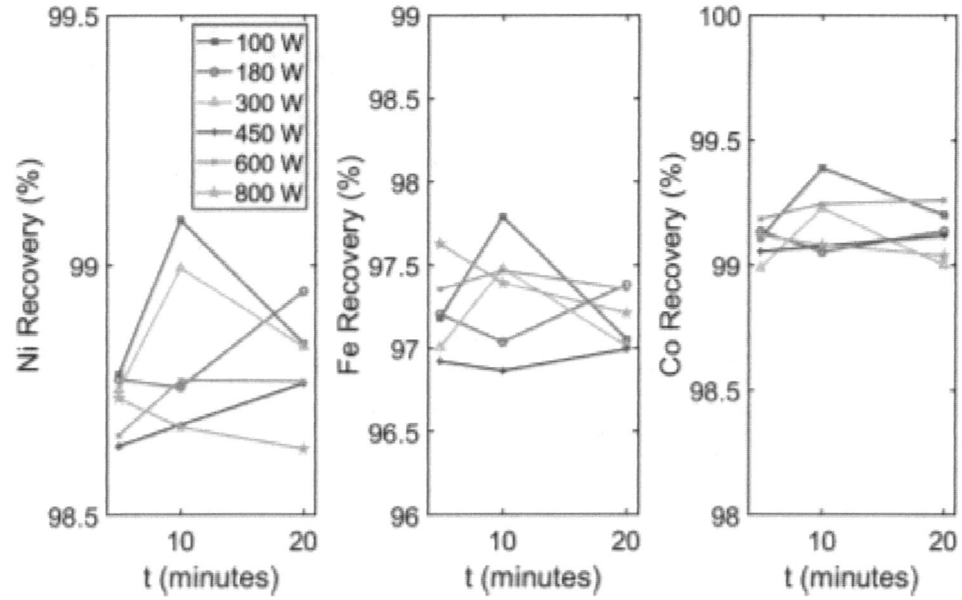

Figure 6. Recovery of Ni, Fe, and Co for the roasting condition at different microwave power (100 W, 180 W, 300 W, 450 W, 600 W, and 800 W) and times (5, 10, and 20 minutes)

Metal sulfides tend to be more soluble in water, while metal oxides generally exhibit lower solubility at room temperature [15]. This aligns with findings from previous studies, which suggest that optimal temperature control is crucial for maximizing metal recovery in hydrometallurgical processes [16,17].

The MAE (Microwave-Assisted Extraction) mechanism for enhancing the recovery of transition metals from nickel laterite ores begins with understanding the microwave heating process. Microwaves generate heat through dielectric heating, where electromagnetic waves (typically at 2.45 GHz) cause polar molecules within the material to oscillate [18]. This oscillation generates heat due to energy losses from molecular friction. In nickel laterite ores, polar molecules like water and hydroxyl groups (OH^-) in minerals absorb this microwave energy, resulting in rapid, localized heating.

A key feature of microwave heating, absent in conventional heating, is selective heating. The metal components in the ore—such as nickel (Ni), iron (Fe), and cobalt (Co)—absorb microwave energy differently due to their distinct dielectric properties, causing them to heat at different rates. This selective heating is more efficient and localized than traditional methods, as microwaves can penetrate the material, heating it from within. Moreover, the various mineral phases in the ore (such as silicates, oxides, and hydroxides) exhibit different dielectric behaviors, leading to differential heating. This effect can selectively break down certain mineral phases, making nickel more accessible during subsequent extraction steps [19-20].

As the nickel laterite absorbs microwave energy, minerals like serpentine, dolomite, goethite, and limonite undergo rapid thermal expansion. This intense, localized heating induces thermal shock, which can fracture or disintegrate the mineral structure, increasing the ore's porosity and surface area. The result is a more reactive and easier for sulfuric acid to penetrate during leaching.

The operational temperature of the sample typically falls within the dehydration or dehydroxylation range of serpentine and dolomite. When heated, dolomite ($CaMg(CO_3)_2$) loses water and transforms into olivine (Mg_2SiO_4) and other phases. These changes make the ore more porous and reactive, enhancing nickel accessibility for acid leaching. Nickel in laterite ores is commonly found in the form of oxides (NiO) or hydroxides ($Ni(OH)_2$), which react with sulfuric acid to form nickel sulfate ($NiSO_4$). Microwave heating accelerates these reactions by increasing the dissolution rate of nickel compounds, making the extraction process more efficient than with conventional heating.

Other metals, such as Fe and Co, may also react with sulfuric acid, forming their respective metal sulfates. While this can complicate the selective extraction of nickel, MAE may help by selectively breaking down the ore matrix, improving the efficiency of nickel recovery. The enhanced heating speeds up chemical reactions and improves the mass transfer of sulfuric acid into the ore, facilitating faster dissolution. Additionally, microwave-induced heating can assist in oxidation-reduction reactions, such as reducing nickel from higher oxidation states (e.g., Ni(III)) to more easily leachable Ni(II), which readily forms nickel sulfate in the presence of sulfuric acid.

4. Conclusion

The findings demonstrate the significant effectiveness of microwave-assisted recovery of transition metals—specifically nickel, iron, and cobalt—from low-grade nickel laterite (Ni ~ 1.2%) through a roasting process with sulfuric acid, followed by leaching at room temperature. Remarkably, recovery rates reached a minimum of 98.6% for nickel, 96.9% for iron, and 99.0% for cobalt, even at low microwave powers of 100 W, 180 W, and 300 W, with a mere 5 minutes of heating time. The solid-to-solution ratio maintained during the process was 1.2 g/mL. Notably, while higher microwave power resulted in a slight decrease in recovery efficiency due to operational temperatures exceeding the boiling point of sulfuric acid, the process remains efficient. The nickel laterite primarily comprises dolomite, calcite, goethite, nepouite, vermiculite, and cristobalite, emphasizing the diverse mineralogy involved. Furthermore, extending the heating duration at lower powers can reduce acid consumption, enhancing the environmental sustainability of the method compared to traditional extraction techniques. Overall, these results not only highlight the potential of microwave technology in mineral processing but also pave the way for more sustainable practices in the recovery of valuable metals from low-grade ores.

Acknowledgments

This research was partially funded by the Directorate General of Higher Education, Ministry of Education, Culture, Research and Technology of the Republic of Indonesia, under Fundamental Research Grant No. 049/E5/PG.02.00.PL/2024. It also resulted from collaborative efforts between Halu Oleo University, Indonesia, and the Research Center for Development of Far-Infrared Region, University of Fukui, Japan.

Journal of Physics: Conference Series **2980** (2025) 012021
doi:10.1088/1742-6596/2980/1/012021

References

[1] Chen, S., Guo, X., Shi, W., and Li, D. (2010). Extraction of valuable metals from low-grade nickeliferous laterite ore by reduction roasting-ammonia leaching method. Journal of Central South University of Technology, 17, 765-769. [https://doi.org/10.1007/s11771-010-0554-9]

[2] Butt, C. W. and Cluzel, D. (2013). Nickel Laterite Ore Deposits: Weathered Serpentinites. Elements, 9, 123-128. [https://doi.org/10.2113/gselements.9.2.123].

[3] Xiao, J. and Zhang, Y. (2019). Recovering Cobalt and Sulfur in Low Grade Cobalt-Bearing V–Ti Magnetite Tailings Using Flotation Process. Processes, 7(8), 536. [https://doi.org/10.3390/pr7080536]

[4] Wang, S. (2006) Cobalt—Its recovery, recycling, and application. JOM 58, 47–50. [https://doi.org/10.1007/s11837-006-0201-y]

[5] Zhao, Y., Geng, S., Cheng, H. Li, G., Lu, X., and Xu, Q. (2018). The role of CaCO3 in the extraction of valuable metals from low-nickel matte by calcified roasting—acid leaching process. IOP Conf. Series: Materials Science and Engineering, 409, 012027 [https://doi.org/10.1088/1757-899X/409/1/012027]

[6] Majzlan, J. (2024). Thermodynamics of the α-FeOOH (goethite)-ScOOH solid solution. Physics and Chemistry of Minerals, 51, 37. [https://doi.org/10.1007/s00269-024-01298-1]

[7] Wang, F. and Dreisinger, D. (2024). Acceleration of Iron-Rich Olivine CO2 Mineral Carbonation and Utilization for Simultaneous Critical Nickel and Cobalt Recovery. Minerals, 14(8), 766. [https://doi.org/10.3390/min14080766]

[8] Eskandarnia, P., Ahmadi Khalaji, A., Kiani, M., Valipor, A. and Tahmasbi, Z. (2023) Mineralogy, Geochemistry, and Origin of Nickel-Bearing Laterites in the Northwestern of Noorabad (Lorestan Province). Journal of Economic Geology, 15(3), 71–102. [https://doi.org/10.22067/ECONG.2023.82772.1079]

[9] Das, G. K. and Li, J. (2023) Iron Removal as Goethite from Synthetic Laterite Leach Solutions. ACS Omega, 8(13), 11931–11940. [https://doi.org/10.1021/acsomega.2c07595]

[10] Su, Y., Berbille, A., Li, X. F., Zhang, Z., PourhosseiniAsl, M. J., Li, H., Liu, Z., Li, S., Liu, J., Zhu, L., and Wang, Z. L. (2024) Reduction of precious metal ions in aqueous solutions by contact-electro-catalysis. Nature Communication, 15, 4196. [https://doi.org/10.1038/s41467-024-48407-w]

[11] Rahmati, S., Adavodi, R., Hosseini, M. R., and Veglio', F. (2024) Efficient Metal Extraction from Dilute Solutions: A Review of Novel Selective Separation Methods and Their Applications. Metals 2024, 14(6), 605- [https://doi.org/10.3390/met14060605]

[12] Agacayak, T., Zedef, V., and Aras, A. (2016) Kinetic study on leaching of nickel from Turkish lateritic ore in nitric acid solution. Journal of Central South University, 23, 39–43. [https://doi.org/10.1007/s11771-016-3046-8]

[13] Batnasan, A., Haga, K., and Shibayama, A. (2014) Recovery of Nickel and Cobalt from a Low Grade Laterite Ore. Resources Processing, 61(2), 100-109. [https://doi.org/10.4144/rpsj.61.100]

[14] Ugwu, I. and Sherman, D.M., (2019). The solubility of goethite with structurally incorporated nickel and cobalt: Implication for laterites. Chemical Geology, 518(4), 1-8. [https://doi.org/10.1016/j.chemgeo.2019.04.021]

[15] Marchioretto, M. M., Rulkens, W., and Bruning, H., (2005). Heavy Metals Precipitation in Sewage Sludge. Separation Sciences and Technology, 40(6), 3393–3405. [https://doi.org/10.1080/01496390500423748]

[16] Torres, R., Segura-Bailón, B., and Lapidus, G. T., (2018). Effect of temperature on copper, iron and lead leaching from e-waste using citrate solutions. Waste Management, 71, 420-425. [https://doi.org/10.1016/j.wasman.2017.10.029]

[17] Suiyi, Z., Jian, W., Yuhong, H., Ying, W., Yuxin, Z., Jiabao, Q, Jiancong, L, Jinlu, Y., and Meichun, J. (2023), Perspective on pH adjustment in hydrometallurgical recycling of valuable metals from waste. Frontiers in Chemistry, 11, 1177173. [https://doi.org/10.3389/fchem.2023.1177173]

[18] Mishra, R. R. and Sharma, K. A. (2016) Microwave–material interaction phenomena: Heating mechanisms, challenges and opportunities in material processing. Composites Part A: Applied Science and Manufacturing, 81, 78-97. [https://doi.org/10.1016/j.compositesa.2015.10.035]

[19] Zhai, X. J., Wu, Q., Fu, Y. , Ma, L. Z., Fan, C. L., and Li, N. J. (2010) Leaching of nickel laterite ore assisted by microwave technique. Transactions of Nonferrous Metals Society of China, 20(1), s77-s81. [https://doi.org/10.1016/S1003-6326(10)60016-7]

[20] Bobicki, E., Liu, Q., and Xu, Z. (2018) Microwave Treatment of Ultramafic Nickel Ores: Heating Behavior, Mineralogy, and Comminution Effects. Minerals, 8, 524. [https://doi.org/10.3390/min8110524]

[21] Wen, T., Zhao, Y., Xiao, Q., Ma, Q., Kang, S., Li, H., and Song, S. (2017) Effect of microwave-assisted heating on chalcopyrite leaching of kinetics, interface temperature and surface energy. Results in Physics, 7, 2594-2600. [https://doi.org/10.1016/j.rinp.2017.07.035]

Prediction of Intercalation Voltage Variation in Ti- and Cr-doped $Na_xV_{1.5}M_{0.5}(PO_4)_3$ (M=Ti, Cr) as Sodium-Ion Battery Cathode

Dhias Pratama Lazuarfy[1,2]*, Fadjar Fathurahman[1,3], Ganes Shukri[1,3], Qory Cipta Scientika[1,3], A.Hafizh Rifa'i[1,2] and Afthon Ilman Huda Isyfi[1,2]**

[1]Nano and Quantum Technology Research Group, Faculty of Industrial Technology, Bandung Institute of Technology, Jalan Ganesa No. 10, Bandung, 40132, Indonesia
[2]PT Perusahaan Listrik Negara (Persero), Jakarta, Indonesia
[3]Research Centre for Nanoscience and Nanotechnology (RCNN), Bandung Institute of Technology, Jalan Ganesa No. 10, Bandung, 40132, Indonesia

*dhias.pratama@pln.co.id
**ganes_tf07@itb.ac.id

Abstract. By means of Density Functional Theory (DFT) calculations, we investigate the effect of Titanium and Chromium doping on the Na intercalation voltage (V_{int}) profile in $Na_xV_{1.5}M_{0.5}(PO_4)_3$ (with M=Ti, Cr and X = 4.0, 3.0, 2.0 and 1.0) for Sodium ion battery cathode. Our findings reveal that both Ti and Cr dopants significantly influence the intercalation voltage profile of the vanadium phosphate-based (NVP-based) cathode material. We find that both Cr and Ti doping increase the overall intercalation voltage range (ΔV) of the doped NVP as compared to the undoped NVP. The largest ΔV is shown by the Cr-doped NVP (~1.25 V), followed by the Ti-doped NVP (~1.15 V) and lastly the undoped (pristine) NVP (~1.00 V). We also find that Cr dopant affect the two end values of V_{int} differently: it decreases (increases) the V_{int} at the X_{Na} = 4.0 (X_{Na} = 1.0) as compared to the undoped NVP. On the other hand, the V_{int} of Ti-doped NVP undergoes simultaneous decrease at the two ends of the V_{int} values (as compared to the undoped NVP). Our findings reveal that incorporating transition metal dopant into the NVP-based cathode is a promising alternative to further boost the working voltage of the NVP cathode.

1. Introduction

Although lithium-ion batteries (LIBs) have revolutionized electronic devices and electric vehicles, the scarcity and high demand for energy raise questions about the ability of LIBs to meet future energy storage needs. To address this issue, much recent research has focused on developing alternative energy storage systems [1]. Sodium-ion batteries (NIBs), which have a similar working principle to LIBs, offer a promising alternative for future energy storage, as sodium is cheaper and more abundant than lithium [2]. In term of safety NIBs can be transported safely at zero voltage, unlike LIBs. They also have a lower risk of fire [3]. Several advantages of NIBs (compared to other type of secondary batteries) are further tabulated in **Table 1**. Overall, the NIB is a promising alternative for a more sustainable and affordable option than its LIBs counterpart [4].

Analog to its LIB counterpart, the overall electrochemical performance of an NIB is highly dependent on the choice of the cathode material being used. Metal oxides, polyanionic compounds, and organic materials have been investigated [4]. Metal oxides are susceptible to structural damage, while organics have slow ion diffusion in general. Polyanionic compounds stand out due to their fast Na^+ diffusion, stable structure, high energy, and good electrochemical performance [1]. $Na_3V_2(PO_4)_3$ (NVP), a polyanionic compound with a rhombohedral structure, is a promising cathode for NIBs. As a member of Na Super Ionic Conductor (NaSICON) material, NVP offers high Na^+ mobility and relatively good structural stability [5]. However, vanadium makes NVP expensive, and its low electronic conductivity hinders its further utilization [6]. Substituting vanadium with a cheaper and more environmentally friendly element is one promising alternative technique for further large-scale NVP applications [7].

Table 1. Selected sustainability considerations for sodium-ion and competing battery technology [3]

	Sodium-ion batteries	Lead-acid	Lithium-ion
Materials	Ubiquitous and abundant	Toxic	Expensive, geographically concentrated and under increasing pressure
Recycling	Limited recycling at present	High recycling rates	Limited recycling at present
Social costs	None	High cost to human health (where regulatory safeguards not followed)	Social and environmental issues in cobalt mining
Safety	Can be shipped and stored in its zero energy state - Excellent safety testing results	Risk of explosion due to hydrogen evolution - Strongly acidic electrolyte	Must be partially charged when shipped

Improving the performance of NVP electrodes has become the focus of various studies. Modifications include carbon coating, doping, and nano-crystallization [1,8]. Carbon coating improves the electronic conductivity of the composite material but does not affect the NVP itself. Doping, especially with transition metal atoms at the Na or V sites, is considered more effective because it can increase the interatomic lattice spacing, intercalation voltage, and electronic conductivity, as well as modulate the anion sites to improve the working voltage and energy density of the material [8]. Titanium is widely distributed across the planet and is found in amounts of about 5.6 grams per kilogram of the Earth's crust. This element is one of the 10 most abundant elements and is more common than chlorine, phosphorus, carbon, and sulfur [9]. In one study, Titanium was used to substitute ($Na_{3-x}V_{2-x}Ti_x(PO4)_3/C$) as a high-performance cathode material for NIBs. As a result, the structural stability and electrochemical performance were greatly improved through Ti substitution. The NVP-$Ti_{0.15}$/C composition showed outstanding rate capability and cyclability with a capacity retention of 60% after 2000 cycles at a high rate of 10 C. Even at 20 C, a reversible capacity of 63 mAh/g was maintained after 300 cycles at a high temperature of 60 °C [10]. A previous density functional theory-based study showed that the superior electrochemical performance of the Ti-doped samples can be attributed to the enhanced intrinsic electronic conductivity, low Na^+ diffusion energy barrier, and high structural stability.

In another study, Chromium was used to dope NVP in the form of $Na_3V_{2-x}Cr_x(PO_4)_3$ (x = 0, 0.02, 0.04, 0.08, 0.10) as a cathode material for sodium-ion batteries (synthesized via the sol-gel method) [11]. Electrochemical tests showed that the cycle stability and specific capacity of sodium-ion batteries can be significantly improved by doping Cr into NVP. Among all Cr-doped cathode materials, $Na_3V_{1.92}Cr_{0.08}(PO_4)_3$ achieved the highest capacity of 112.2 mAh/g and its capacity retention was 97.2% after 50 cycles [11]. Electrochemical impedance spectroscopy measurements showed that Cr doping is an effective method to reduce the interparticle contact resistance by suppressing irreversible phase transformation at low sodium content.

Experimental and theoretical studies show potential for improved electrochemical characteristics in NVP cathodes doped with various elements. However, previous studies lack systematic understanding of how each dopant affects the modified NVP cathode's physical properties. Experimental characterization shows varying physical properties due to differing synthesis methods. Computational approaches using first-principles calculations offer a systematic understanding to explain and predict doping effects on physical properties and electrochemical performance of secondary battery electrodes. [12]. In this study, we use density functional theory (DFT) to investigate the effect of Ti and Cr doping on the change of intercalation voltage (V_{int}) behavior of $Na_xV_{1.5}Ti_{0.5}(PO_4)_3$ and $Na_xV_{1.5}Cr_{0.5}(PO_4)_3$ cathode chemistries. The intercalation voltage is an important quantity in predicting the overall working voltage range of an electrode [13][14]. Knowledge of the impact of dopant variation on this key cathode parameter will be a guide in determining the optimal dopant for further synthesis and experimental efforts.

2. Computational Details

The calculation of intercalation voltage and formation energy was conducted using Density Functional Theory (DFT). VASP, or the Vienna Ab initio Simulation Package, is a highly regarded software package used for performing density-functional theory (DFT) calculations to predict the electronic structure and related properties of materials. It is particularly effective in handling periodic systems, which can include materials like solids with regular lattice structures, surfaces, and even molecules under certain conditions. The core functionality of VASP involves solving the Kohn-Sham equations, which are central to DFT. This involves using wave functions and their corresponding eigenvalues to calculate properties such as the total and local charge densities. [15]

Our present study focused on modifying the $Na_3V_2(PO_4)_3$ cathode by substituting 25% of vanadium (V) atoms with Titanium (Ti) and Chromium (Cr). We manually adjusted the positions of Titanium and Chromium relative to the vanadium atoms in various configurations to optimize the structure. This aimed to identify the most stable parameters and structures of $Na_xV_2(PO_4)_3$, $Na_xV_{1,5}Ti_{0,5}(PO_4)_3$ and $Na_xV_{1,5}Cr_{0,5}(PO_4)_3$ using density functional theory (DFT) calculations as implemented in the Vienna Ab initio Simulation Package (VASP). The projector augmented-wave (PAW) method was employed to solve the Kohn-Sham equation. The exchange-correlation energy was calculated using the Perdew-Burke-Ernzerhof (PBE) functional, with a plane-wave cutoff energy of 520 eV and a Monkhorst-Pack k-point grid of 3 x 3 x 1 for Brillouin zone integration. Convergence criteria for forces and energy were set at 0.05 eV/Å and 10^{-4} eV per atom, respectively. The NVP structure, which was based on the standard crystal database (CCSD), consisted of 126 atoms ($Na_4V_2(PO_4)_3$) in the $R\bar{3}c$ rhombohedral space group and served as the parent structure for various $Na_xV_2(PO_4)_3$ concentrations, where $1 \le x \le 4$. The visualization of Pristine (undoped) and doped NVP cathode shown at **figure 1**.

Journal of Physics: Conference Series **2980** (2025) 012022 doi:10.1088/1742-6596/2980/1/012022

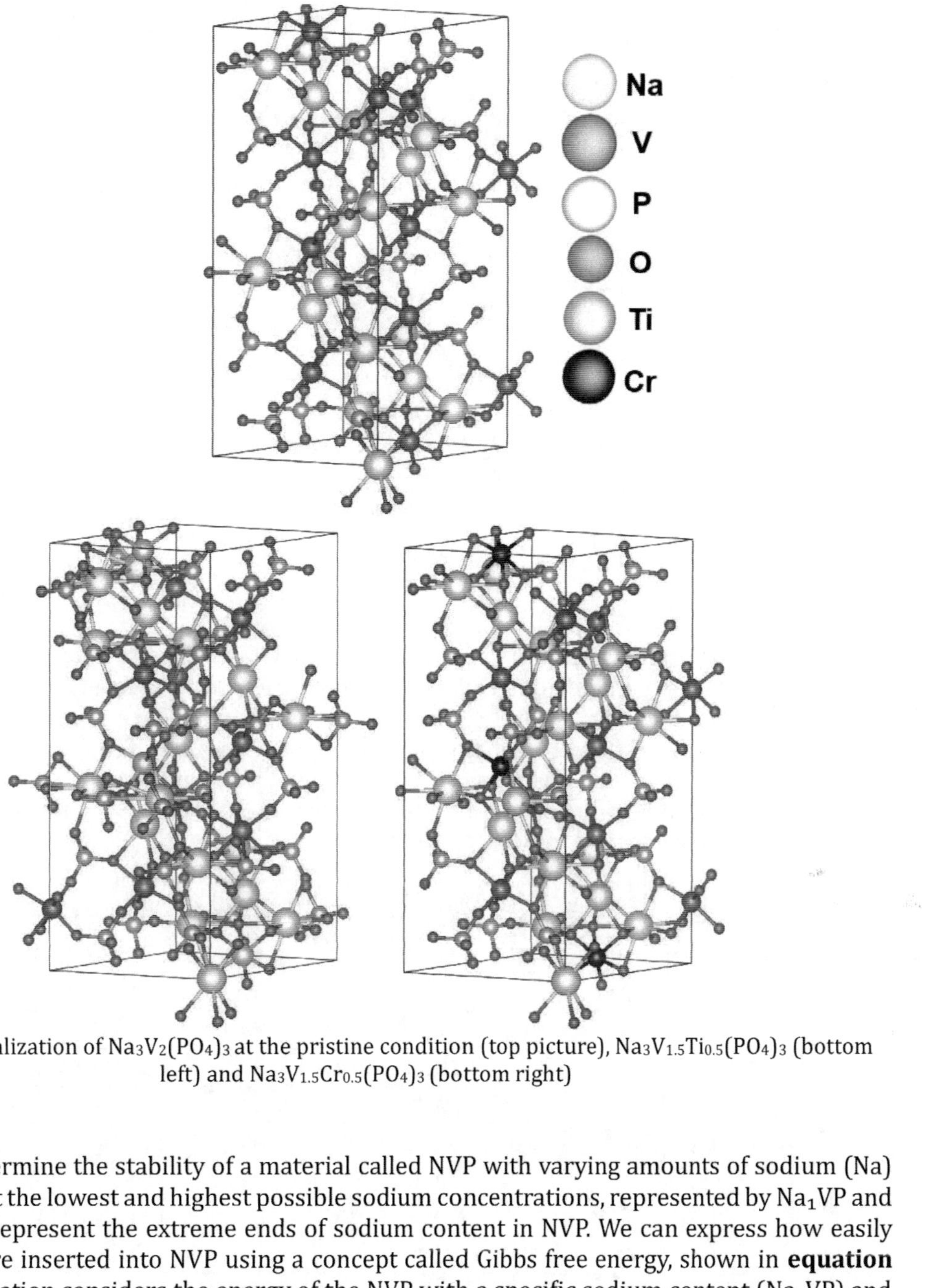

Figure 1 Visualization of $Na_3V_2(PO_4)_3$ at the pristine condition (top picture), $Na_3V_{1.5}Ti_{0.5}(PO_4)_3$ (bottom left) and $Na_3V_{1.5}Cr_{0.5}(PO_4)_3$ (bottom right)

To determine the stability of a material called NVP with varying amounts of sodium (Na) ions, we look at the lowest and highest possible sodium concentrations, represented by Na_1VP and Na_4VP. These represent the extreme ends of sodium content in NVP. We can express how easily sodium ions are inserted into NVP using a concept called Gibbs free energy, shown in **equation (3.1)**. This equation considers the energy of the NVP with a specific sodium content (Na_xVP) and compares it to the energies of the two extreme cases (Na_1VP and Na_4VP).

$$\Delta G_f(Na_xVP) = G[Na_xVP] - \left(\frac{4-x}{3}\right) G[Na_1VP] - \left(\frac{x-1}{3}\right) G[Na_4VP] \qquad (1)$$

In order to simplify the calculation of energy changes associated with the addition or removal of sodium ions from NVP, an approximation method can be employed, as outlined in

Journal of Physics: Conference Series **2980** (2025) 012022 doi:10.1088/1742-6596/2980/1/012022

equation 2. This particular equation utilizes the energies of three distinct NVP variations (Na_xVP, Na_1VP, and Na_4VP), each representing a different sodium concentration within the material. These crucial energy values are determined through computational methods, specifically Density Functional Theory (DFT) calculations.

$$\Delta E_f(Na_xVP) = E[Na_xVP] - \left(\frac{4-x}{3}\right)E[Na_1VP] - \left(\frac{x-1}{3}\right)E[Na_4VP] \qquad (2)$$

The calculation of intercalation voltage follows the formula formulated as follows:

$$V = -\frac{E(Na_xV_2(PO_4)_3) - [E(Na_yV_2(PO_4)_3 + (x-y)\mu_{Na}]}{(x-y)F} \qquad (3)$$

where μ_{Na} is the chemical potential of the Na crystal, F is the Faraday constant with a value of 96845 C/mol or 26801 mAh/mol.

3. Results and Discussions

3.1 Lattice Parameter

Table 2 show the lattice parameters of the unit cell that was simulated, including the values of length, width, and height of the unit cell which are represented by the values a, b, and c. And the angles of the unit cell which are represented by α, β, and γ. From the results of the volume calculation, it can be observed that the doping of titanium and chromium does not cause a significant change in volume when compared to the pristine/ pure condition. This insignificant change in volume indicates that the addition of dopants does not affect the stability of the structure (ΔV still below 8.8%). [16]

Table 2. Lattice parameter (Å and °) and volume across all computed variations

Variation	Nax	a [Å]	b [Å]	c [Å]	α [°]	β [°]	γ [°]	V [Å^3]	ΔV Na1 to Na3
Undoped	Na1	8.49	8.49	21.35	90.35	89.65	120.06	1331.38	5.30 %
	Na2	8.63	8.63	21.34	90.13	89.87	120.47	1368.79	
	Na3	8.75	8.75	21.27	90.25	89.76	120.19	1405.93	
	Na4	8.87	8.87	21.11	90.00	90.00	120.00	1438.49	
Ti-doped	Na1	8.50	8.50	21.32	90.03	89.97	120.04	1332.60	6.17 %
	Na2	8.65	8.65	21.37	90.30	89.72	120.61	1376.10	
	Na3	8.77	8.76	21.39	90.32	89.65	120.20	1420.30	
	Na4	8.89	8.89	21.12	90.00	90.00	120.00	1444.31	
Cr-doped	Na1	8.45	8.45	21.26	90.00	90.00	120.03	1314.45	6.72 %
	Na2	8.61	8.61	21.31	90.33	89.76	120.61	1361.19	
	Na3	8.75	8.75	21.27	90.31	89.68	120.14	1409.20	
	Na4	8.86	8.86	21.09	90.00	90.00	120.00	1433.55	

3.2 Formation Energy

Formation energy is crucial for determining the stability of NaSICON materials. A lower formation energy indicates a more stable structure, which is less likely to degrade during cycling. Stable structures, as indicated by low formation energy, lead to improved cycle life. NaSICON frameworks with stable formation energies are more likely to maintain their integrity over repeated cycles, thereby enhancing cycle life. [17]

Figure 2 shows the results of formation energy calculations based on Density Functional Theory (DFT). The graph displays the relationship between the concentration of Sodium (x) in $Na_xV_2(PO_4)_3$ or $Na_xV_{1.5}M_{0.5}(PO_4)_3$ and its formation energy (in meV/atom). Three types of materials are shown: undoped material (blue line), Titanium-doped material (pink line), and Chromium-doped material (green line).

In general, the formation energy of all three materials decreases with increasing Sodium concentration until it reaches its lowest point at x = 2, then increases again until x = 4. The undoped material has the highest (most positive) formation energy across the entire Sodium concentration range. At x = 2, the undoped material has a formation energy of around -25.88 meV/atom, while the Chromium- and Titanium-doped materials have lower formation energies, around -30.44 meV/atom and -31.19 meV/atom, respectively. At x = 3, the formation energy of the undoped material slightly increases to around -25.18 meV/atom. Meanwhile, the Chromium-doped material has the lowest formation energy compared to x = 2, at around -33.48 meV/atom. The Titanium-doped material also experiences a slight increase in formation energy to around -27.91 meV/atom. It can be concluded that the addition of doping, both titanium and chromium, tends to decrease the formation energy of the material. The effect of Chromium doping is more significant in reducing formation energy compared to Titanium doping.

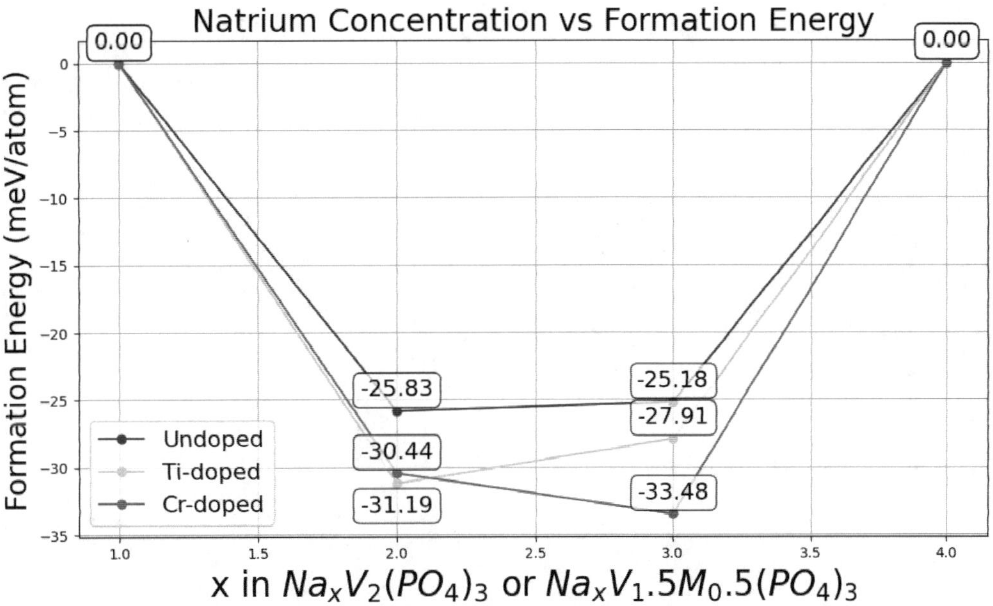

Figure 2 Formation energies of undoped, Ti-doped, and Cr-doped NVP materials vs. Na content (x) in $Na_xV_2(PO_4)_3$ or $Na_xV_{1.5}M_{0.5}(PO_4)_3$.

Journal of Physics: Conference Series **2980** (2025) 012022 doi:10.1088/1742-6596/2980/1/012022

3.3 Intercalation Voltage

Intercalation voltage impacts the energy density of the battery. Higher intercalation voltage generally results in higher energy density. Higher intercalation voltages lead to greater energy density, making these materials suitable for applications requiring efficient energy storage. NaSICON materials with high voltage profiles are promising for achieving high energy densities. [17]

Figure 3 shows the relationship between sodium concentration (x in $Na_xV_2(PO_4)_3$ or $Na_xV_{1.5}M_{0.5}(PO_4)_3$ and intercalation voltage in sodium-ion battery cathode materials. Three types of cathode materials are compared, namely undoped material (blue line), titanium-doped material (pink line), and chromium-doped material (green line). In general, the intercalation voltage tends to decrease with increasing sodium concentration. This can be observed in all three types of cathode materials.

The material with chromium doping shows the highest intercalation voltage at almost every sodium concentration. Titanium doping reduces the intercalation voltage compared to all other material variations (undoped and chromium doped). The difference in intercalation voltage (ΔV) between sodium concentrations 1.0 and 4.0 is also shown. The largest ΔV is shown by the Cr-doped material (1.2480 V), followed by the Ti-doped material (1.1507 V), and lastly the undoped material (0.9943 V).

This graph indicates that doping with titanium and chromium can affect the intercalation voltage in sodium-ion battery cathode materials. The choice of doping type and sodium concentration can be optimized to achieve the desired battery performance. This graph shows that chromium is more optimal in increasing the intercalation voltage in the battery cathode.

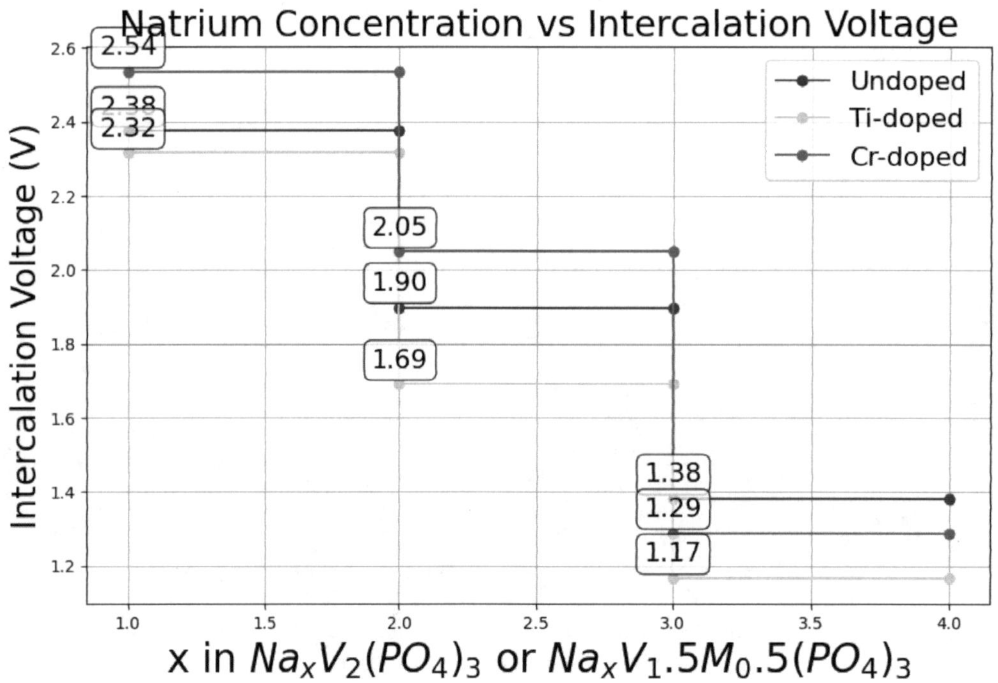

Figure 3 Intercalation Voltage of undoped, Ti-doped, and Cr-doped NVP materials vs. Na content (*x*) in $Na_xV_2(PO_4)_3$ or $Na_xV_{1.5}M_{0.5}(PO_4)_3$.

Journal of Physics: Conference Series **2980** (2025) 012022
doi:10.1088/1742-6596/2980/1/012022

3.4 Magnetic Moment

The arrangement of electrons within an atom dictates its chemical behavior. This arrangement follows certain rules, the Aufbau principle (electrons fill lower energy levels first), Hund's rule (electrons spread out before pairing up), and the Pauli exclusion principle (only two electrons with opposite spins can occupy the same orbital). Vanadium, a transition metal with 23 electrons, has the electron configuration [Ar] $4s^2$ $3d^3$. Interestingly, when vanadium forms ions, electrons are lost from the 4s orbital first, even though the 3d orbital is at a higher energy level. This leads to various oxidation states for vanadium, such as V^{2+} ([Ar] $3d^3$), V^{3+} ([Ar] $3d^2$), and V^{4+} ([Ar] $3d^1$).

Our PBE-based calculations reveal a nearly linear relationship between the magnetic moment of vanadium ions and sodium concentration. Based on the magnetization moment data from **Figure 4**, The three graphs illustrate the magnetization moment of NVP cathodes under three different conditions (undoped, doped with Titanium, and doped with Chromium). The key difference lies in the effect of the dopants on the magnetization moment of the Na atoms.

The undoped NVP exhibits relatively stable Na magnetization moments with slight fluctuations depending on the V atom positions. Na4 shows the highest and most stable moment. In contrast, Ti doping significantly reduces the magnetization moment of all Na atoms, particularly Na1 and Na2, at the Ti substitution sites (atoms 10-12). This suggests that Ti weakens the magnetic interactions within the NVP structure. Conversely, Cr doping markedly increases the magnetization moment of all Na atoms, especially Na1 and Na2, at the Cr substitution sites. This implies that Cr strengthens the magnetic interactions.

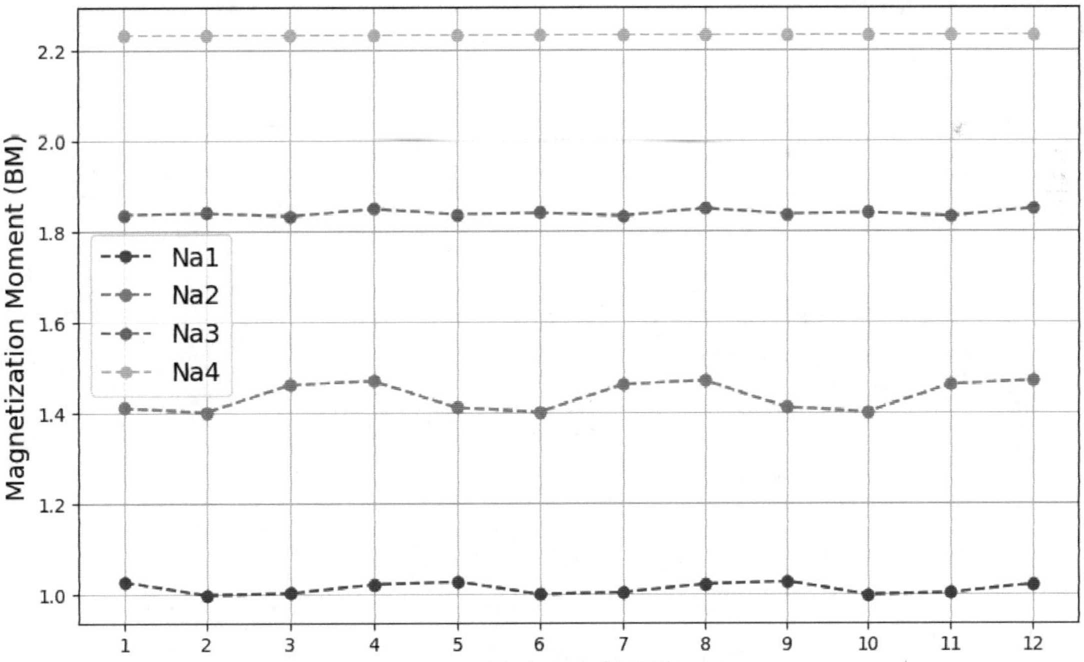

(a)

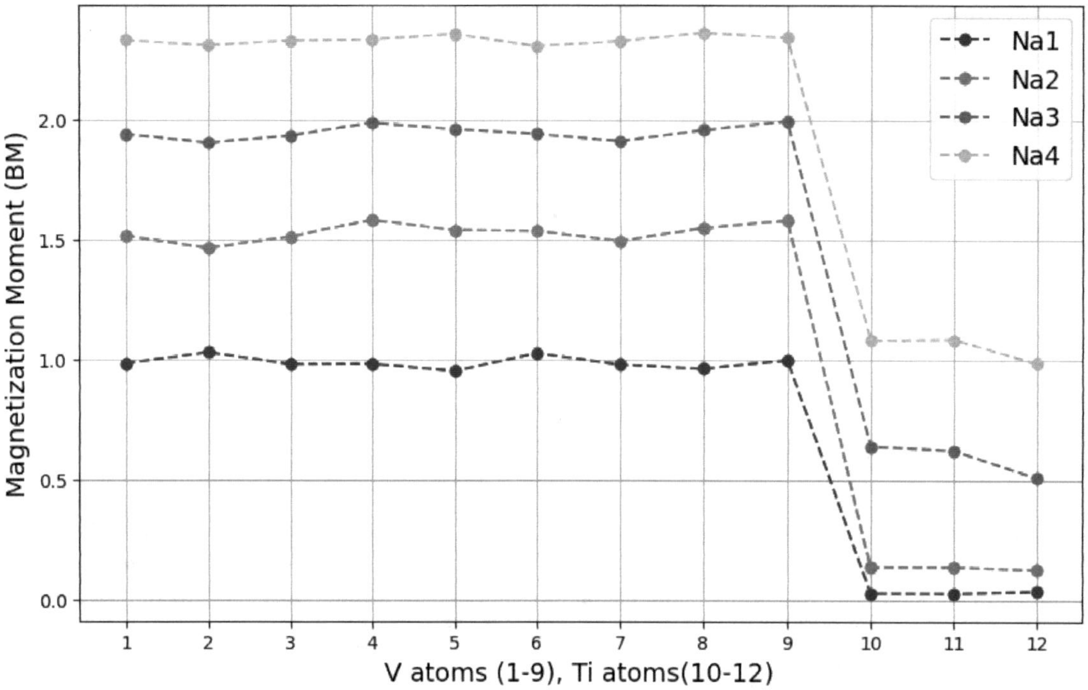

(b)

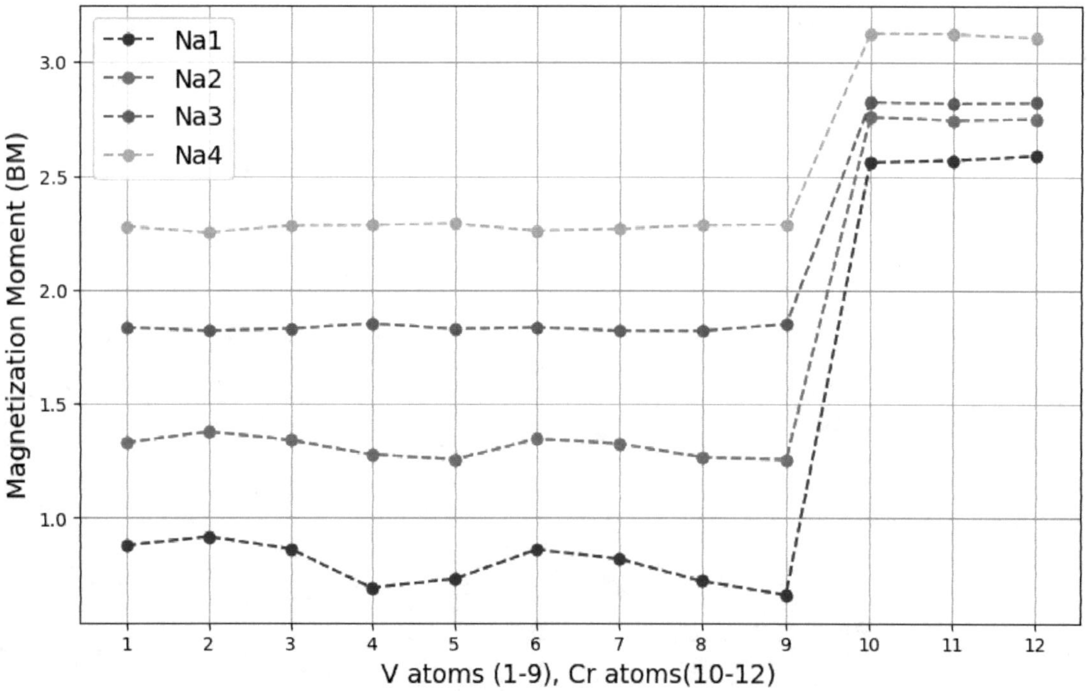

(c)

Figure 4 Magnetization moment profiles of undoped (a), Ti-doped (b), and Cr-doped (c) NVP

3.5 Density of States

Density of States (DOS) analysys was carried out for the transition metal atoms Vanadium, Titanium and Chromium at various stages of charge filling and depletion (Na1, Na2, Na3, Na4) in order to obtain a representation of the electronic condition. The results of the DOS analysis, as shown in Figure 5, indicate that the highest density level change at the Fermi level occurs in vanadium atoms. The variation in density levels in each of these conditions indicates that the main reduction-oxidation (redox) reaction mechanism is dominated by vanadium atoms, not other atoms.

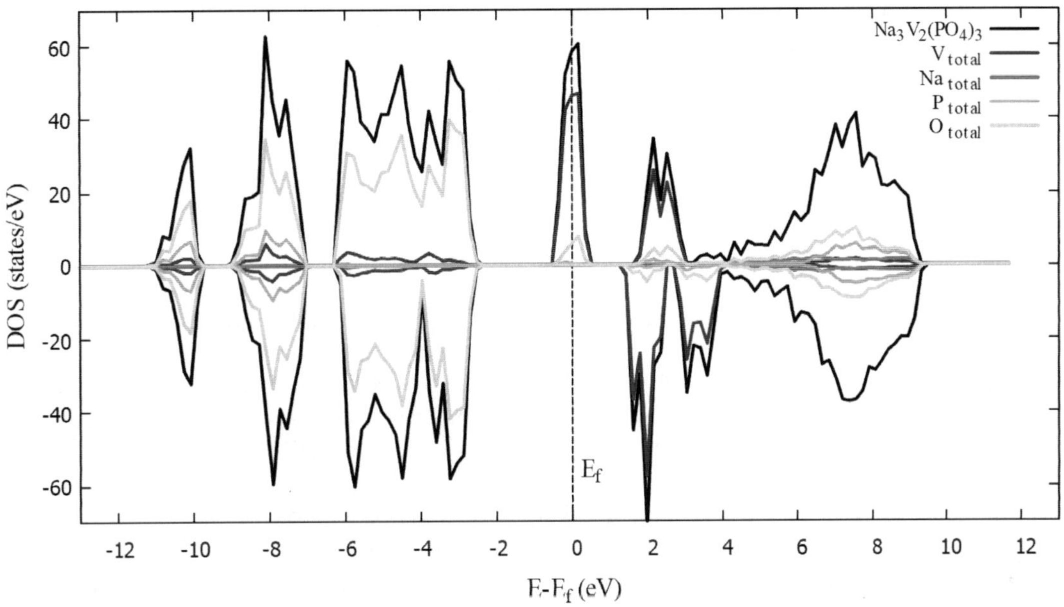

(a)

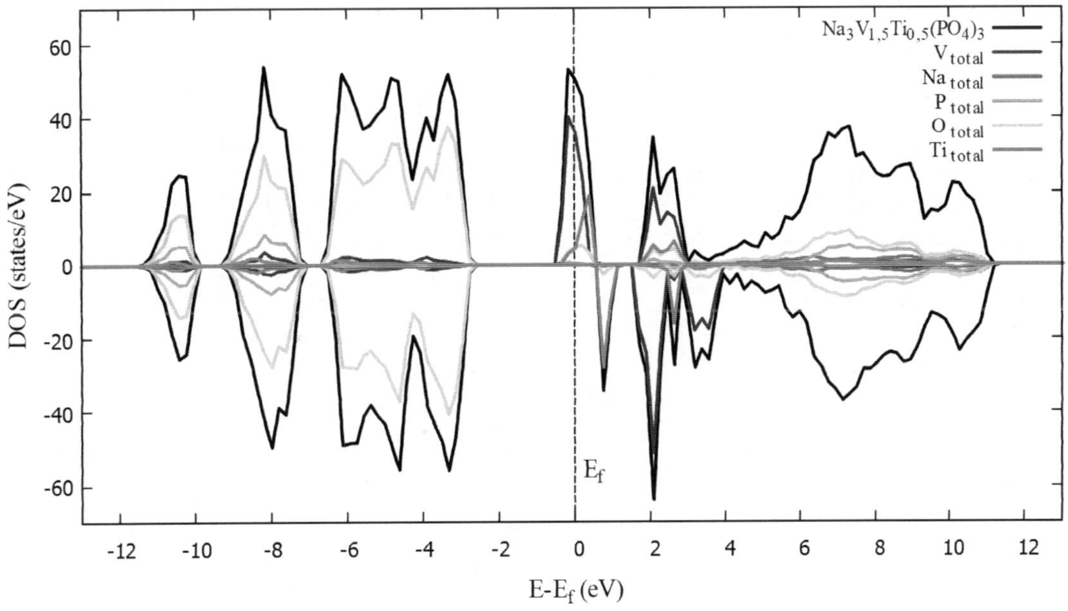

(b)

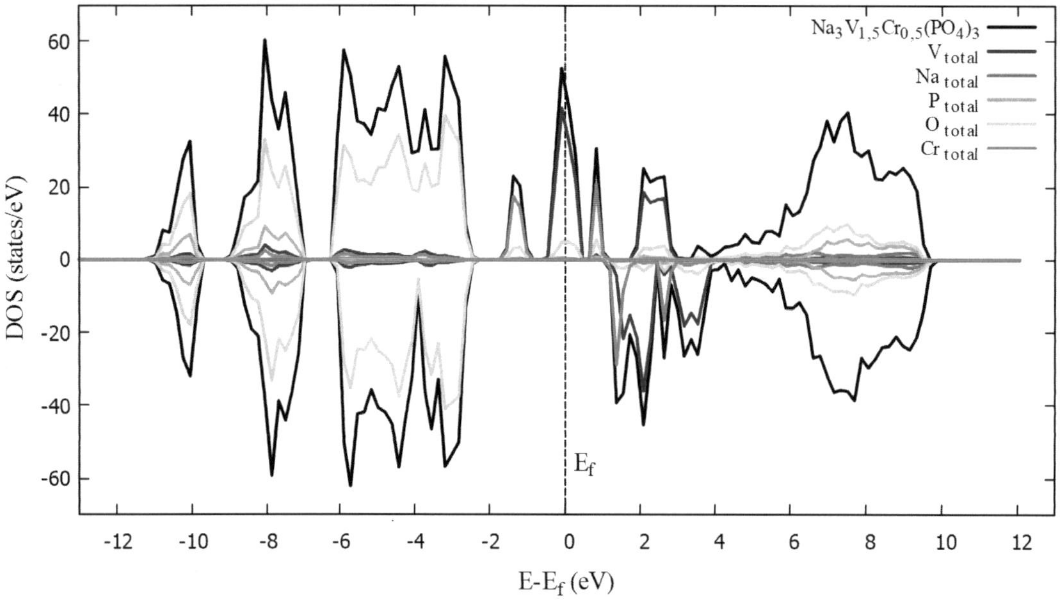

(c)

Figure 5 Density of States of undoped (a), Ti-doped (b), and Cr-doped (c)

The DOS plot in Figure 5 shows that despite the addition of Titanium and Chromium dopants, the V3d state remains dominant around the Fermi level. This indicates that vanadium continues to act as the main redox cation during the desodiation process, especially in the initial stage (Na3.0 to Na2.0). However, there is an increase in the filled 3d Cr state in the energy range of -2.0 eV to -1.0 eV in the Chromium-doped NVP. This filled 3d Cr state is thought to be a contributing factor to the increase in the NVP intercalation voltage. Chromium provides more electrons to the modified NVP, thereby increasing the voltage. When chromium is doped into NVP, there is a change in the electronic structure of the material which is reflected in the DOS graph. This can be seen from the presence of new peaks or an increase in the intensity of existing peaks near the Fermi Energy. This increase in the number of energy states will facilitate electron transfer during the intercalation and deintercalation of Na+ ions.

Titanium doping causes a decrease in the number of energy states available around the Fermi level. This can be seen from the decrease in peak intensity or the appearance of valleys near Ef in the DOS graph. This decrease in the number of energy states will hinder electron transfer during the intercalation and deintercalation of Na+ ions. In other words, it will be more difficult for electrons to move from the electrode to the Na+ ions entering the NVP structure during battery charging, and vice versa. This hindrance in electron transfer will reduce the rate of electrochemical reactions and ultimately decrease the intercalation voltage of the material. This result is consistent with a previous study by Gu et al. (2019) which showed a continuous increase in battery performance with Ti doping variations of 0.02, 0.06, and 0.1 (with the 0.1 variation showing the most optimal battery performance). However, at Ti 0.14, battery performance actually decreased [18].

3.6 Research Insigths

Formation energy indicates the thermodynamic stability of the cathode material. Negative formation energy shows that the mixed state is more stable than the unmixed state, indicating that the material will naturally tend to form a homogeneous mixture. If the formation energy is positive, there is a tendency for the components to separate into different phases, which can affect the performance of the material. This study shows that all variations of doping show a decrease in formation energy, thus increasing thermodynamic stability.

From the perspective of lattice parameter changes, the fact that the volume change for all variations remains below 8.8% is considered as insignificant volume change, it indicates that structural stability can be maintained. It can be concluded that thermodynamically stable material and insignificant volume change is less likely to undergo unwanted chemical reactions during charging and discharging cycles. This can lead to a longer cycle life, meaning the battery can withstand more charge-discharge cycles before its performance degrades significantly.

Intercalation voltage indicates how much energy can be stored in the battery. A higher voltage indicates that more energy can be extracted per unit charge transferred, which leads to a higher energy density. This means the battery can store more energy for the same weight or volume, which is a significant advantage for many applications. Adding more electrons (through doping) changes how the material undergoes reduction and oxidation (redox reactions), which in turn affects its magnetic properties and increases the intercalation voltage. The density of states graph suggests that adding chromium to the material improves its performance, likely by increasing the number of electrons in a specific energy level (3d sub-orbital) of the NVP.

4. Conclusion

Our findings revealed that both Titanium and Chromium doping significantly influence the intercalation voltage and formation energy of the cathode materials. Specifically, Chromium doping was found to increase the intercalation voltage, while Titanium doping decreased it. Additionally, both dopants decreased the formation energy, indicating that doping can enhance the stability of the material. These insights are crucial for the development of improved cathode materials for sodium-ion batteries. By understanding the impact of dopant variation on these key cathode parameters, we can guide the determination of the optimal dopant for synthesis and experimental trials in future research.

Acknowledgements

G.S. would like to acknowledge financial research support granted by the Ministry of Research, Technology and Higher Education (MENRISTEKDIKTI) of the Republic of Indonesia under the 'Riset Fundamental 2024' funding scheme no. 036/E5/PG.02.00.PL/2024. Financial research support granted by the Bandung Institute of Technology under the "Riset Unggulan 2023" scheme no. 308/IT1.B07.1/TA.00/2023 is also acknowledged.

References

[1] Hwang, J.Y., Myung, S.-T., & Sun, Y.-K. (2014). Sodium-ion batteries: present and future. The Royal Society of Chemistry.

[2] Hafiz, N.S.M., Singla, G., & Jha, P. (2022). Next generation sodium-ion battery: A replacement of lithium. *Materials Today: Proceedings.*

[3] Rudola, A., Reene, A.J., Heap, M., Meysami, S.S., Lowbridge, A., Mazzali, F., Sayers, R., Wright, C.J., & Barker, J. (2019). Commercialisation of high energy density sodium-ion batteries: Faradion's journey and outlook. Royal Society of Chemistry (RSC).

[4] Xue, Z., Du, S., Shan, M., Wang, H., Zhang, J., Cui, G., Hu, Z.X., Gordon, K.C., Xu, G., & Zhu, M. (2024). Functional separator materials of sodium-ion batteries: Grand challenges and industry perspectives. *Nano Today.*

[5] Zhang, X., Rui, X., Chen, D., Tan, H., Yang, D., Huang, S., & Yu, Y. (2019). Na3V2(PO4)3: an advanced cathode for sodium-ion batteries. *Nanoscale*, 11, 2556-2576.
[6] Mathato, S., Das, S., Gupta, D., & Biswas, K. (2024). Vanadium substituted Fe, Cr co-doped high performance C/Na3V2(PO4)2F3 cathode for sodium-ion batteries. *Journal of Electroanalytical Chemistry*, 955, 118046.
[7] Lu, F., Wang, J., Chang, S., He, L., Tang, M., Wei, Q., Mo, S., & Kuang, X. (2022). New-type NASICON-Na4FeV(PO4)3 cathode with high retention and durability for sodium ion batteries. *Carbon*, 196, 562-572.
[8] Zhu, Y., Xu, H., Pengdong, C., Chen, Y., Chen, J. (2023). "The recent advances of NASICON-Na3V2(PO4)3 cathode materials for sodium-ion batteries." Journal of Solid State Chemistry, 317, Part A, 123669.
[9] Anke, M., Seifert, M. (2004). "Titanium." In Elements and Their Compounds in the Environment (eds E. Merian, M. Anke, M. Ihnat, M. Stoeppler), Wiley-VCH.
[10] Yang, M., Wang, L., Miao, L., Li, C., Han, J., Huang, Y. (2018). "Superior Na-ion storage achieved by Ti substitution in Na3V2(PO4)3." Energy Storage Materials, 15, 2405-8297.
[11] Ruan, Y., Wang, K., Song, S. et al. (2017). "Improved structural stability and electrochemical performance of Na3V2(PO4)3 cathode material by Cr doping." Ionics, 23, 1097-1105.
[12] Nenni, A. G., Saputro, G. N., Mobarake, N. N., Aprilyanti, F. D., Nuruddin, A., Dipojono, H. K. (2023). "Tuning Na adsorption on the edge of a graphitic nanoribbon by incorporating a functionalized-ligand and single-heteroatom dopant." New Journal of Chemistry, 8, 3773-3782.
[13] Ravanny, W. M., Komalig, G., Shukri, G., Agusta, M. K., Saputro, A. G., Sumboja, A., Dipojono, H. K. (2023). "Enhanced Lithium Diffusivity in Reduced Cerium Oxides: A First-Principles Study." J. Phys. Chem. C, 126, 3328-3338.
[14] Shukri, G., Saputro, A. G., Tarabunga, P. S., Panjaitan, F. V., Agusta, M. K., Nuruddin, H. K. (2022). "Anisotropic Li diffusion in pristine and defective ZnO bulk and (1010) surface." Solid State Ionics, vol. 385, 116205.
[15] Rauch, T., Lodeiro, L. (2022). "DensityTool: A post-processing tool for space- and spin-resolved density of states from VASP." Computer Physics Communications, 277, 108384.
[16] Shen, F., Huang, Q., Tietz, F., Hao, J., Huang, C. T., Hernandha, F. H., Rahman, C. K., Lo, Y. C., Kuei, W. W. (2023). "In Situ Atomic-Scale Investigation of Structural Evolution During Sodiation/Desodiation Processes in Na3V2(PO4)3-Based All-Solid-State Sodium Batteries." Advanced Science, DOI: 10.1002/advs.202301490.
[17] Singh, B., Wang, Z., Park, S., Gautam, G. S., Chotard, J. N., Croguennec, L., Carlier, D., Cheetham, A. K., Masquelier, C., Canepa, P. (2021). "A chemical map of NaSICON electrode materials for sodium-ion batteries." Journal of Materials Chemistry, 9, 2821-292.
[18] Gu, Y., Xiao, F., Xu, K., Pan, X., Li, J., Xu, C., Zhou, X. (2019). "Outstanding electrochemical performance of sodium vanadium phosphate cathode co-modified by carbon-coating and titanium-doping for Na-ion batteries." Ceramics International, 45, DOI: 10.1016/j.ceramint.2019.03.074.

Effect of Iron-Alloyed Concentration Modification on Intercalation Voltage in Sodium-ion Battery Cathodes $Na_xV_{2-y}Fe_y(PO_4)_3$

Afthon Ilman Huda Isyfi[1,2*], Ganes Shukri[1,3] Mohammad Kemal Agusta[1,3***], Qory Cipta Scientika[1], Himawan Widiyanto[1,2], and A.Hafizh Rifa'i[1,2]**

[1]Nano and Quantum Technology Research Group, Faculty of Industrial Technology, Bandung Institute of Technology, Jalan Ganesa No. 10, Bandung, West Java, Indonesia
[2]PT Perusahan Listrik Negara (Persero), Jakarta, Indonesia
[3]Research Centre for Nanoscience and Nanotechnology (RCNN), Bandung Institute of Technology, Jalan Ganesa No. 10, Bandung, West Java, Indonesia

*E-mail: afthon.isyfi@pln.co.id
**E-mail: ganes_tf07@itb.ac.id
***E-mail: kemal@itb.ac.id

Abstract. The NASICON-type $Na_3V_2(PO_4)_3$ (NVP) is a promising cathode material for sodium-ion batteries (SIBs), recognized for its high sodium conductivity, thermal stability, and energy density. The 3D NASICON framework supports efficient Na-ion (de)intercalation during battery cycling. Recent efforts have aimed to enhance NVP's electronic conductivity and operating voltage for improved electrochemical performance. In this study, we employ first-principles calculations to investigate the effects of partially substituting vanadium (V) with iron (Fe) alloy on the intercalation voltage of Fe-alloyed NVP. Our focus is on determining the most thermodynamically stable phase for both pristine NVP and Fe-alloyed $Na_xV_{2-y}Fe_y(PO_4)_3$ (NVFP) at varying Fe levels (x = 0.25, 0.5, and 1) and Na compositions (y = 1, 3, 4). We observe that Fe alloy modifies the electronic environment, impacting the distribution of electronic states and enhancing the intercalation voltage. Pristine NVP, with a lower intercalation voltage range (1.38 to 2.14 V), sees a significant voltage increase with Fe alloying: 1.61–2.37 V at 25% Fe, 1.84–2.95 V at 50% Fe, and 2.53–3.76 V at 100% Fe. The presence of Fe leads to stronger Fe–O bonds, shifts in the crystal field, and an increase in magnetic moment-particularly at higher Fe levels, where a high-spin state for Fe is observed. Additionally, V-O and Fe-O bond lengths expand with higher Fe concentrations, indicating possible lattice expansion. This structural change contributes to an enhanced intercalation voltage, correlating magnetic moment increases with performance gains. This study highlights Fe alloying as a promising strategy to optimize both intercalation voltage and stability in Fe-alloyed NVP, advancing its potential as an SIB cathode material.

1. Introduction

Lithium-ion batteries (LIBs) currently represent the most advanced and widely adopted energy storage technology. The significance of LIBs in the global market is highlighted by the Global Li-ion Battery Market report published by MarketsandMarkets in 2022. This report indicates that the global LIB market was valued at $36.7 billion in 2020 and is projected to grow at an annual rate of 16.6% until 2027 [1].

As a type of rechargeable battery, LIBs have been specifically designed and continuously refined to meet the increasing energy demands of portable electronic devices, electric vehicles, and energy storage power stations. This widespread adoption is primarily attributed to their exceptional properties, including high energy density, high voltage, and environmentally friendly characteristics [2-11]. While the large-scale application of LIBs in energy storage systems may be constrained by the high cost of lithium minerals [2,12,13], it is essential to explore alternative and more sustainable electrochemical battery technologies [2,14,15,16,17].

Sodium-ion batteries (NIBs) have garnered significant attention as a viable alternative to lithium-ion batteries (LIBs), primarily due to the abundance and lower cost of sodium. Sodium is the second most abundant element in the Earth's crust, accounting for approximately 2.6% by weight [18]. Sodium is also significantly cheaper than lithium, with sodium chloride priced at around $100 per ton compared to lithium carbonate at approximately $10,500 per ton in 2020, making sodium around 105 times less expensive than lithium [19]. In addition to cost benefits, NIBs are considered safer than LIBs because Na ions exhibit a lower standard reduction potential (-2.71 V versus -3.04 V for Li), which reduces the risk of thermal runaway and fires due to battery failures [20].

Among the various cathode materials for NIBs, NASICON-type $Na_3V_2(PO_4)_3$ (NVP) has emerged as a leading candidate due to its robust three-dimensional framework, which supports high Na^+ ion mobility and ensures long-term structural stability. The NASICON structure consists of corner-shared VO_6 octahedra and PO_4 tetrahedra linked by oxygen atoms, forming large interstitial spaces that facilitate rapid Na^+ ion intercalation and deintercalation [21]. This structure not only enables high ionic conductivity but also contributes to thermal stability, which is a key requirement for advanced battery systems [22]. However, despite these advantages, NVP's inherently low electronic conductivity remains a primary challenge, limiting its broader adoption in high-performance NIB applications.

To address this limitation, various transition metals, including iron (Fe), have been investigated as dopants to enhance the electronic conductivity and tune the electrochemical properties of NVP. Partial substitution of vanadium (V) with Fe has been shown to significantly affect the structural stability and intercalation voltage of NVP. For instance, Liu et al. (2019) demonstrated that Fe Alloy at the vanadium site enhances electronic conductivity and Na^+ ion diffusion, thus improving the material's overall electrochemical performance [23]. However, the large separation between VO_6 octahedra within the NASICON framework remains a barrier to achieving optimal electronic conductivity [24].

Recent studies have demonstrated that Fe-alloy NVP materials exhibit improved capacity and voltage profiles. For example, Wu et al. (2020) reported a capacity of 91.2 $mAhg^{-1}$ at a 20C rate for Fe-alloy NVP [25], while Novikova et al. (2018) achieved a capacity of 117.6 $mAhg^{-1}$ at a 0.1C rate for Fe-alloy samples synthesized using a sol gel method [26]. Building on these findings, this study systematically investigates the impact of Fe Alloy on the intercalation voltage and Na^+ ion diffusion in NVP, employing first-principles calculations and density functional theory (DFT) to gain deeper insight into these effects.

2. Computational Details

In this work, we conducted all calculations using spin-polarized density functional theory (DFT) within the projector augmented wave (PAW) framework, as implemented in the Vienna *Ab-initio* Simulation Package (VASP) code [27]. We utilized the PAW pseudopotentials for Na ($2s^22p^63s^1$), V ($3s^23p^63d^34s^2$), P ($3s^23p^3$), O ($2s^22p^4$), and Fe ($3d^64s^2$). To examine exchange interactions and

Journal of Physics: Conference Series **2980** (2025) 012023 doi:10.1088/1742-6596/2980/1/012023

electronic correlations, we utilized the Perdew–Burke–Ernzerhof (PBE) form of the generalized gradient approximation (GGA). We investigated four models: pristine Na-based oxides, Na-based oxides with 25% Fe Alloy, Na-based oxides with 50% Fe Alloy, and Na-based oxides with 100% Fe Alloy, as shown in Figure 1.

All self-consistent electronic structure calculations were performed with a convergence limit of 10^{-4} and a kinetic energy cutoff of 520 eV was employed for the expansion of Kohn–Sham wave functions. Convergence for the relaxation of both electronic and ionic structures was achieved with thresholds of 1×10^{-4} and 1×10^{-2} eV/Å, respectively.

To determine how easily Fe alloys could be incorporated into the $Na_3V_2(PO_4)_3$ structure, formation energy calculations were performed. We evaluated the desodiation stability by calculating the formation energy, derived from the development of the Gibbs free energy equation to assess the spontaneity of the reaction at 0 K, which is defined as:

$$\Delta E_f\left(Na_xVP(S)\right) = E[Na_xVP(S)] - \left(\frac{4-x}{3}\right)E[Na_1VP(S)] - \left(\frac{x-1}{3}\right)E[Na_4VP(S)] \tag{1}$$

$E[Na_xVP(S)]$ represents the bulk energy of NVP under partial (de)intercalation conditions, while $E[Na_1VP(S)]$ and $E[Na_4VP(S)]$ correspond to the energy at full intercalation and (de)intercalation states. $E[Na_xVP(S)]$ is calculated based on the lowest energy structure for each x value among various Na^+ vacancy configurations [28].

Additionally, the equilibrium intercalation voltage as a function of Na^+ concentration was calculated using the total energy results obtained from the DFT calculations. A thermodynamic equation was applied to predict the intercalation voltage, which is directly associated with the performance of the cathode during Na^+ ion insertion and extraction. In cathode materials, a high intercalation voltage is desirable; however, it must remain below the electrolyte's potential limit to ensure stability.

The intercalation voltage of Na in NVP is analyzed based on the reaction that represents the reversible process of Na^+ insertion into and extraction from the N_xVP structure.

$$Na_{x_1}V_2(PO_4)_3 + x_2Na^+ + x_2e^- \overset{\Delta G^o}{\longleftrightarrow} Na_{x_1+x_2}V_2(PO_4)_3 \tag{2}$$

With x_1 and x_2 represent the concentration/number of Na^+ ions in the cathode material $Na_xV_{2-1}Fe_x(PO_4)_3$, the intercalation voltage (V) can be expressed as follows.

$$V = -\frac{\Delta G^o}{(x_1-x_2)\,F} \tag{3}$$

$$V = -\frac{E\left(Na_{x_1}V_2(PO_4)_3\right)-\left[E\left(Na_{x_2}V_2(PO_4)_3+(x_1-x_2)\,\mu_{Na}\right]\right.}{(x_1-x_2)\,F}\text{, for NVP} \tag{4}$$

$$\text{And } V = -\frac{E\left(Na_{x_1}V_{2-y}Fe_y(PO_4)_3\right)-\left[E\left(Na_{x_2}V_{2-y}Fe_y(PO_4)_3\right)+(x_1-x_2)\,\mu_{Na}\right]}{(x_1-x_2)\,F}\text{, for NVFP} \tag{5}$$

Where $E\left(Na_{x_1}V_{2-y}Fe_y(PO_4)_3\right)$ represents the total energy of Na at x_1, $E\left(Na_{x_2}V_{2-y}Fe_y(PO_4)_3\right)$ represents the total energy of Na at x_2, μ_{Na} is the chemical potential of the Na crystal, and F is the Faraday constant, with a value of 96,845 C/mol or 26,801 mAh/mol [28].

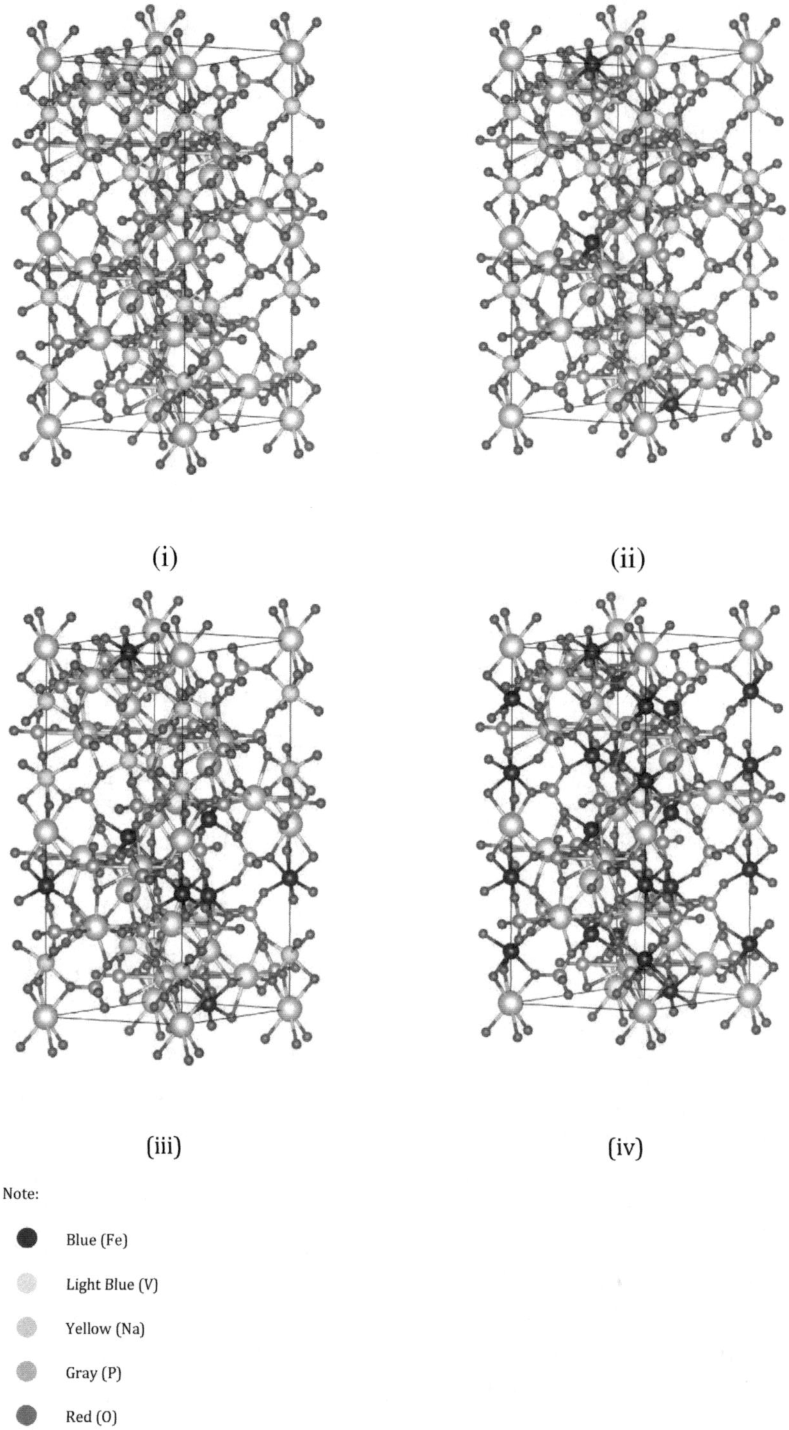

Figure 1. Visualization of $Na_3V_2(PO_4)_3$ at the pristine condition (i), $Na_3V_{1.5}Fe_{0.5}(PO_4)_3$ (ii), $Na_3VFe(PO_4)_3$ (iii), and $Na_3Fe_2(PO_4)_3$ (iv).

3. Results and Discussion

3.1 Structural Properties

The data presented in Table 1 reveal that Fe Alloy significantly influences the lattice parameters and cell volume of Na-based materials (Na_1, Na_3, and Na_4), as calculated using the PBE method. The effects of Fe Alloy on these materials extend beyond observable physical changes in lattice parameters (a, b, c) and involve fundamental mechanisms related to energy stability and interatomic interactions within the crystal structure. The thermodynamic stability of a material largely depends on how atoms interact to achieve their lowest energy configuration [28]. When Fe is introduced as a dopant, it alters the balance of attractive and repulsive forces between atoms. At low to moderate Alloy concentrations (25% and 50%), these changes tend to be localized and controlled, resulting in consistent patterns of expansion or contraction. This suggests that, despite modifications in the distribution of electrostatic forces, the crystal structure can adapt without compromising its overall stability.

However, at full Alloy concentration (Fe 100%), the effects become more pronounced and may induce more significant structural changes. Stronger interactions between Fe atoms and surrounding Na atoms can lead to a more substantial redistribution of forces, resulting in a realignment of atomic positions within the crystal lattice. This can cause the structure to undergo contraction or slight expansion, depending on the specific material type (Na_1, Na_3, and Na_4). Such phenomena indicate a process of structural adjustment aimed at achieving a new state of energy stability, highlighting the complex dynamics of interatomic interactions introduced by Fe Alloy.

Tabel 1. Lattice constants (in Å and °) and volume (in $A^{\circ 3}$) computed with PBE

Na	Alloy	a [Å]	b [Å]	c [Å]	α [°]	β [°]	γ [°]	V [Å^3]
Na_1	Pristine	8.48	8.48	21.34	90.34	89.65	120.06	1331.38
	Fe_25%	8.49	8.49	21.35	90.03	89.97	120.00	1334.64
	Fe_50%	8.48	8.48	21.42	90.00	89.96	120.00	1334.56
	Fe_100%	8.45	8.45	21.39	90.09	89.90	119.92	1325.22
Na_3	Pristine	8.74	8.74	21.26	90.24	89.75	120.19	1405.92
	Fe_25%	8.71	8.70	21.30	89.96	90.00	120.08	1398.50
	Fe_50%	8.75	8.75	21.23	90.21	89.71	120.18	1406.29
	Fe_100%	8.75	8.75	21.27	90.29	89.70	120.16	1408.55
Na_4	Pristine	8.86	8.86	21.11	90.00	90.00	120.00	1438.49
	Fe_25%	8.86	8.86	21.08	90.00	90.00	120.00	1435.90
	Fe_50%	8.84	8.84	21.04	90.00	90.00	120.00	1427.07
	Fe_100%	8.87	8.87	20.98	90.00	90.00	120.00	1432.93

3.2 Electrochemical Stability

The data in Table 2 highlights the significant effect of Fe alloying on the formation energy of Na-based materials. The Na_3 condition in NVFP with 100% Fe alloying exhibits the lowest formation energy, with a minimum value of -40.88 meV/atom. The presence of Fe contributes to a more stable bonding environment within the material's structure by reducing lattice strain and

Journal of Physics: Conference Series **2980** (2025) 012023 doi:10.1088/1742-6596/2980/1/012023

lowering the total system energy. This effect is attributed to the size and electronic interactions of Fe, which are highly compatible with the phosphate framework, resulting in minimal distortion and enhanced ionic interactions. Consequently, the material demonstrates greater resistance to spontaneous phase transitions or structural degradation. This exceptionally low formation energy confirms that NaFP with 100% Fe alloying achieves the highest thermodynamic stability.

The incorporation of Fe in $Na_xV_{2-y}Fe_y(PO_4)_3$ not only enhances the thermodynamic stability of the material but also facilitates its formation process. As the Fe content increases, the energy required to synthesize this compound from its precursors becomes progressively lower. This effect can be attributed to the role of Fe in stabilizing the intermediate structure during the synthesis process. Furthermore, the compatibility of Fe with the phosphate framework reduces the energy barriers for nucleation and crystal growth. Consequently, the formation of this compound becomes easier, more efficient, and more favorable compared to other competing phases.

Table 2. Formation Energy (meV/atom) and intercalation voltage (V) computed with PBE

Alloy	Na Variation	Formation Energy (meV/atom)	Intercalation Voltage (V)
Pristine	Na_1	0.00	2.14
	Na_3	-25.18	2.14
	Na_3	-25.18	1.38
	Na_4	0.00	1.38
25% Fe	Na_1	0.00	2.37
	Na_3	-25.27	2.37
	Na_3	-25.27	1.61
	Na_4	0.00	1.61
50% Fe	Na_1	0.00	2.95
	Na_3	-37.03	2.95
	Na_3	-37.02	1.84
	Na_4	0.00	1.84
100% Fe	Na_1	0.00	3.76
	Na_3	-40.88	3.76
	Na_3	-40.88	2.53
	Na_4	0.00	2.53

The figure 2 illustrates that Fe Alloy in $Na_xV_{2-y}Fe_y(PO_4)_3$ decreases the formation energy, indicating enhanced thermodynamic stability. The pristine material has a higher formation energy (~-25 meV) compared to all Fe-alloyed samples at each x value. Stability increases with higher Fe concentrations, with minimum formation energies of -37 meV (25% Fe), -40 meV (50% Fe), and -45 meV (100% Fe) at x = 3, indicating that 100% Fe Alloy provides the highest stability, while the pristine material has the lowest stability among all variations.

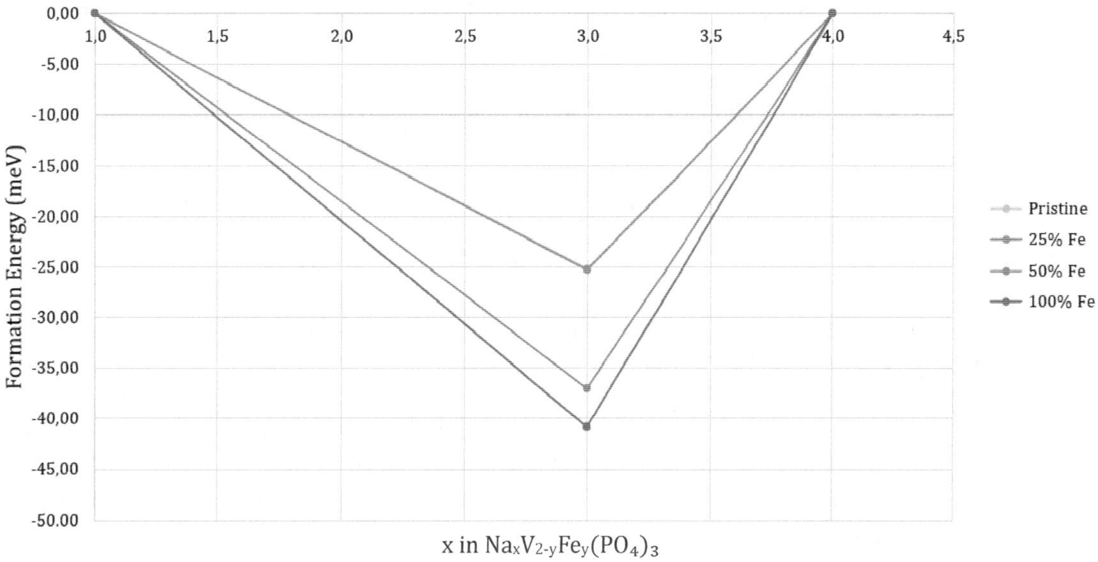

Figure 2. Formation energies of pristine and Fe-alloyed NVP materials for all Na/vacancy ordering vs. Na content (x) in $Na_xV_{2-y}Fe_y(PO_4)_3$

3.3 Intercalation Voltage

The intercalation voltage significantly influences a battery's energy density. Materials with higher intercalation voltages typically deliver greater energy density, making them ideal for applications demanding efficient energy storage [28].

The data in Table 2 and Figure 3 reveal the effect of Fe alloy concentration on the intercalation voltage of $Na_xV_{2-y}Fe_y(PO_4)_3$ at various sodium molar fractions (x). In the pristine material (without Fe alloying), the intercalation voltage remains stable at 2.14 V for x = 1 and x = 3 but decreases to 1.38 V at x = 4. The introduction of Fe alloy significantly enhances the intercalation voltage across all Fe alloy concentrations. The voltage increases to 2.37 V for 25% Fe alloy ($Fe_{0,25}$), 2.95 V for 50% Fe alloy ($Fe_{0,50}$), and reaches a maximum of 3.76 V for 100% Fe alloy (Fe_1) at x = 1 and x = 3.

The widest intercalation voltage range of 2.15 Volts is achieved with 50% Fe alloy. Compared to the $V_{pristine}$ condition, variations in Fe alloy consistently enhance the intercalation voltage at $V_{Fe25\%}$, $V_{Fe50\%}$, and $V_{Fe100\%}$. This indicates that the addition of Fe alloy actively contributes to increasing the intercalation voltage and plays a significant role in the intercalation and deintercalation processes.

A clear trend is observed, where the intercalation voltage increases with higher Fe levels, likely due to modifications in the electronic structure that enhance the material's electrochemical potential. However, this improvement is accompanied by trade-offs: higher Fe concentrations lead to increased formation energy, indicating reduced thermodynamic stability.

This effect arises because Fe substitution introduces lattice strain, which alters interatomic interactions and potentially compromises the structural integrity of the material. Furthermore, the decline in intercalation voltage at higher Na concentrations (x = 4) across all Alloy levels suggests that increased Na insertion induces structural stress, disrupting charge transfer and diminishing electrochemical performance. These findings underscore the necessity of balancing the benefits of enhanced voltage with the challenges of stability when optimizing Fe Alloy in Na-based cathode materials.

Journal of Physics: Conference Series **2980** (2025) 012023 doi:10.1088/1742-6596/2980/1/012023

The analysis shows that at 50% Fe concentration, the material achieves an optimal balance between improved thermodynamic stability and intercalation voltage while avoiding the trade-offs associated with excessive lattice strain typically observed at 100% Fe concentration. Thus, the 50% Fe alloy is ideal for applications requiring both high performance and long-term stability.

This is because, at 50% Fe concentration, the material effectively reduces formation energy through more stable bonding and lower lattice energy, stabilizes the electronic structure to enhance intercalation voltage without inducing instabilities, and maintains structural integrity with moderate lattice strain to prevent distortions during the Na intercalation and deintercalation processes.

The balance between enhancing intercalation voltage and maintaining material stability in Na-based cathodes can be achieved through an integrated approach. This approach includes determining the optimal Fe concentration for balanced stability and performance, engineering the material structure by introducing stabilizing elements or optimizing the NASICON framework, and managing Na insertion during operational cycles to reduce structural stress. Theoretical simulations and experimental validation play a crucial role in designing more efficient materials, while electrolyte optimization can improve system compatibility. With these strategies, Fe-alloyed Na-based cathodes can achieve maximum performance in increasing intercalation voltage without compromising long-term stability.

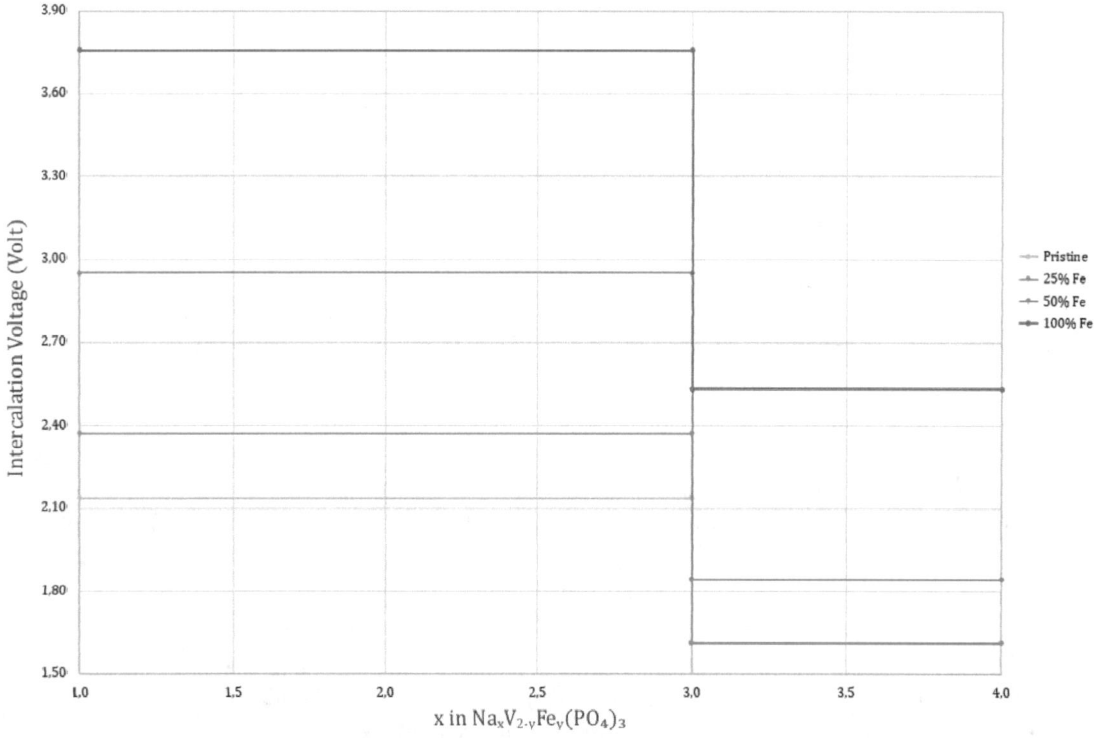

Figure 3. Voltage profiles of the Pristine and Fe-alloyed NVP cathode materials.

Journal of Physics: Conference Series **2980** (2025) 012023 doi:10.1088/1742-6596/2980/1/012023

3.4 Electrochemical Redox Behaviour

The electronic configuration of the material follows the Aufbau principle, Hund's rule, and the Pauli exclusion principle. The Aufbau principle states that electrons fill orbitals starting from the lowest energy level. Hund's rule indicates that electrons are added to orbitals or suborbital singly until each is half-filled before pairing occurs. The Pauli exclusion principle states that electrons, as fermions, cannot occupy the same quantum state; thus, each suborbital (quantum number l) can hold a maximum of two electrons with opposite spins (spin-up and spin-down).

Vanadium, a transition metal with 23 electrons, has the electronic configuration [Ar] $3d^3$ $4s^2$. During ion formation, electrons are first removed from the outer 4s orbital, despite the 3d orbital having a higher energy level. Vanadium exhibits multiple oxidation states, with the ionic electron configurations as follows: (I) V^{2+} is formed by losing both $4s^2$ electrons, (II) V^{3+} is formed by losing an additional electron from the 3d orbital, (III) V^{4+} results from losing two 3d electrons, and (IV) V^{5+} is formed by losing all three 3d electrons, resulting in the [Ar] configuration.

The change in Vanadium ion magnetic moment, as calculated using the PBE method, exhibits a relatively linear trend with varying Na concentration.

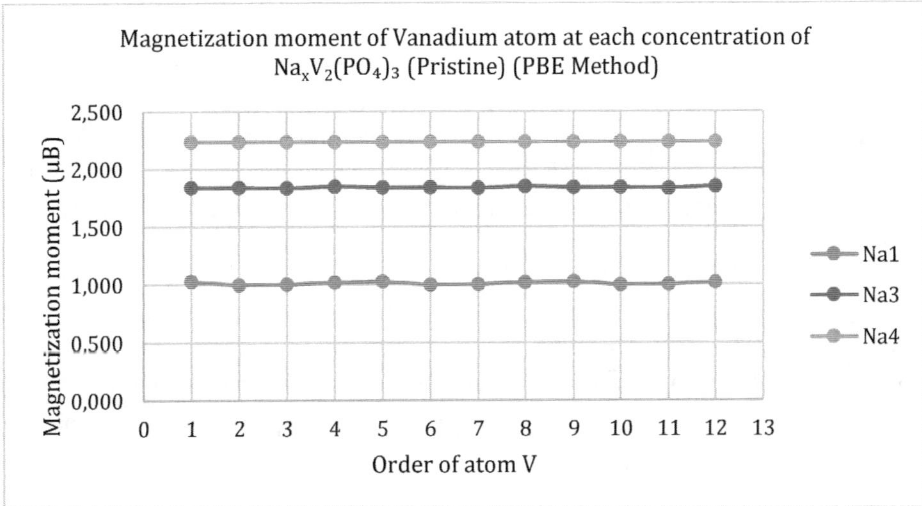

Figure 4. Magnetization moment of V atom at varying Na content (x) in $Na_xV_2(PO_4)_3$ (Pristine)

Figure 4 illustrates the magnetic moment of Vanadium atoms in $Na_xV_2(PO_4)_3$ (pristine NVP) at different Na concentrations (Na_1, Na_3, Na_4) using the PBE method. The magnetic moment values for each V atom range from 0.99 to 1.03 µB for Na_1, 1.83 to 1.85 µB for Na_3, and 2,234 µB for Na_4. These values remain consistent across all V atoms (1-12) without significant fluctuations, indicating stable magnetic configurations in pristine NVP. This stability suggests that pristine NVP has balanced spin interactions, resulting in homogeneous magnetic properties throughout the crystal structure.

Journal of Physics: Conference Series **2980** (2025) 012023 doi:10.1088/1742-6596/2980/1/012023

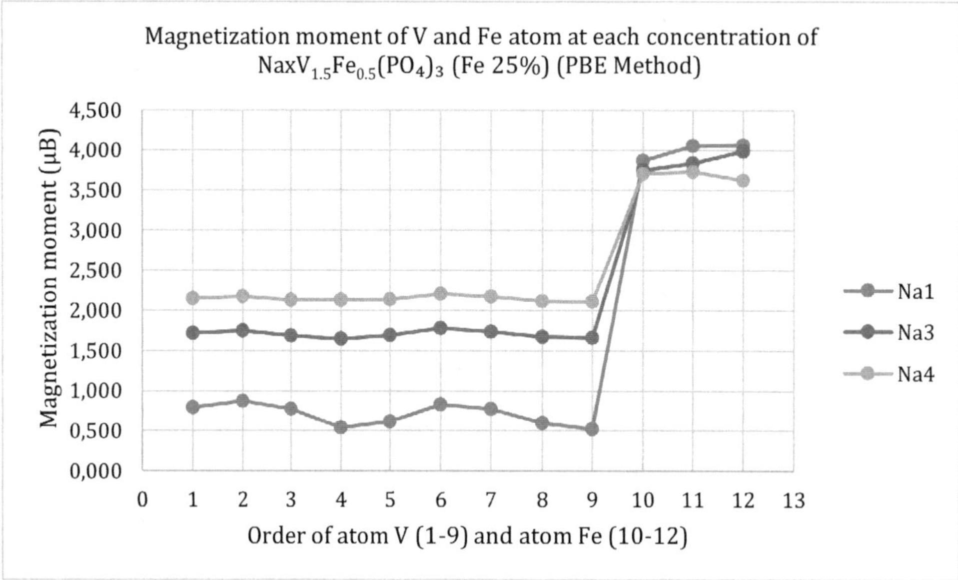

Figure 5. Magnetization moment of V and Fe atom at varying Na content (x) $Na_xV_{1.5}Fe_{0.5}(PO_4)_3$ Fe 25%

Figure 5 shows the changes in the magnetic moment of Vanadium (V) atoms in $Na_xV_{1.5}Fe_{0.5}(PO_4)_3$ with 25% Fe Alloy at three different Na positions (Na_1, Na_3, Na_4) using the PBE method. The results indicate that the magnetic moment of V atoms (positions 1-9) varies significantly at each Na position, with the highest values observed at Na_4. In contrast, the magnetic moment of Fe atoms (positions 10-12) remains high and constant, indicating a stable high-spin configuration. The increase in V magnetic moment after position 9 suggests a strong influence of Fe on the magnetic stability of V.

Overall, 25% Fe Alloy significantly modifies the magnetic moment distribution of V and Fe atoms, particularly enhancing the magnetic moment of Fe atoms and improving the overall magnetic stability of the material.

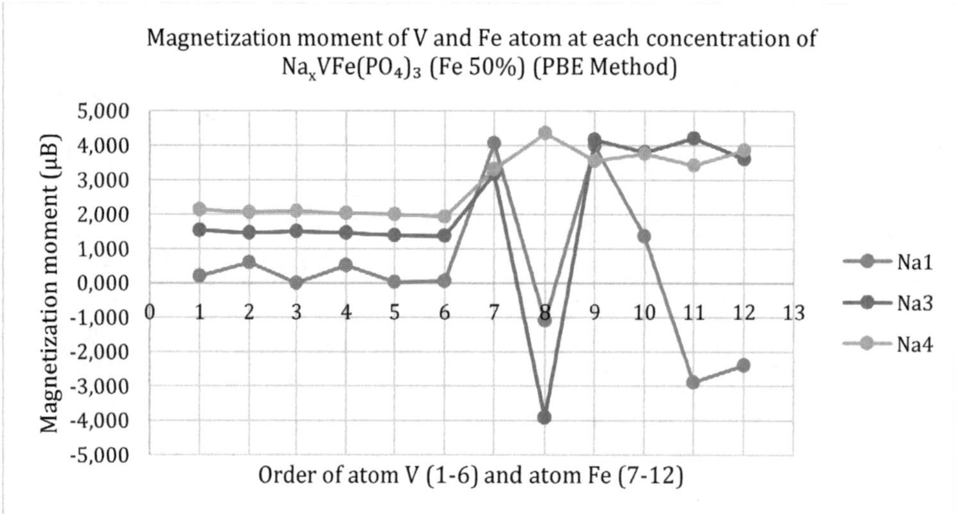

Figure 6. Magnetization moment of V and Fe atom at varying Na content (x) $Na_xVFe(PO_4)_3$ (Fe 50%)

Figure 6 shows that 50% Fe Alloy in $Na_xVFe(PO_4)_3$ causes significant changes in the magnetic moments of Vanadium (V) and Iron (Fe) atoms. The magnetic moment of V atoms (positions 1–6) increases from Na_1 to Na_4, indicating a transition from low-spin to high-spin states. Meanwhile, Fe atoms (positions 7–12) exhibit large variations in magnetic moment at Na_1 and Na_3, indicating spin configuration instability. At Na_4, however, the Fe magnetic moment is more stable with a high-spin configuration, contributing to the overall magnetic stability of the material. The interaction between Fe and V atoms results in an increased magnetic moment contribution at Na_4, providing higher magnetic stability compared to other positions.

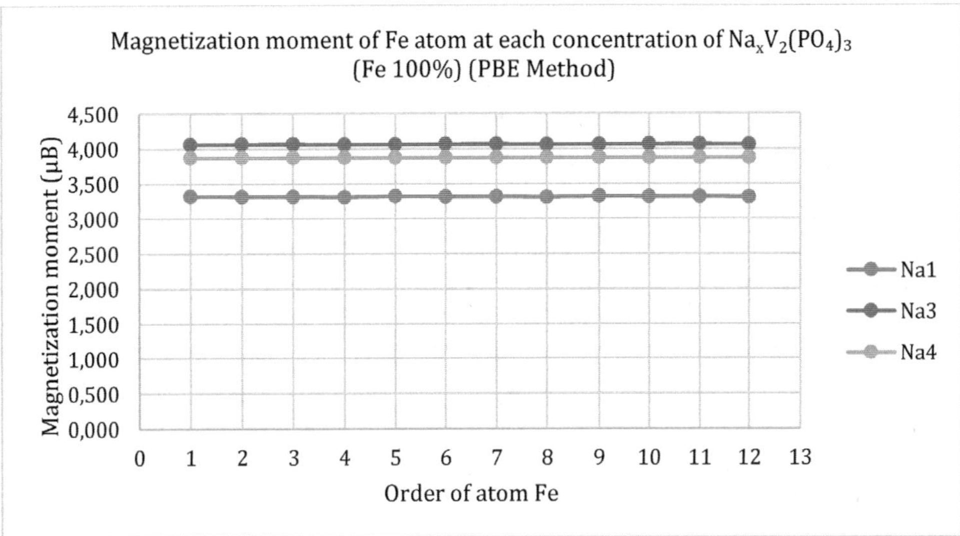

Figure 7. Magnetization moment of Fe atom at varying Na content (x) $Na_xFe_2(PO_4)_3$ (Fe 100%)

Figure 7 shows that the magnetic moment distribution of Fe atoms in $Na_xFe_2(PO_4)_3$ with 100% Fe Alloy exhibits high stability across all Na positions (Na_1, Na_3, Na_4). The magnetic moments remain constant at 3.3 μB (Na_1), 4.06 μB (Na_3), and 3.87 μB (Na_4), indicating a stable high-spin configuration of Fe atoms without significant fluctuations. These results demonstrate that 100% Fe Alloy enhances the overall magnetic stability of the material.

Journal of Physics: Conference Series **2980** (2025) 012023 doi:10.1088/1742-6596/2980/1/012023

Table 3. Electronic structure, magnetic moment and bond length of vanadium ion in NVP (Pristine) calculated with the PBE method for V-ion

V-ion	Orbitals	Variation	NVP	
			Magnetic Moment [μB]	V - O [Å]
V^{2+}/ V^{3+}	t_{2g}^3 t_{2g}^2	Na_4VP	2.23~2.23	2.02 ~ 2.06
V^{3+}	t_{2g}^2	Na_3VP	1.83~1.85	1.93 ~ 2.06
V^{4+}	t_{2g}^1	Na_1VP	0.99~1.03	1.86 ~ 1.97

The table 3 shows that the magnetic moment of V-ions increases with higher Na concentrations, indicating stronger magnetic interactions. Similarly, the V-O bond length increases as Na content rises, suggesting potential lattice expansion or structural distortion. This trend indicates a correlation between Na intercalation and the electronic and structural properties of Vanadium ions in NVP, highlighting the influence of Na concentration on both the magnetic and structural stability of the material.

Table 4. Electronic structure, magnetic moment, and bond length of vanadium ion in NVP and NVFP calculated with the PBE method 25% Fe and 50% Fe for V-ion

V-ion	Orbitals	Variation	$NV_{1.5}Fe_{0.5}P$ (25% Fe)		NVFP (50% Fe)	
			Magnetic Moment [μB]	V - O [Å]	Magnetic Moment [μB]	V - O [Å]
V^{2+}/ V^{3+}	t_{2g}^3 t_{2g}^2	Na_4VP	1.92~2.13	1.20 ~ 2.07	1.92~2.13	1.97 ~ 2.07
V^{3+}	t_{2g}^2	Na_3VP	1.37~1.50	1.91 ~ 2.07	1.37~1.52	1.90 ~ 2.07
V^{4+}	t_{2g}^1	Na_1VP	0.00~0.60	1.94 ~ 2.05	0.00~0.60	1.97 ~ 2.07

Table 5. Electronic structure, magnetic moment, and bond length of vanadium ion in NVP and NVFP. Calculated with the PBE method 25% Fe and 50% Fe for Fe-ion

Fe-ion	Orbitals	Variation	$NV_{1.5}Fe_{0.5}P$ (25% Fe)		NVFP (50% Fe)	
			Magnetic Moment [μB]	Fe – O [Å]	Magnetic Moment [μB]	Fe – O [Å]
Fe^{2+}/ Fe^{3+}	$t_{2g}^4\ e_g^2$; $t_{2g}3\ e_g2$; t_{2g}^5	Na_4VP	3.30~4.35	1.20 ~ 2.07	3.30~4.35	1.96 ~ 2.15
Fe^{3+}	$t_{2g}3\ e_g2$ t_{2g}^5	Na_3VP	3.75~3.99	1.91 ~ 2.07	-3.92~4.21	1.89 ~ 2.21
Fe^{4+}	$t_{2g}^4 e_g^0$	Na_1VP	3.87~4.06	1.94 ~ 2.05	-2.89~4.06	1.88 ~ 2.10

The table 4 and 5 shows that the magnetic moments and bond lengths of V and Fe ions show significant differences between 25% and 50% Fe Alloy levels, indicating changes in electronic configuration and magnetic stability. For Vanadium, the magnetic moments remain relatively stable, with only minor changes in V-O bond lengths, suggesting stable V-O bonds. In contrast, iron exhibits more pronounced changes in magnetic moments at 50% Fe, indicating alterations in spin states and stronger interactions with surrounding ions. The stability of high-spin configurations in Fe enhances the overall magnetic properties of the material, especially in Na_4VP, where the magnetic moments are highest for both Alloy levels.

Table 6. Electronic structure, magnetic moment and bond length of vanadium ion in NVP and NVFP calculated with the PBE method 100% Fe for Fe-ion

Fe-ion	Orbitals	Variation	Magnetic Moment [μB]	Fe – O [Å]
Fe^{2+}/ Fe^{3+}	$t_{2g}^4\ e_g^2$; $t_{2g}3\ e_g2$; t_{2g}^5	Na_4VP	3.872~3.872	1.981 ~ 2.126
Fe^{3+}	$t_{2g}^3\ e_g^2$ t_{2g}^5	Na_3VP	4.058~4.062	1.884 ~ 2.160
Fe^{4+}	$t_{2g}^4 e_g^0$	Na_1VP	3.308~3.317	1.871 ~ 2.098

The table 6 shows that Fe ions exhibit high and consistent magnetic moments across different Na compositions, indicating a stable high-spin configuration. The Fe-O bond lengths vary slightly, reflecting minor structural adjustments with different Fe oxidation states. The highest magnetic moment is observed in the Na_3VP phase, suggesting stronger magnetic interactions in this configuration. These results demonstrate that 100% Fe Alloy maintains stable magnetic properties in NVFP, with minor structural distortions.

Table 7. The compounds formed based on the magnetic moment values and V-O bond lengths in NVP and NVFP calculated with the PBE method

Na concentration, x =	Magnetic Moment [μB]			
	NV_2P	$NV_{1.5}Fe_{0.5}P$	NVFP	NFe_2P
Na =1	0.997~1.026	0.52~0.87	0.00~0.59	3.31~3.32
Na = 3	1.832~1.849	1.65~1.77	1.37~1.52	4,06~4.06
Na = 4	2.234~2.234	2.11~2.20	1.92~2.13	3.87~3.87

Na concentration, x =	Compound			
	NV_2P	$NV_{1.5}Fe_{0.5}P$	NVFP	NFe_2P
Na =1	$Na_1V^{4+}V^{4+}(PO_4)_3$	$Na_1V^{4+}Fe^{4+}(PO_4)_3$	$Na_1V^{4+}Fe^{4+}(PO_4)_3$	$Na_1Fe^{4+}(PO_4)_3$
Na = 3	$Na_3V^{3+}V^{3+}(PO_4)_3$	$Na_3V^{3+}Fe^{3+}(PO_4)_3$	$Na_3V^{3+}Fe^{3+}(PO_4)_3$	$Na_3Fe^{3+}(PO_4)_3$
Na = 4	$Na_4V^{2+}V^{3+}(PO_4)_3$	$Na_4V^{2+/3+}Fe^{2+/3+}(PO_4)_3$	$Na_4V^{2+/3+}Fe^{2+/3+}(PO_4)_3$	$Na_4Fe^{2+/3+}(PO_4)_3$

Table 7 shows that the magnetic moments and resulting compound compositions of NVP and NVFP at varying Na concentrations (x = 1, 3, 4), calculated using the PBE method.

a. At Na = 1, the magnetic moments are lower across all compounds, indicating the dominance of V^{4+} and Fe^{4+} ions. The resulting compounds are mainly composed of $Na_1V^{4+}V^{4+}(PO_4)_3$ and $Na_1V^{4+}Fe^{4+}(PO_4)_3$.

b. At Na = 3 the magnetic moments increase due to the presence of V^{3+} and Fe^{3+} ions, reflecting a shift in valence states. This results in compounds like $Na_3V^{3+}V^{3+}(PO_4)_3$ and $Na_3V^{3+}Fe^{3+}(PO_4)_3$, dan $Na_3Fe^{3+}(PO_4)_3$.

c. At Na = 4, the magnetic moments are highest, with a significant increase in V^{2+} and Fe^{2+} ions, forming compounds such as $Na_4V^{2+/3+}Fe^{2+/3+}(PO_4)_3$ and $Na_4Fe^{2+/3+}(PO_4)_3$.

The table demonstrates that increasing Na concentration (x) leads to higher magnetic moments due to changes in the valence states of V and Fe ions. The shift from V^{4+} and Fe^{4+} at x = 1 to V^{2+} and Fe^{2+} at x = 4 suggests a redistribution of charge and changes in structural stability, affecting the overall magnetic properties of the compounds.

3.5 Implications of NIBs for future energy storage solutions

Sodium-Ion Batteries (NIBs) offer several key advantages over Lithium-Ion Batteries (LIBs). These advantages include the abundant availability of sodium in nature, lower production costs, higher safety due to reduced risk of thermal runaway, and enhanced thermodynamic stability with Fe alloying. Additionally, NIBs are more environmentally friendly as they do not require toxic elements like cobalt and provide sufficient performance for large-scale applications such as renewable energy storage and grid systems. With these benefits, NIBs present a more economical, safer, and eco-friendly alternative to LIBs.

This study demonstrates that the incorporation of Fe alloy into the NVP cathode increases the intercalation voltage (with the highest increase observed in 100% Fe, amounting to 1.23 Volts). Our finding aligns with previous research, which showed that Fe enhances cell voltage, thereby improving charge/discharge capacity [22]. Our results also show that the addition of Fe still maintain the structural stability during the intercalation and deintercalation processes as shown by the relatively low volume change. Thus, our study highlights the significant potential of Fe-alloyed NVP as a cathode material for low-cost rechargeable sodium-ion batteries.

4. Conclusion

Our findings highlight the importance of controlling the alloy concentration of Fe to achieve the desired range of intercalation voltage values, depending on the intended application of the Fe-alloyed NVP-based cathode. The addition of Fe alloys results in an increase in the intercalation voltage level due to the formation of stronger Fe-O bonds. Our results indicate that by precisely optimizing the amount and distribution of Fe within the Fe-alloyed NVP is crucial for optimizing the intercalation voltage profile of the modified NVP cathode. By incorporating Fe into the NVP, we show that the intercalation voltage can be tuned, and it may also enhance the overall structural stability of the modified NVP. These enhancements make Fe-alloyed NVP a promising choice for high-performance sodium-ion battery cathodes.

Acknowledgements

G.S. would like to acknowledge financial research support granted by the Ministry of Research, Technology and Higher Education (MENRISTEKDIKTI) of the Republic of Indonesia under the 'Riset Fundamental 2024' funding scheme no. 036/E5/PG.02.00.PL/2024. Financial research support granted by the Bandung Institute of Technology under the "Riset Unggulan 2023" scheme no. 308/IT1.B07.1/TA.00/2023 is also acknowledged.

References

[1] https://www.marketsandmarkets.com/Market-Reports/lithium-ion-battery-market-49714593.html, November 18, 2024.
[2] Zeng, X., Peng, J., Guo, Y., Zhu, H., and Huang, X., 2020, Front. Chem., 8:635.
[3] Goodenough, J. B., and Park, K. S., 2013, J. Am. Chem. Soc., 135, 1167–1176.
[4] Song, W., Ji, X., Wu, Z., Zhu, Y., Yang, Y., Chen, J., et al., 2014a, J. Mater. Chem. A, 2, 5358–5362.
[5] Han, X., Gui, X., Yi, T. F., Li, Y. W., and Yue, C. B., 2018, Curr. Opin. Solid State Mater. Sci., 22, 109–126.
[6] Yi, T. F., Zhu, Y. R., Tao, W., Luo, S. H., Xie, Y., and Li, X. F., 2018, J. Power Sources, 399, 26–41.
[7] Fang, R., Miao, C., Mou, H. Y., and Xiao, W., 2019, J. Alloys Compd., 818:152884.
[8] Li, R., Xiao, W., Miao, C., Fang, R., Wang, Z. Y., and Zhang, M. Q., 2019, Ceram. Int., 45, 13530–13535.
[9] Fang, R., Xiao, W., Miao, C., Mei, P., Yan, X. M., Zhang, Y., et al., 2020, J. Alloys Compd., 834:155230.
[10] Nie, Y., Xiao, W., Miao, C., Xu, M. B., and Wang, C. J., 2020a, Electrochim. Acta, 334:135654.
[11] Wang, K., Huang, X. B., Zhou, T., Wang, H. Y., Xie, H. S., and Ren, Y. R., 2020, Vacuum, 171:108997.
[12] Jiang, Z., Li, Y. H., Han, C., Huang, Z. Q., Wu, X. W., He, Z. X., et al., 2020, Electrochem. Soc., 167:020550.
[13] Yi, T. F., Wei, T. T., Li, Y., He, Y. B., and Wang, Z. N., 2020, Energy Storage Mater., 26, 165–197.
[14] Yang, Z., Zhang, J., Kintner-Meyer, M. C. W., Lu, X., Choi, D., Lemmon, J. P., et al., 2011, Chem. Rev., 111, 3577–3613.
[15] Yabuuchi, N., Kubota, K., Dahbi, M., and Komaba, S., 2014, Chem. Rev., 114, 11636–11682.
[16] Liu, Y. H., Yu, X. Y., Fang, Y. J., Zhu, X. S., Bao, J. C., Zhou, X. S., et al., 2018, Joule, 2, 725–735.
[17] Mao, M., Cui, C., Wu, M., Zhang, M., Gao, T., Fan, X., et al., 2018, Nano Energy, 45, 346–352.
[18] Carmichael, R. S., 1989, Handbook of Physical Properties of Rocks, Vol. 3, CRC Press.
[19] US Geological Survey, 2021, Mineral Commodity Summaries 2021, U.S. Department of the Interior, https://doi.org/10.3133/mcs2021.
[20] Smith, K., Pesaran, A., & Kim, G.-H., 2012, Battery Safety and Abuse Tolerance, National Renewable Energy Laboratory.
[21] Zhang, Y., 2019, Electrochimica Acta, 300, 458–465.
[22] Rao, Y. B., Achary, K. R., & Patro, L. N., 2022, ACS Omega, 7(51), 48192–48201. https://doi.org/10.1021/acsomega.2c06261.
[23] Liu, X., Feng, G., Wang, E., Chen, H., Wu, Z., Xiang, W., Zhong, Y., Chen, Y., Guo, X., & Zhong, B., 2019, ACS Applied Materials & Interfaces, 11(12), 12421–12430.
[24] Zheng, Y., 2018, Journal of Physical Chemistry C, 122(32), 18353–18359.
[25] Wu, T., Sun, J., Yap, Z. Q. Y., Ke, M., Lim, C. H. Y., & Lu, L., 2020, Materials and Design, 186, 108287.
[26] Novikova, A., Lisovenko, M., & Ivanov, I., 2018, Electrochimica Acta, 273, 416–426.
[27] Hafner, J., 2008, Journal of Computational Chemistry, 29(13), 2044–2078.
[28] Singh, B., et al., 2021, Journal of Materials Chemistry A, 9, 281–292.

A Preliminary Study on Silica/Zinc Ferrite Nanocomposite as Drug Delivery Agent

Anindya Bella Monica[1], Darminto[1], Ahmad Taufiq[2*] and Ainun Nimah[2]

[1] Department of Physics, Faculty of Sciences and Data Analytics, Institut Teknologi Sepuluh Nopember, Surabaya, Indonesia
[2] Department of Physics, Faculty of Mathematics and Natural Sciences, Universitas Negeri Malang, Indonesia

*E-mail: **ahmad.taufiq.fmipa@um.ac.id**

Abstract. Silica/zinc ferrite nanocomposite was prepared using the coprecipitation method to be applied as a candidate for drug delivery agents. The success of nanocomposite synthesis was evidenced by the presence of an amorphous silica phase and zinc ferrite phase in the X-ray diffraction pattern. In addition, in the FTIR characterization, the main functional groups, Si-O, Si-O-Fe, and Zn-O, were detected. Based on SEM characterization, the morphology of silica/zinc ferrite nanocomposite tended to form in a spherical shape with an average particle size of 38.2 nm. The magnetic character of silica/zinc ferrite nanocomposite was evaluated by characterization using VSM, showing a superparamagnetic curve with saturation magnetization of 3.5 emu/g. The drug loading performance showed an absorption peak at a wavelength of 251 nm, confirming the presence of doxorubicin as a drug. Furthermore, the effectiveness of silica/zinc ferrite nanocomposite in drug release was demonstrated by controlled drug release starting at 45 minutes with total release at 160 minutes.
Keywords: Silica/zinc ferrite, nanocomposite, superparamagnetic, doxorubicin, drug delivery agent.

1. Introduction

In the last decade, research in the medical field has experienced rapid development. This development is supported by cutting-edge technology, especially in the utilization of nanotechnology. One technology that has potential in the medical field is drug delivery agents. [1]. This technology is a favorite for researchers because it is a solution to overcome problems such as reducing the development or growth of cancer cells caused by drugs that are not right for cancer cells. Utilizing nano-sized particles to become the mobility of cancer drugs is then directed to the cancer target without damaging normal cells [2]. Furthermore, they function as drug delivery agents, and these nanoparticles can be used as therapeutic agents [3]. Therefore, drug delivery agent technology is very much needed in medical research because it can overcome the limitations of conventional cancer treatment. Cancer drugs used in conventional treatment have very serious weaknesses because they are non-polar substances and are not targeted. A drug often used in cancer treatment is doxorubicin because it is very effective in killing cancer cells but cannot distinguish it from normal cells [4]. With these many problems, drug delivery methods are very effective and efficient in minimizing the side effects of conventional drugs. Until now, a safe and effective drug delivery system remains a challenge for researchers, especially in cancer treatment.

Thus, delivery systems designed to improve the pharmacokinetics of loaded drug molecules, controlled release, and target-specific delivery have received considerable attention in recent years.

Various types of nanoparticles have been explored as drug delivery agents, one of which is silica [5]. The biocompatible nature of silica nanoparticles has attracted the attention of researchers because they can provide therapeutic effects in cancer treatment. In addition, its porous nature, high drug loading efficiency, and suitability to be functionalized with certain ligands are advantages in various treatments [6,7]. The porosity of silica is also believed to be able to encapsulate drugs so that drug release can be controlled within the desired time period [8]. However, one of the important requirements of a targeted drug delivery agent is to have magnetic properties so that it can be controlled by an external magnetic field when applied to the body [6]. Therefore, to improve the application performance of this drug delivery system, many researchers combine materials in the form of composites.

In this study, silica as the main material was composited with zinc ferrite to improve its performance as a drug delivery agent. Zinc ferrite was chosen because it has good superparamagnetic properties so that it can be controlled to the desired target under an external magnetic field [9]. In addition, zinc ferrite is also believed to provide anti-cancer effects by producing reactive oxygen species (ROS) [10,11]. Other researchers, such as Sarala et al., have successfully investigated the toxicity effects of Zinc Ferrite on breast cancer cells [12]. He reported that zinc ions from zinc ferrite can interact directly with cancer cell membranes, causing deformation of the membrane. The combination of these two materials can be a new opportunity in a controlled drug delivery system. Based on the description of the problems and solutions above, this study aims to explore the potential of silica/zinc ferrite nanocomposites that are used as effective drug delivery agents. At the same time, this study focuses on the characterization of physical and chemical properties and effectiveness in loading and releasing doxorubicin drugs.

2. Method

In this study, zinc ferrite was produced from the reaction between iron sand and HCl, while silica was produced from tetraethyl orthosilicate (TEOS) as a silica precursor. In short, as much as 20 g of clean iron sand (already separated) was reacted with 55 mL of HCl by stirring for 30 minutes to obtain a $FeCl_2$ and $FeCl_3$ solution. Through the coprecipitation method, as much as 20 mL of the resulting solution was added with 1.15 g of zinc chloride ($ZnCl_2$), which had previously been calculated using the stoichiometric equation. Furthermore, it was reacted with 20 mL of NH_4OH and stirred for 60 minutes until it became alkaline pH. The basic solution was washed with distilled water until the pH was neutral. Through the sol-gel method, the neutral solution was reacted with 7 mL of TEOS solution using an ultrasonic bath. As much as 5 mL of 5 mL of NH_4OH was added by titration and stirred for approximately 4 hours. The resulting solution was washed repeatedly using distilled water until a neutral pH was obtained. The product was then filtered and dried using an oven at a temperature of 100 °C. The final sample was then evaluated through XRD, FTIR spectroscopy, SEM, VSM, and UV-vis spectroscopy.

3. Result and Discussion

3.1 XRD

The XRD results of silica/zinc ferrite nanocomposite are shown in Figure 1. Based on the figure, it is observed that there is a broad peak at an angle of 2 theta: 20°–30°, which indicates the characteristics of amorphous silica. This result is consistent with previous research by Chen et al., who successfully synthesized amorphous silica using TEOS precursor [13]. Other diffraction peaks are also observed at 2 theta angles: 30.4°, 35.58°, 43.3°, 57.28°, and 62.6° which correspond to hkl planes 022, 113, 004, 115, and 044, respectively. These peaks are characteristic of the zinc ferrite phase, which has a face-centered cubic structure [14,15]. The XRD results of the silica/zinc ferrite nanocomposite were further analyzed using data fitting modeling from AMCSD 0007423 and AMCSD 1010954 for zinc ferrite and silica, respectively. Based on the fitting results using the 2 data models, the lattice parameter values were obtained as 8.39 Å for zinc ferrite and 6.54 Å for silica. Not only that, the grain size of the silica/zinc ferrite nanocomposite was also calculated using the Scherrer equation, obtaining around 80 Å for zinc ferrite and 15.8 Å for silica.

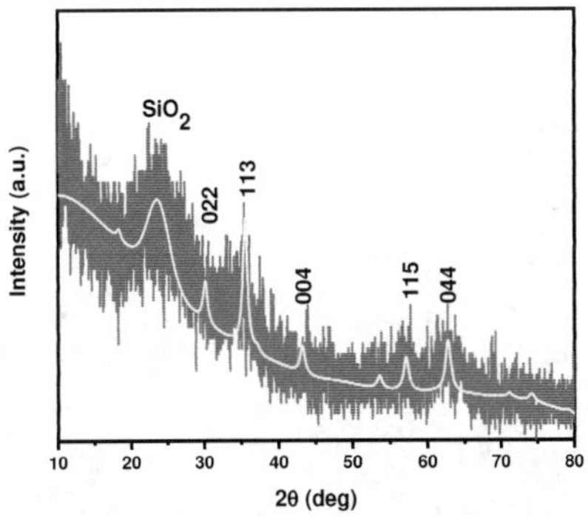

Figure 1. X-ray diffraction pattern of silica/zinc ferrite nanocomposite

3.2 FTIR

FTIR spectrum in the wavenumber range 400–4000 cm^{-1} was carried out to investigate the functional groups of silica/zinc ferrite nanocomposites, as shown in Figure 2. Based on the FTIR spectrum, it was observed that there were Fe-O and Zn-O groups at wavenumbers of 657 cm^{-1} and 611 cm^{-1} in succession [16]. Apart from that, the functional groups of the Si-O-Fe composite were also detected at different wavelengths 931 cm^{-1} [17]. Absorption bands around 161 cm^{-1} are associated with the presence of siloxane groups (Si-O-Si) [18,19]. The absorption band was wide at wavenumber 3100–3600 cm^{-1} and is a stretching vibration of the hydroxyl group, which includes the silanol group -Si (OH), -Si $(OH)_2$, or -Si $(OH)_3$ from the surface of the silica pores [20].

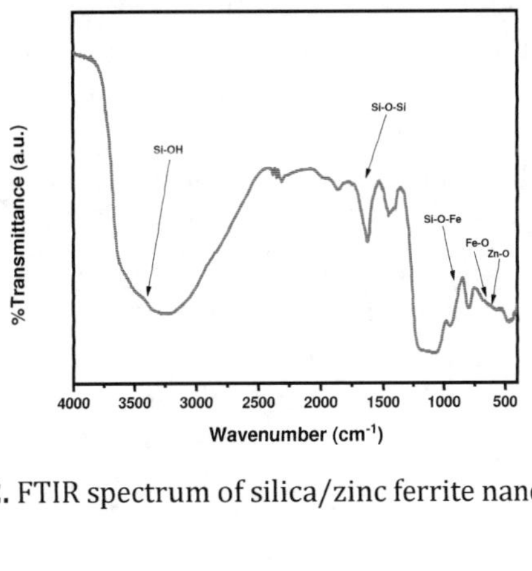

Figure 2. FTIR spectrum of silica/zinc ferrite nanocomposite

3.3 SEM

The surface morphology of the synthesized silica/zinc ferrite nanocomposites was also characterized using scanning electron microscopy (SEM), as shown in Figure 3a. Based on the SEM micrograph, the particles of the silica/zinc ferrite nanocomposites have a spherical shape that tends to agglomerate. This agglomeration is caused by an increase in the surface-to-volume ratio of nanostructured materials, where the very small particle size causes greater surface interactions [21]. The surface morphology of the SEM results also shows that the nanoparticles formed have a relatively uniform size. Based on previous research conducted by Kumar et.al, it was confirmed that the homogeneous particle size proves that the nanoparticles have good crystallinity [22]. In this study, quantitative analysis of the nanocomposite surface was evaluated using analytical data in the form of a histogram as shown in Figure 3b. Based on the analysis results, the average particle size was obtained at around 38.2 nm.

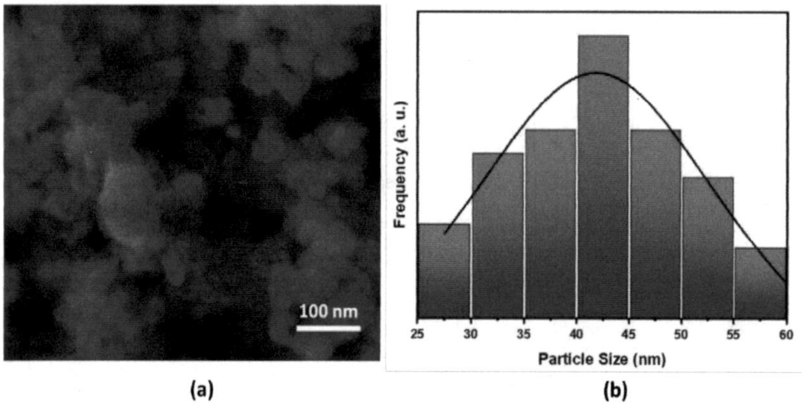

Figure 3. Surface morphology (a) and particle size (b) of the SEM results of silica/zinc ferrite nanocomposite

3.4 VSM

The magnetic properties of the silica/zinc ferrite nanocomposite were investigated using a vibration sample magnetometer (VSM), as shown in Figure 4. Based on the figure, the silica/zinc ferrite nanocomposite has a characteristic S-shaped hysteresis curve, where the curve indicates that the nanocomposite has superparamagnetic magnetic properties [23]. The hysteresis curve of silica/zinc ferrite nanocomposites was also fitted with the Langevin equation to determine the parameters of the saturation magnetization and remanent magnetization of the sample. The values of these parameters were obtained as 3.51 emu/g and 0.00128 emu/g, respectively. Not only that, the value of the square ratio of the saturation magnetization and remanent magnetization was also determined to study the effect of the material domain on the nanoscale size. The results of the square ratio show that values below 0.5 are related to multi-domains, while values above 0.5 are related to single magnetic [24]. These properties indicate the superparamagnetic nature of silica/zinc ferrite nanocomposites, where the magnetization shows an unsaturated type of magnetization, indicating the presence of large particle anisotropy [23].

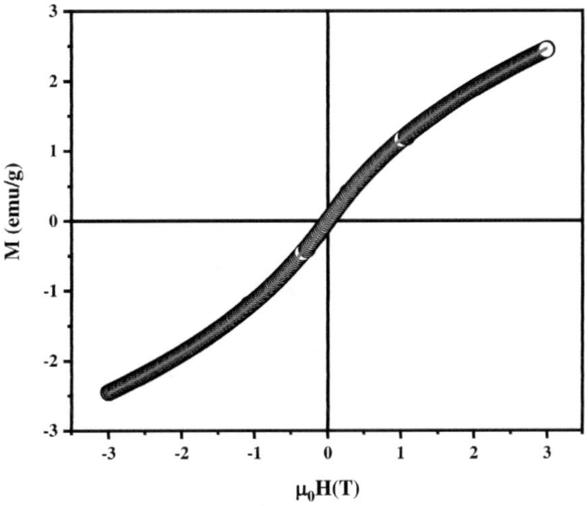

Figure 4. Hysteresis curve of silica/zinc ferrite nanocomposite

3.5 Drug Loading and Drug Release

In this study, doxorubicin was chosen as a model cancer drug to investigate its loading and release process in silica/zinc ferrite nanocomposite. The loading and release of doxorubicin were tested using UV-Vis absorption spectroscopy, as shown in Figure 5. Figure 5a is a graph of dox loading in silica/zinc ferrite nanocomposite, and the results show that the absorption peaks of doxorubicin are at wavelengths of 251 nm and 488 nm [25]. The emergence of this absorption peak indicates that doxorubicin has been well bound to the nanocomposite structure. Based on the literature, the success of loading the doxorubicin drug into the silica/zinc ferrite nanocomposite is due to the interaction between the Dox functional group and the silica/zinc ferrite composite [26]. The silanol groups of the nanocomposite, as confirmed by FTIR results, interact with the amino groups of doxorubicin [27,28]. This interaction provides the advantage of preventing premature release of the drug before reaching the target. The cumulative release characteristics of doxorubicin against silica/zinc ferrite nanocomposite were also analyzed, as shown in Figure 5b. The release profile study of doxorubicin was carried out at a normal body pH of around 7.4 using a PBS buffer solution. Based on the graph, the rapid release profile of the drug from zinc/ferrite nanocomposite began at 40 minutes. This result is consistent with previous research conducted by Cetinkaya et al. [29]. He explained that this rapid initial release is associated with most of the doxorubicin being absorbed on the outer surface of the carrier, not inside its pores. So that doxorubicin bound to the surface of the zinc/ferrite nanocomposite will be more easily released. After the rapid release, the graph shows a slow, even constant release, with only 2% of the drug released after 160 minutes. The slow release of Dox at normal physiological conditions (pH 7.4) is beneficial for cancer treatment because it minimizes the release of drugs in healthy organ cells that have a neutral pH [27]. This helps reduce the side effects of drugs on normal body cells.

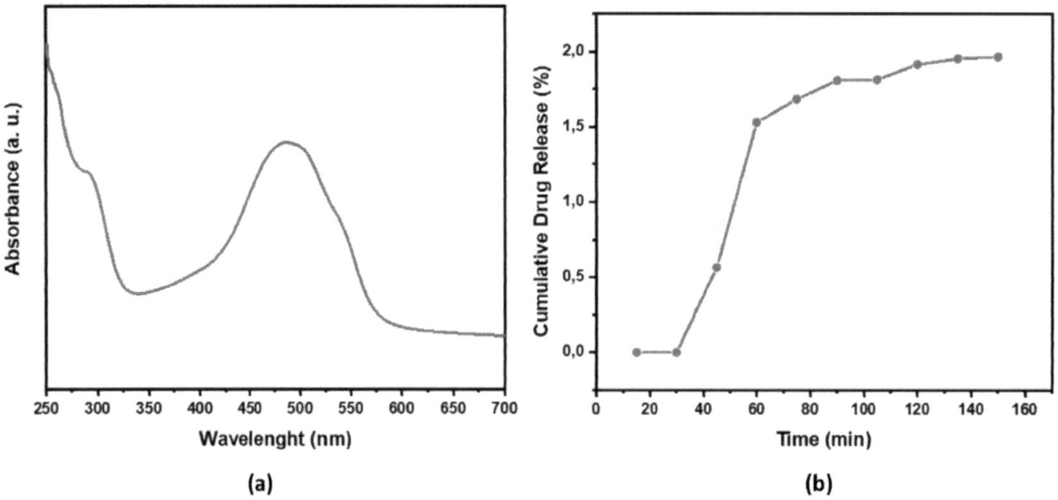

Figure 5. (a) Drug loading and (b) drug release of silica/zinc ferrite nanocomposite against cancer drug doxorubicin.

4. Conclusion

In this study, silica/zinc ferrite nanocomposites have been successfully synthesized using the coprecipitation method. Based on the XRD results, the synthesized sample formed 2 phases, namely the amorphous phase of silica in accordance with the AMCSD 1010954 model data and the crystalline phase of zinc ferrite with the AMCSD 0007423 model data. The FTIR characterization results showed the formation of the main functional groups of the silica/zinc ferrite nanocomposite, including the Si-O, Si-O-Fe, and Zn-O groups. The surface morphology of the SEM results showed that the silica/zinc ferrite nanocomposite had a spherical shape with an average size of 38.3 nm. The magnetic properties of the sample were also investigated using VSM, and the saturation magnetization value was obtained at 3.51emu/g. The potential of silica/zinc ferrite nanocomposites as drug delivery agents was studied using UV-Vis spectroscopy. Loading of doxorubicin into nanocomposite was marked by the presence of doxorubicin absorption peaks at Wavelengths of 251 nm and 488 nm. Meanwhile, the release profile of doxorubicin at normal pH conditions shows a slow release. This is beneficial for cancer treatment because it minimizes the release of drugs in healthy organ cells that have a neutral pH so that side effects of drugs on normal body cells can be avoided.

5. Acknowledgement

This research received support from the State University of Malang for AT through the research funding source of the DRTPM scheme, contract number PT 11.6.34/UN32.14.1/LT/2024

6. References

[1] Jadhav V, Roy A, Kaur K, Rai A K and Rustagi S 2024 Recent advances in nanomaterial-based drug delivery systems *Nano-Structures & Nano-Objects* **37** 101103
[2] Al-Thani A N, Jan A G, Abbas M, Geetha M and Sadasivuni K K 2024 Nanoparticles in cancer theragnostic and drug delivery: A comprehensive review *Life Sciences* **352** 122899
[3] Jain K K 2020 An Overview of Drug Delivery Systems *Drug Delivery Systems* Methods in Molecular Biology vol 2059, ed K K Jain (New York, NY: Springer New York) pp 1–54

[4] Sonam Dongsar T, Tsering Dongsar T, Molugulu N, Annadurai S, Wahab S, Gupta N and Kesharwani P 2023 Targeted therapy of breast tumor by PLGA-based nanostructures: The versatile function in doxorubicin delivery *Environmental Research* **233** 116455

[5] Mohan A, Santhamoorthy M and Lee Y-C 2024 Recent advances in the pH-responsive organic–inorganic mesoporous hybrid silica for targeted drug delivery *European Polymer Journal* **206** 112783

[6] Sopan Sinamo, Slamat Tarigan, Gea S, and Deddi Prima Putra 2022 Utilization of Silica Nanoparticles from Rice Husks for Improving the Mechanical Properties of Dental Materials: A Literature Review *Bioscmed* **6** 2644–50

[7] Stephen S, Gorain B, Choudhury H and Chatterjee B 2022 Exploring the role of mesoporous silica nanoparticle in the development of novel drug delivery systems *Drug Deliv. and Transl. Res.* **12** 105–23

[8] Iranshahy M, Hanafi-Bojd M Y, Aghili S H, Iranshahi M, Nabavi S M, Saberi S, Filosa R, Nezhad I F and Hasanpour M 2023 Curcumin-loaded mesoporous silica nanoparticles for drug delivery: synthesis, biological assays and therapeutic potential – a review *RSC Adv.* **13** 22250–67

[9] Thakur D, Latwal M, Singh J P, Gupta L K and Srivastava R C 2023 Zinc ferrite nanoparticles and their biomedical applications *Oxides for Medical Applications* (Elsevier) pp 233–55

[10] Garg J, Chiu M N, Krishnan S, Kumar R, Rifah M, Ahlawat P, Jha N K, Kesari K K, Ruokolainen J and Gupta P K 2024 Emerging Trends in Zinc Ferrite Nanoparticles for Biomedical and Environmental Applications *Appl Biochem Biotechnol* **196** 1008–43

[11] Anjaneyulu B, Chinmay, Chauhan V, Carabineiro S A C and Afshari M 2024 Recent Advances on Zinc Ferrite and Its Derivatives as the Forerunner of the Nanomaterials in Catalytic Applications *J Inorg Organomet Polym* **34** 1887–907

[12] Sarala E, Madhukara Naik M, Vinuth M, Rami Reddy Y V and Sujatha H R 2020 Green synthesis of Lawsonia inermis-mediated zinc ferrite nanoparticles for magnetic studies and anticancer activity against breast cancer (MCF-7) cell lines *J Mater Sci: Mater Electron* **31** 8589–96

[13] Chen Q, Lei M, Chen Y, Deng Y and Chen M 2023 Preparation of a thick sponge-like structured amorphous silica ceramic coating on 6061 aluminum alloy by plasma electrolytic oxidation in TEOS solution *Ceramics International* **49** 32679–93

[14] Naseri M G, Saion E B, Hashim M, Shaari A H and Ahangar H A 2011 Synthesis and characterization of zinc ferrite nanoparticles by a thermal treatment method *Solid State Communications* **151** 1031–5

[15] Okba E A, Rabea M F, El-Sheikh M Y and Aboelfetoh E F 2024 Design of silver-zinc-nickel spinel-ferrite mesoporous silica as a powerful and simply separable adsorbent for some textile dye removal *Sci Rep* **14** 16481

[16] Fei L, Ali F, Said A, Tariq N, Raziq F, Ali N, Arif U, Akhter M S, Rahdar A and Bilal M 2023 Surface-functionalized spongy zinc ferrite as a robust visible-light driven nanocatalyst for wastewater remediation: characterization, kinetic, and mechanistic insight *Int. J. Environ. Sci. Technol.* **20** 1007–18

[17] Wang B, Yang H, Jin Z, Zhao Q and Fu Y 2023 Effect of Fe/SiO2 ratio and cooling regime on As stability in copper smelting slag *Journal of Non Crystalline Solids* **606** 122190

[18] Christopher I F F, Karuppiah A and Thanapalan V G 2023 Fabrication of silica-cobalt ferrite (SiO2-CoFe2O4) nanocomposites for high electrochemical performances *Ionics* **29** 5055–72

[19] Karami H, Soltanali S, Najafi A M, Ghazimoradi M, Yaghoobpour E and Abbasi A 2023 Amorphous silica-alumina as robust support for catalytic dehydrogenation of propane: Effect of Si/Al ratio on nature and dispersion of Cr active sites *Applied Catalysis A: General* **658** 119167

[20] Shati A A, Alfaifi M Y, Elbehairi S E I, Olegovich B D, Althomali R H, Abdullaev S S, Musad Saleh E A, Hussien B M, Abid M K and Alwave M 2024 Functionalization of porous silica with graphene oxide and polyethyleneimine, containing zinc copper ferrite nanoparticles for water treatment and antibacterial application *Environmental Pollution* **348** 123745

[21] Kumar N, Singh R K and Satyapal H K 2020 Structural, optical, and magnetic properties of non-stoichiometric lithium substituted magnesium ferrite nanoparticles for multifunctional applications *J Mater Sci: Mater Electron* **31** 9231–41

[22] Kumar V, Kumar N, Bhushan Das S, Singh R K, Sarkar K and Kumar M 2021 Sol-gel assisted synthesis and tuning of structural, photoluminescence, magnetic and multiferroic properties by annealing temperature in nanostructured zinc ferrite *Materials Today: Proceedings* **47** 6242–8

[23] Hassan H S, Abdel Maksoud M I A and Attia L A 2020 Assessment of zinc ferrite nanocrystals for removal of 134Cs and 152+154Eu radionuclides from nitric acid solution *J Mater Sci: Mater Electron* **31** 1616–33

[24] El Moussaoui H, Mounkachi O, Masrour R, Hamedoun M, Hlil E K and Benyoussef A 2013 Synthesis and super-paramagnetic properties of neodymium ferrites nanorods *Journal of Alloys and Compounds* **581** 776–81

[25] Taufiq A, Nikmah A, Hidayat A, Sunaryono S, Mufti N, Hidayat N and Susanto H 2020 Synthesis of magnetite/silica nanocomposites from natural sand to create a drug delivery vehicle *Heliyon* **6** e03784

[26] Chen J, Liu J, Hu Y, Tian Z and Zhu Y 2019 Metal-organic framework-coated magnetite nanoparticles for synergistic magnetic hyperthermia and chemotherapy with pH-triggered drug release *Science and Technology of Advanced Materials* **20** 1043–54

[27] Cai W, Guo M, Weng X, Zhang W and Chen Z 2019 Adsorption of doxorubicin hydrochloride on glutaric anhydride functionalized Fe3O4@SiO2 magnetic nanoparticles *Materials Science and Engineering: C* **98** 65–73

[28] Beg M S, Mohapatra J, Pradhan L, Patkar D and Bahadur D 2017 Porous Fe_3O_4-SiO_2 core-shell nanorods as high-performance MRI contrast agent and drug delivery vehicle *Journal of Magnetism and Magnetic Materials* **428** 340–7

[29] Çetinkaya Z, Çiftçi H and Evcin A Synthesis and Characterization of Porous and Spherical Silica Nano-Shells for Controlled Release of Doxorubicin *Available at SSRN 4876529*

Preparation of TiO_2/Fe_3O_4/DMSO Ferrofluid as an Antimicrobial Agent

Fatimah Yamani, ST. Ulfawanti Intan Subadra, Nandang Mufti, Ahmad Taufiq*

Department of Physics, Faculty of Mathematics and Natural Sciences, Universitas Negeri Malang, Malang, Indonesia

*E-mail: ahmad.taufiq.fmipa@um.ac.id

Abstract. TiO_2/Fe_3O_4/DMSO ferrofluid has been successfully synthesized as an antimicrobial agent. The XRD data analysis presented the presence of cubic inverse spinel and tetragonal structures for Fe_3O_4 and TiO_2 phases. The results of SEM analysis showed that the nanocomposite as a filler was composed of spherical particles with a particle size of 37.3 nm. The presence of Fe_3O_4 was confirmed by the appearance of Fe-O bonds on tetrahedral and octahedral sites at 582 cm^{-1} and 443 cm^{-1}. Meanwhile, the C=O bond was detected at 1745 cm^{-1} and 1658 cm^{-1}. The stretching vibration of Ti-O-Ti appeared at 702 cm^{-1}. Furthermore, the S=O vibration of DMSO as a surfactant appeared at 1042 cm^{-1} and 952 cm^{-1}. Furthermore, dispersing liquid was detected by the presence of C-H vibrations at 3007, 2920, and 1461 cm^{-1}. The filler had a superparamagnetic character with a saturation magnetization value of 3.3 emu/g. Interestingly, the antimicrobial performance of the ferrofluid against *Candida albicans*, *Staphylococcus aureus bacteria*, *Bacillus subtilis*, and *Escherichia coli* had inhibition zone values of 8.95, 10.05, 10.17, and 7.72 mm, respectively.
Keywords: TiO_2/Fe_3O_4/DMSO, ferrofluid, antimicrobial agent, inhibition zone, superparamagnetic character.

1. Introduction

Drug-resistant bacterial and fungal infections are a serious threat to the world community because there are around 700,000 deaths due to these infections, and are predicted to reach 10 million deaths by 2050 [1]. *Escherichia coli* (*E. coli*) and *Staphylococcus aureus* (*S. aureus*) are bacteria that can be life-threatening and cause infections such as sepsis and endocarditis [2]. In addition, *Candida albicans* (*C. albicans*) can cause systemic infections in the bloodstream and internal organs [3]. Antibiotics are made to cause the death of microbes. However, microbes evolve by coding genes to synthesize enzymes that fight antibiotics and degrade them [4]. Therefore, it is necessary to develop antimicrobial materials, such as Fe_3O_4 nanoparticles.

Fe_3O_4 nanoparticles were chosen as antimicrobial agents because they are superparamagnetic, non-toxic, highly biocompatible, and bactericidal [5]. These superparamagnetic properties allow direct targeting of antimicrobial agents to areas infected with bacteria using an external magnetic field so that antimicrobial materials can reach the pathogenic zone [6]. Furthermore, a natural surfactant such as *Ziziphus mauritiana* leaf extract is also necessary because it has bioactive such as acyclic diterpene, steroids, flavonol glycosides, and triterpenoid saponins [7]. However, Fe_3O_4 nanoparticles, as one of the antimicrobial materials,

Journal of Physics: Conference Series **2980** (2025) 012025 doi:10.1088/1742-6596/2980/1/012025

have less stable chemical properties and are prone to agglomeration, thus disrupting their antimicrobial performance. Therefore, the addition of TiO_2 is a solution to this problem because they are stable colloidal suspensions [8]. In addition, TiO_2 nanoparticles are superior candidates as composite materials with Fe_3O_4 because they can improve antimicrobial performance. The potential of TiO_2 in nanocomposite systems as an antimicrobial agent is supported by previous studies showing that the antimicrobial performance of Fe_3O_4@TiO_2 is effective against *E. coli* bacteria by producing an inhibition zone of 16.50 mm [9], while pure TiO_2 is only 3 mm [10] and pure Fe_3O_4 is 15.50 mm [9].

In this study, TiO_2/Fe_3O_4 was modified in ferrofluid with almond seed oil as the dispersing medium. Modification in the form of ferrofluid aims to provide steric repulsion between magnetic nanoparticles to overcome the van der Waals force that triggers agglomeration [11]. In addition to magnetic nanoparticles, ferrofluid contains surfactants and a dispersing medium so that the magnetic nanoparticles are suspended by Brownian motion and do not easily settle under normal conditions [12]. Dimethyl sulfoxide (DMSO) surfactant was chosen as a solvent because it can dissolve polar and nonpolar compounds, mix with water, and play a role in the manufacture of bioactive fractions for antimicrobial activity [13]. Interestingly, in this study, almond seed oil was used as the dispersing medium because the Oleic acid (OA) content is higher than olive oil, with an average content of 60.32% [14]. In addition, almond seed oil has advantages such as being proven effective as a dispersing medium in protecting fillers, containing active phenolic compounds that can inhibit bacterial growth, and having antimicrobial properties [15].

2. Materials and Method

This study utilized iron sand, hydrochloric acid (HCl, 12 M, 99%), ammonium hydroxide (NH_4OH, 6.5 M, 99%), titanium trichloride ($TiCl_3$), and dimethyl sulfoxide (DMSO), all obtained from Merck. Deionized (DI) water, *Ziziphus mauritiana*, and almond seed oil were used in pro-analysis grade. The synthesis process began by extracting *Ziziphus mauritiana* using the infusion method by dissolving *Ziziphus mauritiana* powder into DI water at a temperature of 100 °C and stirring using a magnetic stirrer, then filtering to obtain *Ziziphus mauritiana* extract. The synthesis process of Fe_3O_4-*Ziziphus mauritiana* extract using the co-precipitation method by reacting iron sand and HCl to obtain $FeCl_2$ and $FeCl_3$ solutions. Next, $FeCl_2$ and $FeCl_3$ solutions were reacted with *Ziziphus mauritiana* extract, then titrated using NH_4OH and washed until neutral pH and filtered to obtain Fe_3O_4-*Ziziphus mauritiana* extract precipitate [16]. The TiO_2/Fe_3O_4 nanocomposite process using the co-precipitation method began by dissolving $TiCl_3$ and HCl into DI water, followed by a stirring process using a magnetic stirrer, then titrated with NH_4OH. Then, the solution was washed until neutral and filtered. After that, the TiO_2 precipitate was heated at 100 °C then calcined at 600 °C to obtain TiO_2 powder [17]. Next, the TiO_2 powder was added to the Fe_3O_4-*Ziziphus mauritiana* extract precipitate and heated until the TiO_2/Fe_3O_4 nanocomposite powder was formed. The process of making ferrofluids begins by dissolving TiO_2/Fe_3O_4 precipitate into DMSO, then adding almond seed oil and stirring using a magnetic stirrer at a temperature of 90 °C to obtain ferrofluids with TiO_2/Fe_3O_4/DMSO.

3. Result and Discussion

3.1 XRD

The X-ray diffraction pattern of TiO_2/Fe_3O_4/DMSO nanocomposite is presented in Figure 1. The Fe_3O_4 diffraction peaks were identified at 2θ = 30.72°, 35.59°, 43.81°, 53.95°, 57.34°, 63.7°, and 75.09°, which represent the hkl planes (220), (311), (400), (422), (511), (440), and (533), respectively. These results are in line with previous research conducted by Halfadji et al. [18]. Then, the analysis results show Fe_3O_4 has a cubic inverse spinel crystal structure and occupies the Fd-3m space group, where $(Fe^{3+})_A(Fe^{2+}Fe^{3+})_B$ with A at the tetrahedral atom position and B at the octahedral atom position [19]. Furthermore, data analysis shows the presence of anatase TiO_2 diffraction peaks detected at 2θ = 25.24°, 37.92°, 48.03°, 55.20°, 62.75°, 69.17°, 70.45°, and 75.11° which represent the hkl planes (101), (004), (200), (211), (204), (116), (220), and (215), respectively. These results are in accordance with the previous research [20]. TiO_2 has a lattice parameter of $a = b \neq c$ with $\alpha = \beta = \gamma = 90°$, thus forming a tetragonal structure and $I4$/amd space group [8]. In addition, the addition of extract to the nanocomposite did not produce new peaks, indicating that *Ziziphus mauritiana* extract acts as a natural surfactant. In this study, Fe_3O_4 has a crystal size of 11.68 ± 1.55 nm with a lattice parameter of 8.33 ± 0.01 Å and TiO_2 has a crystal size of 14.96 ± 0.03 nm with lattice parameters of 3.79 ± 0.01 Å and 9.45 ± 0.01 Å.

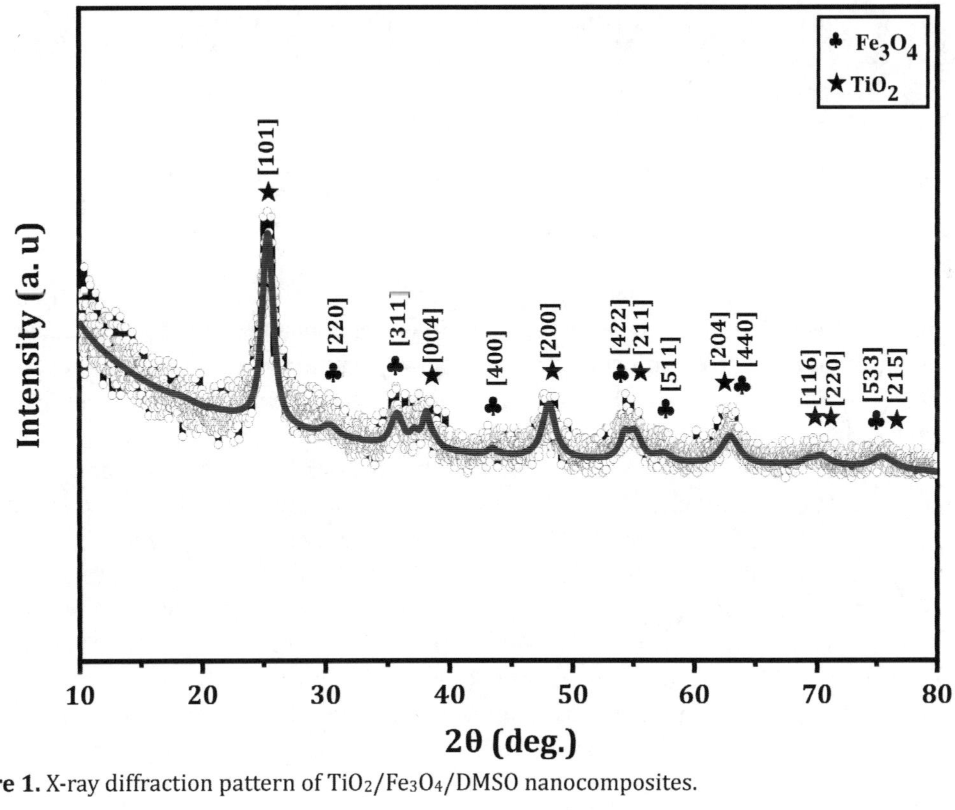

Figure 1. X-ray diffraction pattern of TiO_2/Fe_3O_4/DMSO nanocomposites.

3.2 SEM

The surface morphology and particle size distribution of TiO_2/Fe_3O_4/DMSO nanocomposites characterized using SEM are shown in Figure 2. It can be seen that the nanocomposite is spherical and experiences an agglomeration phenomenon. This agglomeration is caused by the attractive force between the magnetic particles of Fe_3O_4, which are the components of the nanocomposite [21]. The particle size distribution of the nanocomposite is 37.31 ± 0.92 nm. This value is smaller when compared to previous studies that obtained a particle size distribution value of TiO_2/Fe_3O_4 of 43.12 nm [22]. This shows that the addition of natural surfactants of *Ziziphus mauritiana* extract can produce a smaller particle size distribution value than the particle size of non-surfactant nanocomposites.

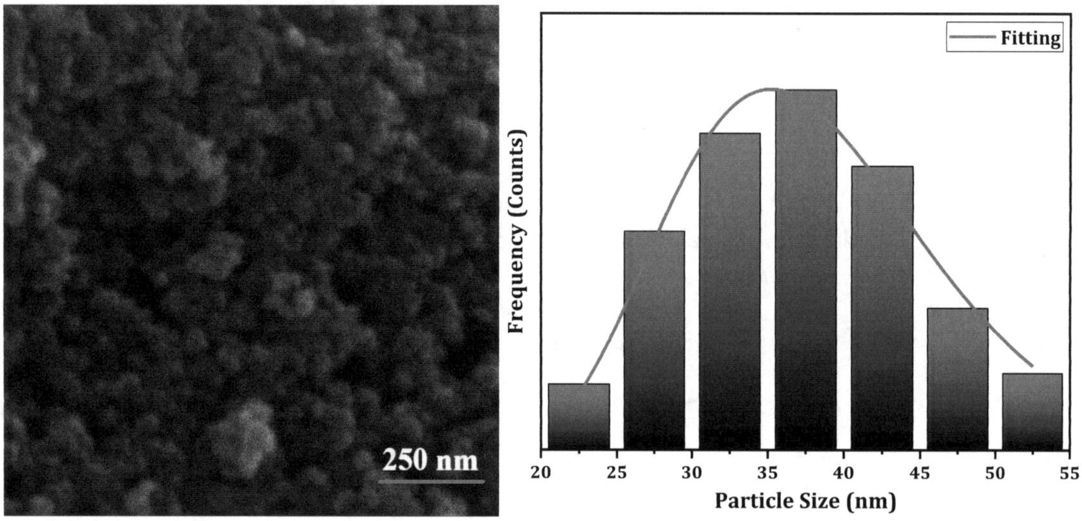

Figure 2. Morphology and particle size distribution of TiO_2/Fe_3O_4/DMSO nanocomposites.

3.3 FTIR

The spectrum of TiO_2/Fe_3O_4/DMSO ferrofluid is shown in Figure 3. The characterization results of the sample showed that the Fe_3O_4 functional group was characterized by the presence of Fe-O vibrations on the tetrahedral and octahedral sites at wavenumbers of 582 cm^{-1} and 443 cm^{-1}, respectively [23]. Furthermore, the O-H stretching vibration was seen at wavenumber 3441 cm^{-1}, indicating the bioactive component of the hydroxyl functional group [24]. The important components of *Ziziphus mauritiana* were detected at wavenumbers 1745 cm^{-1} and 1658 cm^{-1}, indicating the C=O vibration of the carbonyl ester functional group [25]. Furthermore, at 2854 cm^{-1} [26], 1403 cm^{-1}, and 1313 cm^{-1} [27], CH_2 stretching vibrations appear, which originate from the alkyl functional group. Then, the Ti-O-Ti stretching vibration appeared at wavenumber 702 cm^{-1} [28]. The success of surfactant coating by DMSO was marked by S=O vibration at wavenumbers 1042 cm^{-1} and 952 cm^{-1} [29]. Furthermore, CH_3 vibration was seen at wavenumber 1030 cm^{-1} [30]. The FTIR spectrum showed the success of almond seed oil as a dispersing medium with the detection of C-H vibration at wavenumbers 3007 cm^{-1}, 2920 cm^{-1}, and 1461 cm^{-1} [31]. The COO- vibration was identified at wavenumber 1438 cm^{-1} from the carboxyl functional group

[32]. Furthermore, the C-O stretching vibration was detected at wavenumbers 1238 cm^{-1} [33] and 1166 cm^{-1} [31], indicating the presence of ester functional groups.

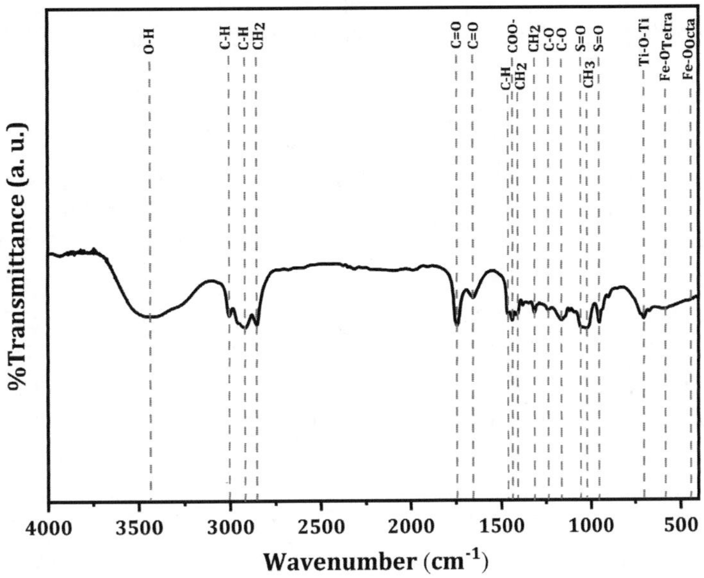

Figure 3. FTIR Spectrum of TiO_2/Fe_3O_4/DMSO ferrofluids.

3.4 VSM

The M-H hysteresis curve of TiO_2/Fe_3O_4/DMSO nanocomposite is shown in Figure 4. The results of the analysis obtained the remanent magnetization (M_r) and coercivity field (H_c) values of the nanocomposite of 0.001 ± 0.0001 emu/g and 0.003 ± 0.0001 T, respectively. The coercivity value is close to zero and the M-H curve formed is similar to the letter S. This indicates that the nanocomposite is in a superparamagnetic state [16]. Furthermore, the saturation magnetization (M_s) is 3.3 ± 0.005 emu/g. This M_s value is much lower than the results of previous studies, which obtained a M_s value of Fe_3O_4 of 31.30 emu/g [34]. This is due to the addition of *Ziziphus mauritiana* and TiO_2, which are diamagnetic materials. As a result, the interaction between Fe_3O_4 particles is reduced and causes a decrease in the overall magnetization [35]. Although the presence of this diamagnetic material causes the M_s to weaken, the resulting magnetization is still large enough to respond to an external magnetic field.

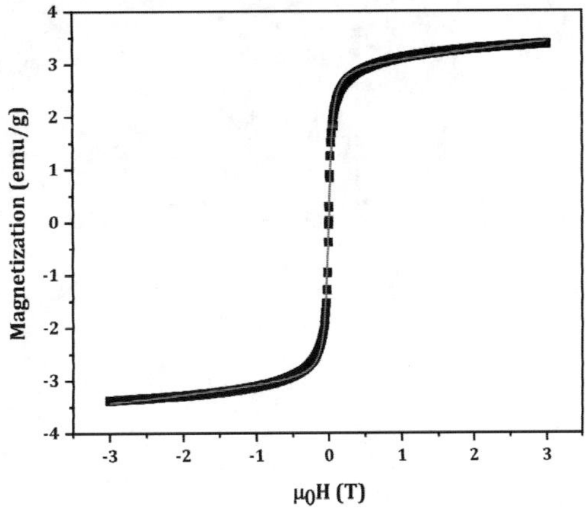

Figure 4. M-H Hysteresis curve of TiO_2/Fe_3O_4/DMSO nanocomposites.

3.5 Antimicrobial Activity

The antimicrobial activity of TiO_2/Fe_3O_4/DMSO ferrofluid was tested using the diffusion method and the results are shown in Figure 5. This study used fungi (*C. albicans*), gram-positive bacteria (*S. aureus* and *B. subtilis*), and gram-negative bacteria (*E. coli*). TiO_2/Fe_3O_4/DMSO ferrofluid had inhibition zones of 8.95 ± 0.97 mm, 10.05 ± 0.82 mm, 10.17 ± 0.48 mm, and 7.72 ± 1.85 mm for *C. albicans*, *S. aureus*, *B. subtilis*, and *E. coli*, respectively. The presence of *Ziziphus mauritiana* as the natural surfactant of Fe_3O_4 plays a role in inhibiting microbial growth to cause microbial death with the presence of phytochemical compounds such as alkaloids, phenolics, flavonoids, polyphenols, and saponins [36]. This is supported by previous research by Rahman et al., which showed that the synthesis of ZnO using *Ziziphus mauritiana* extract as a surfactant showed good antibacterial activity against *S. aureus* [37]. Furthermore, the functionalization of TiO_2 composited with Fe_3O_4-*Ziziphus mauritiana* extract increases antimicrobial performance in ferrofluids. It is shown that Fe_3O_4 can expand the light response of TiO_2 so that photocatalytic reactions can occur in visible light [38], where through this reaction Reactive Oxygen Species (ROS) such as O_2^- (superoxide radical anion), OH (hydroxyl radical), and H_2O_2 [39]. Furthermore, the surface of the nanocomposite containing ROS and covered by DMSO surfactant has a positive charge, while the surface of the microbe has a negative charge, which results in electrostatic attraction between the nanocomposite and the microbe [40]. The ROS produced can interact with the microbial surface, causing damage to the membrane, resulting in leakage of intracellular components, and disrupting cell metabolism to cause microbial death [41]. Interestingly, almond seed oil as a dispersing medium in the TiO_2/Fe_3O_4/DMSO ferrofluid system is able to synergize in disrupting cell structures, which ultimately results in microbial death due to the ability of almond oil to dissolve and damage the outer sheath of the microbial cell wall [42].

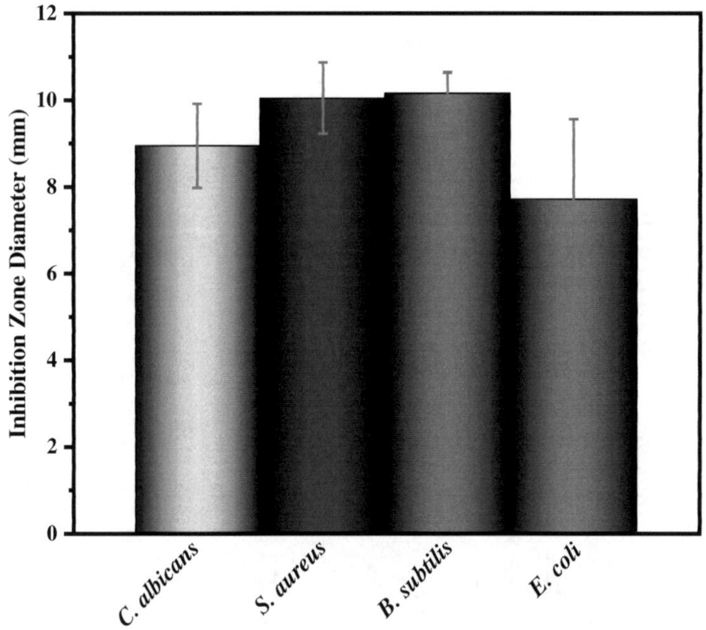

Figure 5. Inhibition zone diameter of antimicrobial activity of TiO_2/Fe_3O_4/DMSO ferrofluids.

4. Conclusion

TiO_2/Fe_3O_4/DMSO ferrofluid has been successfully synthesized. The results of XRD analysis showed the presence of cubic inverse spinel and tetragonal crystal structures with crystal sizes of 11.68 ± 1.55 nm and 14.96 ± 0.03 nm for Fe_3O_4 and TiO_2, respectively. The TiO_2/Fe_3O_4/DMSO nanocomposite was spherical and experienced an agglomeration phenomenon with a particle size distribution of 37.31 ± 0.92 nm. The presence of *Ziziphus mauritiana* was confirmed by the appearance of C=O bonds at 1745 cm^{-1} and 1658 cm^{-1}. The presence of DMSO, as a surfactant, was indicated by S=O bonds at 1042 cm^{-1} and 952 cm^{-1}. Almond seed oil is detected by C-H vibrations at 3007 cm^{-1}, 2920 cm^{-1}, and 1461 cm^{-1}. The nanocomposite has superparamagnetic characteristics with an M_s value of 3.3 ± 0.005 emu/g. Then, the antimicrobial performance of ferrofluid against *C. albicans*, *S. aureus*, *B. subtilis*, and *E. coli* each has an inhibition zone value of 8.95 ± 0.97 mm, 10.05 ± 0.82 mm, 10.17 ± 0.48 mm, and 7.72 ± 1.85 mm.

5. Acknowledgment

Universitas Negeri Malang supported this research through PNBP Grant for AT research group with grant Number 4.4.554/UN32.14.1/LT/2024.

6. References

[1] Samreen, I. Ahmad, H. A. Malak, and H. H. Abulreesh, "Environmental antimicrobial resistance and its drivers: a potential threat to public health," *J. Glob. Antimicrob. Resist.*, vol. 27, pp. 101–111, Dec. 2021, doi: 10.1016/j.jgar.2021.08.001.

[2] J. M. Kwiecinski and A. R. Horswill, "Staphylococcus aureus bloodstream infections: pathogenesis and regulatory mechanisms.," *Curr. Opin. Microbiol.*, vol. 53, pp. 51–60, Feb. 2020, doi: 10.1016/j.mib.2020.02.005.

[3] C. d'Enfert *et al.*, "The impact of the Fungus-Host-Microbiota interplay upon Candida albicans infections: current knowledge and new perspectives," *FEMS Microbiol. Rev.*, vol. 45, no. 3, pp. 1–6, May 2021, doi: 10.1093/femsre/fuaa060.

[4] P. Chatterjee, N. Chauhan, and U. Jain, "Confronting antibiotic-resistant pathogens: Distinctive drug delivery potentials of progressive nanoparticles," *Microb. Pathog.*, vol. 187, no. November 2023, p. 106499, Feb. 2024, doi: 10.1016/j.micpath.2023.106499.

[5] M. Sharaf *et al.*, "Design of lipid-based nanocarrier for drug delivery has a double therapy for six common pathogens eradication," *Colloids Surfaces A Physicochem. Eng. Asp.*, vol. 625, no. April, p. 126662, Sep. 2021, doi: 10.1016/j.colsurfa.2021.126662.

[6] A. B. Seabra, M. T. Pelegrino, and P. S. Haddad, *Antimicrobial Applications of Superparamagnetic Iron Oxide Nanoparticles: Perspectives and Challenges*. Elsevier Inc., 2017. doi: 10.1016/B978-0-323-46152-8.00024-X.

[7] N. Kumar Sishu, U. Das, and C. Immanuel Selvaraj, "Indian jujube a potential fruit tree to improve the livelihood," *Saudi J. Biol. Sci.*, vol. 30, no. 9, p. 103769, Sep. 2023, doi: 10.1016/j.sjbs.2023.103769.

[8] M. Messaoud, F. Trabelsi, P. Kumari, A. Merenda, and L. F. Dumée, "Recrystallization and coalescence kinetics of TiO_2 and ZnO nano-catalysts towards enhanced photocatalytic activity and colloidal stability within slurry reactors," *Mater. Chem. Phys.*, vol. 252, no. May, 2020, doi: 10.1016/j.matchemphys.2020.123235.

[9] N. Rani and B. S. Dehiya, "Magnetic core-shell Fe_3O_4@TiO_2 nanocomposites for broad spectrum antibacterial applications," *IET Nanobiotechnology*, vol. 15, no. 3, pp. 301–308, 2021, doi: 10.1049/nbt2.12017.

[10] U. Q. Satti, S. J. A. Zaidi, A. Riaz, M. A. ur Rehman, C. X. Li, and M. A. Basit, "Simple two-step development of TiO_2/Fe_2O_3 nanocomposite for oxygen evolution reaction (OER) and photo-bio active applications," *Colloids Surfaces A Physicochem. Eng. Asp.*, vol. 671, no. May, p. 131662, 2023, doi: 10.1016/j.colsurfa.2023.131662.

[11] F.-C. Lin, H. L. van de Wouw, O. Campàs, E. M. Sletten, and J. I. Zink, "Synthesis of fluorous ferrofluids and effects of the nanoparticle coatings on field- and temperature-dependent magnetizations," *Chem. Mater.*, vol. 35, no. 19, pp. 7957–7966, Oct. 2023, doi: 10.1021/acs.chemmater.3c01172.

[12] J. E. Król and G. D. Ehrlich, "Using SMART magnetic fluids and gels for prevention and destruction of bacterial biofilms," *Microorganisms*, vol. 11, no. 6, p. 1515, Jun. 2023, doi: 10.3390/microorganisms11061515.

[13] R. Mazumdar, P. P. Dutta, J. Saikia, J. C. Borah, and D. Thakur, "Streptomyces sp. strain PBR11, a forest-derived soil actinomycetia with antimicrobial potential," *Microbiol. Spectr.*, vol. 11, no. 2, Apr. 2023, doi: 10.1128/spectrum.03489-22.

[14] P. K. Ojha *et al.*, "Chemical compositions and essential fatty acid analysis of selected vegetable oils and fats," *Compounds*, vol. 4, no. 1, pp. 37–70, Jan. 2024, doi: 10.3390/compounds4010003.

[15] S. Fabroni *et al.*, "Almond [Prunus dulcis (Mill.) DA Webb] Processing Residual Hull as a New Source of

Bioactive Compounds: Phytochemical Composition, Radical Scavenging and Antimicrobial Activities of Extracts from Italian Cultivars ('Tuono', 'Pizzuta', 'Romana')," *Molecules*, vol. 28, no. 2, p. 605, Jan. 2023, doi: 10.3390/molecules28020605.

[16] A. Taufiq, A. D. Iswatin, R. E. Saputro, and C. I. Yogihati, "Fabrication of New Fe_3O_4/PVA/($C_6H_7O_6Na$)n Nanohybrid Ferrogels for Antibacterial Applications," *Mater. Res.*, vol. 24, no. 2, pp. 6–13, 2021.

[17] S. Sunaryono, S. N. Halizah, S. Zulaikah, H. Susanto, N. Mufti, and A. Taufiq, "Contribution of ZnO/TiO_2 nanocomposite particles towards bacterial growth inhibition," 2021, p. 030013. doi: 10.1063/5.0052530.

[18] A. Halfadji, M. Naous, K. nadia Kharroubi, F. el zahraà Belmehdi, and S. Rajendrachari, "An ultrasonic-assisted Synthesis, Characterization, and application of Nano-Fe_3O_4/TiO_2 as nano-catalyst for the removal of organic dye by Like-Photo-Fenton reactions," *Inorg. Chem. Commun.*, vol. 158, no. August, p. 111686, Dec. 2023, doi: 10.1016/j.inoche.2023.111686.

[19] Sudarmono *et al.*, "Magnetically separable and reusable Fe_3O_4/chitosan nanocomposites green synthesized utilizing Moringa oleifera extract for rapid photocatalytic degradation of methylene blue," *Results Chem.*, vol. 7, no. December 2023, p. 101245, Jan. 2024, doi: 10.1016/j.rechem.2023.101245.

[20] Y. Li, F. Zhang, K. Xu, Z. Yang, and Z. Dong, "The synergistic enhancement of photocatalytic activity by Fe_3O_4–TiO_2 nanosheets for rhodamine B degradation," *Colloids Surfaces A Physicochem. Eng. Asp.*, vol. 700, no. July, p. 134803, Nov. 2024, doi: 10.1016/j.colsurfa.2024.134803.

[21] Z. Li, L. Li, S. Zhang, L. Zhang, Y. Cui, and H. Shi, "Ultrasound assisted TiO_2@Fe_3O_4 nanocomposites photocatalytic degradation of organic pollutants in potato starch processing wastewater," *Sep. Purif. Technol.*, vol. 332, no. September 2023, p. 125799, Mar. 2024, doi: 10.1016/j.seppur.2023.125799.

[22] M. N.Borhani, A. Tavakoli, A. M. Mollaei, and T. N.Borhani, "Visible light photodegradation of methylene blue by ionic liquid based TiO_2/Fe_3O_4 nanophotocatalysts," *Opt. Mater. (Amst).*, vol. 154, no. April, p. 115627, Aug. 2024, doi: 10.1016/j.optmat.2024.115627.

[23] N. Rani and B. S. Dehiya, "Influence of anionic and non-ionic surfactants on the synthesis of core-shell Fe_3O_4@TiO_2 nanocomposite synthesized by hydrothermal method," *Ceram. Int.*, vol. 46, no. 15, pp. 23516–23525, 2020, doi: 10.1016/j.ceramint.2020.06.122.

[24] N. A. Yusuf-Omoloye *et al.*, "Green-synthesis, characterization, and antibacterial activity of Azanza garckeana seed extract silver nanoparticles against vancomycin-resistant Enterococci," *Next Nanotechnol.*, vol. 6, no. March, p. 100035, 2024, doi: 10.1016/j.nxnano.2023.100035.

[25] S. Rajapandi *et al.*, "Chemosphere Ziziphus mauritiana-derived nitrogen-doped biogenic carbon dots : Eco-friendly catalysts for dye degradation and antibacterial applications," *Chemosphere*, vol. 338, no. May, p. 139584, 2023, doi: 10.1016/j.chemosphere.2023.139584.

[26] S. Ozun and M. U. Atalay, "Flotation and adsorption characteristics of albite and quartz with oleic acid-based collector," *Colloids Surfaces A Physicochem. Eng. Asp.*, vol. 672, no. May, p. 131710, Sep. 2023, doi: 10.1016/j.colsurfa.2023.131710.

[27] Y. Yang, J. Liu, M. Zhu, S. Dou, and L. Ding, "Adsorption of glycine at the anatase TiO_2/water interface: Effects of Ca^{2+} ions," *J. Environ. Sci.*, vol. 128, pp. 26–34, Jun. 2023, doi: 10.1016/j.jes.2022.08.016.

[28] N. Rani and B. S. Dehiya, "Influence of anionic and non-ionic surfactants on the synthesis of core-shell Fe_3O_4@TiO_2 nanocomposite synthesized by hydrothermal method," *Ceram. Int.*, vol. 46, no. 15, pp. 23516–23525, Oct. 2020, doi: 10.1016/j.ceramint.2020.06.122.

[29] S. Huang, S. He, Y. Li, S. Wang, and X. Hou, "Hydrogen bond acceptor lined hydrogel electrolyte toward Dendrite-Free aqueous Zn ion batteries with low temperature adaptability," *Chem. Eng. J.*, vol. 464, no. March, p. 142607, May 2023, doi: 10.1016/j.cej.2023.142607.

[30] B. Gao and M. Wang, "Multistep structure evolution during the coagulation of PAN/DMSO solution studied by in-situ FTIR spectroscopy," *Polymer (Guildf).*, vol. 283, no. May, p. 126221, Sep. 2023, doi: 10.1016/j.polymer.2023.126221.

[31] K. D. Mekonnen, "Fourier transform infrared spectroscopy as a tool for identifying the unique characteristic bands of lipid in oilseed components: Confirmed via Ethiopian indigenous desert date fruit," *Heliyon*, vol. 9, no. 4, p. e14699, Apr. 2023, doi: 10.1016/j.heliyon.2023.e14699.

[32] S. Pandey, Y. Premjit, A. Khuntia, A. Kadival, and J. Mitra, "Development of probiotic-loaded calcium alginate-maltodextrin microparticles based on electrohydrodynamic technique," *Powder Technol.*, vol. 428, no. July, p. 118808, Oct. 2023, doi: 10.1016/j.powtec.2023.118808.

[33] N. Mohammadi, N. Ostovar, and D. Granato, "Pyrus glabra seed oil as a new source of mono and polyunsaturated fatty acids: Composition, thermal, and FTIR spectroscopic characterization," *LWT*, vol. 181, no. April, p. 114790, May 2023, doi: 10.1016/j.lwt.2023.114790.

[34] M. Sajid, S. Shuja, H. Rong, and J. Zhang, "Size-controlled synthesis of Fe_3O_4 and Fe_3O_4@SiO_2 nanoparticles and their superparamagnetic properties tailoring," *Prog. Nat. Sci. Mater. Int.*, vol. 33, no. 1, pp. 116–119, Feb. 2023, doi: 10.1016/j.pnsc.2022.08.003.

[35] L. Jestha *et al.*, "Photodegradation enhancement of organic dyes using magnetically separable and reusable Fe_3O_4/rGO nanocomposites green-synthesized utilizing plant leaves extract," *Case Stud. Chem. Environ. Eng.*, vol. 10, no. August, p. 100915, 2024, doi: 10.1016/j.cscee.2024.100915.

[36] E. El, Y. El, N. A. S. Qarah, A. Khalid, K. Moustaid, and B. Nasser, "Genus Ziziphus : A comprehensive review on ethnopharmacological, phytochemical and pharmacological properties," *J. Ethnopharmacol.*, vol. 259, no. September 2019, p. 112950, 2020, doi: 10.1016/j.jep.2020.112950.

[37] A. Rahman, M. H. Harunsani, A. L. Tan, N. Ahmad, and M. M. Khan, "Antioxidant and antibacterial studies of phytogenic fabricated ZnO using aqueous leaf extract of Ziziphus mauritiana Lam," *Chem. Pap.*, vol. 75, no. 7, pp. 3295–3308, Jul. 2021, doi: 10.1007/s11696-021-01553-7.

[38] Z. Zhang, Y. Fang, L. Zhuo, H. Yuan, and L. Zhang, "Reduced graphene oxide wrapped Fe_3O_4@TiO_2 yolk–shell nanostructures as a magnetic recyclable photocatalytic antibacterial agent," *J. Alloys Compd.*, vol. 904, p. 164001, 2022, doi: 10.1016/j.jallcom.2022.164001.

[39] R. Hatefi, H. Younesi, A. Mashinchian-moradi, and S. Nojavan, "A facile decoration of anatase Fe_3O_4/TiO_2 nanocomposite with graphene quantum dots : Synthesis , characterization , and photocatalytic activity," *Adv. Powder Technol.*, vol. 32, no. 7, pp. 2410–2422, 2021, doi: 10.1016/j.apt.2021.05.020.

[40] A. Taufiq *et al.*, "Synthesis of Fe_3O_4/Ag nanohybrid ferrofluids and their applications as antimicrobial and antifibrotic agents," *Heliyon*, vol. 6, no. 12, 2020, doi: 10.1016/j.heliyon.2020.e05813.

[41] A. Halfadji, L. Bennabi, S. Giannakis, A. G. Marrani, and S. Bellucci, "Sono-synthesis and characterization of next-generation antimicrobial ZnO/TiO_2 and Fe_3O_4/TiO_2 bi-nanocomposites, for antibacterial and antifungal applications," *Ceram. Int.*, vol. 50, no. 20PB, pp. 39097–39108, 2024, doi: 10.1016/j.ceramint.2024.07.276.

[42] O. M. Kolawole, A. T. Adesegun, S. C. Isreal, R. O. Ayorinde, and B. O. Silva, "Comparative Evaluation of the Physicochemical, Antimicrobial and Stability profile of Olive oil, Almond oil and Coconut oil-based Emulgels," *Trop. J. Phytochem. Pharm. Sci.*, vol. 3, no. 3, pp. 234–239, Jul. 2024, doi: 10.26538/tjpps/v3i3.4.

Isolation and Characterization of Chitosan from Nacre of the Oyster Pearl Shell (*Pinctada Maxima*) as Biopolimer

Susi Rahayu[1, 2], Masruroh[1*], D.J.D.H Santjojo[1], Nonik Septiani[2], Rahadi Wirawan[2]

[1] Physics Department, Faculty of Mathematics and Natural Sciences, Universitas Brawijaya, Malang, 65145, Indonesia
[2]Physics Department, Faculty of Mathematics and Natural Sciences, University of Mataram, 83125, Indonesia

*E-mail: ruroh@ub.ac.id

Abstract. Chitosan is derived from the deacetylation of chitin using a strong base. Chitin, a biopolymer commonly found in oyster pearl shells (*Pinctada maxima sp)*, is transformed into chitosan. This study aims to investigate the properties of chitosan extracted from oyster pearl shells, focusing on the effect of variations in NaOH concentration on the chitosan produced. The isolation of chitosan involved deproteinization, demineralization, and deacetylation. Proximate content, functional groups, degree of deacetylation (DD), crystal structure and crystallinity, and particle size were analyzed using proximate testing, FTIR, XRD, and PSA. The study found that both chitosan samples had low water and fat content, displayed typical chitosan functional groups (OH and NH_2), and possessed an orthorhombic crystal structure with consistent lattice parameters of 8.24 Å. Chitosan treated with 80% NaOH achieved the highest DD of 88%, the greatest crystallinity of 72.07%, and a particle size of 44.306 µm. In comparison, chitosan treated with 70% NaOH had a DD of 82.61%, crystallinity of 39.65%, and a particle size of 55.509 µm. Thus, variations in NaOH concentration are proposed to have differences in DD, crystallinity, and particle size of chitosan. The research successfully isolated chitosan from oyster pearl shells, but further optimization of process parameters is necessary to enhance its potential as a biomaterial.

1. Introduction

The Pinctada maxima crustacean has significant amounts of chitin (69.65%) and calcium carbonate ($CaCO_3$) (98.7%) [1] . $CaCO_3$ can be converted into hydroxyapatite through synthetic methodologies. At the same time, chitin can be altered into chitosan using deacetylation [1–3]. Chitosan is a linear polysaccharide that features reactive amino and hydroxyl groups. Its polymer chain is characterized by a structure composed of β-(1-4)-2-acetamido-D-glucose units (acetylation units) and β-(1-4)-2-amino-D-glucosamine units (deacetylation units). Chitin is often found in blood cockle shells [4], shrimp shells [5], and oyster shells [6,7]. Chitin and chitosan are biopolymers with diverse functions and extensive applications across different industries. With the increasing demand for chitosan, there is a need for a cost-effective and sustainable source of chitin and chitosan [8]. Chitosan is primarily derived from the deacetylation of chitin, which can be achieved through either enzymatic or chemical methods [9].

Chitosan is chemically produced by isolating it through deproteinization, demineralization, and deacetylation processes. Converting acetyl groups into amine groups during deacetylation necessitates using a strong base [10]. Deacetylation plays a crucial role in producing high-quality chitosan. When chitin from Pinctada maxima is deacetylated to form chitosan, it achieves a deacetylation degree (DD) of 77.21%. In contrast, gradual deacetylation of the same chitin results in a DD of 57.50% [6,7].

Meanwhile, isolating chitosan from oyster pearl shells resulted in DD reaching 81.50%. In this study, a modified version of the isolation method described [10] was employed, which includes deproteinization, demineralization, and deacetylation. The resulting chitosan is creamy white, odorless, in powder form, yielding 20.63%, contains OH and NH_2 groups, and has a crystallinity of 36.94% [10]. According to these characteristics, the chitosan meets the quality standards set by the National Standardization Agency (BSN) in the SNI (7949: 2013) chitosan quality standard.

Using a strong base for the deacetylation of chitin transforms acetyl groups into amine groups directly related to the DD or chitosan purity. Increasing the concentration of the strong base enhances the purity of chitosan. This study has yet to explore using Pinctada maxima pearl shells as a source of chitin. Consequently, further research is needed to improve chitosan quality and examine the potential for shellfish waste. This research will focus on analyzing the properties of chitosan isolated with different amounts of strong bases.

2. Method

2.1. Material Synthesis

Chitosan-based production was carried out leveraging cutting-edge approaches. The experiment utilized isolation techniques to produce chitosan from oyster pearl shell waste. Extraction of the chitosan powder adopted the method used [10]: deproteinization, demineralization, and deacetylation. However, this research modified the NaOH concentration in the deacetylation. The percentage of NaOH were 70% (Cs-1) and 80% (Cs-2). The two analysis results have been compared with the reference chitosan, NaOH 60% (CS-ref). Regarding chitosan production, the preliminary step began by dissolving 80 g of oyster pearl shell powder in 800 mL of 4% NaOH. The solution was homogenized with a magnetic stirrer for one hour at 80°C. In addition, the pH of the solution was neutralized using distilled water using a washing technique. Once neutral, the solution was precipitated and then the precipitate was separated and dried at 80°C for three hours in an oven. It generated powder, which was the final result of the deproteinization stage.

The deproteinized powder was prepared for demineralization. The demineralization involved dissolving the deproteinized powder with 1 M HCl solution using a ratio of 1:15 (w/v). The solution was then homogenized by using a magnetic stirrer for one hour. Then, the homogeneous solution was neutralized and left undisturbed until it fully settled. This residue was dried at 80°C for three hours to obtain chitin powder. After that, chitin powder was transformed into chitosan through a deacetylation reaction. Chitin was mixed with NaOH (70% concentration is the Cs-1 sample and 80% is the Cs-2 sample) using a 1:15 (w/v) ratio for three hours at 120°C. The residue from this process was neutralized and dried at 50°C for five hours to produce chitosan powder.

2.2. Material Characterization and Analysis

The testing of chitosan properties was conducted using various methods, including proximate analysis, FTIR, XRD, and PSA. Proximate analysis determined the chemical composition of chitosan, such as moisture, ash, protein, and fat content (AOAC 2011). Moisture content was analyzed by weighing 1 g of chitosan into a Petri dish, heating it in an oven at 105°C for 24 hours until the weight stabilized, and calculating the weight difference as the moisture content (% DM). Ash content analysis began by drying the dishes at 105°C for 3 hours, followed by burning 1-2 g of chitosan in a kiln at 600°C for 18-24 hours. The ash sample was then cooled in a desiccator for 1 hour before weighing. Fat content was measured using the gravimetric method. A 1 g sample of chitosan was dissolved in 15 mL of 25% HCl and heated for 3 hours. The resulting residue was dried, extracted with hexane, and weighed to determine the fat content. Protein content analysis was performed using the Kjeldahl method with a nitrogen-to-protein conversion factor of 5.33. A 1 g sample was mixed with selenium and concentrated H_2SO_4, then digested at 420°C. The digested sample was distilled using 40% NaOH, and the vapor was collected in a 4% H_3BO_3 solution. Finally, the solution was titrated with 0.2 N HCl until the color changed from green to red, indicating the protein content [11].

Water content determined the quality of chitosan. High levels of moisture could affect its structural and compositional makeup. As indicated by the ash residue measurement, it verified information regarding inorganic substances remaining from combustion. Low ash-forming means it is deficient in minerals, which might effectively raise the level of chitosan purification. Protein content investigation ascertained the quality of the chitosan, and low-fat influenced the physical properties and stability of chitosan [12].

Secondly, the FTIR examination was proper to specify the functional groups and DD of chitosan. Data from FTIR demonstrated chitosan functional groups, such as NH, OH, CC, CH and C=C. Whereas, DD demonstrated the purity of the isolated chitosan by analyzing the absorbance ratio between two specific wave numbers: A1655, which corresponded to the amine absorption band, and A3450, which aligned to the hydroxyl group. Calculation of the DD proportion was measured using the baseline method in equation below [13].

$$\%DD = 100 - \left[\frac{A_{1655}}{A_{3450}} x \frac{100}{1,33}\right] \qquad (1)$$

The DD standard for chitosan used in the healthcare sector follows the BSN recommendation of exceeding 75%. It also complied with the standards set by the European Pharmacopeia 6.0 (Eur: Ph 6.0) and the United States Pharmacopeia 34-NF 29 (USP 34-NF 29) [14,15].

The other physicochemical properties can be determined using XRD analysis, which provides both qualitative and quantitative crystal structure data. This analysis revealed information about the crystal structure, lattice parameters, phase transitions, and degree of crystallinity. Phase transitions were identified through the formation of characteristic peaks. These sample peaks were compared with the JCPDS Card No.39-1894 database to determine the chitosan spectrum formed at 2θ = 13.612°-37.934°, serving as a reference for identifying the phase of pure chitosan. The degree of crystallinity of a compound can be calculated using the following equation:

$$crystallinity = \frac{crystal\ area\ fraction}{fraction\ L\ crystal\ +\ fraction\ L\ amorphous} \quad (2)$$

In addition to determine the degree of crystallinity, the calculation of dislocation density was also important. Dislocations are irregularities in the crystal structure that significantly influence and alter many of a material's properties. Dislocation density referred to the number of dislocations per unit volume of material and was measured by the length of dislocation lines per unit volume of crystal. The dislocation density can be calculated using the following equation:

$$Dislocation\ density\ (\delta) = \frac{1}{D^2} \quad (3)$$

with D being the crystal diameter of the sample [16]. The properties and characteristics of a particle were closely linked to its shape, size, and size distribution. A common technique for assessing these properties was Particle Size Analyzer (PSA). The PSA results revealed the size distribution of chitosan particles. Testing the particle size distribution was essential for understanding the physical characteristics and performance of chitosan. Additionally, determining the size of chitosan provided insights into the production process, potential applications, and the stability of the resulting particles [17,18].

3. Result and Discussion

Oyster pearl shell powder has been effectively transformed into chitosan through a series of processes including deproteinization, demineralization, and deacetylation. During the deacetylation step, chitin was converted to chitosan using the strong base NaOH. Different concentrations of NaOH resulted in varying levels of water, ash, fat, and crude protein content, as shown in Table 1.

Table 1. Proximate test analysis of chitosan from pearl oyster shells

Parameters	Cs-Ref (%)	Cs-1 (%)	Cs-2 (%)
Water Content	0.724±0.0381	1.138±0.0286	8.051±0.3018
Ash Content	71.432±0.105	74.217±0.146	60.267±0.470
Fat Level	0.226±0.0306	0.176±0.0102	0.291±0.0049
Crude Protein Content	2.377±0.0069	3.054±0.0018	2.773±0.0049

According to the proximate analysis (table 1), the water content of chitosan complied with the BSN standards for powder preparations. Lower water content was associated with the storage duration of the material, with high-quality chitosan maintaining its properties when stored for longer periods. Chitosan quality was further evaluated based on its crude protein, fat, and ash content. Both specimens exhibited low fat percentages and contained crude protein. The lower fat content in chitosan improved its ability to reduce body fat [10,12]. The protein in the sample was a by-product of the deproteinization process. This protein in chitosan was valuable as it serves as an organic matrix. Based on these parameters, it can be inferred that chitosan is suitable for use as a biomaterial. Nevertheless, the proximate analysis indicated that the ash content in chitosan was an upward trend, surpassing 60%. This showed considerable minerals, resulting from an incomplete demineralization phase during its isolation.

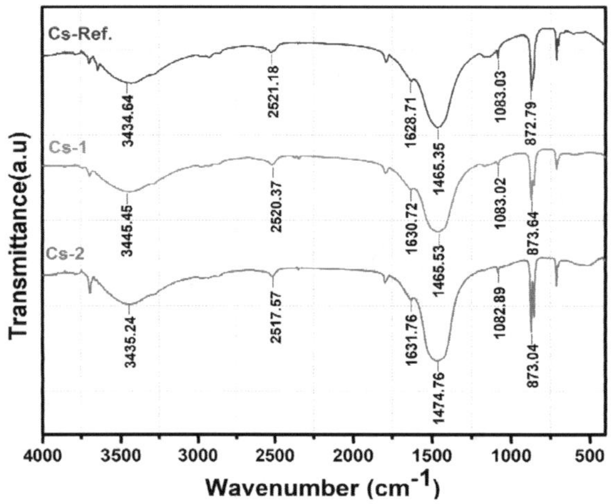

Figure 1. FTIR Spectrum Analysis of Chitosan

Due to proximate interpretation, chitosan can be classified as a biomaterial. This classification was supported by identifying chitosan's characteristic functional groups, specifically the OH and NH_2 spectra (Fig. 1). The functional groups of chitosan exhibited stretching and bending vibration modes (Table 2). Chitosan's high purity and quality were confirmed by analyzing its DD (Table 3) and crystal structure (Table 4). The DD was calculated using the baseline method in equation (1), with the highest DD value of 88.00% achieved at an 80% NaOH concentration (Cs-2).

Table 2. Absorbance analysis of chitosan functional groups

Functional Group	Wave Number (cm^{-1})		
	Cs-Ref	Cs-1	Cs-2
O-H *(stretch)*	3434.64	3445.45	3434.7
C-H *(stretch)*	2521.18	2520.37	2517.57
$N-H_2$ *(bend, primary amine)*	1628.71	1630.72	1631.76
$C-H_3$ *(bend)*	1465.35	1465.53	1474.76
C-O-C *(glucosamine ring)*	1083.03	1083.02	1082.89
β-1,4- glycosidic	872.29	873.64	873.04

The DD reflected the removal of acetyl groups from chitin, with a higher DD value signifying greater purity and quality. A high concentration of NaOH contributed to the significant degradation of acetyl groups. Additionally, elevated temperatures during the deacetylation process were crucial as they enhanced the solution's kinetic energy, accelerating the degradation of acetyl groups. As more acetyl groups were degraded, more amine groups were generated, resulting in an increased DD of chitosan.

Table 3. Characteristics of chitosan from pearl oyster shells

Samples	DD(%)	Crystallinity (%)	Particle Size (μm)
Cs-Ref	81.50	36.94	-
Cs-1	82.61	39.65	48.420
Cs-2	88.00	72.07	59.141

The results of crystallinity calculations using equation (2) exhibited that DD affected the crystallinity of chitosan (Table 3). Calculation of crystallinity based on the results of XRD analysis (Fig. 2) which shows the emergence of chitosan compounds at the 2θ 23.82° (220); 29.58° (123); 31.50° (241); 36.43° (104) on Cs-1 and 21.07° (200); 23.69° (220); 29.75° (123); 31.42° (241); 36.29 (104) on Cs-2 (Table 4).

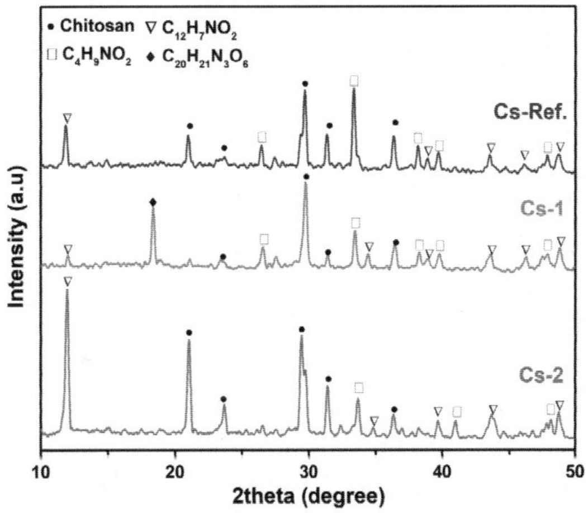

Figure 2. XRD analysis spectrum of chitosan

As the DD increases, the degree of chitosan crystallinity also rises (Table 3). The higher the DD, the more acetyl groups are lost. The fewer acetyl groups that are owned will impact the acceleration of crystal growth so that the crystallinity increases (Fig. 3) [19]. Other studies indicated that DD also impacted the crystallinity level, drug release within the chitosan matrix, and the extent of chitosan cross-linking [20].

Table 4. Qualitative and Quantitative Analysis of chitosan crystal structure

Parameter	Cs-Ref	Cs-1	Cs-2
Indeks miller	102; 220; 202; 241; 104	220; 123; 241; 104	200; 220; 123; 241; 104
angle 2θ (°)	20.98; 23.74; 27.42; 31.34; 36.52	23.82; 29.58; 31.50; 36.43	21.07; 23.69; 29.75; 31.42; 36.29
crystal structure	Orthorombic	Orthorombic	Orthorombic
Parameter kisi (Card No: 00-039-1894)	8.24 Å	8.24 Å	8.24 Å
FWHM (°)	-	0.197; 0.177; 0.472; 0.315	0.216; 0.157; 0.118; 0.197; 0.157

Density (ρ)	-	1.52	1.52
crystal size (nm)	-	41.21; 46.42; 17.50; 26.55	37.41; 51.70; 69.66; 41.89; 53.24
Dislocation density (nm^{-2})	-	0.590; 0.464; 3.270; 1.420	0.714; 0.374; 0.206; 0.570; 0.353
Microstain	-	4.075; 2.925; 7.302; 4.180	5.070; 3.267; 1.938; 3.056; 2.090
unit cell volume	-	1410.91	1410.91

The graph shows (Fig. 3) a positive linear relationship between NaOH concentration and the degree of deacetylation (DD). As the NaOH concentration increases from 60% to 80%, the DD rises from 81% to 88%. With an (R^2) value of 0.8737, the model demonstrates a significant correlation, indicating that higher NaOH concentrations directly enhance the degree of deacetylation.

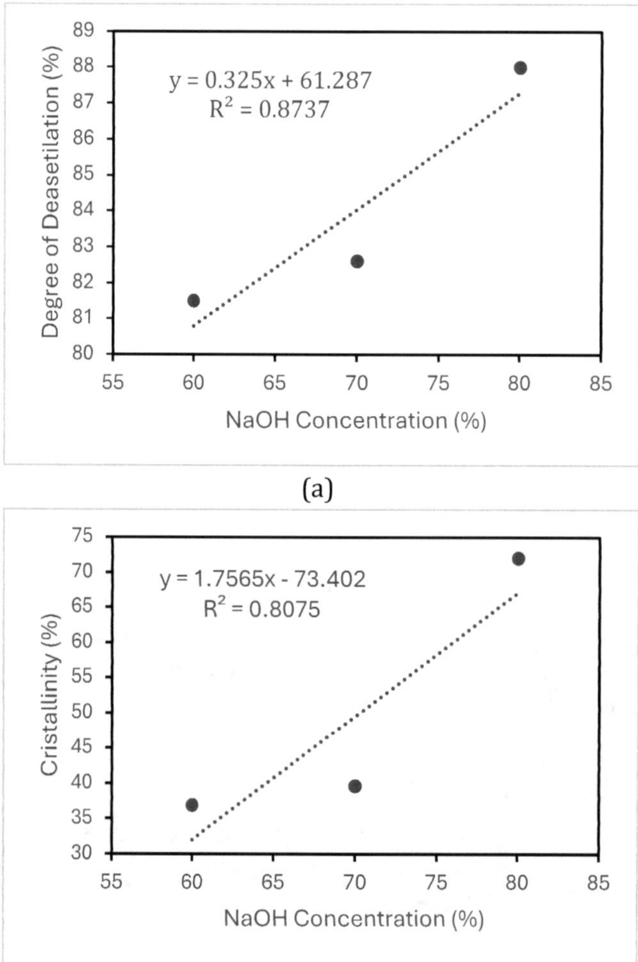

(b)

Figure 3. Graphical illustration of the relationship between (a) NaOH concentration dengan Degree of Deacetilation (b) NaOH concentration and crystallinity.

The graph illustrates a positive linear relationship between NaOH concentration and crystallinity. As the NaOH concentration increases from 60% to 80%, the crystallinity rises significantly from approximately 35% to 70%. The regression equation with an (R^2) value of 0.8075 indicates that 80.75% of the variation in crystallinity is explained by the NaOH concentration. This suggests that higher NaOH concentrations effectively enhance the crystallinity. In addition, variances in crystallinity statistics were affected by factors such as the source organism from which chitin was derived and the polymer extraction process [12]. The degree of crystallinity played a crucial role in medical applications, impacting the physical and chemical properties of materials, the formulation process, the characteristics of drug release from the chitosan matrix, and the rate and kinetics of drug delivery [21].

The concentration of NaOH shaped the average diameter of chitosan (Table 3). According to Figure 4, the Dv10, Dv50, and Dv90 values were recorded. For each sample, the Dv10 values were 25.850 and 35.177, while the Dv50 values, representing the median particle size, were 44.306 µm and 55.509 µm. The Dv90 values for the specimen were 75.892 and 87.404. This data indicated that the particle size of the Cs-2 was more prominent than Cs-1.

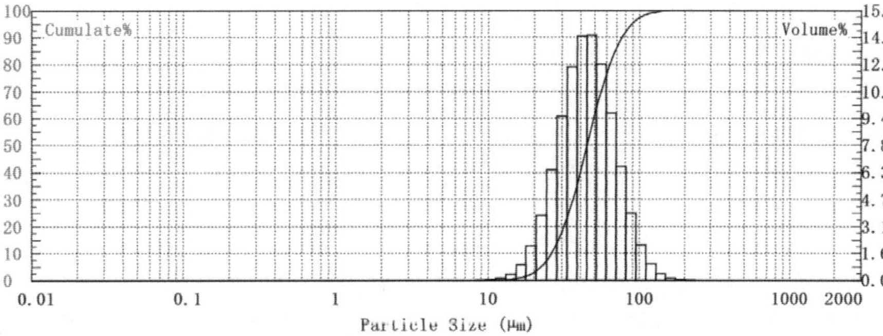

(a)

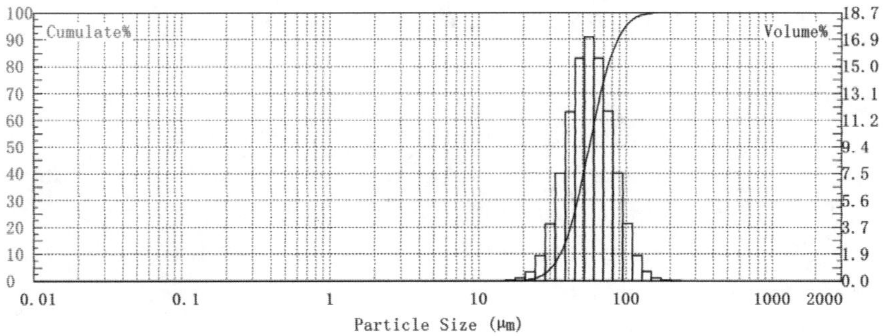

(b)

Figure 4. Particle size distribution (a) Cs-1 (b) Cs-2

Based on PSA analysis, it can be seen that the higher concentration of NaOH indicates a smaller particle size. Changes in particle size are caused by breaking the polymer chain, increasing crystal regularity, and reducing aggregation. These small particles exhibit better reactivity and interact more effectively with other materials. Additionally, the size of the particles played a crucial role in improving the material's performance and suitability for various applications, as it affected its physical, chemical, and mechanical properties [17].

The concentration of NaOH in the deacetylation process of chitosan significantly influences its chemical transformation and physical properties. Higher NaOH concentrations enhance the removal of acetyl groups, resulting in the formation of more reactive amine groups and an increased degree of deacetylation (DD) in chitosan. This increase in DD alters the chemical characteristics of chitosan, making it more polar, improving its solubility in acidic environments, and enhancing its antimicrobial properties and hydrogen-bonding capacity. Additionally, higher NaOH concentrations affect the crystallinity of chitosan; more intense alkaline conditions lead to polymer degradation, which reduces particle size and impacts polymer chain interactions. These changes make chitosan more adaptable for biomedical applications, as its physical and chemical properties can be tailored by adjusting the NaOH concentration.

4. Conclusion

Chitin has been successfully deacetylated into chitosan, meeting the quality standards recommended by BSN, with a DD exceeding 75%, as well as the standards set by the European Pharmacopoeia 6.0 (Eur: Ph 6.0) and the United States Pharmacopeia 34-NF 29 (USP 34-NF 29). Chitosan was characterized using 80% NaOH, achieving the highest DD of 88%, crystallinity of 72.07%, and a particle size of 44.306 µm. In contrast, chitosan prepared with 70% NaOH had a DD of 82.61%, crystallinity of 39.65%, and a particle size of 55.509 µm. However, the analysis revealed that chitosan produced with excessively high base concentrations is unsuitable for food and pharmaceutical applications. The ability to control the degree of deacetylation, crystallinity, and particle size enables the production of chitosan with targeted characteristics, such as improved solubility, bioactivity, and structural adaptability. These enhanced characteristics make chitosan highly valuable for drug delivery systems, wound dressings, and tissue engineering scaffolds in the biomedical field, where specific physical and chemical properties are essential for biocompatibility and effectiveness. Therefore, innovative isolation techniques that avoid using high base concentrations are required.

Acknowledgments

Thanks to all who support this research, especially the Fundamental Physics Laboratory and Material Physics Research Laboratory, Faculty of Mathematics and Natural Sciences, University of Mataram, for the research facilities. This research funding was also assisted by an internal grant from the University of Mataram for fiscal year 2024 and Hibah Doktor Lektor kepala FMIPA, DRPM Universitas Brawijaya.

References

[1] Alabaraoye E, Achilonu M, Hester R. Biopolymer (Chitin) from various marine seashell wastes: isolation and characterization. J Polym Environ 2018;26:2207–18.

[2] Kurniawidi DW, Alawiyah G, Rahayu S, Masruroh, Wirawan R, Destrianingtyas AS, Septiani N, Ardianto T, Illahi RR. Modification of Chitosan Isolation Method From Pearl Oyster Shell (*Pinctada Maxima*) as a Source of Natural Polymer. Journal of Physics: Conference Series 2024 Vol. 2866 (012015).

[3] Kurniawidi DW, Alaa S, Nurhaliza E, Safitri DO, Rahayu S, Ali M, et al. Synthesis and Characterization of Nano Chitosan from Vannamei Shrimp Shell (Litopenaeus vannamei). Jurnal Ilmiah Perikanan Dan Kelautan 2022;14:380–7.

[4] Siswoyo E, Zahra RN, Mai NHA, Nurmiyanto A, Umemura K, Boving T. Chitosan of Blood Cockle Shell (*Anadara granosaai)* as a natural Coagulant for Removal of Total Suspended Solid (TSS) and Turbidity of Well-water. Egyptian Journal of Aquatic Research : 49(2023) 283-289.

[5] Ilyas HN, Zia KM, Rehman S, Ilyas R, Sultana S. Utilization of shellfish industrial waste for isolation, purification, and characterizations of chitin from crustacean's sources in Pakistan. J Polym Environ 2021;29:2337–48.

[6] Nurlaili N, Alaa S, Rahayu S. Modifikasi Teknik Isolasi Biopolimer Kitosan Dari Cangkang Kerang Mutiara (Pinctada Maxima) Sebagai Adsorben Zat Warna Metilen Blue. ORBITA: Jurnal Pendidikan Dan Ilmu Fisika 2022;8:268–73.

[7] Hadi AG, Saddam NS, Ahmed DS, Yousif E. Extraction and Characterization of Chitosan Gained from Oyster and Fish Shells and Evaluation Its Application as Antioxidant. AIP Conference Proceedings (2022) 2386, 030016.

[8] Namboodiri MMT, Pakshirajan K. Valorization of waste biomass for chitin and chitosan production. Waste biorefinery, Elsevier; 2020, p. 241–66.

[9] Kou SG, Peters LM, Mucalo MR. Chitosan: A review of sources and preparation methods. Int J Biol Macromol 2021;169:85–94.

[10] Nurmaulida SE, Rahayu S, Alawiyah G, Hidayatullah K, Kurniawidi DW, Ali M. Fabrication of Chitosan Biopolymer from Pearl Oyster Shells (Pinctada maxima) for Medical Applications. Indonesian Physical Review 2023;6:240–9.

[11] Klenden YL, Mukkun L, Ndapamuri MH, and Pramatana F. The Migratory Locust (*Locusta migrotoria*) as a Potential Source Protein and Biopolymer Compounds In The Future. Biodiversitas 2023; 24(11): 5979-5987

[12] da Silva Lucas AJ, Oreste EQ, Costa HLG, López HM, Saad CDM, Prentice C. Extraction, physicochemical characterization, and morphological properties of chitin and chitosan from cuticles of edible insects. Food Chem 2021;343:128550.

[13] Fatima B. Quantitative analysis by IR: determination of chitin/chitosan DD. Modern Spectroscopic Techniques and Applications 2020;107.

[14] Revision USPConventionC of. The United States Pharmacopeia, United States Pharmacopeial Convention, Incorporated; 1984.

[15] Keitel S. Pharmacopoeial Standards: European Pharmacopoeia. Encyclopedia of Pharmaceutical Science and Technology, 2013:2691–703.

[16] Sutapa IW, Wahid Wahab A, Taba P, Nafie NL. Dislocation, crystallite size distribution and lattice strain of magnesium oxide nanoparticles. J Phys Conf Ser, vol. 979, IOP Publishing; 2018, p. 012021.

[17] Sidauruk H, Munandar A, Haryati S. The production of nanoparticles chitosan from crustaceans shell using the top-down method. E3S Web of Conferences, vol. 147, EDP Sciences; 2020, p. 03025.

[18] LUTHFIYANA N, BIJA S, NUGRAENI CD, LEMBANG MS, ANWAR E, LAKSMITAWATI DR, et al. Characteristics and antibacterial activity of chitosan nanoparticles from mangrove crab shell (Scylla sp.) in Tarakan Waters, North Kalimantan, Indonesia. Biodiversitas 2022;23.

[19] Alemu D, Getachew E, Mondal AK. Study on the Physicochemical Properties of Chitosan and their Applications in the Biomedical Sector. Int J Polym Sci 2023;2023:5025341.

[20] Safdar R, Omar AA, Arunagiri A, Regupathi I, Thanabalan M. Potential of Chitosan and its derivatives for controlled drug release applications–A review. J Drug Deliv Sci Technol 2019;49:642–59.

[21] M. Ways TM, Lau WM, Khutoryanskiy V V. Chitosan and its derivatives for application in mucoadhesive drug delivery systems. Polymers (Basel) 2018;10:267.

Journal of Physics: Conference Series **2980** (2025) 012027 doi:10.1088/1742-6596/2980/1/012027

OPTIMIZATION OF GREEN SYNTHESIS OF NANOGOLD USING RED GINGER RHIZOME EXTRACT AND POLYVINYLPYRROLIDONE (PVP)K60

Titta H Sutarna[1], Gladis K Pratiwi[1], Fikri Alatas[1], Hestiary Ratih[1], Alifiani M Hermawati[1]

[1] Pharmaceutical Technology, Faculty of Pharmacy, Universitas Jenderal Achmsd Yani, Cimahi, West Java, Indonesia

*E-mail: titta.hartyana@lecture.unjani.ac.id

Abstract. Green synthesis of nanoparticles is an environmentally friendly approach that utilizes plant-derived reducing agents to reduce metal ions to their elemental form. One of the bioreductants is red ginger (*Zingiber officinale* Rosceo) rhizome which contains flavonoids and polyphenols. This research aimed to optimize the concentration of red ginger extract and PVP-K60 for stable nanogold synthesis. The study began with determining the antioxidant activity of red ginger extract (IC50 = 29.63 μg/mL) and synthesizing nanogold using 600 ppm extract. Different concentrations of PVP-K60 (0.5%, 1%, 2%, and 5%) were used as stabilizers. The synthesized nanoparticles were characterized using visual observation, UV-vis spectrophotometry, and particle size analysis. Red ginger rhizome extract effectively reduces gold ions at 600 ppm, and 2% PVP-K60 stabilizes nanoparticles for 28 days.

Keywords: Nanogold, Red ginger (*Zingiber officinale* Rosceo), PVP-K60

INTRODUCTION

Nanotechnology is a science that focuses on nanometer-scale particles that are increasingly being developed byscientists. Nanotechnology is used invarious fields, one of which is pharmaceutical science. The gold nanoparticle or nanogold that have been successfully synthesized are between 10-100 nm in size [1]. Two main approaches for synthesizing gold nanoparticles are top-down and bottom-up [2]. Top-down synthesis involves breaking down larger materials into nanoparticles through mechanical processes like milling. Bottom-up synthesis, on the other hand, builds nanoparticles from smaller units, such as atoms or molecules, using techniques like sol-gel, chemical precipitation, and gas phase agglomeration. [3,4].

Journal of Physics: Conference Series **2980** (2025) 012027 doi:10.1088/1742-6596/2980/1/012027

Unfortunately, these methods suffer from limitations including the use of toxic solvents, hazardous waste, and high energy requirements. [5]. A sustainable and unconventional method for nanoparticle synthesis is bioreduction, which employs plant extracts to facilitate the reduction of metal ions into nanoparticles [6–8]. Red ginger (*Zingiber officinale* Roscoe) rhizome extract can be employed to synthesize nanogold particles. While the ideal nanoparticle concentration is 600 ppm, aggregation consistently occurs. To address this issue, a stabilizing agent is required to maintain the stability of the nanoparticle size [9–11]. Polyvinylpyrrolidone (PVP) is a polymer utilized as a stabilizer in the manufacture of nanogold particles [12–15]. The function of the stabilizer in addition to preventing uncontrolled particle growth, preventing aggregation due to surface energy of large nanoparticles, can also control the color changes that occur [16–18].

While the synthesis of gold nanoparticles using red ginger rhizome extract has been explored, the long-term stability of these nanoparticles remains a concern. Aggregation is a common issue that can significantly impact the properties and applications of nanoparticles. The optimal concentration of PVP-K60 as a stabilizer for gold nanoparticles synthesized using red ginger rhizome extract has not been definitively determined. The aim of this study was to synthesize stable gold nanoparticles using red ginger rhizome extract as a reducing agent and to optimize the concentration of PVP-K60 as a stabilizer to increase the stability of the synthesized nanoparticles.

MATERIALS AND METHOD

Materials

The main materials used in this research are gold metal purchased from PT. Antam Indonesia. Red ginger (*Zingiber officinale* Rosceo) rhizome was obtained from Balitro, Bogor, Indonesia. Nitric acid and hydrochloric acid purchased from Merck Indonesia, while polyvinylpyrrolidone K60 was obtained from BASF, Indonesia.

Production of Red Ginger Rhizome Extract (*Zingiber officinale* Rosceo)

The red ginger rhizomes were initially cleaned with running water and then cut into smaller pieces. Subsequently, the pieces were dried in a drying cabinet at a temperature of 50°C. The dried material was then ground into a fine powder and sieved through a 60-mesh sieve. To extract the bioactive compounds, a maceration process using 96% ethanol was employed. The resulting ethanol extract was concentrated to a thick consistency using a rotary evaporator at a temperature of 50°C.

Determination IC $_{50}$ Red Ginger Rhizomes with the DPPH method

Approximately 25 mg of red ginger extract was accurately weighed and dissolved in 50 mL methanol. The solution was then homogenized and diluted into six different concentrations. Two milliliters of each concentration were mixed with two mL of DPPH solution and incubated in the dark for a duration of 30 minutes. The absorbance of the resulting mixture was measured at the maximum wavelength of DPPH using UV-Visible spectrophotometry. The percentage inhibition

Journal of Physics: Conference Series **2980** (2025) 012027 doi:10.1088/1742-6596/2980/1/012027

of DPPH was calculated using equation 1.

$$\% \, Inhibition = \frac{\text{Absorbance Blanko} - \text{Absorbance Sample x 100\%}}{\text{Absorbance Blanko}} \quad \text{(equation 1)}$$

Five milligrams of DPPH were weighed and dissolved in 100 mL of methanol in a dark flask. The solution was shaken until complete dissolution. Four milliliters of this DPPH solution were pipetted into test tubes and incubated for 30 minutes at 37°C in the dark.

Preparation of 0.5 mM Chloroauric Acid (HAuCl4) Solution

A 0.05 g gold sample was dissolved in a 3:1 mixture of HCl and HNO3. The solution was heated until complete dissolution of the gold, then diluted to 500 mL with deionized water to obtain a 0.5 mM HAuCl4 solution [19].

Synthesis of Nanogold

A 2000 ppm stock solution of red ginger rhizome extract was prepared and diluted to concentrations of 600, 900, and 1200 ppm. Three mL of each concentration was transferred to separated amber vial, followed by the addition of three mL of 0.5 mM HAuCl4 solution. The gold nanoparticles were characterized using visual inspection, UV-Visible spectroscopy, and particle size analyzer.

Addition of PVP-K60 into Nanogold

A total of 3 mL of 600 ppm red ginger rhizome extract solution containing 0.5%, 1%, 2%, or 5% Polyvinylpyrrolidone-K60 (PVP-K60) was heated to a temperature of 60°C for one minute. Subsequently, 3 mL of 0.5 mM HAuCl4 solution was added to each solution. The resulting gold nanoparticle solutions were characterized using visual observation, UV-Visible spectrophotometry, and Particle Size Analysis (PSA).

Characterization Nanogold by Visual

Observation of Color Change of NanogoldVisual observation of the color change of nanogold was carried out at 24 hours.

Characterization Nanogold UV-Visible Spectrophotometry

A mL sample was placed in a cuvette and measured using a UV-Visible spectrophotometer against a deionized water blank. The extract solution mixed with 0.5 mM chloroauric acid was analyzed at 0 and 24 hours within the wavelength range of 400-600 nm. Nanogold typically absorbs light in the 500-600 nm range. [10].

Journal of Physics: Conference Series **2980** (2025) 012027 doi:10.1088/1742-6596/2980/1/012027

Characterization Nanogold by Particle Size Analyzer (PSA)

Samples were collected and placed into the cuvette, which was subsequently positioned in the cuvette holder for particle measurement.

Zeta Potential Measurement of Nanogold

Zeta potential measurements were performed using a Horiba SZ-100V2 particle size analyzer. Samples were diluted in 10 mL of deionized water and measured in a cuvette.

Stability Test

The stability test of nanogold after the addition of Polyvinyl Pyrolidone (PVP-K60) stabilizer of 0.5%; 1%; 2% and 5% respectively was carried out by visual observation, UV-visible spectrophotometer, and Particle Size Analyzer (PSA) after storage for 0, 1, 7, 14, 21, and 28 days.

RESULTS AND DISCUSSION

The initial step involves determining the water content of simplisia. To prevent microbial growth, the water content was reduced to 3.1%, meeting the Indonesian Herbal Pharmacopoeia Edition II requirement of less than 11% [20]. The next step involved determining the total ash content, water-soluble ash content, and acid-soluble ash content to assess the inorganic and mineral content and identify potential impurities like sand. The results are summarized in Table 1.

Table 1. Results of Determination of Ash Content of Red Ginger Rhizome Simplicia

Characterization Type	Determination of Ash Content
Total Ash Content	4.91±0.17%
Water Soluble Ash Content	4.63±0.32%
Acid Soluble Ash Content	0.35%±0.35%

n=3

To identify the bioreductive compounds in red ginger rhizomes, a phytochemical screening was conducted, revealing the presence of flavonoids [21–23]. The rhizomes were extracted using 96% ethanol through maceration, a non-heat-based method to preserve the bioactive compounds. The extraction yield was 22.33%. The antioxidant potential of the red ginger rhizome thick extract was assessed using the DPPH (2,2-diphenyl-1-picrylhydrazil) method[24–26]. This method measures the ability of a compound to scavenge DPPH free radicals. The flavonoids present in the red ginger rhizome extract can donate hydrogen atoms to DPPH radicals, thereby stabilizing them. The decrease in DPPH radical concentration was monitored using a UV-Visible spectrophotometer at a maximum wavelength of 520 nm [27]. The calculated IC50 value of 29.63 µg/mL for the red

Journal of Physics: Conference Series **2980** (2025) 012027 doi:10.1088/1742-6596/2980/1/012027

ginger rhizome extract indicates strong antioxidant activity.

Prior to nanogold synthesis, UV-visible spectrophotometry was used to determine the optimal wavelengths of the red ginger rhizome extract and HAuCl4 solutions. Red ginger extract concentrations of 20, 30, and 40 g/mL were used for the synthesis. Figure 1 shows the visual appearance of nanogold synthesized using red ginger bioreductant extract with concentrations of 600, 900, and 1200 ppm.

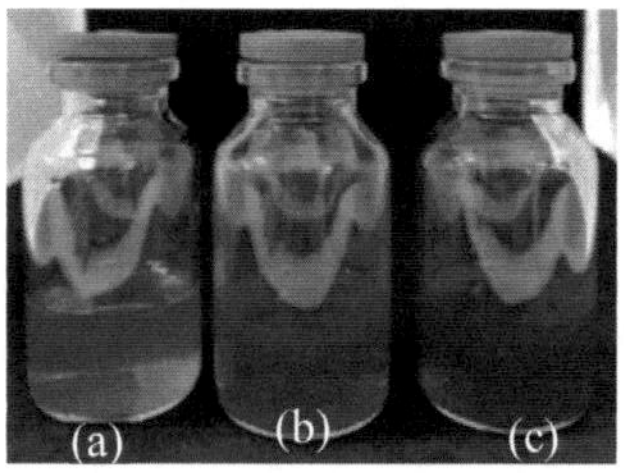

Figure 1. Visual appearance of nanogold synthesized using red ginger rhizome extract with concentrations (a) 600. (b) 900, and (c) 1200 ppm.

$HAuCl_4$, being a weak acid, establishes an equilibrium in solution. To ensure efficient nanogold formation, a 1:1 ratio of HAuCl4 and red ginger extract was employed. Qualitative observations, manifested as color changes from clear to pink to purple, indicated successful nanoparticle formation within 60 minutes at all concentrations. The color change is attributed to the growth of gold clusters, resulting from the interaction of Au atoms through metallic bonds. The reduced concentration of $AuCl4^-$ ions facilitate their reduction to neutral Au atoms. In their ionic form, $AuCl4^-$ ions repel each other due to electrostatic repulsion. However, upon reduction to neutral Au atoms, the electrostatic repulsion is eliminated, allowing the Au atoms to interact and form nano-sized clusters [10].

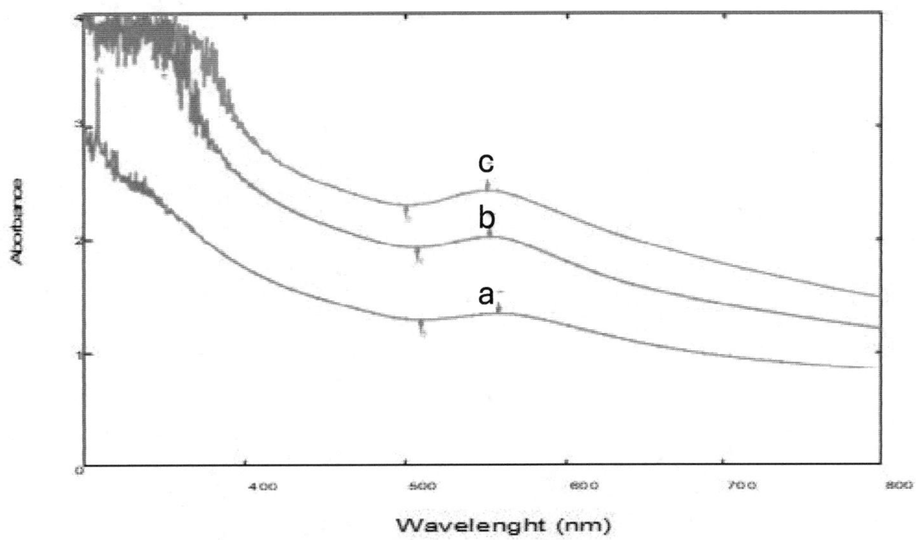

Figure 2. UV-vis spectrum of nanogold synthesized using red ginger rhizome extract with concentrations (a) 600. (b) 900, and (c) 1200 ppm.

UV-vis absorption spectra of 600, 900, and 1200 ppm red ginger extract solutions confirmed the formation of nanogold. Figure 2 shows the UV-Vis spectrum of nanogold synthesized using red ginger extract. The maximum wavelengths of 557.7 nm, 552 nm, and 550 nm, respectively, fall within the typical nanogold absorption range of 500-600 nm. Further characterize the synthesized nanogold, a concentration of 600 ppm red ginger rhizome extract was chosen, as previous research had demonstrated successful nanoparticle formation at this concentration [10,28,29].

To prevent nanoparticle aggregation and maintain their size and number, stabilizers like excipients are added. These stabilizers reduce interparticle forces, preventing the formation of larger clusters. [30]. One of the stabilizers is a hydrophilic polymer, namely PVP- k60 , this stabilizer as a binder is used at a concentration of 0.5-5% [16]. Gold nanoparticles were stabilized using four concentrations of PVP-K60: 0.5%, 1%, 2%, and 5%. Particle size, polydispersity index (PI), and zeta potential were measured using PSA. The particle sizes of PVP-K60 stabilized nanogold are shown in Figure 3. Initial particle size measurements, performed at time zero, revealed nanogold particle sizes in the range of 200–300 nm at all PVP-K60 concentrations with polydispersity index (PI) values below 0.5. Lower PI values indicate a more homogeneous particle size distribution and, consequently, greater stability [29–31]. However, a marked increase in particle size beyond 1000 nm was observed after 21 days of storage at 0.5 and 1% of PVP-K60 additions, while at 2 and 5% PVP-K60 additions the particle sizes remained below 1000 nm. The addition of 0.5% and 1% concentrations proved insufficient to inhibit aggregation beyond a 14-day period. This may be attributed to the lower viscosity of the solution, which is conducive to particle aggregation and subsequent growth [31].

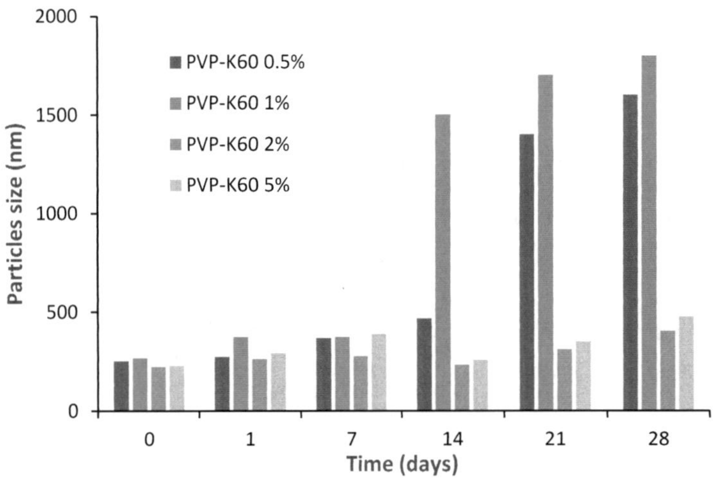

Figure 3. Particle size of nanogold after storage with addition of PVP-K60 stabilizer

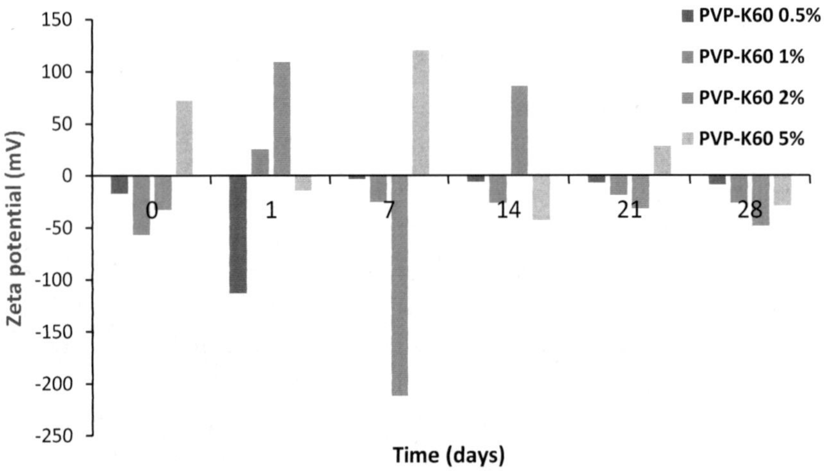

Figure 4. Zeta potential of nanogold with addition of PVP-K60 stabilizer

Zeta potential is a crucial parameter for predicting the stability of nanoparticle dispersions [32,33]. Figure 4 shows the zeta potential value of nanogold with the addition of PVP K-60 as a stabilizer. The 5% PVP-K60 concentration exhibited lower zeta potential values of 28.17 mV and -29.13 mV at 21 and 28 days, respectively, indicating instability. In contrast, the 2% concentration demonstrated superior stability with a zeta potential of -48.43 mV at 28 days.

CONCLUSION

Red ginger rhizome extract effectively reduces gold ions to form nanoparticles at a concentration of 600 ppm. A 2% concentration of PVP-K60 stabilizer maintains nanoparticle stability for up to 28 days.

ACKNOWLEDGEMENT

The research team thanks LPPM Universitas Jenderal Achmad Yani for their financial support.

REFERENCES

[1] Hammami I, Alabdallah NM, Al A. Gold nanoparticles : Synthesis properties and applications. J King Saud Univ – Sci 2021;33:101560. https://doi.org/10.1016/j.jksus.2021.101560.

[2] Jiang Z, Li L, Huang H, He W, Ming W. Progress in Laser Ablation and Biological Synthesis Processes: "Top-Down" and "Bottom-Up" Approaches for the Green Synthesis of Au/Ag Nanoparticles. Int J Mol Sci 2022;23. https://doi.org/10.3390/ijms232314658.

[3] Harish V, Ansari MM, Tewari D, Gaur M, Yadav AB, García-Betancourt ML, et al. Nanoparticle and Nanostructure Synthesis and Controlled Growth Methods. Nanomaterials 2022;12:1–30. https://doi.org/10.3390/nano12183226.

[4] Lin HK, Yan TH, Bashir S, Liu JL. Synthesis of nanomaterials using bottom-up methods. Adv Nanomater Their Appl Renew Energy, Second Ed 2022:61–110. https://doi.org/10.1016/B978-0-323-99877-2.00003-5.

[5] Thakkar KN, Mhatre SS, Parikh RY. Biological synthesis of metallic nanoparticles. Nanomedicine Nanotechnology, Biol Med 2010;6:257–62. https://doi.org/10.1016/j.nano.2009.07.002.

[6] Mofrad R, Hadi R, Tahmasebi B, Farhoudian S, Mehravar M, Nasiri R. Green synthesis of gold nanoparticles using plant extract: Mini-review. Nanochem Res 2017;2:8–19. https://doi.org/10.22036/ncr.2017.01.002.

[7] Elia P, Zach R, Hazan S, Kolusheva S, Porat Z, Zeiri Y. Green synthesis of gold nanoparticles using plant extracts as reducing agents. Int J Nanomedicine 2014;9:4007–21. https://doi.org/10.2147/IJN.S57343.

[8] Bharadwaj KK, Rabha B, Pati S, Sarkar T, Choudhury BK, Barman A, et al. Green synthesis of gold nanoparticles using plant extracts as beneficial prospect for cancer theranostics. Molecules 2021;26. https://doi.org/10.3390/molecules26216389.

[9] Štrba P, Šebesta M, Kratošová G, Urík M, Vávra I, Peikertová P, et al. Nanogold Biosynthesis Mediated by Mixed Flower Pollen Grains 2019;19:2983–8. https://doi.org/10.1166/jnn.2019.15853.

[10] Aslan K, Lakowicz JR, Geddes CD. Nanogold-plasmon-resonance-based glucose sensing

2004;330:145–55. https://doi.org/10.1016/j.ab.2004.03.032.

[11] Wang N, Cao P, Ma H, Lin M. How Stabilizers and Reducing Agents A ff ect the Formation of Nanogold Amalgams 2021. https://doi.org/10.1021/acs.langmuir.1c00618.

[12] Sutarna TH, Gozali D, Panantarani C. Box Behnken Design Optimization for Synthesis of Dexibuprofen Nanocrystals Using Anti- solvent Precipitation Method and Oral Bioavailability Testing 2023;34:5626–43.

[13] Ibrahim AM, Alzahrani HAA, Abd El-Latif MM, Selim MM. Influence of different stabilizers on the morphology of gold nanoparticles. Bull Natl Res Cent 2019;43:0–4. https://doi.org/10.1186/s42269-019-0070-5.

[14] Nalawade P, Mukherjee T, Kapoor S. Green synthesis of gold nanoparticles using glycerol as a reducing agent. Adv Nanoparticles, 2013;2:78–86.

[15] Bandulasena M V., Vladisavljević GT, Odunmbaku OG, Benyahia B. Continuous synthesis of PVP stabilized biocompatible gold nanoparticles with a controlled size using a 3D glass capillary microfluidic device. Chem Eng Sci 2017;171:233–43. https://doi.org/10.1016/j.ces.2017.05.035.

[16] Sutarna TH, Gozali D, Panantarani C, Alatas F. Nanocrystals of Poorly Soluble Drugs : Polyvinylpyrrolidone (PVP) as Stabilizer 2022;20:4655–68. https://doi.org/10.14704/nq.2022.20.5.NQ22746.

[17] Madkour M, Bumajdad A, Al-Sagheer F. To what extent do polymeric stabilizers affect nanoparticles characteristics? Adv Colloid Interface Sci 2019;270:38–53. https://doi.org/10.1016/j.cis.2019.05.004.

[18] Kim EJ, Yeum JH, Choi JH. Effects of Polymeric Stabilizers on the Synthesis of Gold Nanoparticles. J Mater Sci Technol 2014;30:107–11. https://doi.org/10.1016/J.JMST.2013.11.012.

[19] Husen A. Gold Nanoparticles from Plant System : Synthesis , Characterization and their Gold Nanoparticles from Plant System : Synthesis , Characterization and their Application 2018. https://doi.org/10.1007/978-3-319-46835-8.

[20] Kementerian Kesehatan Republik Indonesia, editor. Farmakope Herbal Indonesia. 2nd ed. Jakarta: Kementerian Kesehatan Republik Indonesia; 2017.

[21] Irfan M, Moniruzzaman M, Ahmad T, Mandal PC, Bhattacharjee S, Abdullah B. Ionic liquid based extraction of flavonoids from Elaeis guineensis leaves and their applications for gold nanoparticles synthesis. J Mol Liq 2017;241:270–8. https://doi.org/10.1016/j.molliq.2017.05.151.

[22] Sierra JA, Vanoni CR, Tumelero MA, Plá Cid CC, Faccio R, Franceschini DF, et al. Biogenic approaches using citrus extracts for the synthesis of metal nanoparticles: the role of flavonoids in gold reduction and stabilization. New J Chem 2016;40:1420–9. https://doi.org/10.1039/C5NJ02128F.

[23] Tsai C, Lin L. Heliyon DPPH scavenging capacity of extracts from Camellia seed dregs using polyol compounds as solvents. Heliyon 2019;5:e02315.

https://doi.org/10.1016/j.heliyon.2019.e02315.

[24] Foti MC, Biomolecolare C, Gaifami P. Use and Abuse of the DPPH • Radical 2015. https://doi.org/10.1021/acs.jafc.5b03839.

[25] Ionita P. The Chemistry of DPPH · Free Radical and Congeners. Int J Mol Sci 2021;22:1–15. https://doi.org/https://doi.org/10.3390/ijms22041545.

[26] Irivibulkovit KS, Ouanthavong SN, Ameenoi YS. Paper-based DPPH Assay for Antioxidant Activity Analysis 2018;34:795–800.

[27] Tjaša, Prevc and Nataša Šegatin, Nataša Poklar Ulrih BC. Talanta DPPH assay of vegetable oils and model antioxidants in protic and aprotic solvents 2013;109:13–9. https://doi.org/10.1016/j.talanta.2013.03.046.

[28] Aslan K, Lakowicz JR, Geddes CD. Nanogold Plasmon Resonance-Based Glucose absorption bands due to electron oscillations induced by 2014;77:2007–14.

[29] Barchiesi D, Kessentini S, Guillot N, Chapelle L De, Grosges T. Localized surface plasmon resonance in arrays of nano-gold cylinders : inverse problem and propagation of uncertainties Abstract : 2013;21:2245–62.

[30] Rajapriya M, Aruna S, Raju S, Ravichandran B. Correction to : Synthesis and Characterization of Zinc Oxide Nanoparticles Using Cynara scolymus Leaves : Enhanced Hemolytic , Antimicrobial , Antiproliferative , and Photocatalytic Activity. J Clust Sci 2019;1:10876. https://doi.org/10.1007/s10876-019-01734-1.

[31] Leiva A, Bonardd S, Pino M, Saldías C, Kortaberria G, Radić D. Improving the performance of chitosan in the synthesis and stabilization of gold nanoparticles. Eur Polym J 2015;68:419–31. https://doi.org/10.1016/j.eurpolymj.2015.04.032.

[32] Clogston JD, Patri AK. Chapter 6 Zeta Potential Measurement 2011;697. https://doi.org/10.1007/978-1-60327-198-1.

[33] Pandey PK, Maheshwari R, Raval N, Gondaliya P, Kalia K, Tekade RK. Journal of Colloid and Interface Science Nanogold-core multifunctional dendrimer for pulsatile chemo- , photothermal- and photodynamic- therapy of rheumatoid arthritis. J Colloid Interface Sci 2019;544:61–77. https://doi.org/10.1016/j.jcis.2019.02.073.

Green synthesis of ZnO-red betel nanocomposites (*Piper crocatum Ruiz & Pav*) and their antibacterial activity

Salma Aridha Muflihah[1], Akfiny Hasdi Aimon[1,2*], Arie Wibowo[3,4], Neni Surtiyeni[1], Mikrajuddin Abdullah[1] and Muhammad Ilham Gusti Yuliantoro[1]

[1] Department of Physics, Faculty of Mathematics and Natural Sciences, Bandung Institute of Technology, Bandung 40132, Indonesia
[2] Collaboration Research Center for Advanced Energy Materials, National Research and Innovation Agency – Institut Teknologi Bandung, Bandung 40132, Indonesia
[3] Materials Science and Engineering Research Group, Faculty of Mechanical and Aerospace Engineering, Institut Teknologi Bandung, Bandung 40132, Indonesia
[4] Research Center for Nanoscience and Nanotechnology, Institut Teknologi Bandung, Bandung 40132, West Java, Indonesia

*E-mail: akfiny@itb.ac.id

Abstract. Antibacterial resistance has been a major global health problem, requiring the development of alternative antimicrobial agents. This research studied the synthesis and properties of zinc oxide (ZnO)-red betel nanocomposites, mainly focused on its antibacterial properties. ZnO, known for its antimicrobial efficacy, is combined with red betel leaves that rich of bioactive compounds to enhance antibacterial activity against resistant bacterial strains. The goal is to create a sustainable and effective antimicrobial agent that maximize the synergistic effects of ZnO nanoparticles and compounds in red betel leaves. Characterization techniques including XRD and SEM were employed to confirm the structural and morphological properties of the nanocomposites. Antibacterial assays demonstrated significant activity against *S. aureus* and *E. coli*, suggesting the potential of these nanocomposites as innovative solution in the fight against antibacterial resistance. This work contributes to the growing field of nanotechnology-driven solutions for healthcare, highlighting a promising approach to address one of the most pressing issues in modern medicine.

Keywords: antibacterial, nanocomposites, red betel, ZnO.

1. Introduction

Nanomaterials research has been one of the leading fields of study in this century. Researchers are continually optimizing their properties, such as their electrical conductivity, mechanical strength, chemical reactivity, and biocompatibility. These remarkable properties give nanomaterials great potential for a wide range of applications. They are being used in biomedicine, automotive industries, infrastructure development, battery and energy storage systems, demonstrating their versatility and transformative impact across various sectors [1].

Among various applications, nanomaterials have shown promise in addressing critical health challenges, particularly in the development of antibacterial agents.

Nanoparticles and nanocomposites are two notable types of nanomaterials. Nanocomposites typically combine two or more materials to harness the advantageous properties of each. One widely used material is zinc oxide (ZnO), known for its excellent optical, chemical, and electronic properties [1]. Nano-ZnO has gained widespread interest due to their biocompatibility, low toxicity, and multifunctionality. Known for their photocatalytic activity, UV-blocking capabilities, and strong antibacterial properties, ZnO has been utilized in various industries, including cosmetics, sunscreens, and coatings[2]. Its ability to generate reactive oxygen species (ROS) in response to light or environmental conditions allows it to disrupt bacterial cell walls, making it a potent antibacterial agent [3].

ZnO is often combined with other materials; for instance, it has been paired with titanium dioxide (TiO_2) to serve as a reducing agent and to degrade water pollutants. The significance of ZnO nanocomposites lies in their diverse applications and potential to enhance performance in areas such as environmental remediation, catalysis, and biomedical devices[4]. ZnO can also be integrated with biological materials, including plants, microorganisms, and bio-regenerable substances [1,3,5]. Biological materials act as reducing agents, stabilizers or capping agents in nanomaterial synthesis [1]. Plants are preferred over microorganisms for nanoparticle synthesis because various parts such as leaves, stems, and roots can be used [1]. Moreover, plants do not require cell culture preparation, incubation periods, or extensive examination [1]. The phytogenic synthesis of nano-ZnO typically involves mixing plant extracts rich in phytochemicals with a zinc precursor solution, with adjustments to concentrations, pH levels, and temperatures [1]. Many plant leaves exhibit inherent antibacterial activity due to the presence of bioactive compounds such as flavonoids, tannins, and alkaloids. These compounds have been used for centuries in traditional medicine, offering a natural and sustainable approach to combating bacterial infections. Recent research has focused on combining these natural antibacterial agents with nanotechnology to enhance their efficacy.

Antimicrobial resistance, a phenomenon where microbial such as bacteria, virus, and fungi change and the usual antimicrobial medicine unable to kill them anymore [6,7]. This issue has been a global concern currently. As pathogens evolve, they develop resistance mechanisms that render standard treatments ineffective, leading to prolonged infections and increased mortality rates. This can be a wider problem, affecting many life aspects such as in economy [7]. The urgency to develop novel antibacterial agents that can overcome this resistance is greater than ever, as the current arsenal of antibiotics dwindles in effectiveness.

In response to this critical need, this research aims to develop an alternative antibacterial agent by synthesizing nanocomposites of ZnO with natural leaves. We selected red betel (*Piper crocatum Ruiz & Pav*), a plant widely used in Asia, known for its beneficial properties. This plant contains bioactive compounds with potent antibacterial properties, such as phenol, flavonoids, saponins, terpenoids, alkaloids, and tannins [8]. This research could provide a sustainable, low-cost solution to the growing problem of antibacterial resistance while minimizing environmental impact.

Journal of Physics: Conference Series **2980** (2025) 012028 doi:10.1088/1742-6596/2980/1/012028

2. Materials and Methods

2.1 Materials

The materials used in this experiment are the same as our latest research. They are $Zn(NO_3)_2 . 6 H_2O$, NaOH, demineralized water, and fresh red betel leaves. All the chemicals used were of analytical grade and applied without further purification.

2.2 Synthesis Methods

To prepare the plant material, we used the maceration method as described in reference [5]. Initially, red betel leaves were thoroughly washed twice with tap water and once with distilled water to remove any dirt. Following this, 6 grams of the red betel leaves were boiled in 150 mL of distilled water for 8 minutes. The resulting solution was allowed to cool to room temperature before being filtered through Whatmann no. 1 filter paper to eliminate any residual leaf fragments. The next step involved synthesizing ZnO-red betel nanocomposites. We began by mixing 4.46 grams of $Zn(NO_3)_2{\cdot}6H_2O$ with the prepared red betel solution for one hour. This mixture was then heated at 60 °C for an hour. Then, NaOH was added to the solution until the pH of solution reach 7. The sample was subsequently centrifuged at 3500 rpm for 15 minutes to separate the supernatant from the pellet. The pellet was washed several times with demineralized water and then dried at 220 °C for 90 minutes. The resulting products were further characterized as detailed in the following section.

2.3 Characterization Methods

Scanning Electron Microscopy (SEM) is utilized to analyze the surface morphology of samples. SEM SU 3500 that we used operates by scanning a focused electron beam over the sample to produce detailed images. X-ray Diffraction (XRD) is primarily used to study crystallinity and determine the crystalline size of the samples; in this research, we employed a Rigaku Miniflex for XRD analysis. The antibacterial activity was assessed using the Total Plate Count (TPC) method, with *Staphylococcus aureus* and *Escherichia coli* as the test bacterium. The TPC method begins with weighing the sample and diluting it in distilled water to prepare 10 ppm solutions. The samples are then mixed with *Staphylococcus aureus* and *Escherichia coli* bacterial suspensions with known populations. The mixtures are left to contact for 5 minutes. Afterward, the Total Plate Count (TPC) is tested on the resulting mixtures of the samples and the bacteria. These are incubated at 37 °C for 24 hours. Finally, the number of colonies that grow is counted and compared with the initial population of each bacterium.

3. Result and Discussions

The experiment produced a white-greyish powder of ZnO-red betel nanocomposite, which was characterized using various methods. XRD characterization revealed intensity peaks over a 2θ graph, indicating a pattern that could be analyzed further. SEM images qualitatively showed the surface morphology and size of the samples. Lastly, the antibacterial test revealed the total bacteria colony count on each plate for different test bacteria.

The XRD pattern was shown in Figure 1. It was analyzed using Match! 3 software and compared with the (JCPDS) 036-1451, confirming a hexagonal crystal system with ZnO wurtzite structure [9]. The identified peaks and their corresponding hkl indices were 31.88° (100), 34.55° (002), 36.37° (101), 47.69° (102), 56.73° (110), 62.99° (103), and 68.11° (112).

The crystalline size of the obtained ZnO can be predicted from XRD result using Debye-Scherrer equation [3], which is shown in Equation 1 as follow:

$$D = K\lambda/\beta \cos\theta \quad (1)$$

In this context, D is the size of the nanoparticles' crystals, K is the Scherrer constant (0.9), λ is the wavelength (1.54), and β is the full width at half maximum (FWHM). Average crystallite size was obtained as 12.845 nm. This small size provides a higher surface area, likely attributed to the use of low temperatures during the synthesis process. [10].

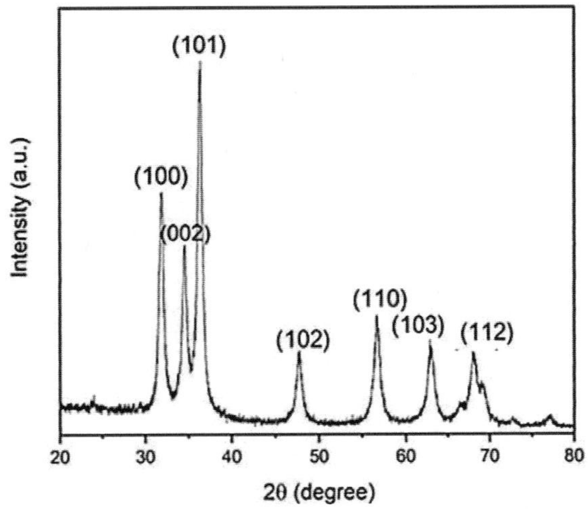

Figure 1. XRD spectra of the sample.

The mechanisms involved in the synthesis of nanoparticles using plant extract are described as follows. Phytochemicals (bioactive compounds found in plants) initially coordinate with Zn^{2+} ions, primarily through donor oxygen atoms, leading to the creation of a Zn-phytochemical complex [1]. Karnan et al. using *Nephelium lappaceum L.* extract in the synthesis process [11]. Here, polyphenolic ellagic acid interacts with zinc ions through its aromatic hydroxyl groups, forming a zinc-ellagate complex [11] In a separate study conducted by Thi et al., organic compounds such as flavonoids, limonoids, and carotenoids from natural sources form complex

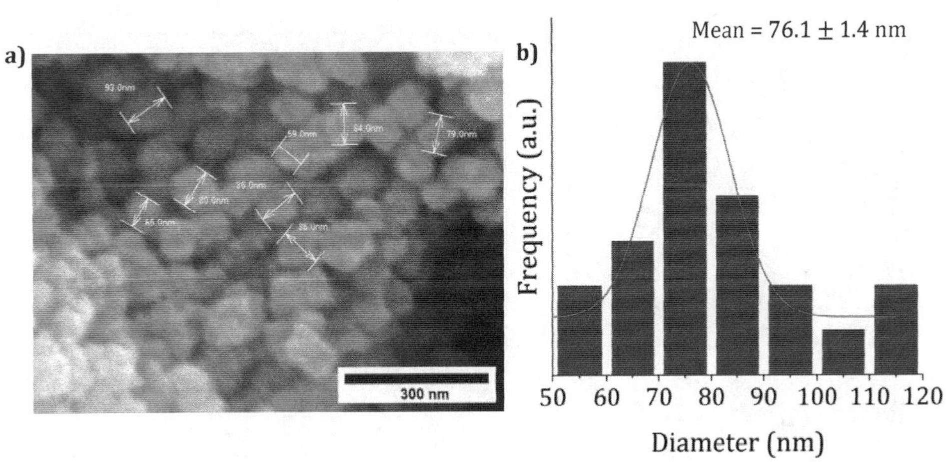

Figure 2. a) SEM image and b) particle size distribution of ZnO-red betel composites.

ligands with zinc ions through their hydroxyl aromatic ring groups, which facilitates the nucleation and stabilization of ZnO nanoparticles [3].

The SEM images in Figure 2 reveal that the particle sizes are uniform with average size 76.2 ± 1.4 nm and exhibit the formation of clusters, consistent with Karnan's findings [11]. These clusters, referred to as aggregation or agglomeration, could be attributed to several factors. The red betel extract used in the synthesis has a pH of 5, and the acidity of the extract may contribute to the agglomeration process[12]. The particles morphology exhibits a uniform semi-spherical shape, which also tend to create clusters[12].

S. aureus is bacteria that causes skin infections, pneumonia, and sepsis [13]. This bacterium resistants to various antibiotic agent such as methicillin [14] while *E. coli* resistants to betalactams and quinolones [15,16]. Yu et al's research also stated that *E. coli* multiresistant to multiantibiotic penicillin (100%), acetylspiramycin (100%), lincomycin (98.8%), oxacillin (98.8%), and sulphamethoxazole (53%) [17].

The antibacterial properties of red betel has been examined against several bacteria. Heliawati et al. studied its antibacterial properties against bacteria in dental caries. Due to its rich antibacterial biocompound, red betel extract is efficient against Gram-positive bacteria (*S. mutans, S. sanguinis*), and Gram-negative bacteria (*Veillonella parvula*) [8]. While Edikresnha et al. studied antibacterial activities of hydrogel loaded by *Piper crocatum* against *S. aureus* and *P. aeruginosa*[18]. This research performed a strong antibacterial effect against *S. aureus* [18].

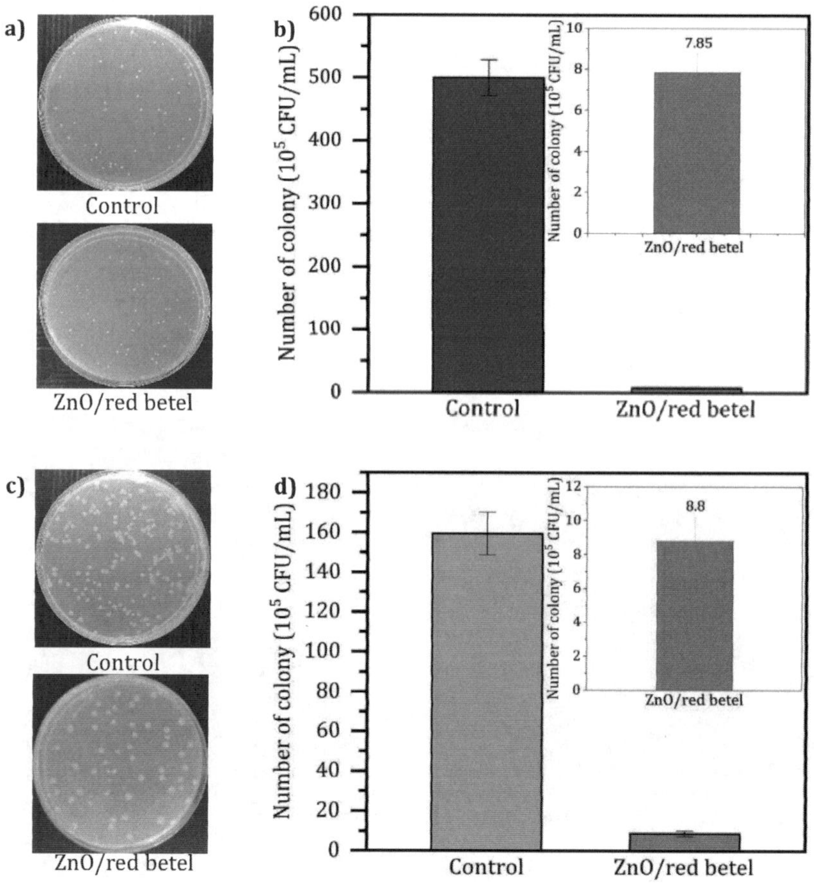

Figure 3 Comparison of the antibacterial effects of control and ZnO/red betel (10 ppm): visual representation of a) *S. aureus* and (c) *E. coli* colonies; and quantitative analysis of colony counts for b) *S. aureus* and (d) *E. coli.*

The antibacterial properties of this sample were evaluated by TPC methods. This method measured the initial amounts of bacteria and the final amount bacteria after being contacted with the sample. The antibacterial activity then is calculated using the equation (2) below:

$$Antibacterial\ activity\ (\%) = \left[\frac{a-b}{a}\right] \times 100 \qquad (2)$$

Figure 3 shows antibacterial activity of ZnO-red betel against *S. aureus* and *E. coli* bacteria. As we can see in Figure 3, the ZnO/red betel sample in 10 ppm concentration possess great antibacterial against Gram-positive (*S. aureus*) and Gram-negative (*E. coli*) bacteria. In detail, they were 98,4% for *S. aureus* and 94,5% for *E. coli.* Meanwhile, another research showed that Zn^{2+} with concentration of 10^{-5} – 10^{-4} M considered as an effective concentration that has antibacterial rate over 80% [19]. This result implies that incorporating ZnO with red betel successfully enhance its antibacterial efficacy rather than using ZnO itself.

In another side, metal oxides were known for their effectivity in their strong properties against Gram-positive and Gram-negative bacteria. Thi et al. claims that ZnO nanoparticles (NPs) produce reactive oxygen species (ROS) in water suspensions, which inhibit the growth of bacteria and fungi [3]. The study found that higher synthesis pH levels increased the formation of ROS, enhancing the bactericidal activity of ZnO NPs. Additionally, smaller ZnO NPs and those containing orange-peel extract demonstrated higher antibacterial effects against *S. aureus* and *E. coli*, with the effect attributed to both ROS and organic compounds in the extract[3].

Gram-negative bacteria, like *E. coli*, are more resistant to antibacterial agents due to their lipopolysaccharide-rich outer membrane[18]. In contrast, Gram-positive bacteria, such as *S. aureus* have a porous peptidoglycan layer that allows easier penetration of agents like bioactive compounds, disrupting their metabolism[18,20]. However, this research demonstrated strong antibacterial properties against both Gram-negative and Gram-positive bacteria, even at the small concentrations (10 ppm). ZnO's positive surface charge (+24 mV) interacts with the negatively charged bacterial membranes, creating an electrostatic gradient that damages the cell membrane[21]. This mechanism is effective against both Gram-positive *Staphylococcus aureus* and Gram-negative *Escherichia coli*, despite their structural differences, as the charge interaction disrupts membrane stability in both types[21].

Antibacterial mechanisms in ZnO-red betel nanocomposite needs to be studied further. It may combine ROS production from ZnO and different mechanisms from bioactive compound of red betel. These bioactive compounds may function as toxins within the protoplasm, destabilizing bacterial balance, inhibiting organelle formation, and breaking down the cell membrane [8].

4. Summary

We successfully synthesized ZnO-red betel nanocomposite that exhibits excellent antibacterial properties. Further characterization is needed to analyze the ratio and distribution of each component within the nanocomposite, as well as the properties of individual molecules. Techniques such as zeta-potential analysis, and thermogravimetric analysis should be considered for comprehensive characterization to measure the repulsion force for avoiding agglomeration and analyze the mass stability. We hope this study can be one solution for antibacterial resistance issue.

Acknowledgement

We gratefully acknowledge the financial support provided by Riset Unggulan ITB 2024 for funding this research. Without their support, this work would not have been possible. We would also like to extend our sincere gratitude to Electron Microscope Laboratory – RCNN ITB for their assistance in characterizing the samples, providing access to advanced instrumentation, and offering invaluable technical support throughout the research process.

References

1. Sivagami M and Asharani I V 2022 *Inorg. Chem. Commun.* **145** 110054
2. Bai X, Yan J and Gilchrest B A 2024 *J. Invest. Derm.* **144(2)** 430-434.e1
3. Doan Thi T U, Nguyen T T, Thi Y D, Ta Thi K H, Phan B T, and Pham K N 2020 *RSC Adv.* **10(40)** 23899–907
4. Tripathy N and Kim D H 2018 *Nano Converg.* **5(1)** 27
5. Muflihah S A, Sustini E, Aimon A H, Permatasari F A and Abdullah M 2024 *J. Phys. Conf. Ser.* **2734(1)** 012045
6. Murray C J L et al. 2022 *The Lancet* **399(10325)** 629–55
7. Jonas O B, Irwin A, Berthe F C J, Le Gall F G and Marquez P V (The World Bank Team) 2017 *Drug-resistant infections : a threat to our economic future (Vol. 2) : final report (English)* Washington, D.C.: The World Bank
8. Heliawati L, Lestari S, Hasanah U, Ajiati D and Kurnia D 2022 Molecules **27(9)** 2861
9. Shatnawi M et al. 2016 *Results Phys.* **6** 1064–71
10. Ebin B, Arıg E, Özkal B and Gürmen S *Int. J. Miner. Metall. Mater.* **19(7)** 651–6
11. Karnan T and Selvakumar S A S 2016 *J. Mol. Struct.* **1125** 358–65
12. Khairunnisa S, Wonoputri V and Samadhi T W 2021 *IOP Conf. Ser. Mater. Sci. Eng.* **1143(1)** 012006
13. Cheung G Y C, Bae J S and Otto M 2021 *Virulence* **12** 547–69
14. Turner N A et al. 2019 *Nat. Rev. Microbiol.* **17** 203–18
15. Puvača N and Frutos R de L 2021 *Antibiotics* **10** 1–19
16. Kilani H, Ferjani S, Mansouri R, Boutiba-Benboubaker I and Abbassi M S 2020 *J. Glob. Antimicrob. Resist.* **20** 50–5
17. Yu Z N, Wang J, Ho H, Wang Y T, Huang S N and Han R W 2020 *J. Glob. Antimicrob. Resist.* **22** 94–101
18. Edikresnha D, Suciati T, Suprijadi, Khairurrijal K 2021 *J. Mater. Res. Technol.* **15** 17–36
19. Ning C et al. 2015 *Chem. Res. Toxicol.* **28(9)** 1815–22
20. Edikresnha D, Suciati T, Munir M M, Khairurrijal K 2019 *RSC Adv.* **9(45)** 26351–63
21. Mendes C R et al. 2022 *Sci. Rep.* **12** 2658

Study of the Effect of Anodizing Times on the Degradation Rate of Mg-5CAp Biodegradable Composites as Bone Implants

Iwan Setyadi[1], Ana Karina Bano[2], Adhinda Septhia Nur Rizky[1], Mirza Wibisono[1], Suryadi[1], Arif Hidayat[1], Krisna Adhitya[1], Tika Mustika[1], Muslim Efendi Harahap[1], Giri Wahyu Alam[1], Joni Sah[1], I Nyoman Jujur[1], Ni Nyoman Rupiasih[2]*

[1] Advanced Materials Research Center, BRIN, BJ Habibie Science and Technology (KST) Area, Building 224, South Tangerang, Banten 15314, Indonesia
[2] Department of Physics, Faculty of Mathematics and Natural Sciences, Udayana University, Bali 80361, Indonesia

*E-mail: rupiasih@unud.ac.id

Abstract. Currently, significant efforts are being made to develop biodegradable Mg-based materials as alternatives to bone implants. Beyond enhancing strength, it is crucial to slow down the degradation rate to align with the bone healing process. Magnesium-carbonate apatite (Mg-xCAp) composites were developed through a combination of warm compaction and extrusion processes. Carbonate Apatite (CAp), a calcium phosphate-based bio-ceramic, not only serves as a reinforcement but also as a stimulus for the formation of osteoblast cells to expedite bone healing. This research focuses on improving the degradation rate of Mg-5CAp through an anodizing process conducted for varying durations of 5, 10, and 15 minutes at a voltage of 10 V and comparing the results with the non-anodized (NA) conditions. Characterization includes electrochemical corrosion tests such as EIS and Tafel, observations of layer thickness, and oxide element formation using SEM-EDX mapping. The results indicate that the 15 minutes anodizing process significantly improves the degradation rate, especially compared to NA conditions. The degradation rate decreased from 2.4 to 1.1 mm/y, and SEM mapping observation revealed at the best condition, the anodic oxide layer thickness is ranging from 5.4 - 8.7 μm.

Keywords: Composites, Magnesium-5Carbonate Apatite (Mg-5CAp), anodizing times, degradation rate, bone implants.

1. Introduction

In the past decade, a paradigm shift has emerged among orthopaedic doctors to develop biodegradable implants as an alternative for treating traumatic bone injuries (1). Biodegradable implant materials, such as magnesium (Mg), have been developed in recent years (2). Magnesium is a lightweight metal with bioabsorbable properties and an elastic modulus similar to bone (3). However, for its development as an alternative biodegradable implant material, it is essential to slow down the degradation rate of magnesium and enhance its strength. One approach involves the formation of new structures by developing a magnesium-carbonate apatite (Mg-CAp) composite (4). Carbonate apatite (CAp) is considered superior to hydroxyapatite (HA) due to its closer resemblance to bone, ease of absorption, and good osteoconductivity. The composition of CAp used in this research is 5 wt%, processed through extrusion, resulting in the Mg-5CAp composite (1).

The development of Mg-CAp composites has led to a significant increase in material strength; however, further efforts are necessary to reduce the degradation rate of magnesium. One method to achieve this is by applying protective coatings to the Mg-5CAp composite, such as through anodizing techniques (5). One crucial parameter in anodizing is the anodizing time, as it influences the formation of the surface layer on the material (6). This study focuses on observing the degradation rate of Mg-5CAp composites subjected to anodizing in a Na_3PO_4 solution with three different anodizing durations: 5, 10, and 15 minutes.

2. Material and methods

2.1 Material

The material used in this study was powder-based Mg-5CAp rods, which were developed by PRMM-BRIN through a combination of warm compaction and hot extrusion processes with an extrusion ratio of 4.

2.2 Sample preparation

The Mg-5CAp rod (Ø = 9 mm) was initially sectioned into disc-shaped specimens, each with a diameter of 9 mm and a thickness of 3 mm. Each sample was carefully soldered to a conductive wire to facilitate the anodizing and subsequent electrochemical corrosion testing. The specimens were mounted in epoxy resin to ensure that only its exposed surface was involved in film growth during the anodizing process. The mounted samples underwent a sequential grinding process using grit #400, #800, #1000, #2000, #5000, & #10000 abrasive papers and then polished using DiaPro 3 μm diamond suspension to achieve a uniform, smooth, and reflective surface finish. The etching procedure began by immersing the sample in an HNO_3 solution for 10 seconds, followed by immersion in an 80 °C NaOH solution for 1 minute, rinsing with distilled water for 1 second, and drying with a hairdryer.

2.3 Anodizing process

The anodizing process was carried out in a 0.215 M Na_3PO_4 solution at a constant voltage of 10 V for 5, 10, and 15 minutes. During the procedure, the Mg-5CAp sample functioned as the cathode, while a platinum (Pt) plate served as the anode. Upon completion of the anodizing process, the samples were rinsed with distilled water and dried using a hairdryer. Non-anodized Mg-5CAp was prepared as a reference.

2.4 Electrochemical corrosion test

The degradation rate analysis was performed using the electrochemical corrosion test. The test was conducted with a PARSTAT 4000A potentiostat/galvanostat using the three-electrode system. The Mg-5CAp sample ($A = 0.636\ cm^2$) served as the working electrode (WE) and was immersed in simulated body fluid (SBF). An Ag/AgCl reference electrode (3.5 M KCl, 0.205 V vs

Standard Hydrogen Electrode (SHE)) and a platinum counter electrode (CE) were employed at room temperature. The working electrode was initially submerged for 30 minutes to attain the open circuit potential (OCP) and assess its condition. The Electrochemical Impedance Spectroscopy (EIS) test was then performed using an AC signal of 10 mV and a frequency range of 10 mHz to 10 kHz with Electrical Equivalent Circuit (EEC). Tafel measurement was later conducted with a scan rate of 5 mV/s.

2.5 SEM-EDX

The thickness and composition of the anodic layer were analyzed by observing the cross-sections of both anodized and non-anodized Mg-5CAp samples using a scanning electron microscope and energy dispersive X-ray (SEM-EDX (JSM-IT200) at a magnification of 3000x and an accelerating potential difference of 15 kV.

3. Results and discussion

3.1 Degradation behaviour

The Tafel test results of the Mg-5CAp composite samples measured in SBF solution are presented in Figure 1. Through the extrapolation of the Tafel curve, Figure 1, the corrosion rate values can be obtained as listed in Table 1.

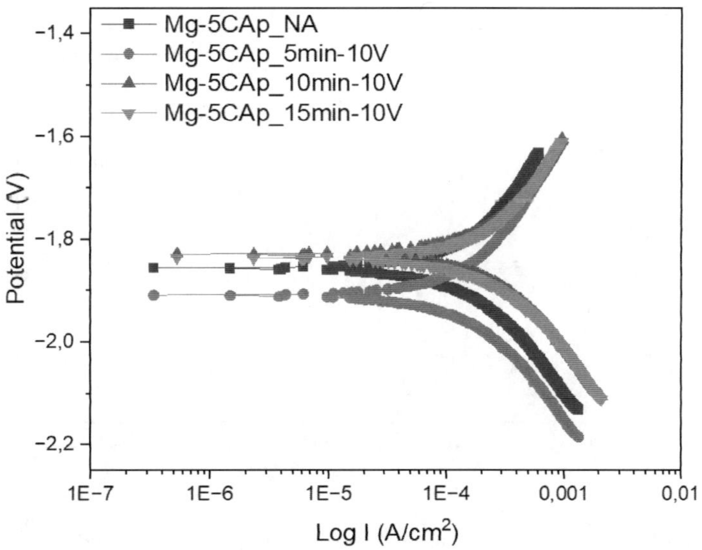

Figure 1. The Tafel plot for non-anodized (NA) and Anodized Mg-5CAp in SBF with anodizing times: 5, 10, and 15 minutes.

Table 1 shows the Mg-5CAp anodized for 15 minutes exhibited the lowest corrosion rate (1.046 mm/y) compared to the Mg-5CAp anodized for 10 minutes (1.1308 mm/y), 5 minutes (2.0027 mm/y), and the non-anodized sample (2.3769 mm/y). This result indicates that increasing the anodizing time can reduce the degradation rate of Mg-5CAp as shown in Table 1. The addition of anodizing time will increase the thickness of the oxide layer formed on the surface of the Mg-5CAp alloy which acts as a barrier that slows the penetration rate of corrosive ions. This

Journal of Physics: Conference Series **2980** (2025) 012029 doi:10.1088/1742-6596/2980/1/012029

is discussed further in the coating thickness sub-chapter. Fitriana & Anawati also reported that increasing the *anodizing* time on AZ31 alloy can improve the corrosion resistance (6).

Table 1. The corrosion rate of non-anodized (NA) and anodized Mg-5CAp in SBF with anodizing times: 5, 10, and 15 minutes.

Anodizing time (min)	Corrosion rate (mm/y)
0 (NA)	2.3769
5	2.0027
10	1.1308
15	1.046

The Nyquist curve from the EIS test, as shown in Figure 2, provides insight into the corrosion resistance of the samples. Generally, a larger semicircle diameter indicates better corrosion resistance [7]. The fitting of the EIS data, performed using the ZView software and modelled with an EEC as shown in Figure 3, yielded the charge transfer resistance (Rct) values, which are presented in Table 2.

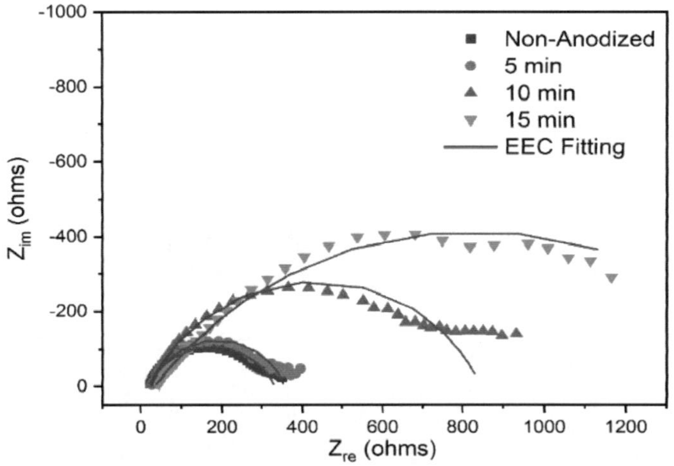

Figure 2. Nyquist curve for non-anodized (NA) and anodized Mg-5CAp in SBF with anodizing times: 5, 10, and 15 minutes.

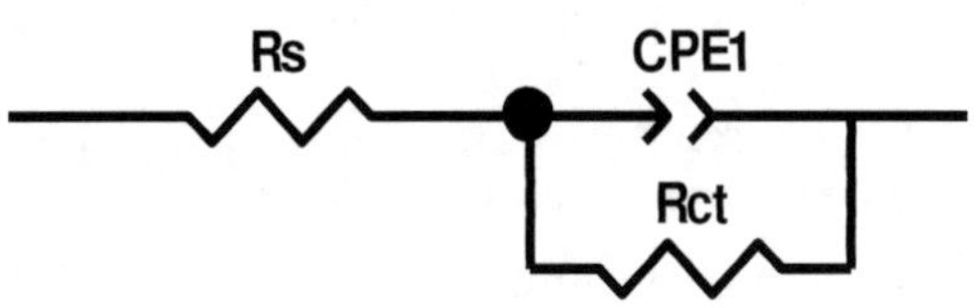

Figure 3. The EEC for fitting the EIS results.

From data Table 1 and 2 can be plotted curves of anodizing times vs corrosion rate (mm/y) and anodizing times vs Rct as shown in Figures 4a and 4b. The figure shows a higher Rct value corresponds to higher corrosion resistance (7).

Table 2. The charge transfer resistance of non-anodized (NA) and anodized Mg-5CAp in SBF with anodizing times: 5, 10, and 15 minutes.

Anodizing time (min)	Rct (Ω)
0 (NA)	327.5
5	309.8
10	821.3
15	1584

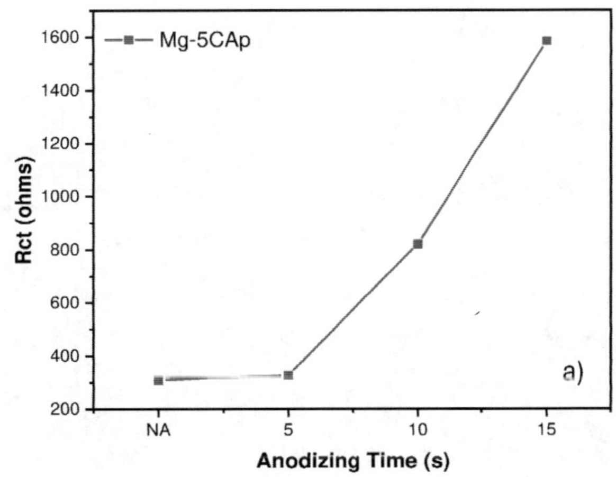

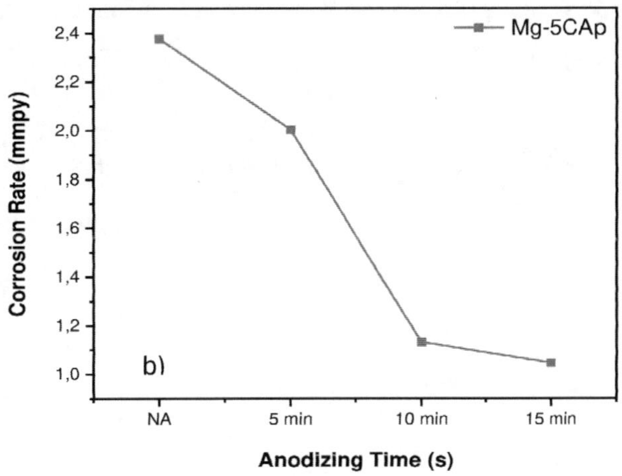

Figure 4. The relation between anodizing time and (a) charge transfer resistance (Rct) and (b) corrosion rate.

Table 2 shows the Mg-5CAp sample anodized for 15 minutes exhibited the highest Rct value (1584 Ω), followed by the samples anodized for 10 minutes (821.3 Ω), 5 minutes (327.5 Ω), and the non-anodized sample (309.8 Ω). These findings are consistent with the earlier Tafel test results. The increase in R_{ct} occurs along with the increasing anodizing time. That may happen because the increases in anodizing time significantly increase the thickness of the oxide layer, thereby reducing the charge transfer resistance and slowing down the degradation process. This is in line with the research report on anodizing AZ31B alloy (8).

3.2 Coating thickness

The SEM observations in Figure 5 demonstrate the differences in the thickness of the oxide layer formed on the anode on the surface of non-anodized and anodized Mg-5CAp with the corresponding values presented in Table 3.

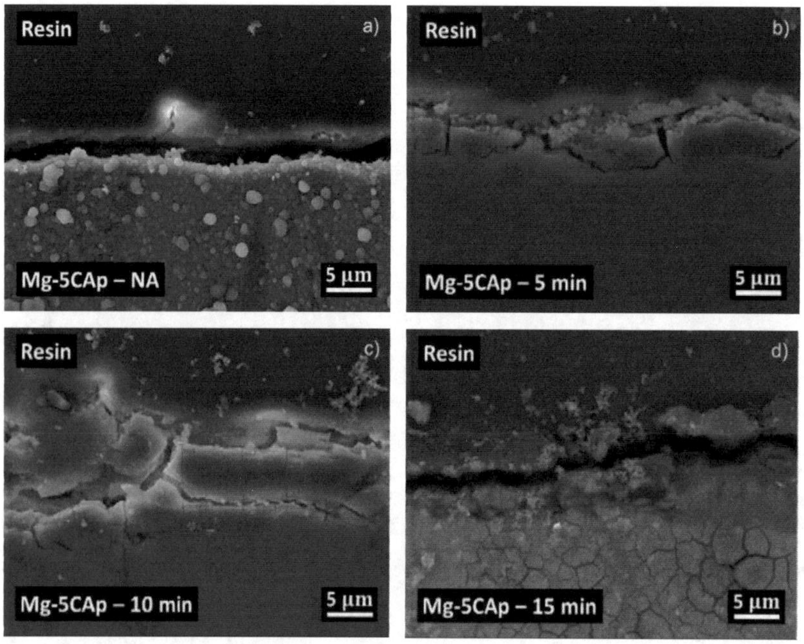

Figure 5. SEM results for (a) non-anodized Mg-5CAp and Mg-5CAp anodized for (b) 5, (c) 10, and (d) 15 minutes.

Table 3. The film thickness of non-anodized and anodized Mg-5CAp in SBF with anodizing times: 5, 10, and 15 minutes.

Anodizing time (min)	Film thickness (µm)
0 (NA)	-
5	1.7-6.8
10	2.4-9.1
15	5.4-8.7

Upon examination, Table 3 shows that the anodic layer formed on the Mg-5CAp substrate anodized for 15 minutes exhibited a higher minimum thickness of 5.4 µm, followed by samples anodized for 10 and 5 minutes e.g. 2.4 and 1.7 µm respectively. According to Lin & Fu (9), the presence of a high oxygen content, a constituent of MgO and $Mg_3(PO)_4$, indicates the successful formation of the anodic film on the sample surface (9–12). Therefore, the oxygen element mapping shown in Figure 6 further confirms the existence of the anodic layer on the substrate surface, which is indicated by the highly saturated yellow color on the surface.

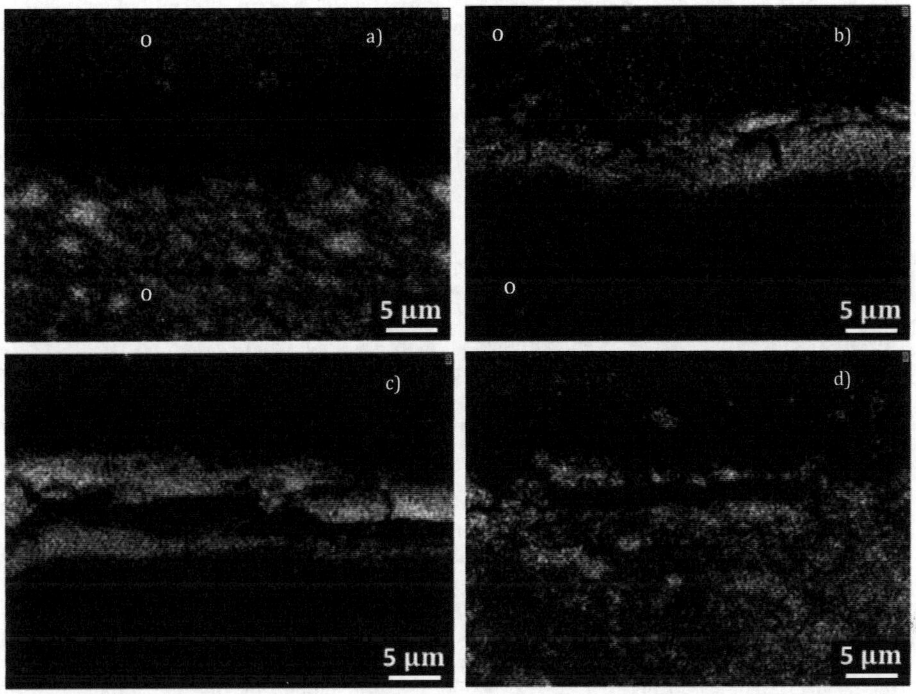

Figure 6. EDX-Mapping results that show the oxygen (O) element for (a) non-anodized Mg-5CAp and Mg-5CAp anodized for (b) 5, (c) 10, and (d) 15 minutes.

4. Conclusions

This study shows that the anodizing process is quite effective in reducing the degradation rate of Mg-5CAp. This significant reduction in degradation rate was proven from electrochemical testing and surface morphology characterization. The increase in anodizing time also influences the reduction in the degradation rate and layer thickness. The time variations used are 5, 10, and 15 minutes with a fixed voltage of 10 V. Electrochemical testing shows a decrease in the degradation rate occurs with increasing anodizing time. This reduction in degradation rate is reviewed by comparing time-varying anodizing materials and non-anodizing materials. The results of the degradation rate were obtained from the extra polarization of the Tafel curve, it can be seen that increasing the anodizing time had an effect on reducing the degradation rate. The results of decreasing the degradation rate are also confirmed by the EIS curve obtained, which shows that the larger the diameter of the curve, the better the reduction in the degradation rate. The SEM-EDS mapping characterization shows that increasing the time of the anodizing process has an effect on the thickness of the layer formed. The highest minimum thickness results were shown

Journal of Physics: Conference Series **2980** (2025) 012029 doi:10.1088/1742-6596/2980/1/012029

at an anodizing time of 15 minutes of 5.4 - 8.7 μm compared to other conditions. The formation of a high oxygen content indicates the successful formation of an anodic layer on the sample surface.

References

[1] Setyadi I, Suryadi, Jujur IN, Harahap ME, Tiwibowo B, Damisih, et al. Observation of mechanical properties improvement of powder-based magnesium-carbonate apatite (Mg-10CA) through deformation strengthening process. In: AIP Conference Proceedings. American Institute of Physics; 2024.

[2] Li S, Ren J, Wang X, Ding Y, Li P, Hu Y, et al. Dilemmas and countermeasures of Fe-based biomaterials for next-generation bone implants. Vol. 20, Journal of Materials Research and Technology. Elsevier Editora Ltda; 2022. p. 2034–50.

[3] 3. Asaduzzaman Chowdhury M, Helal Hossain MD, Hossain N, Hossen Z, Arefin Kowser M, Masud Rana M. Advances in coatings on Mg alloys and their anti-microbial activity for implant applications. Arabian Journal of Chemistry. 2022 Nov 1; 15(11).

[4] Setyadi I, Marsetio AF, Kamal AF, Rahyussalim, Supriadi S, Suharno B. Microstructure and microhardness of carbonate apatite particle-reinforced Mg composite consolidated by warm compaction for biodegradable implant application. Mater Res Express. 2020 May 1; 7(5).

[5] Nugroho F, Teknik J, Stt M, Jl A, Janti BR, Lanud A, et al. Effect of Current Meeting and Anodizing Time on the Thickness of Aluminum Oxide Layer on Aluminum Alloy AA 2024-T3.

[6] Fitriana MF, Anawati A. Remarkable improvement in corrosion resistance of anodized AZ31 alloy by sealing with beeswax-colophony resin. In: Journal of Physics: Conference Series. Institute of Physics Publishing; 2019.

[7] Abdo HS, Seikh AH, Mohammed JA, Luqman M, Ragab SA, Almotairy SM. Influence of chloride ions on electrochemical corrosion behavior of dual-phase steel over conventional rebar in pore solution. Applied Sciences (Switzerland). 2020 Jul 1; 10(13).

[8] Kang Y, Yan S, Li Z, Wang Z, Yang A, Ma W, et al. Influence of Anodic Oxidation on the Organizational Structure and Corrosion Resistance of Oxide Film on AZ31B Magnesium Alloy. Coatings. 2024 Mar 1; 14(3).

[9] Lin CS, Fu YC. Characterization of Anodic Films on AZ31 Magnesium Alloys in Alkaline Solutions Containing Fluoride and Phosphate Anions. J Electrochem Soc. 2006; 153(10): B417.

[10] Tsakiris V, Tardei C, Clicinschi FM. Biodegradable Mg alloys for orthopedic implants – A review. Vol. 9, Journal of Magnesium and Alloys. National Engg. Reaserch Center for Magnesium Alloys; 2021. p. 1884–905.

[11] Xiang E, Gómez-Cerezo MN, Ali Y, Ramachandra SS, Yang N, Dargusch M, et al. Surface Modification of Pure Zinc by Acid Etching: Accelerating the Corrosion Rate and Enhancing Biocompatibility and Antibacterial Characteristics. ACS Appl Mater Interfaces. 2022 May 18; 14(19): 22554–69.

[12] Liu X, Yang H, Liu Y, Xiong P, Guo H, Huang HH, et al. Comparative Studies on Degradation Behavior of Pure Zinc in Various Simulated Body Fluids. JOM. 2019 Apr 20; 71(4): 1414–25.

Journal of Physics: Conference Series **2980** (2025) 012030 doi:10.1088/1742-6596/2980/1/012030

Analysis Of The Use Of Agar Phantom In The Spatial Resolution Test Of Ultrasound Images (USG)

Nadiah Fakhira[1*] and Rena Widita[1]

[1] Department of Physics, Bandung Institute of Technology, Bandung, Indonesia

*E-mail: rena@itb.ac.id

Abstract. The research has been conducted in Radiology Installation of Santo Boromeus Hospital regarding the analysis of agar-based Phantom Ultrasound in testing the spatial resolution of ultrasound images. In addition to aiming to replace factory-made phantoms which are relatively expensive, this research also aims to determine the accuracy of agar-based ultrasonic phantoms in testing the quality control of more effective ultrasonic aircraft spatial resolution. The research was carried out experimentally, namely by making an agar-based ultrasound phantom with a composition of 10 g agar powder and 1000 ml water, using an object in the form of 3 straws containing a sugar-water solution. The phantom scanning process with GE brand 2D ultrasound type Logic 7 is carried out using 2 types of transducers, namely linear array transducers and curved array transducers. The results obtained show that the spatial resolution value of the image using a linear array transducer is 65.679 and 65.7 pixels/cm with a distance between the three objects of (1.995 ± 0.25) cm and (1.996 ± 0.2) cm from the actual distance of 2 cm. Meanwhile, the spatial resolution value of the image using a curved array transducer is 65.366 and 64.7 pixels/cm with a distance between the three objects of (1.985 ± 0.75) cm and (1.966 ± 1.7) cm from the actual distance of 2 cm.

Keywords: Agar Phantom, Curved Array Transducer, Linear Array Transducer, Spatial Resolution, Ultrasound

1,INTRODUCTION

Hospitals and clinics have used ultrasound to diagnose abnormalities in many organs or tissues in patients, but generally without a quality assurance process, either at the time of purchase or during the use of the ultrasound itself. Quality assurance should be performed routinely before each ultrasound examination is performed[1,2].

Quality control must be carried out to ensure that the ultrasound machine used is operating properly and to ensure that the operation is carried out according to plan so that if damage and deviations occur, the source of the problem can be found and identified and then corrected so that what is expected can be achieved[3].

One of the quality control tests in ultrasound is spatial resolution. Spatial resolution refers to the ability to see the smallest details, namely how small an object can be seen on a particular imaging system[4]. Small details in ultrasound images will be very difficult to analyze using the human eye because they are on the order of pixels (the smallest elements of the image), so with these limitations a method is needed to analyze spatial resolution.

To carry out quality control tests on ultrasound machines, a tool is needed, namely an ultrasound phantom. In 2008, A. J. D. Frederick Ayers studied the fabrication and characterization of silicon-based phantom tissue with optical properties in the visible and near infrared domains. Optical phantom simulation networks play a central role in the development, validation, and application of optics-based biomedical technologies[5].Factory-made ultrasonic phantoms are also known to be quite expensive so that not all institutions can provide phantoms for learning/practice activities. Therefore, an ultrasonic phantom test tool is needed that is affordable, easy to construct, reusable, and durable .

Several previous studies have made and used ultrasonic phantoms made from gelatin, agar, paraffin gel, and silicone rubber. The gelatin-based phantom has the disadvantage that it does not last long so it cannot be used for the long term, in addition, the gelatin phantom requires a refrigerator or cooler in the manufacturing process[6]. The paraffin gel-based ultrasonic phantom also has the disadvantage that it is difficult to change the speed of sound from the material used[7].

Meanwhile, in previous studies, there have been studies that have conducted spatial resolution quality control tests on ultrasound using silicon rubber phantoms[8]. So in this study, a spatial resolution quality control test will be carried out on ultrasound using an agar-based ultrasound phantom.

The agar-based phantom is considered quite promising as an ultrasound phantom because agar has a very affordable price when compared to other materials such as silicon rubber. Agar is also easy to make and maintain because it does not require cooling, does not melt easily, and can be used repeatedly. In addition, agar-based phantoms also have structural similarities to human tissue, especially soft tissue[9]

The purpose of this research is to study the accuracy of ultrasound-based phantoms in order to carry out quality control tests on the spatial resolution of ultrasound which is more effective and affordable and has output results that match the real value.

The design and implementation of this research used digital image processing software that has been installed on a computer.

2.METHODOLOGY

The instrument used to test the quality of spatial resolution of ultrasound images is an ultrasound phantom which in this research will be made from agar-based materials. Phantom agar is made using a composition of 1000 ml of water and 10 g of agar powder with the object used being 3 straws, each of which has the same diameter. These four straws are filled with sugar water at the same concentration.

In this case, the sugar water is made using 60 ml of water and then 30 g of sugar is added. Inside the phantom, an object is also placed in the form of a 1 cm iron needle which is used as a reference scale for changing the size of pixels to cm. The phantom design contains an object in the form of an arrangement of 3 straws of the same diameter with a distance between straws of 2 cm that it can also be seen in Figure 1.

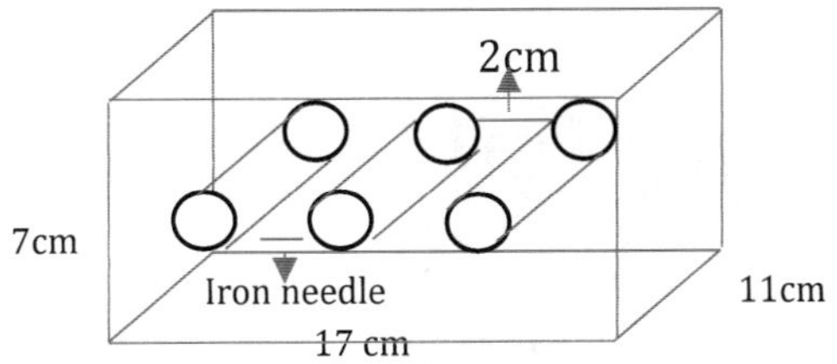

Figure 1. The phantom design contains an object in the form of an arrangement of 3 straws of the same diameter with a distance between straws of 2 cm and there is an iron needle with a length of 1 cm

The step for analyzing spatial resolution begins with scanning the phantom with depth and gain settings. Depth is one of the tools on the ultrasound machine that functions to change the image plane displayed in one centimeter gradation increments. Increasing depth will reduce image resolution. At higher depths, the transducer needs to travel a greater distance, therefore the frame rate and resolution are both lower, while gain is a tool used to adjust the overall brightness of the real-time ultrasound image (B-mode). Changes in gain can make an image that is too dark brighter or vice versa, so that subtle differences in texture can be seen.

In this research . the phantom scanning process with GE brand 2D ultrasound type Logic 7 is carried out using 2 types of transducers, namely linear array transducers and curved array transducers. The linear array transducer produces a rectangular field of view with uniform beam density throughout. Meanwhile, the curved array transducer has a curved array that allows a wider field of view.

To analyze the spatial resolution, a phantom was scanned with depth and gain settings, namely a depth of 11 cm then a gain of 70 and 50.

Next, image processing is carried out from the obtained ultrasound image. The ultrasound image is read using a computer that has installed image processing software where the image format is grayscale. The last stage of the spatial resolution analysis process is carried out by measuring the distance between adjacent objects.

3.RESULT

In this study, a quality control analysis of spatial resolution analysis of Ultrasonography (USG) images was carried out using an agar-based ultrasound phantom. The phantom scanning image obtained was then read using a laptop that had image processing software installed where the image format must be grayscale.

One of the spatial resolution analysis processes is carried out by measuring the distance between adjacent objects. Based on the phantom design made, the arrangement between objects is at the same depth with the difference in distance in the horizontal direction.

Measurement of the distance between objects can be done by directly measuring the distance between 2 sides of adjacent circular objects. Measurement of the distance between objects in this study was carried out using a digital image processing application that had been installed on the computer. After measuring the image, the distance value in pixels listed in the application can be

Journal of Physics: Conference Series **2980** (2025) 012030 doi:10.1088/1742-6596/2980/1/012030

seen. computer. After measuring the image, the distance value in pixels listed in the application can be seen.

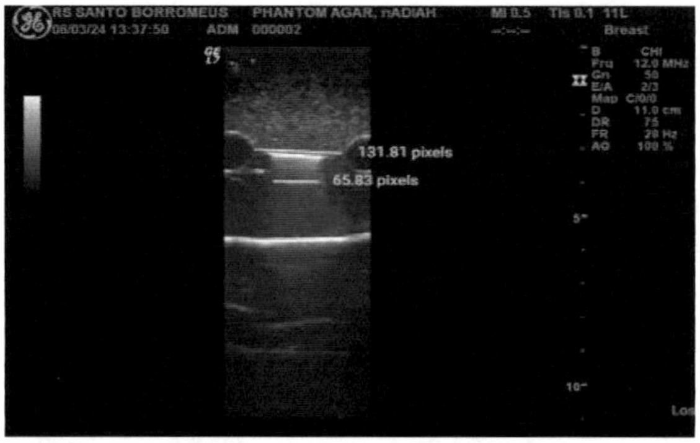

Figure 2. Calculation of the length of the distance between object 1 and object 2 and the length of the iron needle in the image using a linear array transducer.

Previously, an additional object had been placed on the phantom in the form of a 1 cm iron needle which would be used as a scale reference for converting the pixel size to cm units. In Figure 2 it can be seen that the pixel size for a 1 cm iron needle is 65.83 pixels. This indicates that for a scale of 1 cm it is equal to 65.83 pixels. Then it can also be seen in Figure 2 that the distance between object 1 and object 2 is 131.81 pixels.

The phantom scanning process using a linear array transducer was carried out 5 times, likewise the phantom scanning process using a curved array transducer was also carried out 5 times. And to make it easier to repeat, in this case object 1 is named 01, as well as objects 2 and 3 are named 02 and 03.

Table 1. The distance between the three objects on the agar-based phantom using a linear array transducer.

Experiment	The distance between 01 and 02 (pixels)	The distance between 02 and 03 (pixels)
1	131.81	130.33
2	130.33	130.09
3	133.22	130.13
4	131.33	133.22
5	130.10	133.23
Average	**131.358**	**131.4**

It can be seen in Table 1 that the average distance between object 1 and object 2 is 131.358 pixels where this number if converted into cm units is equal to 1.995 cm. While the distance between objects 2 and 3 has an average value of 131.4 pixels or 1.996 cm.

In Figure 3, it can also be seen that the size of the pixel unit for a 1 cm iron needle is 65.85 pixels. This value is almost the same as the value of the iron needle size in the image results using a linear array transducer. This also indicates that for a scale of 1 cm it is equal to 65.83 pixels. Then it can also be seen in Figure 3 that the distance between object 1 and object 2 is 131.67 pixels.

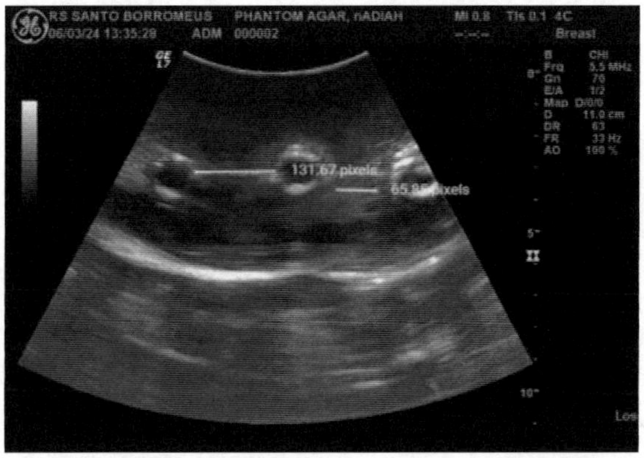

Figure 3. Calculation of the distance between the three objects in the image using a curved array transducer in Experiment 1.

Table 2. The distance between the three objects on the agar-based phantom using a curved array transducer.

Experiment	The distance between O1 and O2 (pixels)	The distance between O2 and O3 (pixels)
1	131.67	130.10
2	130.09	130.09
3	130.13	128.53
4	131.67	130.13
5	130.10	128.52
Average	**130.732**	**129.474**

It can be seen in Table 2 that the average distance between object 1 and object 2 on the curved array transducer is 130.732 pixels, where this number if converted into cm units is equal to 1.985 cm. While the distance between objects 2 and 3 has an average value of 129.474 pixels or 1.966 cm.

Next, an analysis of the spatial resolution of the image resulting from scanning an agar-based phantom using a linear array transducer and curved array transducer was carried out.

$$\text{Spatial resolution} = \frac{number\ of\ pixels}{actual\ distance\ (\ cm)} \quad (1)$$

Table 3. Calculation of Image Spatial Resolution from Image Results Using Linear Array Transducers.

Object	Average number of pixels	Distance	Spatial Resolution
1 and 2	131.358	2	65.679
2 and 3	131.4	2	65.7

Table 4. Calculation of Image Spatial Resolution from Image Results Using Curved Array Transducers.

Object	Average number of pixels	Distance	Spatial Resolution
1 and 2	130.732	2	65.366
2 and 3	129.474	2	64.7

Based on the calculation results of Table 3 and Table 4, it can be seen that the spatial resolution value obtained is accurate and in accordance with the actual object distance. Then the standard deviation and % error of the spatial resolution value of the image obtained are calculated.

$$\sigma = \sqrt{\frac{\sum_{i=1}^{n}(X_i - \bar{X})^2}{n-1}} \quad (2)$$

$$\%\,(error) = \left(\frac{\substack{actual\ distance\ value- \\ average\ value\ of\ calculated\ distance}}{actual\ distance\ value}\right) \quad (3)$$

Table 5. Calculation of Standard Deviation and % Error from the distance value between objects in ultrasound images using linear array transducers.

Object	Average value of calculated distance (cm)	Actual distance value (cm)	Standard Deviation	% Error
1 and 2	1.995	2	1.2561	0.25
2 and 3	1.996	2	1.6684	0.2

Table 6. Calculation of Standard Deviation and % Error from the distance value between objects in ultrasound images using curved array transducers

Object	Average value of calculated distance (cm)	Actual distance value (cm)	Standard Deviation	% Error
1 and 2	1.985	2	0.8563	0.75
2 and 3	1.966	2	0.8664	1.7

Based on the calculation results of Table 5 and Table 6, it can be seen that the % error value for the distance between objects in the image using a linear array transducer has a smaller number compared to the % error value for the distance between objects in the image using a curved array transducer, this also indicates that in this study the spatial resolution value obtained in the image using both transducers is equally good and accurate and in accordance with the actual distance between objects. The error number also shows a small number and the values obtained still meet the standards set by the American Association of Physicists in Medicine (AAPM).

4.CONCLUSION

Agar-based ultrasound phantom can be used as a simple and accurate ultrasound phantom to conduct ultrasound quality control tests in terms of spatial resolution by measuring the distance between objects replacing factory-made phantoms because they have a spatial resolution value that matches the actual distance between objects and has a small error value and the values obtained still meet the standards set by the American Association of Physicists in Medicine (AAPM). Increasing the composition of the phantom material can still be attempted to further increase spatial resolution, especially for clinical applications, namely by increasing the concentration of sugar water in the object placed in the phantom so that the image results obtained will also be clearer and more accurate.

5. ACKNOWLEDGMENT

This work is partly supported by the Ministry of Education, Culture, Research and Technology of the Republic of Indonesia through the 2024 research program and ITB research program year 2024. We also thank Prof. Yudi Darma for useful discussions in data analysis.

6.LIST OF REFERENCES

[1] Coulam, E., Rollo, J., (1981). The Physical Basis of Medical Imaging, New York.

[2] Kremkau, F. W., (1989). Diagnostic Ultrasound principle and instrument, 5th ed. WB Saunders, Philadelphia.

[3] Amitasari, R., Adi, K., Fisika, J., & Diponegoro, U. (2015). Pengaruh Gain Terhadap FWHM Dan FWTM Pada Citra B-Mode Ultrasonogram (USG), 18(2), 59–66.

Journal of Physics: Conference Series **2980** (2025) 012030 doi:10.1088/1742-6596/2980/1/012030

[4] M. Mahesh, "The Essential Physics of Medical Imaging", Third Edition, Med. Phys., vol. 40, no. 7, pp. 077301, 2013.

[5] Frederick Ayers, Alex Grant, Danny Kuo, David J. Cuccia, A. J. D. (2008). Fabrication and characterization of silicone-based tissue phantoms with tunable optical properties in the visible and near infrared domain. SPIE. 6870. hlm. 1–9.

[6] Morrow, S. D., Cupp J. A., Broder J. S. (2016). Versatile, Reuseable and Inexpensive Ultrasound Phantom Procedural Trainers. Journal of Ultrasound in Medicine 35(4): 831-841.

[7] Vieira, S. L., Pavan, T. Z., Junior, J. E., and Carneiro, A. A. O. (2013). Paraffin-Gel Tissue-Mimicking Material for Ultrasound-Guided Needle Biopsy Phantom. Ultrasound in Medicine and Biology. 39 (12). hlm. 2477-2484.

[8] Ratar, V. I., Trihandaru, S. S. , and Maslebu, G. (2020). Analisis Resolusi Spasial Citra Ultrasonografi (USG) pada Arah Tangensial Radias Citra menggunakan Phantom Berbasis Silicon Rubber. Jurnal Fisika dan Aplikasinya. Vol. 16. No.1. 1 hlm.

[9] Putri, C. F. R., Maslebu, G. , Setiawan, A. (2021). Analisis Densitas Larutan Berbasis Citra Phantom Agar dalam Kendali Mutu USG. POSITRON Vol. 11. No. 1. hlm. 26-30.

Journal of Physics: Conference Series **2980** (2025) 012031 doi:10.1088/1742-6596/2980/1/012031

Design, Simulation, and Miniaturization of All-Optical AND and XOR Logic Gates using 2D Photonic Crystals

Jonathan Sugijanto[1*], Melinda Alberta[1], Arfat Pradana[2] and Akhmadi Surawijaya[3,4]

[1] Department of Physics - Faculty of Mathematics and Science, Institut Teknologi Bandung, Bandung, Indonesia
[2] Center for Nanoscience and Nanotechnology, Institut Teknologi Bandung, Bandung, Indonesia
[3] Department of Electrical Engineering - School of Electrical Engineering and Informatics, Institut Teknologi Bandung, Bandung, Indonesia
[4] Center for Microelectronic, Institut Teknologi Bandung, Bandung, Indonesia

*E-mail: 10222007@mahasiswa.itb.ac.id

Abstract. This study focuses on the design, simulation, and miniaturization of all-optical AND and XOR logic gates using 2D photonic crystals. The proposed logic gates use light interference in photonic crystal waveguide junctions to amplify or annihilate the gate's output power. The waveguides are made by carving a path-like defect from a 2D triangular lattice of silicon nanorods suspended in air. The pristine photonic crystal has a bandgap from 1462 nm to 1845 nm to support the commonly used 1550 nm light. Using the FDTD simulation method, contrast ratios of up to 8.34 dB for cell sizes as small as 149 μm^2 have been demonstrated. A small dimension means a high switching frequency of up to 1.43 THz is possible. Lastly, efforts in migrating from 2D to 3D implementation are also discussed, with the most efficient result shown by translating the 2D design into a slab and sandwiching it with two thin Aluminum films as reflective layers.

1. Introduction

Photonic Integrated Circuit (PIC) is a microchip that integrates many photonic components in a single chip. Unlike a conventional photonics circuit with many lenses and mirrors, a PIC can offer a low-cost, compact, and fast optical circuit [1]. Such a small and versatile device can find application not only in data and telecommunication but also in other sectors such as photonic computing.

PIC consists of many components such as lasers, resonators, waveguides, and photonic crystals (PhC). PhC is an optical structure with periodic modulation of refraction index [2]. It has been shown in theory that such a periodicity can introduce a dead zone (band gap) in the frequency domain where no photon with said frequency can pass [3]. Said Photonic Band Gap (PBG) may depend on the polarization of the light in 2 and 3-dimensional PhC cases. For example, rectangular and triangular 2D lattices of dielectric rods suspended in air or air holes in a dielectric medium typically have large TM or TE PBG, respectively.

By introducing defects in the PhC, one can engineer a waveguide-like structure for the light in the PBG [4]. By introducing more complex defects, several authors have shown the possibility of logical functions through FDTD analysis. FDTD itself is short for Finite Difference Time-Domain.

FDTD is a method for solving Maxwell equations where space and time are divided into grids so the electric and magnetic fields can be solved sequentially [5].

Currently, there are four ways one can make a fast PhC logic gate: self-collimation effect, Multi-Mode Interference (MMI), waveguide interference, and Kerr effect. The self-collimation effect describes the absence of diffraction in certain structures of PhC. By utilizing the reflection and transmission of two perpendicular self-collimated beams, Y. Zhang, et al have managed to build OR and XOR functionality with a Contrast Ratio (CR) between the output and input above 17 dB [6]. On the other hand, by allowing the possibility of multiple modes in their combined waveguide, W. Liu, et al could design XOR, XNOR, NAND, and OR functionality with above 25 dB of CR [4]. Next, Y. Fu has shown that even a simple combination of two waveguides can interfere and produce OR, XOR, NOT, XNOR, and NAND functionality with as high as 20 dB CR [7]. Lastly, a CR of approximately 10 dB was achieved by M. F. Yanik, et al by introducing Kerr nonlinearity in the PhC [8].

In all of the existing logic gate designs, it was apparent that minimum effort has been made to miniaturize the gates for compact PIC applications. Although some methods are theoretically unaffected by cell size, the proposed mechanism may leak the confined light or perform differently when the dimension is minimized. In addition, no 3D simulation has been made for dielectric rods in air PhC logic gates. While air holes in dielectric medium PhC waveguides have seen 3D simulation [9] and even fabrication in recent years [10], few studies have discussed the 3D implementation of dielectric rods in air [11], the setup used by most 2D PhC logic gates.

Therefore, in this study, we propose the design, simulation, and miniaturization of PhC logic gates with a high contrast ratio. Then, the final design would be further characterized for comparison with other studies. We also analyze the possible 3D implementation of the gates especially for dielectric rods in air type PhC. This research would be limited to only the AND and XOR logic gates via the waveguide interference method.

2. Methodology

All the designs in this study are based on triangular silicon nanorods in air PhC. On the other hand, all the light sources used in this study are Gaussian with 1550 nm wavelength in TM mode. The choice of material and wavelength are taken with conventional Integrated Circuit (IC) and Optical Communication in mind. Therefore, ease of integration should be expected.

Light interference in a waveguide effect is chosen as the basic principle used in this study. To simulate the waves' interference, the Finite-Difference Time-Domain (FDTD) method offered by ANSYS Lumerical 2022 R1.4 is used. A lattice constant of half a wavelength is chosen for ease of interference.

An optimization of the rods' radii is employed to minimize waveguide loss. Two methods are cascaded to find it. First, a range of possible radii is found by profiling the PBG against radius (Figure 1a). A circuit similar to Figure 2 is then made for various possible radii. The radius showing the highest power throughput from A to C is then chosen, namely 219.19 nm. The full band structure for the chosen parameters can be seen in Figure 1b.

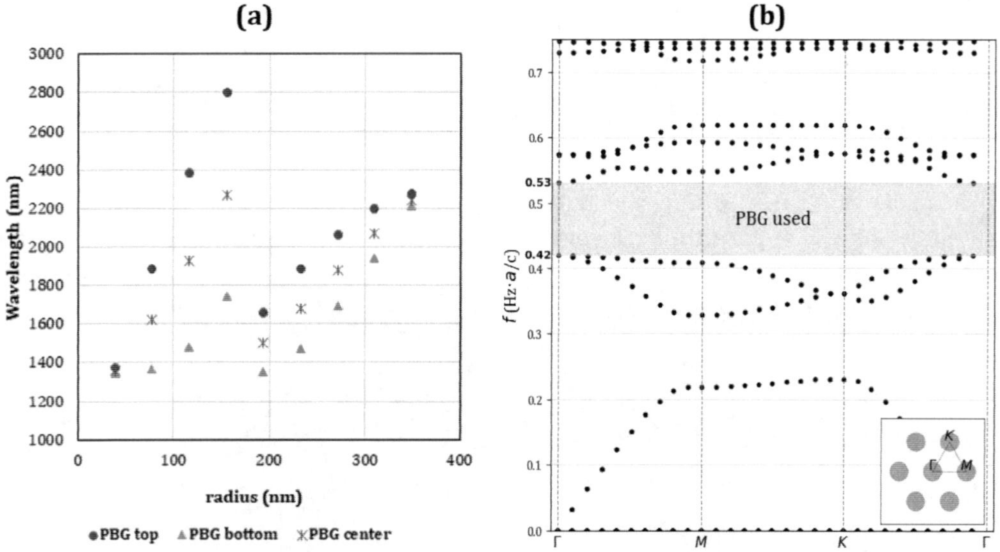

Figure 1. (a) TM Photonic Band Gap (PBG) of triangular silicon in air Photonic Crystal (PhC) with lattice constant a=775 nm for various rod radiuses. (b) TM photonic bandstructure of the PhC used in this study with lattice constant a=775 nm and radius r=219 nm. The normalized frequency is plotted against a wavevector along the Brillouin zone of a triangular lattice shown on the bottom right. The band structure shows a band gap (PBG) for 1462 – 1845 nm wavelength.

Defects of certain patterns are introduced to the pristine PhC to create the functionality needed in this study. In particular, a variation of single-row defect is used because said defect is commonly used to make PhC waveguides. By controlling the defect path length for two coherent inputs, one can simulate AND and XOR logic gate functionality. Contrast Ratio (CR) defined as the ratio between the highest logical "0" and lowest logical "1" is then calculated as the study's figure of merit. A reduction in the number of rods in the X and Y axes is also evaluated to find the highest CR.

Next, we discuss potential ways to migrate the 2D PhC design to 3D. Multiple slab-based waveguide structures are simulated and compared to find the best method to translate 2D PhC waveguide-based design to 3D. The PhC parameter used here is the same as in the 2D counterpart.

Journal of Physics: Conference Series **2980** (2025) 012031 doi:10.1088/1742-6596/2980/1/012031

3. Results and Discussions

3.1. AND Logic Gate

AND logic gate is realized using constructive interference according to the PhC structure presented in Figure 2. A and B are the Input ports for the light source, whereas C is the output port where power measurement is done.

Because of the linearity of silicon's dielectricity, the performance of the gates is not affected by the input powers. Therefore, the performance of the gates can be tested by switching the inputs on (digital high) or off (digital low) as presented in Figure 3. Due to the symmetric nature of the

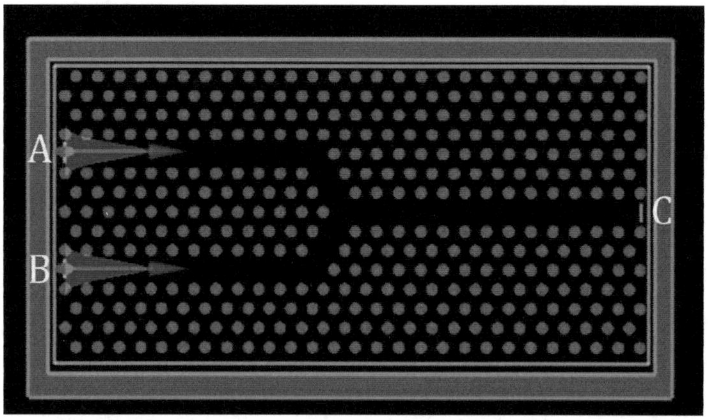

Figure 2. Final silicon PhC AND logic gate design with ports A and B as the input ports (1550 nm Gaussian), and C as the output port. The rectangle surrounding the PhC structure is the FDTD PML boundary condition.

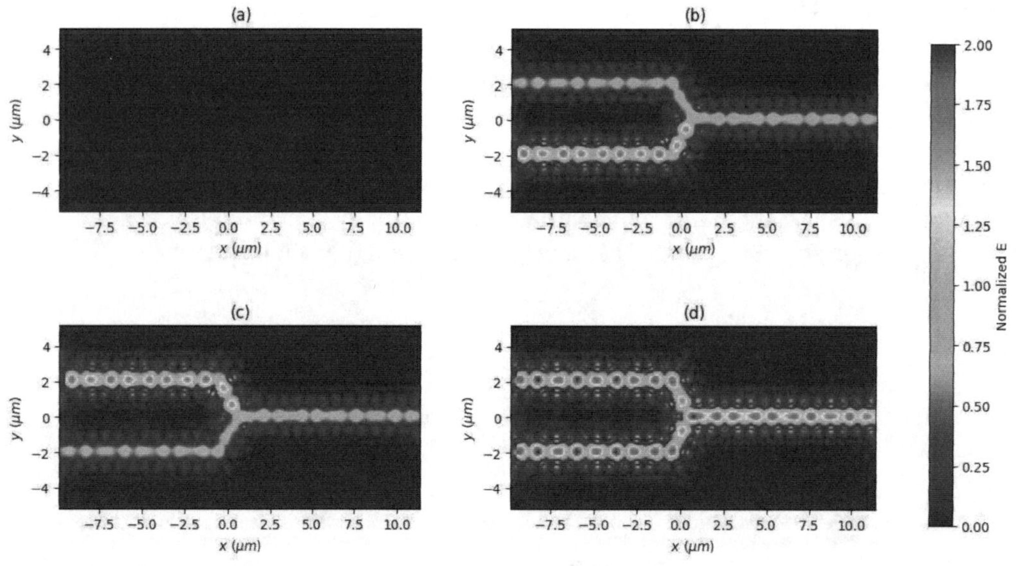

Figure 3. Electric field profile of the final PhC AND Logic Gate design when (a) there is no input, (b) input B is digital high, (c) input A is digital high, and (d) both inputs are digital high. The input lights are Gaussian with 1550 nm wavelength.

Journal of Physics: Conference Series **2980** (2025) 012031 doi:10.1088/1742-6596/2980/1/012031

design, there is no difference between the single excitation of port A and port B. In either case, the light from the input is distributed at the intersection into three equal parts. Two parts are transmitted to ports B and C while the rest is reflected to port A. On the other hand, when both inputs are high, constructive interference can be seen at the junction, resulting in more light arriving at port C.

The resulting table of truth and the measured CR is presented in Table 1. By measuring the power at port C using a power monitor, we can see that port C receives the highest power when both port A and B are excited, the lowest when both input ports are off, and a small amount of power when either port A or B is excited. The behavior of the input-output is similar in function to an AND logic gate by defining throughput (normalized power) more than 1 as the logical "1" and throughput less than 0.35 as the logical "0". CR is then calculated by taking into account logical "1" is 1.43 and logical "0" is 0.34 to obtain the value of 6.22 dB.

With a working design in hand, we proceed with our investigation to miniaturize the AND logic gate cell by reducing the number of nanorods in the Y-axis (Ny) while maintaining the number of nanorods in the X-axis (Nx). The result samples of the miniaturization are presented in Table 2. The highest CR is observed at Nx = 27 and Ny = 15 which correspond to cell size of 20.15 x 9.40 μm^2. We then take the cell size with the highest CR as our final AND gate design. Note that the field and table of truth presented in Figure 2, Figure 3, and Table 1 are calculated using the final design.

Table 1. Table of truth for the Final PhC AND logic gate design. Port A and B are the input ports while port C is the output. Throughput is defined as the ratio of output and single input power. Contrast Ratio is taken from the ratio between the lowest digital "1" throughput and the highest digital "0" throughput.

A	B	C	Throughput at C	Contrast Ratio (CR), dB
0	0	0	0.00	-
1	0	0	0.34	-
0	1	0	0.34	-
1	1	1	1.43	6.22

Table 2. PhC AND logic gate miniaturization result samples. The number of rods in the Y axis (Ny) is reduced from 27 to 7 rods while keeping the number of rods in the X axis constant (Nx). The first, best, and final iteration is shown in the table.

Nx	Ny	CR (dB)
27	27	6.06
27	15	6.22
27	7	6.02

3.2. XOR Logic Gate

XOR gate is constructed by utilizing destructive interference between both inputs. The phase shift required for the interference is realized by adding half a wavelength length to the path taken by input B. Such a path length addition can be easily realized because of the chosen half-a-wavelength lattice constant. XOR logic gate PhC structure is presented in Figure 4.

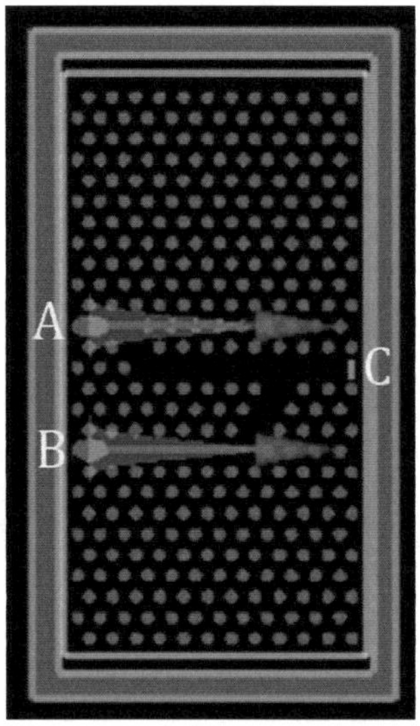

Figure 4. Final silicon PhC XOR logic gate design with ports A and B as the input ports (1550 nm Gaussian), and C as the output port. The rectangle surrounding the PhC structure is the FDTD PML boundary condition.

The inputs of the XOR logic gate can then be tested just like the AND gate. Figure 5 shows the electrical field distribution of said input variations.

Similar to the AND logic gate, a single input light is distributed along the three available paths, reducing the output power at port C. However, the asymmetric structure concentrates a little more light on port C compared to the AND counterpart, yielding 50% higher throughput. When both ports A and B are excited, an almost perfect destructive interference occurs, resulting in an almost zero output. Throughput measurements in Table 3 show XOR gate-like characteristics for the design. By defining logical "0" as less than 0.1 throughput and logical "1" as more than 0.4 throughput, we obtained a CR of 8.34 dB.

With a working design in hand, we proceed with the miniaturization of the cell using the same procedure as the previous logic gate. However, in this part of the study, the number of Nx is reduced while keeping Ny constant. The result samples of the miniaturization are presented in Table 4. The highest CR is observed for Nx=12 and Ny=27 which correspond with cell size of 8.52 x 17.45 μm^2.

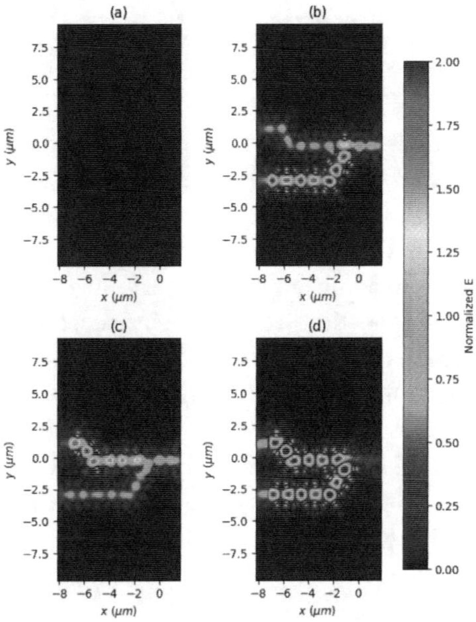

Figure 5. Electric field profile of the final PhC AND Logic Gate design when (a) there is no input, (b) input B is digital high, (c) input A is digital high, and (d) both inputs are digital high. The input lights are gaussian with 1550 nm wavelength.

Table 3. Table of truth for the Final PhC XOR logic gate design. Ports A and B are light inputs, while port C is the measured output. Throughput is defined as the ratio of output and single input power. Contrast Ratio is taken from the ratio between the lowest digital "1" throughput and the highest digital "0" throughput.

A	B	C	Throughput at C	Contrast Ratio (CR), dB
0	0	0	0,00	-
1	0	1	0,45	8.34
0	1	1	0,45	8.34
1	1	0	0,07	-

Table 4. PhC XOR logic gate miniaturization result samples. The number of rods in the Y axis (Ny) is reduced from 27 to 7 rods while keeping the number of rods in the X axis constant (Nx). The first, middle, and final iteration is shown in the table.

Nx	Ny	CR (dB)
20	27	2.37
15	27	4.36
12	27	8.34

Journal of Physics: Conference Series **2980** (2025) 012031 doi:10.1088/1742-6596/2980/1/012031

3.3. Results of AND and XOR PhC-based logic gates

For the AND and XOR PhC logic gates, we found that the optimum value of Nx is 27 for the AND gate and 12 for the XOR gate, while the optimum value of Ny is 15 for the AND gate and 27 for the XOR gate. A comparison of our design with previous designs in PhC-based logic gates for 1550 nm wavelength is summarized in Table 5 and Table 6.

For the AND gate, we can see that this study exhibits an unremarkable CR. However, in terms of cell size, this study has managed to achieve its goal of minimizing the cell size of the logic gate while keeping good CR. The cell size of this study is only preceded by Rani's work. Rani could achieve such a small cell size because the lattice constant she used corresponds to the lowest bandgap point (at a normalized frequency of around 0.25 if applied for this study based on Figure 1b). While this study doesn't optimize the lattice constant used for the lowest bandgap possible, more research is underway to combine this study's and Rani's methods in optimizing the cell size.

On the other hand, a similar situation can be seen for the XOR logic gate. While having an average CR, the design proposed in this study manages to excel in cell size if compared with studies that used similar lattice constant. Therefore, this study presents the possibility of PhC logic gate cell size reduction without sacrificing much CR.

Table 5. Comparison of cell size and Contrast Ratio (CR) for PhC AND logic gates with previous studies. The studies are limited to logic gates for light with 1550 nm wavelength.

References	Material	Cell size (μm^2)	CR (dB)
Raghda (2014), simulation [16]	Silicon	472.74	6.00
Rani (2015), simulation [17]	Silicon	18.13	8.76
Pedraza (2022), simulation [9]	GaAs/AlGaAs	598.08	6.19
This study (2024), simulation	Silicon	189.34	6.22

Table 6. Comparison of cell size and Contrast Ratio (CR) for PhC XOR logic gates with previous studies. The studies are limited to logic gates for light with 1550 nm wavelength.

References	Material	Cell size (μm^2)	CR (dB)
Yulan Fu, (2013) simulation [7]	Silicon	499.94	21.1
Rani (2015), simulation [17]	Silicon	18.13	8.49
This study (2024), simulation	Silicon	148.76	8.34

3.4. Time response of the PhC logic gates

Although much faster than an electron, PhC logic gates' switching speed is still limited by light speed and all the scattering interactions involved. The time responses of the two logic gates proposed are graphed in Figure 6.

Because of the AND PhC logic gate's big size, a delay of up to 0.28 ps can be observed. The resulting interactions would then create ripples and stabilize at 0.35 ps. On the other hand, the XOR PhC logic gate is short horizontally, so the delay is limited to just 0.1 ps. However, the difference in the optical path for input A and B results in A arriving before B. This discrepancy initializes a spike from 0.1 to 0.22 ps before reaching a stable low intensity at 0.4 ps. Such a spike could be detrimental for some applications, so further research is necessary to reduce it.

All of the materials used in the simulation are linear, so the time required to go from "0" to "1" is the same as that from "1" to "0". Therefore, the maximum switching frequency of these logic gates is 1.43 THz and 1.25 THz for the AND and XOR gates, respectively.

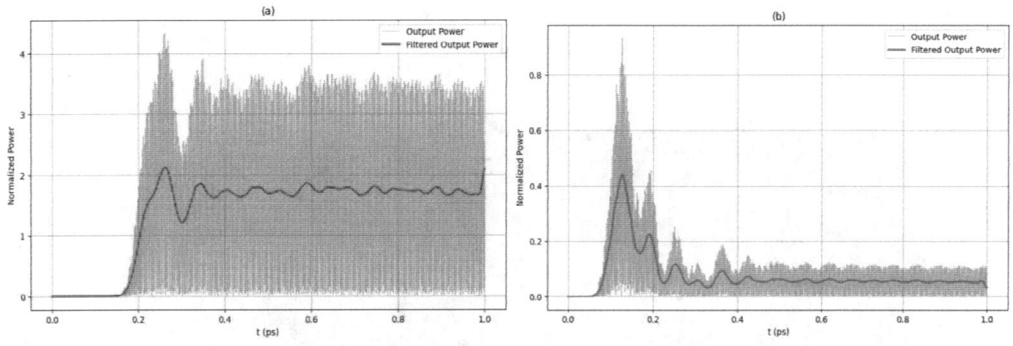

Figure 6. Normalized power intensity at output port C for both logic gates against the time interval after all inputs are set to digital high. (left) AND PhC logic gate and (right) XOR PhC logic gate. The raw output (blue) is low pass filtered (red) to show the average power over time.

Table 7. Relation between PhC slab waveguide height to light source height ratio and the average output power. The output power is expressed in its ratio against the 2D simulation result for comparison.

Ratio of PhC and source height	Mean output power against 2D (%)
1	75
3	8.21
5	6.57

Journal of Physics: Conference Series **2980** (2025) 012031 doi:10.1088/1742-6596/2980/1/012031

3.5. Potential 3D implementation of PhC logic gates

Unlike air holes in dielectric medium type PhC, dielectric medium in air type PhC does not naturally exhibit total internal reflection. A slab model would immediately disperse the light in the vertical direction. On the other hand, an infinitely tall slab model would also immediately disperse the light to the infinitely tall vertical direction. While an infinitely tall PhC cannot be simulated, the result of an increasingly tall PhC is presented in Table 7. The table shows a steep decline in output intensity corresponding to the increase in height. Thus, neither of the above methods work.

To solve this problem, we propose to put two reflective films on either side of the slab. A new simpler prototype as seen in Figure 7 was used to characterize a few common reflectors. Both PhC and light source were kept at half a lattice constant in height to emphasize the effect of the reflector used. The result presented in Table 8 suggests that thin Aluminum film is the best reflector to realize slab PhC.

Although only having an efficiency of 53.86%, a higher than 50% efficiency suggests that light confinement is indeed possible for dielectric rods in air PhC slab structures. Since most 2D PhC logic gate designs use this type of 2D PhC, further work in this efficiency may unlock the possibility of fabrication and even applications for this field in the future.

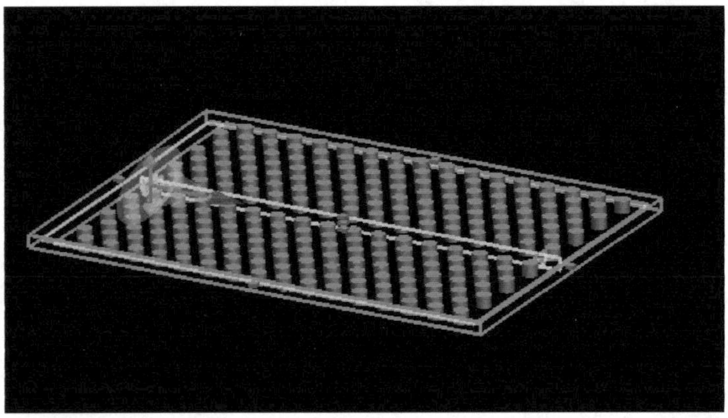

Figure 7. The 3D PhC waveguide slab base model used in this study. The best reflector for a given waveguide-based slab photonic crystal structure is found using this. The thickness of the slab used is 775 nm.

Table 8. Normalized output power (Pout/Pin) for various 3D PhC waveguide slab models sandwiched by a particular reflector type. The thickness of the slab used is 775 nm.

Reflector	Pout/Pin (%)
2D	95.89
Ideal reflector	96.16
Si – SiO_2 Bragg's reflector	5.08
Silver film (0.3875 μm)	44.01
Aluminum film (0.1 μm)	53.86

Journal of Physics: Conference Series **2980** (2025) 012031 doi:10.1088/1742-6596/2980/1/012031

3.6. Design limitation: back output and future directions

In the presented designs, the simulation results (Figures 3 and 4) indicate a phenomenon where input signals partially reflect or scatter into unintended paths, a behavior referred to as "back output". In the AND gate, when only one input (e.g., A or B) is active, the output power at port C is low. However, significant light intensity is observed propagating back into the inactive input port or adjacent waveguides. A similar issue occurs in the XOR gate, where stray light contributes to unwanted residual power in port C even in logical "0" conditions. This behavior can degrade performance in cascaded configurations, as the reflected or stray light may interfere with subsequent logic gates.

The reflected light can distort the signal, leading to reduced signal fidelity by altering the amplitude and phase of the light, making it difficult for the system to maintain correct logic states [12]. Additionally, back output can reduce the contrast ratio (CR) between logical "1" and "0" states, which is crucial for ensuring clear and distinguishable outputs. As the unwanted reflected light interferes with the expected signal, it effectively lowers the difference in output intensities, which is especially problematic in cascaded systems where multiple logic gates rely on precise outputs [13]. Furthermore, this effect can result in mode coupling, where energy unintentionally couples into other waveguides or modes, leading to inefficient signal transmission and power losses [14].

Back output is a well-documented challenge in photonic crystal logic gate designs. For example, Li, Qing et al. and Sharma et al. identified that unintended reflections and mode coupling in photonic waveguides can lead to reduced contrast ratios and signal fidelity [14] [15]. Moreover, Zhong et al. highlighted the need for improved light confinement and directional propagation to mitigate these effects in cascaded systems [10]. By addressing these challenges, cascading multiple photonic logic gates becomes feasible, paving the way for complex photonic computational systems.

4. Conclusion

This study has presented the design, simulation, and miniaturization of PhC AND and XOR logic gates using silicon rods in air triangular lattice with a lattice constant $a = 775$ nm and radius $r = 219.19$ nm. A CR of 6.22 dB has been achieved for an AND logic gate with a cell size of 189.34 μm^2. A similar result can be found for the XOR logic gate with an 8.34 dB contrast ratio and 148.76 μm^2 cell size. Both logic gates exhibit maximum theoretical switching frequencies of 1.43 and 1.25 THz, respectively. Additionally, we have designed and simulated potential slab structures to implement said 2D logic gates in 3D. The best result was found by sandwiching the PhC slab with two thin Aluminum films, resulting in an efficiency of 53.86%.

While the proposed design in this study does not offer a very high contrast ratio, both designs exhibit relatively small cell sizes, which proves the effectiveness of the miniaturization employed in this study. Further studies are underway to increase the size further by optimizing the lattice constant used. Furthermore, additional studies on translating 2D design to 3D using other materials and the possibility of cascading multiple logic gates using a photonic diode are underway to solve the issues presented in this study. Such solutions would be necessary to create Very Large Scale Integration (VLSI) in Photonic Integrated Circuits (PIC), thereby realizing fast and efficient photonic computing.

Acknowledgment

We thank the Center for Nanoscience and Nanotechnology ITB for the ANSYS Lumerical 2022 R1.4 FDTD used in this study.

References

[1] G. Roelkens, et al, "III-V/Si photonics by die-to-wafer bonding," *Materials Today,* vol. 10, no. 7-8, pp. 36-43, 2007.

[2] Toshihiko, Photonic Crystal –An Introduction, Yokohama National University.

[3] S. G. Johnson and J. D. Joannopoulos, Introduction to Photonic Crystals: Bloch Theorem, Band Diagrams, and Gaps (But No Defects), 2003.

[4] W. Liu, D. Yang, G. Shen, H. Tian and Y. Ji, ""Design of ultra compact all-optical XOR, XNOR, NAND and OR gates using photonic crystal multi-mode interference waveguides," *Optics & Laser Technology,* vol. 50, no. 1, pp. 55-64, 2013.

[5] A. Taflove and S. C. Hagness, Computational electrodynamics: The finite-difference time-domain method, Boston: Artech House, 2000.

[6] Y. Zhang, Y. Zhang and B. Li, "Optical switches and logic gates based on self-collimated beams in two-dimensional photonic crystals," *Optics Express,* vol. 15, no. 15, pp. 9287-9292, 2007.

[7] Y. Fu, X. Hu and Q. Gong, "Silicon photonic crystal all-optical logic gates," *Physics Letters A,* vol. 377, no. 3-4, pp. 329-333, 2013.

[8] M. F. Yanik, S. F. Marin, Soljaˇcic´ and J. D. Joannopoulos, "All-optical transistor action with bistable switching in a photonic crystal cross-waveguide geometry," *OPTICS LETTERS,* vol. 28, no. 24, pp. 2506-2508, 2003.

[9] C. LP, P. ML, R. JC, G. PSS and V. N. OP, "Photonic crystal integrated logic gates and circuits," *Opt Express,* vol. 30, no. 2, pp. 1976-1993, 2022.

[10] Chi-O Cho et al, "Photonic Crystal Slab Waveguides Fabricated by the Combination of Holography and Photolithography," *Japanese Journal of Applied Physics,* vol. 43, p. 1384, 2004.

[11] S. G. Johnson, et al, "Guided modes in photonic crystal slabs," *Physical Review B,* vol. 60, no. 8, pp. 5751-5758, 1999.

[12] H. Zhong, T. He, Y. Meng and Q. Xiao, "Photonic Bound States in the Continuum in Nanostructures," *MDPI,* vol. 16, no. 22, pp. 1-20, 2023.

[13] K. Zheng, Y. Yuan, J. He, G. Gu, F. Zhang, Y. Chen, J. Song and J. Qu, "Ultra-high light confinement and ultra-long propagation distance design for integratable optical chips based on plasmonic technology," *Nanoscale,* vol. 11, no. 10, pp. 1-20, 2019.

[14] Q. Li, G. Moille, H. Taher, A. Adibi and K. Srinivasan, "Improved coupled-mode theory for high-index-contrast photonic platforms," *Physical Review A,* vol. 102, pp. 1-9, 2020.

[15] E. K. Sharma, J. P. Nath and N. Dhingra, "Coupled mode theory and coupled mode photonic devices: A Review," *Asian Journal of Physics,* vol. 30, no. 5, pp. 1-18, 2021.

[16] R. M. Younis, N. F. F. Areed and S. S. A. Obayya, "Fully Integrated AND and OR Optical Logic Gates," *IEEE Photonics Technology Letters,* vol. 26, no. 19, pp. 1900-1903, 2014.

[17] P. Rani, Y. Kalra and R. Sinha, "Design of all optical logic gates in photonic crystal waveguides," *Optik,* vol. 126, no. 9-10, pp. 950-955, 2015.

Amino Groups Effect on the Phosphorescence Carbon Dots: DFT Study

Reza Umami[1], Fitri Aulia Permatasari[1,2,3*], and Ferry Iskandar[1,2,3*]

[1] Department of Physics, Faculty Mathematics and Natural Sciences, Institut Teknologi Bandung, Jalan Ganesha 10, Bandung, West Java 40132, Indonesia.
[2] Research Center for Nanoscience and Nanotechnology, Institut Teknologi Bandung, Jalan Ganesha 10, Bandung, West Java 40132, Indonesia.
[3] Collaboration Research Center for Advanced Energy Materials, National Research and Innovation Agency - Institut Teknologi Bandung, Jalan Ganesha 10, Bandung 40132, West Java, Indonesia

*E-mail: ferry@itb.ac.id, fauliap@itb.ac.id

Abstract. Carbon dots (CDs) that exhibit room-temperature phosphorescence (RTP) are attracting interest due to their broad potential applications, such as anti-counterfeiting, chemical sensing, bioimaging, and photocatalysis. The process of designing and synthesizing RTP CDs has been challenging because it often involves finding an appropriate solid material or adding heavy atoms to enhance intersystem crossing (ISC) and suppress vibrational dissipation between singlet and triplet states. Nitrogen functional groups, specifically Amino (NH_2), might effectively enhance RTP in CDs by facilitating the ISC process. In this study, we proposed the use of surface-functionalized amino-doped CDs to modulate afterglow properties by altering the energy gap (ΔE_{ST}) between the lowest excited singlet (S_1) and triplet state (T_1) through time-dependent density functional theory (TD-DFT) calculations. We also analyzed the spin-orbit coupling (SOC) constant, which governs radiative recombination in the triplet state. We conducted computational simulations using DFT, utilizing optimized model systems to support experimental findings.

Introduction

Phosphorescent carbon dots (CDs) have emerged as an innovative nanomaterial, rapidly gaining interest among researchers due to their unique properties and broad applications. CDs present several advantages, including their ability to be synthesized from various precursors, high stability, excellent biocompatibility, low toxicity, and cost-effectiveness[1,2]. CDs have already been utilized in multiple fields, such as biomedicine, optoelectronics, sensors, and catalysis[3]. Recently, room-temperature phosphorescent (RTP) CDs have become a significant focus of research due to their distinctive characteristics, such as longer emission lifetimes and high selectivity, making them ideal for applications in areas like anticounterfeiting, information encryption, and optoelectronic devices[4].

RTP in CDs originates from the direct radiative transition of triplet excitons from the first triplet state to the ground state, enabling light emission at room temperature. These attributes make RTP CDs highly desirable for a wide range of advanced applications. Surface functionalization, particularly with amino groups, can additionally alter the charge distribution

Journal of Physics: Conference Series **2980** (2025) 012032 doi:10.1088/1742-6596/2980/1/012032

and electronic energy levels of CDs, leading to enhancements in their optical properties, such as redshifts in absorbance spectra, changes in energy gaps, and increased spin-orbit coupling (SOC)[5–7]. These modifications are essential for understanding key processes like intersystem crossing (ISC) and reverse intersystem crossing (RISC), which are critical for achieving long-lived excited states and enhanced phosphorescence[8].

This study investigates CDs' optical and electronic properties functionalized with amino groups. Using theoretical calculations, Density Functional Theory (DFT), we explore the effects of amino group concentrations on the CDs' absorbance, energy gaps, and spin-orbit coupling constants. Through this work, we aim to provide deeper understanding the effect of surface modification on the properties of CDs, enabling their optimized use in optoelectronic and photonic applications.

Computational Method

The carbon dot (CDs) model was developed using pyrene, a representative polycyclic aromatic hydrocarbon (PAHs) compound. Thus, pristine pyrene will be referred to as pristine CDs, while pyrene decorated with amino groups will be termed CDs functionalized with amino groups. Subsequent calculations were performed to assess the impact of surface functionalization with amino groups ($-NH_2$) at varying concentrations relative to the pristine state. Geometry optimizations were carried out using Density Functional Theory (DFT) in the Gaussian 09 software [9], ensuring convergence without imaginary frequencies. The absorbance spectra of the CDs were then calculated using Time-Dependent Density Functional Theory (TD-DFT) with the B3LYP functional (three-parameter) and the 6-311G(d,p) basis set. The hole-electron distributions were analyzed and visualized using the Multiwfn 3.8 package alongside Chemcraft software[10,11]. Additionally, the energy levels associated with the singlet and triplet states were determined at each optimized geometry using B3LYP/DEF2-SVP, providing vertical excitation energies from the singlet to the triplet state. Spin-orbit coupling (SOC) constants were also analyzed at the same level of theory using the ORCA 5.0 package to complement the ΔE_{ST} data further [12].

Result and Discussion

The CDs model utilizes pyrene, a representative polycyclic aromatic hydrocarbon (PAH) compound. This structure is considered highly representative due to its similar absorbance characteristics with experimental results and has been reported [5,13]. In this work, we investigate the incorporation of amino functional groups (NH_2), experimentally confirmed by our work [14–16]. These amino groups are introduced on the surface of CDs via nitrogen-containing precursors such as urea, o-phenylenediamine, and ethylenediamine. The influence of varying amino group concentrations is examined by specifying the number of functional groups involved. To simplify the analysis, the geometric structure shown in Fig. 1 is selected to represent CDs and their surface modifications. The optimized structure is planar and has successfully converged.

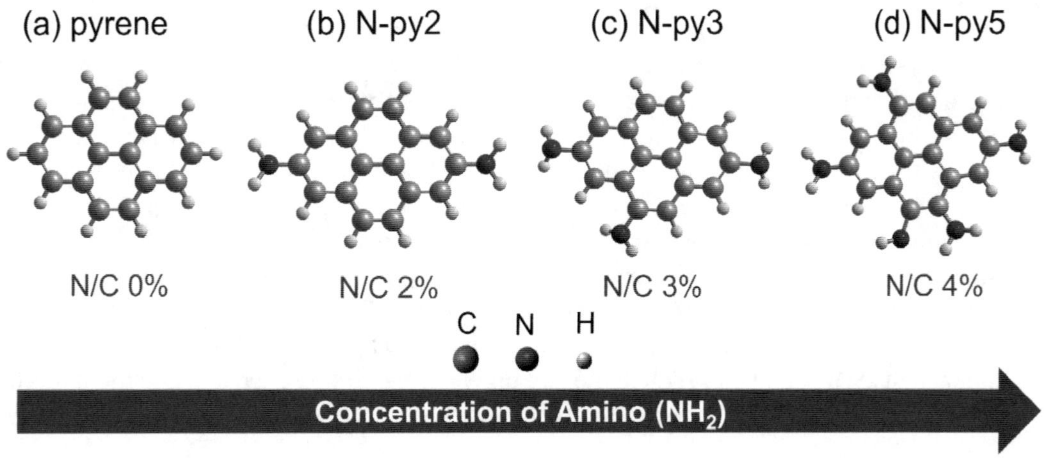

Figure 1. Optimized structure of CDs-based pyrene (a) pristine pyrene and (b-d) pyrene with varying amino (-NH_2) concentration.

To investigate the absorbance behavior resulting from the amino functional groups on the surface of CDs, TD-DFT calculations were performed with the Gaussian 09 software. The absorbance spectrum of the CDs is presented in Fig. 2. The pristine CDs exhibit characteristic peaks around ~230 nm and 340 nm (Fig. 2a), corresponding to transitions typically observed in CDs synthesized via a bottom-up method[17–20]. The peak at ~230 nm is assigned to a $\pi-\pi*$ transition (C=C bond), while the peak at 340 nm is indicative of an n–π* transition (C-N, C=N bonds)[14,17]. The introduction of amino groups (Fig. 2b-d) results in a redshift in the absorbance spectrum, with the wavelengths shifting by approximately 5–50 nm. The shifting has been observed in previous reports due to the symmetry breaking caused by the amino groups. This symmetry disruption permits electronic transitions previously forbidden in pristine CDs, thereby modifying their absorbance characteristics [21–23].

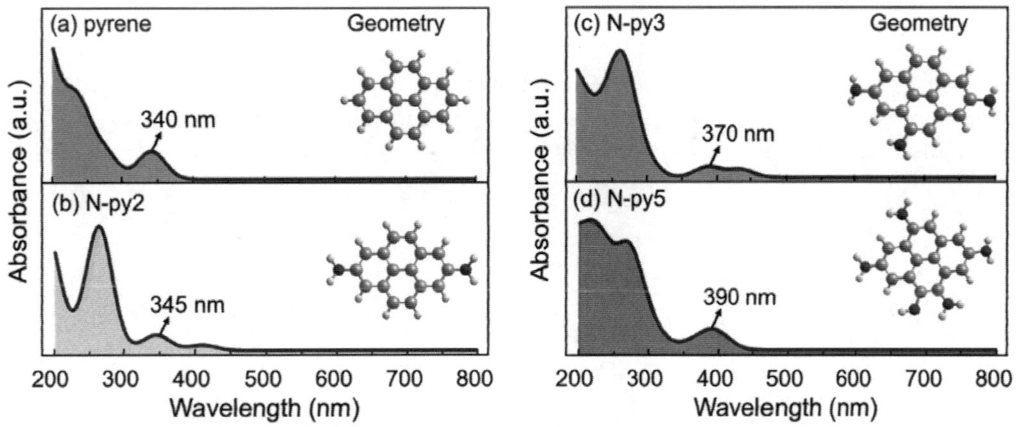

Figure 2. Absorbance spectra of CDs (a) pristine pyrene and (b-d) pyrene with varying amino (-NH_2) concentration.

The TD-DFT calculations performed using the ORCA 5.0 package are presented in Fig. 3. The energy gap decreases from 3.81 eV to 3.50, 3.49, and 3.46 eV, respectively, following the addition of amino incorporation of functional groups onto the surface of the CDs. The spin-orbit coupling (SOC) value between the triplet (T_1) and first excited singlet state (S_1) is initially 0.00 cm^{-1} in pristine CDs but increases to 0.26 cm^{-1} as the concentration of amino groups increases. The band gap energy between the first excited singlet state and the first triplet state (ΔE_{ST}) decreases from 2.09 eV in pristine CDs to 1.83 eV in CDs with the highest amino group concentration. According to Marcus rate theory, the reverse intersystem crossing (RISC) process becomes more favorable in amino-functionalized CDs due to increased SOC and the narrower ΔE_{ST} [24]. This can be attributed to nitrogen configurations, such as amino, which donate electrons to the π-system of the CDs structure, facilitating excited-state electron transitions. In pristine CDs, where localized (π, π*) transitions dominate, the intersystem crossing (ISC) and RISC processes are less likely due to weaker SOC and a larger ΔE_{ST} [21,23]. In amino-functionalized CDs, the additional amino groups enhance the SOC value and reduce the gap between the first excited singlet state and the first triplet state (ΔE_{ST}), facilitating phosphorescence with a prolonged lifetime.

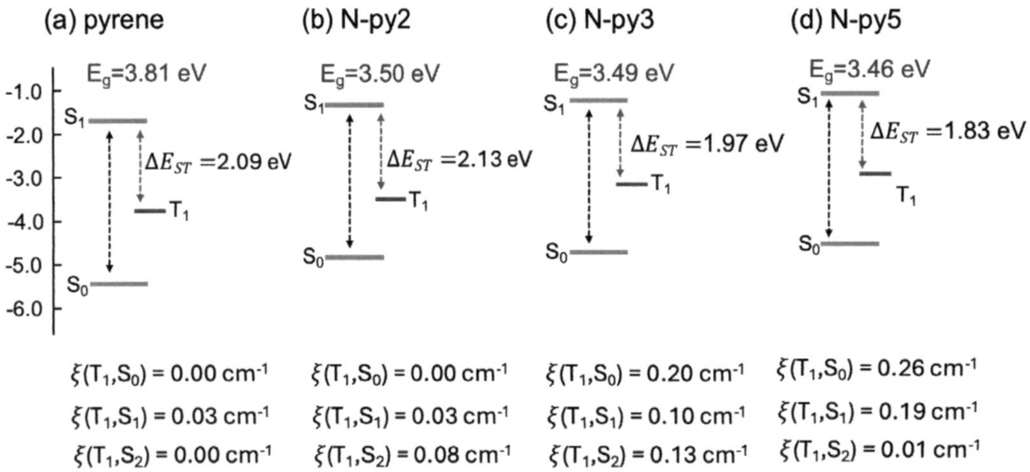

Figure 3. Energy bandgap E_g (eV), ΔE_{ST} (eV), and SOC index ξ (cm^{-1}) of CDs (a) pristine pyrene and (b-d) pyrene with varying amino ($-NH_2$) concentration in singlet and triplet levels. The red and green lines represent the singlet and triplet levels.

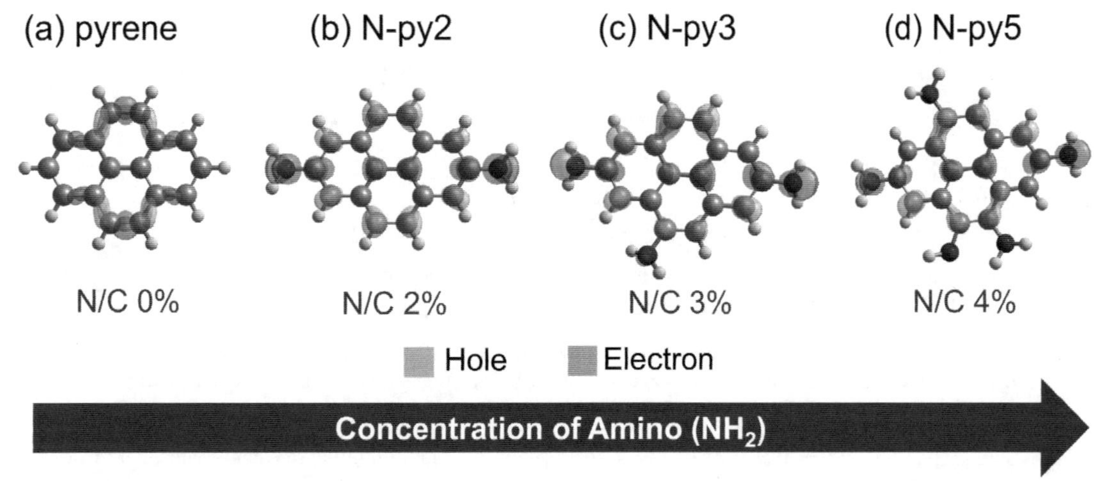

Figure 4. Distribution of charge electron and hole in CDs structure (a) pristine pyrene and (b-d) pyrene with varying amino (-NH_2) concentration. The calculations use isosurface 0.000244 a.u. Inset, the purple and red balloons represent electrons and holes.

To elucidate the optical properties of GQDs, the electron and hole distributions are depicted in Fig 4. The electron-hole distributions corresponding to the strong absorption peaks of the CDs are shown, where green represents holes and red represents electrons[25]. Figure 4 highlights the impact of surface functionalization on the charge distribution within CDs. It is observed that the electrons and holes are delocalized onto the amino functional groups as a result of the surface modification. Moreover, amino groups on the CDs surface lead to a more symmetrical distribution of electrons and holes in the functionalized CDs.

Conclusion

In brief, the surface functionalization of CDs with amino groups significantly alters their optical and electronic properties. The introduction of amino groups causes a redshift in absorbance due to symmetry breaking and enhanced electronic transitions. Additionally, the higher spin-orbit coupling (SOC) and smaller energy gap (ΔE_{ST}) between the singlet and triplet states encourage reverse intersystem crossing (RISC), enabling longer phosphorescence lifetimes. The delocalization of electrons and holes onto the amino functional groups further leads to a more symmetrical charge distribution, highlighting the importance of surface modification in tuning the electronic characteristics of CDs for optoelectronic applications.

Acknowledgment

This work is fully supported by International Research Grant 2023: Research Grant and Staff Exchange Program, Institut Teknologi Bandung Grant No. 8693/IT1.B07.1/TA.00/2023. R.U. would like to thank the Ministry of Fiscal Indonesia Endowment Fund for Education (LPDP) for her doctoral scholarship.

Journal of Physics: Conference Series **2980** (2025) 012032 doi:10.1088/1742-6596/2980/1/012032

References

[1] Iravani S and Varma R S 2020 Green synthesis, biomedical and biotechnological applications of carbon and graphene quantum dots. A review *Environmental Chemistry Letters 2020 18:3* **18** 703–27

[2] Liu M 2020 Optical Properties of Carbon Dots: A Review *Nanoarchitectonics* **1** 1–12

[3] Nguyen K G, Huš M, Baragau I A, Bowen J, Heil T, Nicolaev A, Abramiuc L E, Sapelkin A, Sajjad M T and Kellici S 2024 Engineering Nitrogen-Doped Carbon Quantum Dots: Tailoring Optical and Chemical Properties through Selection of Nitrogen Precursors *Small* 2310587

[4] Bai J, Yuan G, Chen X, Zhang L, Zhu Y, Wang X and Ren L 2022 Simple Strategy for Scalable Preparation Carbon Dots: RTP, Time-Dependent Fluorescence, and NIR Behaviors *Advanced Science* **9** 1–9

[5] Kundelev E V., Tepliakov N V., Leonov M Y, Maslov V G, Baranov A V., Fedorov A V., Rukhlenko I D and Rogach A L 2019 Amino Functionalization of Carbon Dots Leads to Red Emission Enhancement *Journal of Physical Chemistry Letters* **10** 5111–6

[6] Zhang Q, Wang R, Feng B, Zhong X and Ostrikov K (Ken) 2021 Photoluminescence mechanism of carbon dots: triggering high-color-purity red fluorescence emission through edge amino protonation *Nature Communications 2021 12:1* **12** 1–13

[7] Cao N, Wang Q, Zhou X, Gao Y, Feng Y, Li H, Bai P, Wang Y and Zhou G 2021 Where is the best substitution position for amino groups on carbon dots: a computational strategy toward long-wavelength red emission *J Mater Chem C Mater* **9** 14444–52

[8] Sun J, Sun Z, Wang Z, Wang N, Han Y, Zhang L, Zhang B and Zhang X 2024 Ultra-Long-Lived Red TADF-CDs: Solid-State Synthesis, Time-Dependent Phosphorescence Color And Luminescent Mechanism *Adv Opt Mater* **12** 1–9

[9] M.J. Frisch, G.W. Trucks, H.B. Schlegel, G.E. Scuseria, M.A. Robb, J.R. Cheeseman, G. Scalmani, V. Barone, G.A. Petersson, H. Nakatsuji, X. Li, M. Caricato, A. Marenich, J. Bloino, B.G. Janesko, R. Gomperts, B. Mennucci, H.P. Hratchian, J.V. Ortiz, A.F. I D J F 2016 Gaussian 09, Revision D.01 *Gaussian, Inc., Wallingford CT (2016)*

[10] Lu T and Chen F 2012 Multiwfn: A multifunctional wavefunction analyzer *J Comput Chem* **33** 580–92

[11] Zhurko G A 2005 Chemcraft - graphical program for visualization of quantum chemistry computations. Ivanovo, Russia, 2005. https://chemcraftprog.com

[12] Neese F 2022 Software update: the ORCA program system, version 5.0 *WIRES Comput. Molec. Sci.* **12** e1606

[13] Liu B, Aquino A J A, Nachtigallová D and Lischka H 2020 Doping Capabilities of Fluorine on the UV Absorption and Emission Spectra of Pyrene-Based Graphene Quantum Dots *Journal of Physical Chemistry A* **124** 10954–66

[14] Permatasari F A, Fukazawa H, Ogi T, Iskandar F and Okuyama K 2018 Design of pyrrolic-N-rich carbon dots with absorption in the first near-infrared window for photothermal therapy *ACS Appl Nano Mater* **1** 2368–75

[15] Ogi T, Aishima K, Permatasari F A, Iskandar F, Tanabe E and Okuyama K 2016 Kinetics of nitrogen-doped carbon dot formation via hydrothermal synthesis *New Journal of Chemistry* **40** 5555–61

[16] Permatasari F A, Nakul F, Mayangsari T R, Aimon A H, Nuryadin B W, Bisri S Z, Ogi T and Iskandar F 2021 Solid-state nitrogen-doped carbon nanoparticles with tunable emission prepared by a microwave-assisted method *RSC Adv* **11** 39917–23

[17] Wang B, Yu J, Sui L, Zhu S, Tang Z, Yang B and Lu S 2021 Rational Design of Multi-Color-Emissive Carbon Dots in a Single Reaction System by Hydrothermal *Advanced Science* **8**

[18] Xian Y and Li K 2022 Hydrothermal Synthesis of High-Yield Red Fluorescent Carbon Dots with Ultra-Narrow Emission by Controlled O / N Elements **2201031** 1–13

[19] Permatasari F A, Aimon A H, Iskandar F, Ogi T and Okuyama K 2016 Role of C-N Configurations in the Photoluminescence of Graphene Quantum Dots Synthesized by a Hydrothermal Route *Sci Rep* **6** 1–8

Journal of Physics: Conference Series **2980** (2025) 012032 doi:10.1088/1742-6596/2980/1/012032

[20] Cao M, Li Y, Zhao Y, Shen C, Zhang H and Huang Y 2019 A novel method for the preparation of solvent-free, microwave-assisted and nitrogen-doped carbon dots as fluorescent probes for chromium(VI) detection and bioimaging *RSC Adv* **9** 8230–8

[21] Santika A S, Permatasari F A, Umami R, Muyassiroh D A M, Irham M A, Fitriani P and Iskandar F 2022 Revealing the synergetic interaction between amino and carbonyl functional groups and their effect on the electronic and optical properties of carbon dots *Physical Chemistry Chemical Physics* **24** 27163–72

[22] Permatasari F A, Umami R, Sundari C D D, Mayangsari T R, Ivansyah A L, Muttaqien F, Ogi T and Iskandar F 2023 New insight into pyrrolic-N site effect towards the first NIR window absorption of pyrrolic-N-rich carbon dots *Nano Res* **16** 6001–9

[23] Umami R, Permatasari F A, Muyassiroh D A M, Santika A S, Sundari C D D, Ivansyah A L, Ogi T and Iskandar F 2022 A rational design of carbon dots: Via the combination of nitrogen and oxygen functional groups towards the first NIR window absorption *J Mater Chem C Mater* **10** 1394–402

[24] Park I S, Yang M, Shibata H, Amanokura N and Yasuda T 2022 Achieving Ultimate Narrowband and Ultrapure Blue Organic Light-Emitting Diodes Based on Polycyclo-Heteraborin Multi-Resonance Delayed-Fluorescence Emitters *Advanced Materials* **34** 2107951

[25] Sun L, Chen Y, Sun M and Zheng Y 2022 Physical Mechanism of Fluorescence and Chirality of Functionalized Graphene Quantum Dots *Journal of Physical Chemistry C* **126** 12845–59

Material Characterization using Magnetic Induction Based Measurement Technique at Low-Frequency Range

Kurnia Nugraha[1,2,3], Winarto Winarto[1*], Didied Haryono[3,4], Amalia Sholehah[3,4], Harisman Nugraha[5]

[1] Department of Metallurgical and Material Engineering, Faculty of Engineering, Universitas Indonesia, Depok 16424, Indonesia
[2] Department of Mechanical Engineering, Faculty of Engineering, Universitas Sultan Ageng Tirtayasa, Cilegon 42435, Indonesia
[3] Advanced Materials and Tomography Laboratory, Faculty of Engineering, Universitas Sultan Ageng Tirtayasa, Cilegon 42435, Indonesia
[4] Department of Metallurgical Engineering, Faculty of Engineering, Universitas Sultan Ageng Tirtayasa, Cilegon 42435, Indonesia
[5] C-Tech Labs Edwar Technology, Tanggerang 15141, Indonesia

*E-mail: winarto.msc@ui.ac.id

Abstract. Material characterization is the process of measurement and analysis aimed at understanding the physical, chemical, mechanical, or structural properties of a material. The primary objective is to determine how the material will behave or perform under specific conditions. In this study, material characterization was carried out using a magnetic-induction-based measurement technique (MIT), which is a technique for obtaining information about the conductivity or magnetic permeability properties of an object by utilizing a magnetic field. This material characterization is part of the development of a non-destructive testing device for detecting defects in steel welds. The purpose of this study is to determine the impedance values of various materials using the MIT system developed by C-Tech Labs Edwar Technology was performed at a low-frequency range. Experiments were conducted by measuring each material in the 100 - 200 kHz frequency range. The materials measured include steel, steel with artificial defects, copper, aluminum, acrylic, and air. The results show that air and aluminum have the lowest and highest values in the frequency range. In addition, the measurement of steel with artificial defects resulted in lower impedance values compared to normal (non-defect) steel material. Thus, the MIT system that has been developed can characterize various types of materials, especially for the detection of artificial defects in steel material.

Keywords: low-frequency, magnetic induction, material characterization, non-destructive testing, steel welds.

1. Introduction

Welding is the process of joining two or more materials into a single unit by heating them to the welding temperature, either with or without the application of pressure, and with or without the use of filler metal [1, 2]. Welding is a commonly used method for steel in industrial applications. However, welding can cause various issues, such as defects, distortion, residual stress, and metallurgical changes, which affect the quality and performance of the welded joints [3, 4]. Therefore, it is important to evaluate welded steel using Non-destructive Testing (NDT), which can detect and characterize defects without damaging the material or affecting its future use [3, 5, 6]. NDT is a testing method that does not damage or alter the physical, chemical, or mechanical properties of the material being tested [5, 7]. NDT can be used for various purposes, such as quality inspection, process control, and maintenance. NDT includes various techniques, such as visual testing, penetrant testing, radiographic inspection, ultrasonic testing, magnetic particle inspection, eddy current testing, and acoustic emission [6, 7]. These techniques have different principles, equipment, and applications, depending on the type, size, and location of the defects to be detected. The appropriate method selection depends on the specific application, fluid type, pipe configuration, and other specific requirements [8].

Eddy current testing (ECT) and Electromagnetic Tomography (EMT) are NDT methods that use electromagnetic induction to detect flaws in conductive materials [6, 7]. This method is also known as the magnetic induction-based measurement technique (MIT). In recent years, MIT has been expanded for medical and industrial applications, including stroke detection [9], Hemorrhage [10], monitoring of molten metal flow [11, 12], non-destructive testing (NDT) for material characterization [13], and as a potential detection tool for National Nuclear Security [14]. The development of MIT, both in terms of instrumentation and software, has been noted [15]. MIT works by passing an alternating electric current through a coil of wire placed near the surface of the material being tested. This electric current generates a changing magnetic field around the coil and penetrates the material. This magnetic field then induces an electric current in the material called eddy current. This eddy current also generates its own magnetic field that opposes the magnetic field of the coil. This magnetic field affects the current and voltage flowing in the coil. The changes in current and voltage can be detected using either the same coil or a separate detector coil, forming the basis of the eddy current testing inspection technique. Any changes in the conductivity of the material being tested, such as surface defects, crack or variations in thickness, will distort the magnitude of the eddy currents [16-21].

The use of magnetic induction based measurement technique (MIT) in inspection and testing of materials can be applied to a wide range of industrial applications [22] [13] has several benefits: 1) MIT does not damage the material being tested [23]. This is crucial for saving costs and time, as well as maintaining the quality and functionality of the material [24]. 2) MIT can provide testing results with a high degree of accuracy because the technology used in this method is highly sensitive to changes in the electrical conductivity of the material. 3) MIT can be performed quickly and easily, as it only requires a probe or coil placed near the surface of the material being tested. MIT does not require special preparations, such as surface cleaning, lubrication, or marking. 4) MIT can be used for various types of conductive materials, such as metals, alloys, composites, and ceramics. MIT can also be used for materials of various shapes and sizes, such as pipes, plates, wires, rods, and so on. MIT can be adjusted with various parameters, such as frequency, amplitude, phase, and the geometry of the probe or coil, to enhance sensitivity and selectivity towards specific defects [25]. 5) MIT does not use radiation sources that are hazardous to human health and the environment, as used in radiographic testing. Magnetic induction-based measurement techniques only use harmless magnetic fields and do not require protective equipment or special supervision. Magnetic induction-based measurement techniques also do not leave radiation traces on the tested material, thus avoiding the risk of contamination or pollution. However, the limitations of MIT include sub-optimal sensor sensitivity for detecting very small defects, which requires high resolution. Additionally, material characteristics such as conductivity and permeability can affect detection results, making materials less responsive to magnetic fields and reducing the effectiveness of MIT. The previous MIT system is also less effective in detecting defects in non-metallic or composite materials with different electromagnetic properties. Frequency variation is also an important factor, as each frequency may be optimal only for certain defect types or materials, thereby increasing the complexity of calibration and adjustment.

MIT's potential in industrial applications demonstrates versatility across various sectors. In the oil and gas industry, MIT can enhance pipeline integrity monitoring by identifying early signs of corrosion, blockages, or structural weaknesses, thus preventing costly leaks and environmental risks. In aerospace and automotive manufacturing, MIT can ensure component quality by detecting subsurface defects without damaging the material. In construction, MIT can monitor concrete structures for internal flaws or degradation, allowing for early intervention to prevent structural failures. Additionally, in the pharmaceutical and medical industries, MIT can be utilized to inspect the purity of chemical substances or detect anomalies in biological materials. In the energy sector, MIT can support condition monitoring of turbines and generators, enhancing the reliability of power plants. Overall, MIT can significantly improve safety, quality control, and operational efficiency across industries, offering a non-invasive alternative where traditional methods may be less effective.

In this research, material characterization using the MIT system developed was performed at a low-frequency range. Characterization was carried out on various types of materials, including steel, aluminum, copper, acrylic, and air. The impedance value of various types of materials were further analyzed afterward. The potential of this MIT system being used for welded-steel defect detection was also discussed.

2. Methods

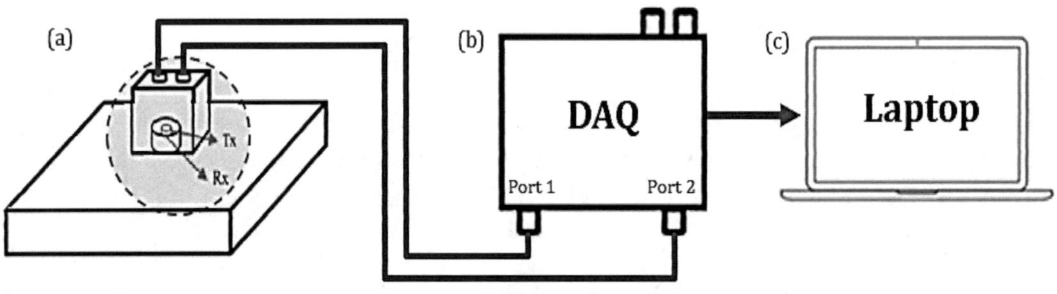

Figure 1. Schematic diagram of experiment procedure, (a) magnetic induction sensor, (b) data acquisition, and (c) laptop.

The experiment was conducted using the MIT system developed by C-Tech Labs Edwar Technology. The experimental steps are as follows: 1) device calibration and 2) material measurement. The calibration process of the device confirms the accuracy of the measuring instrument by resetting it to its original settings without additional adjustments and addressing any measurement errors from internal or external influences to ensure precise and reliable results. Calibration was performed sequentially on port 1 and port 2 using the standard method, involving measurements in three modes: short, open, and load. The short mode was conducted by directly connecting the ports to check for minimum resistance. The open mode was performed by disconnecting the ports to measure maximum impedance. The load mode was carried out by connecting a known reference load to validate the accuracy of the measurements.

Table 1. Specifications of solenoid coil sensor.

Parameter	Value
Number of turns Transmitter (T_x)	115
Number of turns Receiver (R_x)	1150
Coil diameter Transmitter (T_x)	0.5 cm
Coil diameter Receiver (R_x)	1.5 cm

The data acquisition system (DAQ) is connected, with the transmitter (T_x) cable of the sensor attached to port 1 and the receiver (R_x) cable to port 2. For material measurement, each material was measured in the frequency range of 100-200 kHz with five repetitions to obtain the electromagnetic impedance value influenced by the material's permittivity, conductivity, and permeability. Data from each measurement were saved on a computer in .csv format and used for further analysis. The materials used in this study were normal steel, steel with artificial defects, copper, aluminum, acrylic, and air. The artificial defects were made to resemble porosity defects with a diameter of 3 mm. The MIT system used consisted of: 1) a magnetic induction sensor, 2) data acquisition system, and 3) personal computer. The schematic diagram of the experiment and the sensor specifications can be seen in Figure 1 and Table 1, respectively.

3. Discussion

Electromagnetic impedance is a concept that encompasses how electromagnetic waves interact with specific media or materials in the form of the resistance encountered by the electromagnetic wave when interacting with a medium [16]. Impedance is influenced by several factors, including permittivity, conductivity, and permeability of the material. Permittivity is a measure of how easily a material allows electric field lines to pass through it, and it is related to the formation of an electric field. Conductivity, on the other hand, measures how well a material can conduct electric current. Permeability is a measure of how easily a material allows magnetic field lines to pass through it, and it is related to the formation of a magnetic field. The relationship between these properties and impedance is complex and depends on the frequency.

Figure 2 shows the characterization of the impedance signal as a function of frequency on various types of materials. The colors on the graph indicate the materials tested, i.e., air (red line), steel (green line), steel with artificial defect (gray line), copper (blue line), aluminum (pink line), and acrylic (purple line). In general, the trend in impedance measurements for all materials decreases up to a frequency of 140 kHz, then increases to 150 kHz, and tends to stagnate up to a frequency of 200 kHz. Apart from that, the impedance value of each material relative to other materials also tends to be constant, except at a frequency of 100 kHz. At a frequency of 100 kHz, overlapping occurs for several materials. Based on the raw data analyzed, the 100 kHz frequency tends to have higher noise compared to other frequencies. Thus, analysis at a frequency of 100 kHz can be ignored.

Journal of Physics: Conference Series **2980** (2025) 012033 doi:10.1088/1742-6596/2980/1/012033

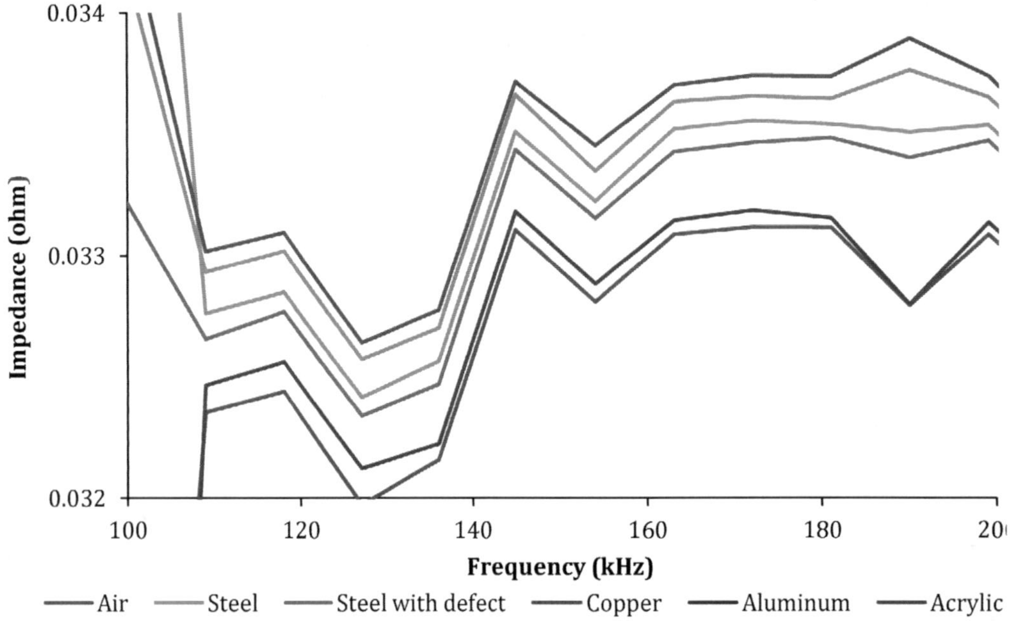

Figure 2. Graph of impedance signal characterization as a function of frequency.

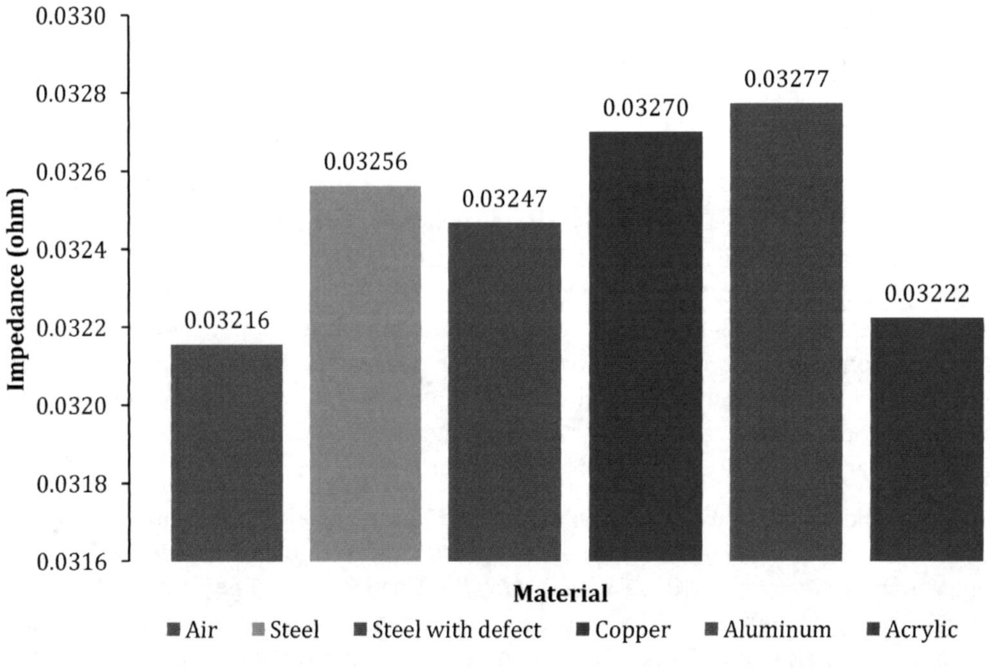

Figure 3. Bar Diagram of Material Impedance

From the results of the signal characterization in Figure 2, the optimal condition of one of the frequencies is found so that a bar chart is plotted as in Figure 3. The determination of optimal conditions is based on changes in the impedance signal between steel and steel with artificial defect because in the future this MIT system will be used for defect detection in steel welds. As can be seen in Figure 3, the impedance values from lowest to highest are air, acrylic, steel with artificial defects, steel, copper, and aluminum. The impedance value of air is 32.16 mΩ, whilst the impedance value of aluminum is 32.77 mΩ. In addition, steel and steel with artificial defect materials have a difference of impedance value of 9.5 x 10^{-2} mΩ, where steel with artificial defect has a lower impedance value compared to normal steel. This trend is in accordance with the measurement results of the base material, where the measured values for air are lower compared to steel. Thus, if there is a defect such as porosity, the impedance measurement will decrease at this frequency range. A summary of impedance measurements at the frequency of 136 kHz for all materials using the MIT system developed by C-Tech Labs Edward Technology.

Study of material characterization using the MIT system at low frequencies have been carried out on various types of materials. From the study, it can be concluded that air and aluminum are the materials with the lowest and highest impedance with values of 32.16 mΩ and 32.77 mΩ respectively. In addition, the difference in impedance measurement of steel and steel with artificial defect has value of 9.5 x 10^{-2} mΩ. The difference in impedance values between steel and steel with artificial defect materials is a good indication in the MIT system development process. In which, the MIT system is currently capable of detecting artificial defect with a diameter of 3 mm as concluded in this study. When compared to conventional non-destructive testing (NDT) methods such as ultrasonic testing (UT) and radiographic testing (RT), the MIT system offers several significant advantages. Unlike UT, which requires direct contact with the material surface and a coupling medium like gel, MIT can operate without physical contact, making it ideal for applications in hard-to-reach environments. Additionally, unlike RT, which involves ionizing radiation that may pose health risks to operators, MIT uses a safe and environmentally friendly electromagnetic field. However, the MIT system also has several limitations that should be considered. Its sensitivity is influenced by the size, shape, and orientation of defects, which can affect detection accuracy. Materials with similar electromagnetic properties can also pose challenges in distinguishing different types of defects. Moreover, the current MIT system is more effective on metallic materials with high conductivity, while its effectiveness on non-metallic materials such as composites still requires improvement. Therefore, future research will focus on enhancing the measurement sensitivity of the MIT system by reducing the size of artificial defects to resemble real defects, as well as expanding testing to include various types of materials. Through these advancements, the MIT system is expected to become a more reliable and competitive NDT technology for broader industrial applications.

4. Conclusion

In this research, material characterization using the MIT system at low frequencies has been carried out using various types of materials. From the study, it can be concluded that air and aluminum are the materials with the lowest and highest impedance, with values of 32.16 mΩ and 32.77 mΩ, respectively. In addition, the difference in impedance measurement of steel and steel with artificial defect has a value of 9.5 x 10^{-2} mΩ. The difference in impedance values between steel and steel with artificial defect materials is a good indication of the MIT system development process. The MIT system is currently capable of detecting artificial defects with a diameter of 3 mm as concluded in this study. Further studies will focus on the sensitivity of MIT measurements by reducing the size of artificial defects to real defects.

Journal of Physics: Conference Series **2980** (2025) 012033 doi:10.1088/1742-6596/2980/1/012033

References

[1] ANSI, *Standard Welding Terms and Definitions Including Terms for Adhesive, Bonding, Brazing, Soldering, Thermal Cutting, and Thermal Spraying*, in *American Welding Society*. 2001, American Welding Society: N.W. LeJeune Road, Miami, Florida 33126. p. 42.

[2] Weman, K., *Welding processes handbook*. Vol. 2. 2012, Daryaganj, New Delhi - 110002 India: Woodhead Publishing Limited. 267.

[3] Deepak, J., et al., *Non-destructive testing (NDT) techniques for low carbon steel welded joints: A review and experimental study.* Materials Today: Proceedings, 2021. **44**: p. 3732-3737.

[4] Messler Jr, R.W., *A Practical guide to welding solutions: overcoming technical and material-specific issues.* 2019: John Wiley & Sons.

[5] Shull, P.J., *Nondestructive evaluation: theory, techniques, and applications*. 2002: CRC press.

[6] Shaloo, M., et al., *A review of non-destructive testing (NDT) techniques for defect detection: application to fusion welding and future wire arc additive manufacturing processes.* Materials, 2022. **15**(10): p. 3697.

[7] Gupta, M., et al., *Advances in applications of Non-Destructive Testing (NDT): A review.* Advances in Materials and Processing Technologies, 2022. **8**(2): p. 2286-2307.

[8] Datta, S. and S. Sarkar, *A review on different pipeline fault detection methods.* Journal of Loss Prevention in the Process Industries, 2016. **41**: p. 97-106.

[9] Park, G.S. and D.S. Kim, *Development of a magnetic inductance tomography system.* IEEE transactions on magnetics, 2005. **41**(5): p. 1932-1935.

[10] Zhu, Z., et al. *Hemorrhage Detection by Magnetic Induction Tomography with Helmholtz Coil.* in *2023 42nd Chinese Control Conference (CCC)*. 2023. IEEE.

[11] Ma, X., et al., *Electromagnetic techniques for imaging the cross-section distribution of molten steel flow in the continuous casting nozzle.* IEEE Sensors Journal, 2005. **5**(2): p. 224-232.

[12] Muttakin, I., T. Wondrak, and M. Soleimani, *Magnetic induction tomography sensors for quantitative visualization of liquid metal flow shape.* IEEE Sensors Letters, 2020. **4**(7): p. 1-4.

[13] Wei, H.-Y. and M. Soleimani, *A magnetic induction tomography system for prospective industrial processing applications.* Chinese Journal of Chemical Engineering, 2012. **20**(2): p. 406-410.

[14] Darrer, B.J., et al., *Magnetic imaging: a new tool for UK national nuclear security.* Scientific reports, 2015. **5**(1): p. 7944.

[15] Ma, L. and M. Soleimani, *Magnetic induction tomography methods and applications: A review.* Measurement Science and Technology, 2017. **28**(7): p. 072001.

[16] Levy, S., D. Adam, and Y. Bresler, *Electromagnetic impedance tomography (EMIT): a new method for impedance imaging.* IEEE transactions on medical imaging, 2002. **21**(6): p. 676-687.

[17] Sophian, A., et al., *Electromagnetic and eddy current NDT: a review.* Insight, 2001. **43**(5): p. 302-306.

[18] García-Martín, J., J. Gómez-Gil, and E. Vázquez-Sánchez, *Non-destructive techniques based on eddy current testing.* Sensors, 2011. **11**(3): p. 2525-2565.

[19] Kasai, N., et al., *Crack detection using a cutting-edge flexible eddy current sensor with voltage and phase measurement techniques.* Mechanical Systems and Signal Processing, 2024. **219**: p. 111613.

[20] Wei, H.-Y., *Magnetic Induction Tomography for Medical and Industrial Imaging: Hardware and Software Development.* 2012, University of Bath.

[21] Piscitelli, G., et al. *Magnetic induction tomography via the monotonicity principle*. in *Journal of Physics: Conference Series*. 2023. IOP Publishing.

[22] Hansen, J. and R.B. Peoples, *Using eddy current testing to solve industrial problems.* Materials evaluation, 2006. **64**(6): p. 543-546.

[23] Griffiths, H., *Magnetic induction tomography.* Measurement science and technology, 2001. **12**(8): p. 1126.

[24] Elia, S.J. and W.H.A. Putra, *Analisis Kemampuan Pendeteksian Pengujian Eddy Current Testing (ECT) terhadap Crack pada Bollard dengan Bahan Casting.* Jurnal Teknik ITS, 2021. **10**(1): p. G36-G43.

[25] Udpa, S.S. and P.O. Moore, *Electromagnetic testing.* Nondestructive testing handbook, 2004. **5**: p. 230.

Journal of Physics: Conference Series **2980** (2025) 012034 doi:10.1088/1742-6596/2980/1/012034

Molecular Dynamics Simulations of Temperature-Induced Changes in Water Nanodroplet-Graphene Nanoribbon

Mauludi Ariesto Pamungkas[1*], Ghozi Zhorif Althof[1], Farid Surya[1], Gancang Saroja[1], Agus Dardiri[1]

[1] Department of Physics, Universitas Brawijaya, Indonesia

*E-mail: m_ariesto@ub.ac.id

Abstract. Wettability is a critical factor in diverse fields such as oil recovery, microfluidics, and nanotechnology. The behavior of liquid droplets on solid surfaces, influenced by surface properties, has been extensively investigated. Understanding these interactions is essential for controlling droplet dispersion. This study investigates the impact of temperature on the interaction between water nanodroplets and graphene nanoribbons (GNRs) using molecular dynamics (MD) simulations. The results demonstrate that the potential energy, pressure, and enthalpy increase with the addition of 1-3 water nanodroplets to the GNR system at varying temperatures. Furthermore, the simulations confirm the hydrophilic nature of GNRs.

1. Introduction

The interactions between water and graphene are of paramount importance for a wide range of applications, including graphene-based membranes, electrodes, and catalysts[1]. Despite extensive research, the detailed structure of water hydrogen bonds at the graphene-water interface remains elusive. The hydrophilicity or hydrophobicity of graphene has been a subject of ongoing debate, with the consensus that it is strongly influenced by the number of graphene layers[2,3][3]. Hojat Zahedi and Masumeh Foroutan have performed molecular dynamics simulations to explore the interactions between water droplets and graphene oxide at elevated temperatures[4]. In this study, we aimed to elucidate the effects of varying the number of water droplets and temperature on the interfacial properties of water-graphene systems. Specifically, we sought to determine the hydrophilicity or hydrophobicity of graphene under different conditions through molecular dynamics simulations.

2. Computational Methods

Molecular dynamics simulations were performed using LAMMPS to investigate the interactions between water nanodroplets and graphene nanoribbons (GNRs)[5]. The GNRs were modeled as 10 x 5 nm systems with 2016 carbon atoms in a honeycomb lattice, while water droplets consisted of 521 oxygen and 1042 hydrogen atoms. To assess the impact of the number of water droplets, we varied the droplet number from 1 to 3. Periodic boundary conditions were employed in the horizontal direction to eliminate finite size effects and simulate a bulk system. Both zigzag and armchair GNR structures were studied[6]. To explore the temperature effects, simulations were

Journal of Physics: Conference Series **2980** (2025) 012034 doi:10.1088/1742-6596/2980/1/012034

carried out at 300 K, 400 K, and 500 K. Given that contact angle measurements are usually carried out at room temperature, this simulation aimed to investigate the effects of temperature increments of 100 K starting from ambient conditions. A hybrid AIREBO-Lennard-Jones potential was employed to accurately describe the interactions between atoms. The AIREBO potential, which accounts for bonded and non-bonded interactions, was combined with the Lennard-Jones potential for van der Waals interactions[7][8].

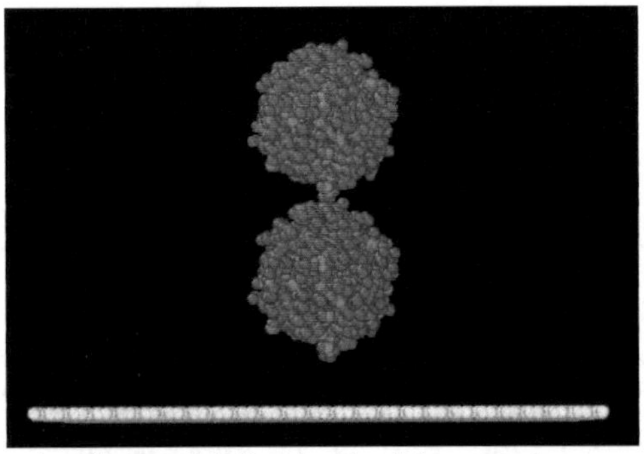

Figure 1. Initial model of water droplet on graphene nanoribbon

3. Results and Discussion

Figure 2 illustrates the temporal evolution of potential energy for a single, double, and triple water droplet resting on a graphene nanoribbon (GNR) at temperatures of 300 K, 400 K, and 500 K . The results indicate that the pattern of potential energy fluctuations occurs earlier at higher temperatures. This higher kinetic energy allows water molecules to move faster and more freely, resulting in more frequent and stronger interactions between water molecules and the graphene surface. Higher temperatures can also cause softening of bonds between water molecules or between water molecules and the graphene surface. Weaker bonds are more easily broken and reformed, leading to faster fluctuations in potential energy

Beyond the temperature effects, Figure 2 further reveals a positive correlation between the number of water droplets deposited on graphene and the system's potential energy. As the droplet count increases, so does the frequency of interactions involving water molecules and graphene carbon atoms. These interactions, predominantly hydrogen bonding, van der Waals forces, and electrostatic forces, contribute to a cumulative increase in the system's potential energy.

Journal of Physics: Conference Series **2980** (2025) 012034 doi:10.1088/1742-6596/2980/1/012034

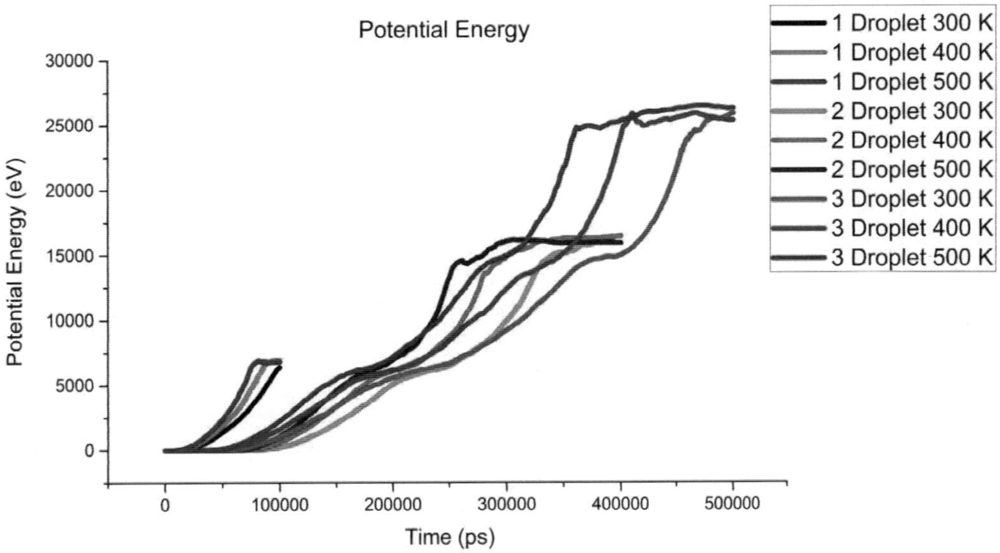

Figure 2. Time-dependent potential energy profile of a single, double, and triple water droplet deposited on graphene nanoribbon (GNR) at temperatures of 300 K, 400 K, and 500 K

In the preceding sections, the effects of temperature on the potential energy of a water droplet on graphene nanoribbons (GNR) were discussed. Additionally, Figure 3 illustrates the influence of temperature and the number of water droplets on the pressure exerted on graphene nanoribbons (GNR). With different temperature, the patterns of the pressure of the system are similar. However, the earlier onset of pressure fluctuations at higher temperatures was observed. This suggests that water molecules at higher temperatures possess greater kinetic energy, leading to faster movement and more frequent interactions with the graphene surface. Consequently, the interactions between water droplets and graphene occur more rapidly and intensely, resulting in earlier and more pronounced pressure fluctuations. The increased number of water droplets results in a higher proportion of water-graphene interactions. High temperatures can also induce changes in the graphene surface structure, such as increased vibrations of carbon atoms. These structural changes can affect the interactions between water molecules and graphene. In addition, higher temperatures can decrease the viscosity of water. Lower viscosity allows water molecules to move more freely and interact with graphene more quickly.

Journal of Physics: Conference Series **2980** (2025) 012034 doi:10.1088/1742-6596/2980/1/012034

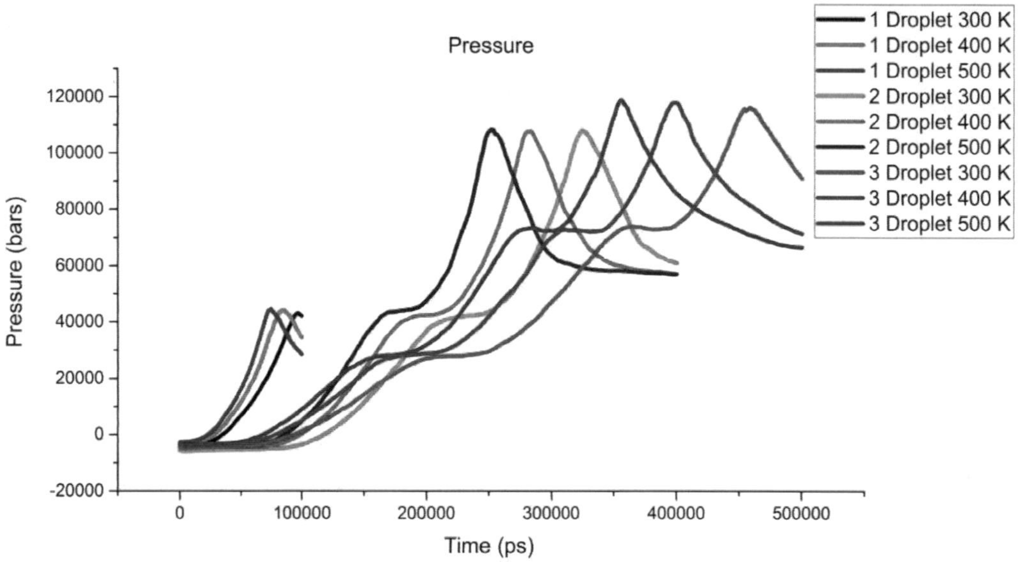

Figure 3. The relationship between the number of water droplets and temperature with the pressure exerted on graphene nanoribbons (GNR). Elevated temperature and a greater number of water droplets lead to earlier onset and increased magnitude of pressure fluctuations

In Figure 4 we found that relationship between temperature and the number of water droplets with enthalpy was similar to that of enthalpy. At higher temperatures, the frequency of collisions between water molecules and the graphene surface increases significantly. More frequent collisions cause faster changes in the system's potential energy, leading to a clearer and earlier onset of enthalpy fluctuations.

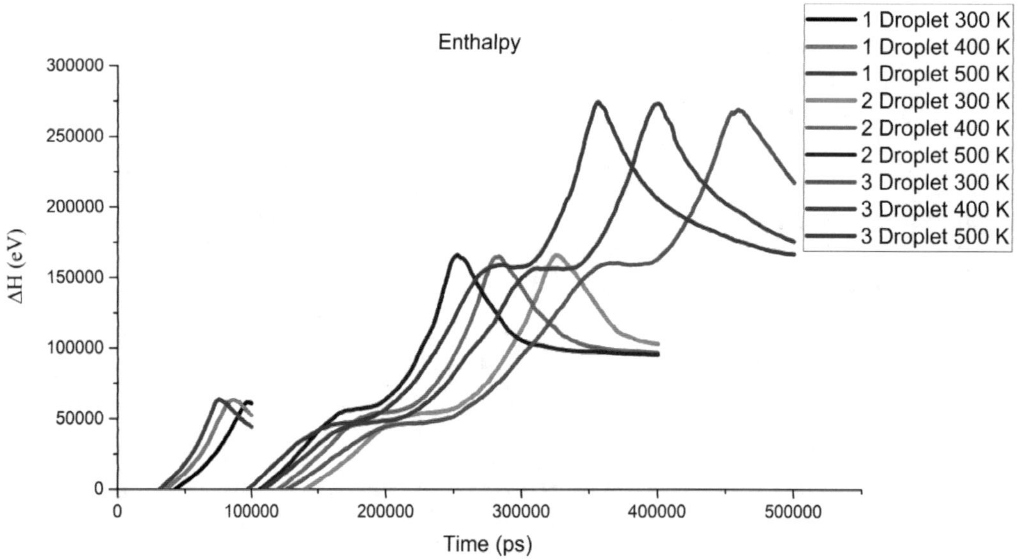

Figure 4. Temporal evolution of enthalpy as a function of temperature and the number of water droplets. Higher temperature and a larger droplet count result in more frequent and pronounced fluctuations in enthalpy

Finally, we calculated the contact angle of water droplets on graphene nanoribbons (GNR), as shown in Figure 5. The addition of more droplets increased the contact angle. Notably, three droplets exhibited a contact angle of only 15 degrees. While one and two droplets did not penetrate the GNR, three droplets were able to penetrate the material. These simulation results reinforce the conclusion that graphene is a hydrophilic material[9]. While some studies suggest that graphene is hydrophobic, it is becoming increasingly evident that its hydrophobicity or hydrophilicity is influenced by the number of layers. Monolayer graphene exhibits lower hydrophobicity compared to bilayer or multilayer graphene[3]. However, there are also studies suggesting the opposite, with monolayer graphene exhibiting higher hydrophobicity than multilayer graphene[10]. This discrepancy highlights the complex nature of graphene's interaction with water, influenced by numerous factors.

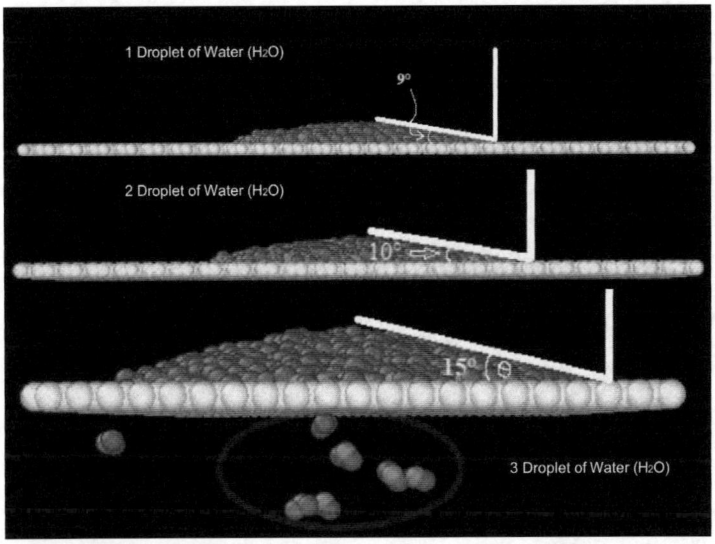

Figure 5. Contact angle of water nanodroplet on the GNR surface

4. Conclusion

Increasing the number of water droplets on graphene led to an increase in potential energy, pressure, and enthalpy. At various temperatures, the time-dependent profiles of potential energy, pressure, and enthalpy exhibited similar fluctuation patterns. However, these fluctuations occurred earlier at higher temperatures. These findings suggest that temperature plays a crucial role in the dynamics of water-graphene interactions. The simulations confirm the hydrophilic nature of GNRs.

Acknowledgments

The research was funded by Universitas Brawijaya (grant number: 00140.2/UN10.A0501/B/PT.01.03.2/2024)

Journal of Physics: Conference Series **2980** (2025) 012034 doi:10.1088/1742-6596/2980/1/012034

References

[1] E. Kim, D. Kim, K. Kwak, Y. Nagata, M. Bonn, M. Cho, Wettability of graphene, water contact angle, and interfacial water structure, Chem 8 (2022) 1187–1200. https://doi.org/10.1016/j.chempr.2022.04.002.

[2] Sebastián R. Accordino; Joan Manuel Montes de Oca; J. Ariel Rodriguez Fris; Gustavo A. Appignanesi, Hydrophilic behavior of graphene and graphene-based materials, Journal of Chemical Physics 143 (2015) 154704.

[3] C.E.G.R.L.M.-W.D.K.G.O.K. Martin Munz, Thickness-Dependent Hydrophobicity of Epitaxial Graphene, ACS Nano 9 (2015) 8401–8411.

[4] H. Zahedi, M. Foroutan, Molecular dynamics simulation of a water nano-droplet on graphene oxide surface at high temperature: Evaporation or spreading?, Appl Surf Sci 455 (2018) 789–794. https://doi.org/10.1016/j.apsusc.2018.06.046.

[5] A.P. Thompson, H.M. Aktulga, R. Berger, D.S. Bolintineanu, W.M. Brown, P.S. Crozier, P.J. in 't Veld, A. Kohlmeyer, S.G. Moore, T.D. Nguyen, R. Shan, M.J. Stevens, J. Tranchida, C. Trott, S.J. Plimpton, LAMMPS - a flexible simulation tool for particle-based materials modeling at the atomic, meso, and continuum scales, Comput Phys Commun 271 (2022) 108171. https://doi.org/10.1016/j.cpc.2021.108171.

[6] T. Enoki, S. Fujii, K. Takai, Zigzag and armchair edges in graphene, Carbon N Y 50 (2012) 3141–3145. https://doi.org/10.1016/j.carbon.2011.10.004.

[7] A.B.T.J.A.H. Steven J. Stuart, A reactive potential for hydrocarbons with intermolecular interactions, J Chem Phys 112 (2000) 6472–6486.

[8] J.B. Adams, Bonding Energy Models, Encyclopedia of Materials: Science and Technology (2001) 763–767.

[9] L.A. Belyaeva, P.M.G. van Deursen, K.I. Barbetsea, G.F. Schneider, Hydrophilicity of Graphene in Water through Transparency to Polar and Dispersive Interactions, Advanced Materials 30 (2018). https://doi.org/10.1002/adma.201703274.

[10] R. Seki, H. Takamatsu, Y. Suzuki, Y. Oya, T. Ohba, Hydrophobic-to-hydrophilic affinity change of sub-monolayer water molecules at water–graphene interfaces, Colloids Surf A Physicochem Eng Asp 628 (2021). https://doi.org/10.1016/j.colsurfa.2021.127393.

DFT Study of H_2 Adsorption and Dissociation on Supported PdZn Clusters

Marleni Wirmas[1], Muhammad Haris Mahyuddin[2,3*], Mohammad Kemal Agusta[2,3], and Hermawan Kresno Dipojono[2,3]

[1] Doctoral Program of Engineering Physics, Faculty of Industrial Technology, Institut Teknologi Bandung, Jl. Ganesha 10 Bandung 40132, Indonesia
[2] Quantum and Nano Technology Research Group, Faculty of Industrial Technology, Institut Teknologi Bandung, Jl. Ganesha 10 Bandung 40132, Indonesia
[3] Research Center for Nanoscience and Nanotechnology, Institut Teknologi Bandung, Jl. Ganesha 10 Bandung 40132, Indonesia

*E-mail: mahyuddin133@itb.ac.id

Abstract. The adsorption and dissociation of hydrogen molecules are essential for various hydrogenation reactions and utilization in hydrogen storage. Herein, by utilizing Density Functional Theory (DFT), we introduce Zn atoms to a Pd subnanometer cluster to investigate H_2 dissociation. Two cluster models with ZrO_2 support are considered, namely Pd_4Zn/ZrO_2 and Pd_3Zn_2/ZrO_2. We find that H_2 molecule strongly adsorbed on the Pd clusters, where Pd_3Zn/ZrO_2 gives slightly lower adsorption energy. Both configurations were found to facilitate the H_2 dissociation into two H atoms thermodynamically, which takes place on top of Pd atoms with activation barriers of 0.63 and 0.67 eV on the Pd_4Zn/ZrO_2 and Pd_3Zn_2/ZrO_2., respectively. Overall, this study provides new insights into the effect of cluster forms of Pd-based catalyst on the reaction mechanism of H_2 dissociation.

Keywords: subnanometer cluster, H_2 dissociation, Density Functional Theory, heterogeneous catalyst

1. Introduction

Transition metals have become significant tools for organic synthesis catalysts with vast attention and have been investigated experimentally and theoretically in the past decades for industrial purposes. One of the specific metals, palladium-based catalyst has been widely used for reactions of hydrogenation, oxidation, reduction, and many other organic syntheses on various reactants and products. Palladium has been demonstrated to have predictable and controllable catalytic properties with good susceptibility to wide range of functional groups in various reactions, as well as capability on adsorbing and dissociating hydrogen molecules. Many studies reported that several forms of Pd catalysts exhibit exceptional activities compared to other transition metals because of their high stability and high affinity of H_2 spillover [1–3]. However, as a noble metal, Pd with its high cost stimulates extensive search for effective and efficient systems where modification of its form in small numbers with good performance are needed.

Some strategies, including modification of active sites into supported single atom and subnanometer cluster catalysts, have been introduced in past years. In single atom catalyst (SAC), 100% atom utilization is one of its biggest advantages, since SACs resemble homogenous metal complexes, clearly can utilize metal atoms to their fullest potential and have well defined structures. However, since the metal atoms' surface free energy are higher, the drop in particle size is associated by an increased tendency of aggregation [4]. Furthermore, in many essential industrial processes, the single atomic active site is unable to adsorb and activate several substrates simultaneously, particularly in the catalytic reactions of multiple substrates [5]. This would result in a reduction in catalytic efficiency or even causing inactivity for the reaction [6]. On the other hand, two or more atoms in the form of supported atomic clusters can provide enough sites for the adsorption and activation of several substrates, potentially resulting in the creation of a new catalytic route that lowers the reaction barrier and increases catalytic activity. Moreover, introducing another metal species on the cluster structure often leads to different geometric and also the electronic structure due to the overlapping orbitals between two metal atoms, thus giving enhancement to unique catalytic behaviors in heterogeneous catalytic reactions [7].

Hydrogen molecules are the most abundant chemical element, making up approximately 75% of all normal matter, and essential part of natural gas. Hydrogen can be utilized as resources of fuels in clean energy, ammonia manufacturing, as well as hydrogenation of CO, CO_2, and other organic compounds into value-added materials. The adsorption and dissociation of hydrogen molecule is one of the important factors for hydrogen storage and hydrogenation reaction in heterogeneous catalysis. It can also be a key step in heterogeneous catalysis, especially if the dissociation of one of the reactants is activated, making it the rate-limiting step of the overall reaction [8]

In hydrogen catalyst, as well as storage applications, it is generally required for the catalyst to have a proper adsorption energy (thermodynamic factor) and low activation energy (kinetic factor). Thermodynamically, hydrogen storage and catalyst should have a suitable bond interaction strength. Kinetically, lower activation energy enables the reaction to proceed at lower temperature and pressure. In Pd cluster catalyst, H_2 adsorption and dissociation on various small Pd clusters had been investigated. On Pd_3 and Pd_4 clusters, H_2 dissociation occurs with energy barrier of 0.31 and 0.46 eV [9]. In a more recent study, DFT calculation of H_2 dissociation on Pd_6 cluster showed that barrier energy of 0.47 eV is required for the formation of the intermediate 2H, where one H atom bounds to the single Pd atom and the second H bridging the Pd–Pd edge [10]. As for supports for the cluster, ZrO_2 is one of the versatile metal oxide structures with robust interaction with other metals, which leads to a stronger molecule binding [11,12]. Furthermore, alloying active sites such Pd clusters with additional metal atom and combining with metal oxide support would alter the electronic structure that affects the interaction between the catalyst and intermediates. However, only some Pd clusters have been synthesized, and combinations of metals that could be more effective in catalyzing chemical reactions related to hydrogenation are still unclear. Furthermore, there is a lack of clarity on the electronic structure and the role of electronic interactions between the sites and molecules. As such, it is fundamentally significant to systematically understand the distinctive properties and superiority of cluster.

In this work, we propose a new model of supported cluster catalyst to be performed in density functional theory (DFT) calculations of the H2 dissociation reaction over two types of clusters, namely Pd_4Zn/ZrO_2 and Pd_3Zn_2/ZrO_2, to understand the mechanism of how these structures enhance the reaction in terms of thermodynamics and kinetics. This work intends to understand the characteristics and gain insights for improvement in the catalyst design of H_2 dissociation for a sustainable pathway. Theoretical methods, however, have some limitations such as approximations for exchange-correlation functionals, which can lead to inaccuracies in binding energy and reaction barrier calculations. These discrepancies may result in over, or underestimation of adsorption strengths, especially for systems with complex electronic interactions. Nevertheless, DFT simulations can identify the preferred adsorption sites and mechanism in reaction pathways, guiding the rational design of more advanced catalyst [13].

2. Computational Details

All calculations were performed using Vienna ab-initio simulation package (VASP 5.4.4) [14,15] based on the Kohn–Sham formulation [16,17]. The calculations were non-spin polarized as there is no magnetization for the supported PdZn cluster. The projector augmented wave (PAW) was used to treat the core-electron interactions [18,19]. The electron exchange-correlation was calculated using the generalized gradient approximation by Perdew-Burke-Ernzrnhof (GGA-PBE) [20]. The Grimme's D3 scheme with zero damping was used to account for the van der Waals dispersion correction [21]. Bader charge analysis was used to calculate atomic charge and spin density [22]. A cutoff energy of 500 eV was used in all calculations and the plane wave expansion with the total-energy convergence was set to 1×10^{-5} eV for the electronic self-consistent iteration. The force convergence of every ionic iteration was below 0.03 eV $Å^{-1}$. The Brillouin zone was sampled with 3×3×1 Monkhorst-Pack *k*-points meshes [23]. The climbing image nudged elastic band (CI-NEB) method was used to locate transition states (TS) [24,25]. The Image-Dependent Pair Potentials method was employed to generate initial NEB images [26]. The optimized structures were visualized by using VESTA [27].

The selection of specific supported PdZn clusters is driven by their unique structural and electronic properties, which play a critical role in catalytic performance. The configurations were chosen based on the optimization of Pd-Zn interaction and ZrO_2 support, creating favorable and effective sites for H_2 adsorption, dissociation, and spillover [28-31]. Previous research shows that alloying Pd with Zn alters the electronic density of states, which affects hydrogen binding energies and contributes to the activation energies for bond dissociation. This enhances reaction selectivity and efficiency. Additionally, ZrO_2, often used as a support, provides strong metal-support interactions that stabilize these alloys and prevent agglomeration under reaction conditions, improving their performance in hydrogen reaction processes [32]. These rational designs aim to incorporate activity and stability, which is particularly promising for hydrogen applications.

For the surface system, a 3 × 3 supercell of Pd_4Zn/t-ZrO_2(101) and Pd_3Zn_2/t-ZrO_2(101) slab models with five atomic layers and a vacuum space of 20 Å along the *z*-direction were constructed. This model allows one to minimize the effect of interactions between the periodically repeated adsorbates due to the large enough separation distance in both *x* and *y* directions. During calculations, the bottom two layers were fixed to their bulk positions while the remaining layers were free to relax. The adsorption energy (E_{ads}) was calculated by using the following equation.

$$E_{ads} = E_{slab+adsorbate} - (E_{slab} + E_{adsorbate})$$

where $E_{slab+adsorbate}$, $E_{adsorbate}$, E_{slab} are the total energies of the adsorbate-surface system, the slab surface, and an isolated adsorbate, respectively. Using energy in transition and initial state, Activation energy for the dissociation reaction was determined by subtracting the total energy of the transition state (TS) and initial state (IS), as follows:

$$E_a = E_{TS} - E_{IS}$$

Journal of Physics: Conference Series **2980** (2025) 012035 doi:10.1088/1742-6596/2980/1/012035

3. Results and Discussion

We show in Figure 1 the energy diagram of H_2 dissociation on Pd_4Zn/ZrO_2 and Pd_3Zn_2/ZrO_2. H_2 molecule is adsorbed on both systems to start the reaction. For all possible adsorption configurations, the most stable adsorption site is on top of the PdZn clusters for both Pd_4Zn and Pd_3Zn_2 with adsorption energy E_{ads} = -0.48 and -0.53 eV. These numbers are found to be more stable than one in PdZn(111) surface reported by Brix et al. [31] and comparably higher than on Pd_4 and Pd_6 clusters (-0.81, -0.67 eV) [33]. Following the adsorption, the H-H bond length is slightly elongated from 0.74 Å in the gas phase to 0.82 Å on Pd_4Zn/ZrO_2 and 0.83 Å on Pd_3Zn_2/ZrO_2, suggesting that the adsorbed H_2 has already been activated even at this early stage of the reaction. As for Pd-H bond length, the results are similar on both clusters (r_{H-Pd} = 1.80 Å). Compared to existing results, we find quite a similarity on PdZn surface (r_{H-Pd} = 1.81 Å) [31] and on small Pd clusters (r_{H-Pd} = 1.78 Å)[29].

Adsorption of H_2 is followed by subsequent H–H bond scission via transition state (TS), where the bond further elongates to 2.0 and 1.70 Å on Pd_4Zn/ZrO_2 and Pd_3Zn_2/ZrO_2, respectively. The two H atoms interact with one (on Pd_3Zn_2) and two (on Pd_4Zn) of the Pd atoms (see Figure 1) with Pd–H bond lengths being similar for both catalysts (about 1.6 Å). The H-H bond length discrepancy is probably due to different step on the transition state. On Pd_4Zn/ZrO_2 case, the TS is found to be late-barrier step (Figure 1, red lines) where 2 hydrogen atoms separation is quite far and interacts with more atoms in the cluster. The activation energy required for the dissociation in this system is 0.62 eV, slightly lower than on Pd_3Zn_2/ZrO_2. With activation barrier energy around 0.6 eV, this number is found to be higher than one on Pd clusters (0.4 eV on Pd_4 and Pd_6) [33] and PdZn surface (0.02 eV on PdZn(111)) [31].

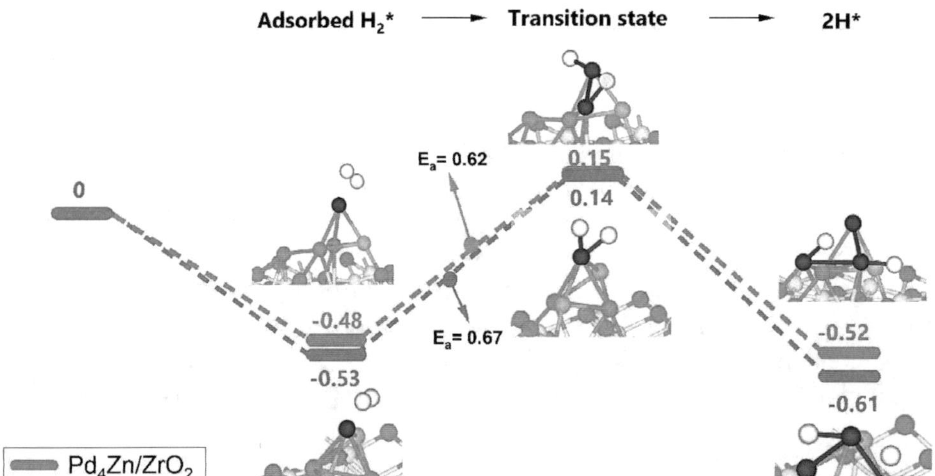

Figure 1. Energy diagrams (in eV) of H_2 dissociation reaction taking place on Pd_4Zn/ZrO_2 (red lines) and Pd_3Zn_2/ZrO_2 (green lines). H_2 adsorption occurs on top of Pd atom of the clusters on both systems, which is followed by the elongated H-H bond in transition state located in the cluster, to finally dissociates into 2H* in the second layer of the PdZn clusters.

The reaction is followed by the formation of adsorbed 2H* on both catalyst systems (Figure 1). On Pd_4Zn/ZrO_2, H atoms are separated with far H-H distance 3.43 Å, each H interacts with 2 atoms, resulting in adsorption energy E_{ads} = -0.52 eV. One H atom is located on the hollow site of the cluster while another H atom is located on the interfacial site of cluster-support. Meanwhile on Pd_3Zn_2/ZrO_2, adsorption of 2H* has similar H-H distance with Pd_4Zn/ZrO_2, and both H atoms are adsorbed on the hollow site of the cluster with lower adsorption energy (E_{ads} = -0.61 eV), in agreement with adsorption on PdZn(111) surface reported by Brix et., al. [31] with E_{ads} = -0.62 eV.

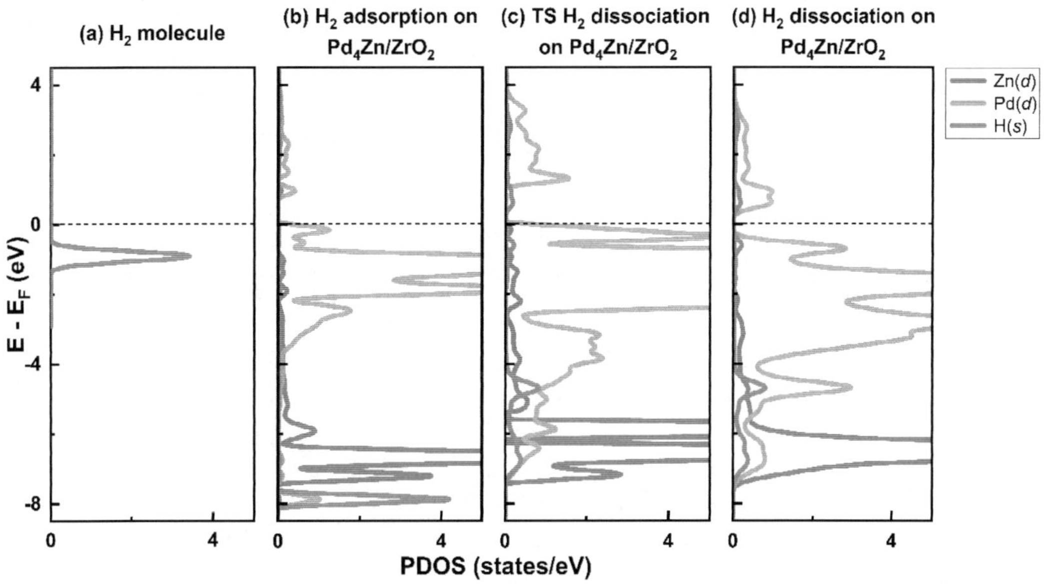

Figure 2. Calculated projected density of states (PDOS) for (a) isolated H_2 molecule, (b) H_2 adsorption, (c) transition state, and (d) 2H* on Pd_4Zn/ZrO_2. The Fermi level is set to zero. The stability of hydrogen interactions with the catalyst resulting from the H-Pd-Zn interaction is depicted by overlapping PDOS.

For a deeper understanding of the performance, electronic properties, and interactions, we also perform a projected density of states (PDOS) analysis for elementary steps in this reaction. First, the isolated H_2 molecule has electrons accumulating at energy level E = -0.95 eV. Upon adsorption of H_2 molecule on both systems, we observe that electrons in H atoms shifted to a lower energy level to interact with Pd atoms, which can be seen at visible peak located around E = -7.8 eV (Figure 2b, 3b), showing that this adsorption has interaction between H-Pd and explains the early H-H bond activation. Following transition state in Figures 2c and 3c, electrons on hydrogen atoms are distributed in the occupied (above Fermi level) and unoccupied (below Fermi level) state. On Pd_3Zn_2/ZrO_2 transition state (Figure 3c), H, Pd, and Zn atoms are interacting at energy level E = -4.76 eV, also H-Pd interaction at E = -4.16 eV showing a more stable system than on Pd_4Zn/ZrO_2. On the adsorption of 2H* on Pd_4Zn/ZrO_2, H-Pd interactions are shown in several energy E = -0.75 and -4.65 eV, while H-Pd-Zn interaction is also yielded at E = -6.56 eV, showing the role of Zn atom in hydrogen adsorption. In line with Pd_4Zn/ZrO_2, 2H* on Pd_3Zn_2/ZrO_2 also show H-Pd and H-Pd-Zn interactions (Figure 3d). We found that the interaction between H and the cluster metal atom dominantly contributes to H_2 adsorption and dissociation on both Pd_4Zn/ZrO_2 and Pd_3Zn_2/ZrO_2.

Journal of Physics: Conference Series **2980** (2025) 012035 doi:10.1088/1742-6596/2980/1/012035

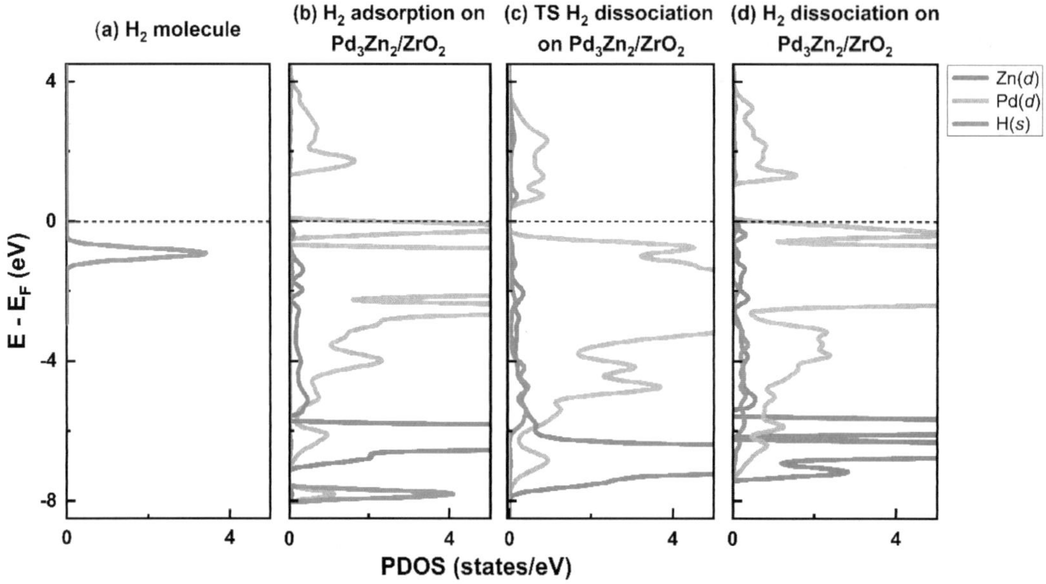

Figure 3. PDOS of: (a) isolated H_2 molecule, (b) H_2 adsorption, (c) transition state, and (d) 2H* on Pd_3Zn_2/ZrO_2. The Fermi level is set to zero. The stability of hydrogen interactions with the catalyst resulting from the H-Pd-Zn interaction is depicted by overlapping PDOS.

4. Conclusions

In this study, we explore the catalytic performance of Pd_4Zn/ZrO_2 and Pd_3Zn_2/ZrO_2 clusters for the H_2 dissociation. Using DFT calculation, we identify the process in thermodynamics and kinetics aspects. We find that the reaction is exothermic in both systems. However, the activation barriers are relatively high with similar results, E_a = 0.62 and 0.67 eV on Pd_4Zn/ZrO_2 and Pd_3Zn_2/ZrO_2, indicating the possibility of needing a better structure and atomic numbers in the cluster for a proper catalyst. Electronic structure analysis reveals the adsorption strength of H_2* on supported clusters comes from the interaction between hydrogen atoms with Pd and Zn. These findings offer valuable insights into a better catalyst system with good performance. In our following study, we will pursue deeper investigations on optimizing cluster composition and performing ab initio molecular dynamics simulations to understand temperature effects and reaction dynamics, which offer a more comprehensive understanding of catalytic behavior under realistic conditions.

Acknowledgments
This research was partly supported by the Osaka Gas Foundation of International Cultural Exchange (OGFICE) with grant no.: 8620/IT1.B07.1/TA.00/2024. M. W. gratefully acknowledges a doctoral scholarship support from the Indonesia Endowment Fund for Education, the Ministry of Finance of Indonesia (LPDP) Indonesia.

References

[1] Lin L, Hou C, Zhang X, Wang Y, Chen Y and He T 2018 *Appl Catal B* **221** 312–9
[2] Bredig G and Carter S R 1914 *Berichte der deutschen chemischen Gesellschaft* **47** 541–5
[3] Kanega R, Ertem M Z, Onishi N, Szalda D J, Fujita E and Himeda Y 2020 *Organometallics* **39** 1519–31
[4] Yang X, Wang A, Qiao B and Li J U N 2013 *Acc Chem Res* **46**

[5] Wang H, Liu J X, Allard L F, Lee S, Liu J, Li H, Wang J, Wang J, Oh S H, Li W, Flytzani-Stephanopoulos M, Shen M, Goldsmith B R and Yang M 2019 *Nat Commun* **10** 1–12
[6] Rong H, Ji S, Zhang J, Wang D and Li Y 2020 *Nature Communications 2020 11:1* **11** 1–14
[7] Sankar M, Dimitratos N, Miedziak P J, Wells P P, Kiely C J and Hutchings G J 2012 *Chem Soc Rev* **41** 8099–139
[8] Kroes G J 1999 *Progress in Surface Science* **60** 1-85
[9] Moc J, Musaev D G and Morokuma K 2000 *J Phys Chem A* **104** 11606–14
[10] Wang Y, Cao Z and Zhang Q 2003 *Chem Phys Lett* **376** 96–102
[11] Zhang Z, Zhang L, Hülsey M J and Yan N 2019 Z *Molecular Catalysis* **475**
[12] Wang X, Zhao J, Sun W, Pei Y, An J, Li Z and Ren J 2019 *Comput Mater Sci* **159** 210–21
[13] Parr R G and Weitao Y 1995 *Int Series of Mono* 169-194
[14] Kresse G and Furthmüller J 1996 *Comput Mater Sci* **6** 15–50
[15] Kresse G and Furthmüller J 1996 *Phys Rev B* **54** 11169–86
[16] Hohenberg P and Kohn W 1964 *Physical Review* **136** B864–71
[17] Kohn W and Sham L J 1965 *Physical Review* **140** A1133–8
[18] Blöchl P E 1994 *Phys Rev B* **50** 17953–79
[19] Kresse G and Joubert D 1999 *Phys Rev B* **59** 1758–75
[20] Perdew J P, Burke K and Ernzerhof M 1996 *Phys Rev Lett* **77** 3865–8
[21] Grimme S 2006 *J Comput Chem* **27** 1787–99
[22] Henkelman G, Arnaldsson A and Jónsson H 2006 *Comput Mater Sci* **36** 354–60
[23] Monkhorst H J and Pack J D 1976 *Phys Rev B* **13** 5188–92
[24] Henkelman G, Uberuaga B P and Jónsson H 2000 *J Chem Phys* **113** 9901–9904
[25] Henkelman G and Jónsson H 2000 *J Chem Phys* **113** 9978–9985
[26] Smidstrup S, Pedersen A, Stokbro K and Jónsson H 2014 *J Chem Phys* **140** 214106
[27] Momma K and Izumi F 2011 *J Appl Crystallogr* **44** 1272–6
[28] Jung K D and Bell A T 2000 *J Catal* **193** 207–23
[29] Pelzer A W, Jellinek J and Jackson K A 2013 *J Phys Chem A* **117** 10407–15
[30] Nozawa K, Endo N, Kameoka S, Pang Tsai A and Ishii Y 2011 *J Physical Soc Japan* **80** 064801
[31] Brix F, Desbuis V, Piccolo L and Gaudry É 2020 *J Phys Chem Lett* **11** 7672–8
[32] Tauster S J, Fung S C and Garten R L 1978 *J Am Chem Soc* **100** 170–5
[33] Liu X, Tian D and Meng C 2015 *J Mol Struct* **1080** 105–10

Journal of Physics: Conference Series **2980** (2025) 012036 doi:10.1088/1742-6596/2980/1/012036

Solvent effect on the adsorption of lithium polysulfides on single-atom Ni on N-doped graphene: a first-principles study

Jessie Manopo, Muhammad Khaishar Mahardhika, Charlie Ofiyen, and Yudi Darma*

Quantum Semiconductor and Devices Laboratory, Department of Physics, Institut Teknologi Bandung, Bandung, Indonesia
*E-mail: yudi@itb.ac.id

Abstract. We study the effect of solvent (dimethyl ether [DME] and 1,3-dioxolane [DOL]) on the adsorption of lithium polysulfides (Li_2S_4, Li2S_6, and Li_2S_8) on single-atom Ni anchored on N-doped graphene by using density functional theory (DFT). All DFT calculations were carried out using Quantum ESPRESSO software. We found that lithium polysulfides (LiPSs) have relatively higher binding energy when they interact with DME solvent due to weaker overlapping between Li 2p and O 2p states. The binding energy difference of LiPSs ranges from 22-54 meV. In both solvents, the oxygen atom in the solvent interacts with the lithium atom in the LiPSs, thus weakening the bond between LiPSs and the catalysts. This research can better understand the sulfur reduction reaction for lithium-sulfur batteries applications.

1. Introduction

The need for long-term and highly efficient batteries is increasing due to the massive development of electric vehicles. One of the main issues in electric vehicles is the driving range, which is limited by the energy density of the batteries [5]. As the space in electric vehicles is limited, the size of the battery cannot be further increased to achieve a longer driving range. One of the best solutions is to develop a battery with higher energy density. Lithium-sulfur (Li-S) batteries are one of the most promising candidates for replacing lithium-ion (Li-ion) batteries. Li-S possesses a theoretical energy density of ~2600 Wh/kg, which is much larger compared to that of Li-ion (387 Wh/kg) [2, 4]. However, there are several issues to be tackled in Li-S batteries, one of them being the lithium polysulfides (LiPSs) shuttle effect [3, 15].

To solve this problem, we can develop highly efficient catalysts to promote LiPS conversion. Developing a catalyst not only enhances the reaction kinetics of LiPSs but also suppresses the shuttle effect [16, 17]. To better understand the catalytic reaction, we must consider the interaction between the LiPSs and the solvent [1]. To avoid LiPSs dissolution, the LiPSs should bind the catalysts stronger than to bind the solvent molecule. Therefore, studying the interaction between LiPSs and the solvent molecule is important.

Single-atom catalysts have gained tremendous interest for energy conversion applications, such as fuel cells [12], metal-air batteries [13], and lithium-sulfur batteries[7, 20]. Among single-atom catalysts, metal-nitrogen-carbon (M-N-C) catalysts are widely used due to their adjustable electronic properties and have promising electrocatalytic activity [18]. Previous studies have revealed the high activity of Ni-based single-atom catalysts for lithium-sulfur batteries [6].

However, this study has neglected the solvent effect which plays an important role in the catalytic processes [8]. Therefore, in this study, we study the solvent effect Ni as the metal center due to.

In this paper, by using density functional theory, we investigate the interaction between Li_2S_x (x=4, 6, 8) and the solvent because they are soluble in commonly used electrolytes such as

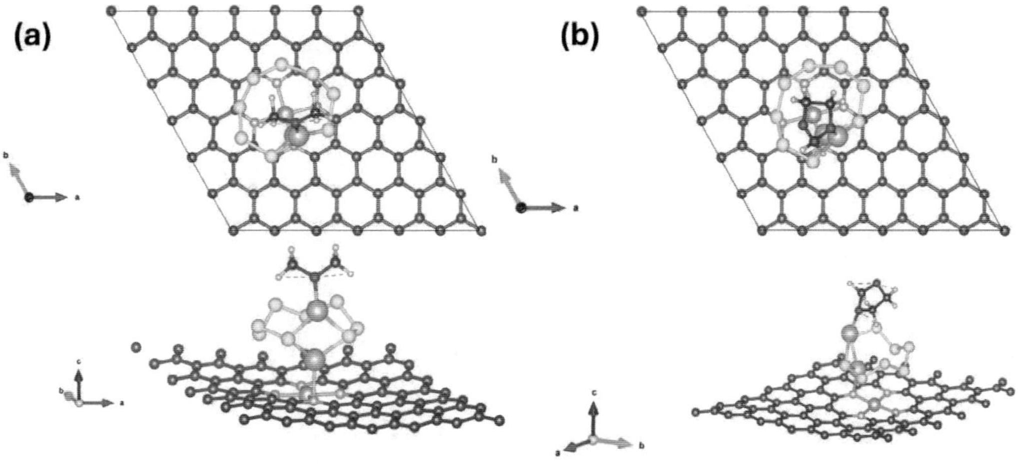

Figure 1. Top view and side view of adsorbed Li_2S_8 on Ni-N-C materials in (a) DME solvent and (b) DOL solvent

dimethyl ether (DME) and 1,3-dioxolane (DOL). According to previous research, the DME and DOL solvent suits these applications, LiPSs are found to be soluble in solvents and weakly bind to the substrate in other solvents such as ethylene carbonate and polyethylene carbonate [11].

2. Computational Methods

To model the catalyst, we used a $Ni-N_4$ active site placed on a 6×6×1 graphene supercell. To simplify the solvent model, we only focused on the interaction between one DME/DOL molecule and the adsorbed LiPSs. In this case, we neglect the interaction between solvent molecules; to get better insight, we should consider the solvent in actual density to accurately mimic the solvent-solvent interaction. We only focused on Li_2S_4, $Li2S_6$, and Li_2S_8 owing to their high binding energy to DME and DOL solvents. The models used in this proceeding are depicted in **Figure 1.** All DFT calculations were carried out using Quantum ESPRESSO software.[9, 10] We used Pymatgen[14] to model the initial adsorption structure and relax the structure using the Broyden–Fletcher–Goldfarb–Shanno (BFGS) algorithm implemented in Quantum ESPRESSO. Only the gamma point is used for all calculations. To account for van der Waals interactions, we used vdW-DF2 functional which was previously used to describe the interaction between graphene and LiPSs [19]. The

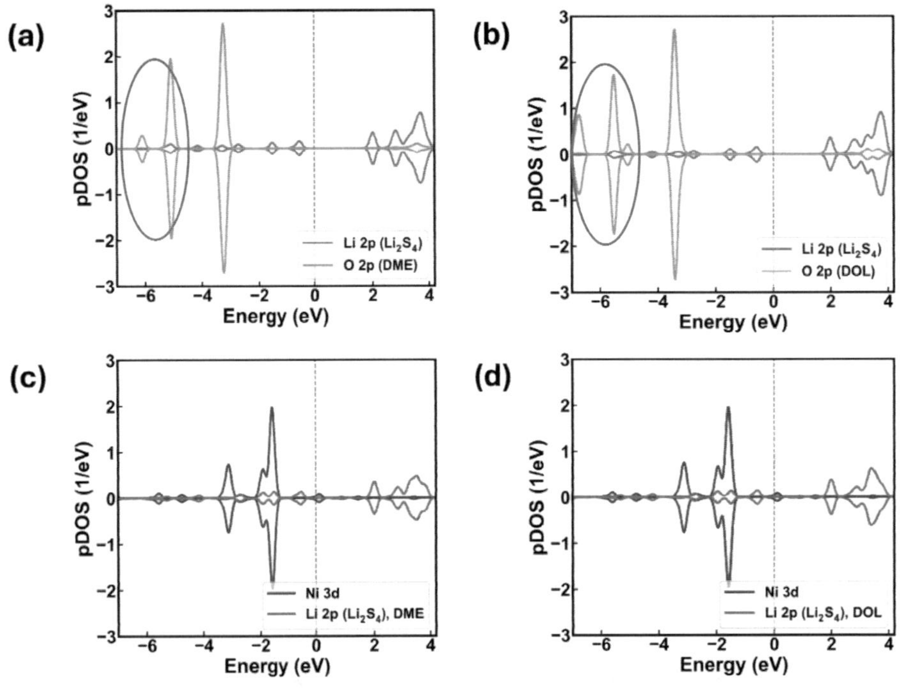

Figure 2. Density of states of Li 2p states of Li_2S_4 with the O 2p in (a) DME molecule and (b) DOL molecule, density of states of Li 2p states of Li_2S_4 which interact with the Ni 3d in (c) DME solvent and (d) DOL

Table 1. Binding energies of LiPSs in DME and DOL solvent

Solvent	LiPSs	Binding energy (eV)
DME	Li_2S_4	-1.470
	Li_2S_6	-1.742
	Li_2S_8	-2.020
DOL	Li_2S_4	-1.416
	Li_2S_6	-1.764
	Li_2S_8	-1.999

research shows that the interaction between graphene and lithium polysulfides is dominated by van der Waals interaction. We used ultrasoft pseudopotentials. The cutoff energy used was 60 Ry.

3. Results and Discussions

To investigate the interaction between the solvent and the LiPSs, we calculated the binding energy using the following equation:

$$\Delta E = E_{system} - (E_{adsorbate} - E_{solvent+substrate})$$

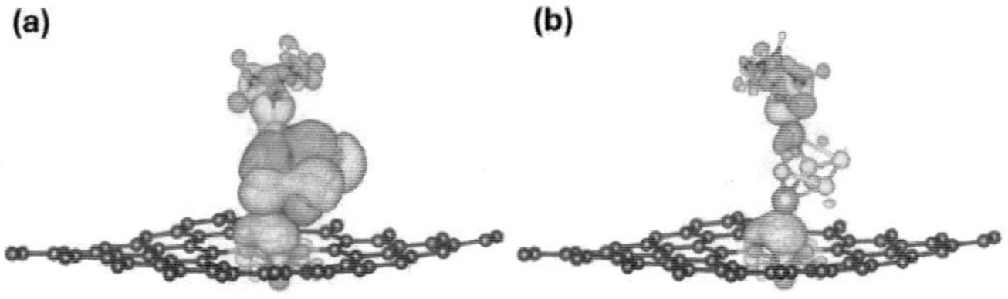

Figure 3. Charge density difference of adsorbed Li_2S_4 on Ni-N-C catalyst with (a) DME solvent and (b) DOL solvent

where E_{system} is the total energy of the Ni-N-C substrate with adsorbed LiPSs and the solvent, $E_{adsorbate}$ is the total energy of isolated LiPSs, $E_{solvent+substrate}$ is the total energy of the solvent and the LiPSs.

The binding energy of LiPSs to the system is shown in Table 1. Generally, the binding energy of the LiPSs is larger in the DME solvent. To investigate the reasoning behind this fact, we plot the density of states (DOS) of Li 2p states of Li_2S_4 with the O 2p in the DME molecule and DOL molecule, which is shown in Figure 2 (a)-(b). From Figure 1, we can see the LiPSs interact with the solvent through the Li-O bond, therefore we investigated the interaction between the Li 2p states in LiPSs and the O 2p states of the solvent. It is shown that the overlapping states are larger in DOL compared to those in DME. Thus, LiPSs will have stronger interactions with DOL compared to those with DME. Therefore, LiPSs will have stronger binding energy to Ni-N-C in DME solvent.

To further confirm the argument, we also plot the charge density differences in Figure 3. Charge density differences show that the charge depletion area (depicted in cyan) between the LiPSs and the solvents is larger in DME compared to that in DOL, which further supports our argument that LiPSs will have stronger binding energy to Ni-N-C in DME solvent.

From Figure 1, it is shown that Ni-N-C interacts with the LiPSs through the N-Li bond. Therefore, we plot the DOS of Li 2p states of Li_2S_4 which interact with the Ni 3d orbital in Figure 2(c)-(d). From the DOS plot, we cannot see a clear distinction between these two environments (DME and DOL), which supports the fact that the binding energy in DME and DOL does not exhibit significant differences.

4. Conclusions

We have examined the influence of solvents (specifically dimethyl ether [DME] and 1,3-dioxolane [DOL]) on the adsorption characteristics of lithium polysulfides (Li_2S_4, Li_2S_6, and Li_2S_8) on single-atom nickel supported on nitrogen-doped graphene, employing density functional theory (DFT) as the methodological framework. Our findings indicate that the inclusion of solvents diminishes the interaction strength between the lithium polysulfides (LiPSs) and the catalytic surfaces. In both solvent environments, the oxygen atom within the solvent engages with the lithium atom of the LiPSs. We found that LiPSs have relatively higher binding energy when they interact with DME solvent due to weaker overlapping between Li 2p and O 2p states. This study could be further improved by accounting for solvent in actual density at room temperature. This research endeavor contributes to an enhanced comprehension of the sulfur reduction reaction pertinent to applications in lithium-sulfur battery technology.

Journal of Physics: Conference Series **2980** (2025) 012036 doi:10.1088/1742-6596/2980/1/012036

Acknowledgments

This work is partly supported by the Ministry of Education, Culture, Research and Technology of the Republic of Indonesia through the 2024 research program and ITB research program year 2024. J.M. acknowledges PMDSU scholarship batch VI from the Ministry of Education, Culture, Research and Technology of the Republic of Indonesia. M.K.M. and C.O. thank the Indonesian Endowment Fund for Education (LPDP) for their scholarship. All calculations in this research were computed using the High-Performance Computing (HPC) facility provided by the National Research and Innovation Agency, Indonesia.

References

[1] Boteju T, Abraham A M, Ponnurangam S and Thangadurai V 2023 Theoretical Study on the Role of Solvents in Lithium Polysulfide Anchoring on Vanadium Disulfide Facets for Lithium–Sulfur Batteries *The Journal of Physical Chemistry C* **127** 4416-24

[2] Bruce P G, Freunberger S A, Hardwick L J and Tarascon J-M 2012 Li–O2 and Li–S batteries with high energy storage *Nature Materials* **11** 19-29

[3] Cao R, Xu W, Lv D, Xiao J and Zhang J-G 2015 Anodes for Rechargeable Lithium-Sulfur Batteries *Advanced Energy Materials* **5** 1402273

[4] Chen R, Zhao T and Wu F 2015 From a historic review to horizons beyond: lithium–sulphur batteries run on the wheels *Chemical Communications* **51** 18-33

[5] Deng J, Bae C, Denlinger A and Miller T 2020 Electric Vehicles Batteries: Requirements and Challenges *Joule* **4** 511-5

[6] Dong W, Gu H, Zhu X, Zhao M, Chang Q, Yang F, Shen D, Tang S, Hong X, Dong Z and Yang S 2024 First-principles study of sulfur-host materials utilizing single-atom catalysts embedded within g-C3N4 *Surfaces and Interfaces* **48** 104315

[7] Du Z, Chen X, Hu W, Chuang C, Xie S, Hu A, Yan W, Kong X, Wu X, Ji H and Wan L-J 2019 Cobalt in Nitrogen-Doped Graphene as Single-Atom Catalyst for High-Sulfur Content Lithium–Sulfur Batteries *Journal of the American Chemical Society* **141** 3977-85

[8] Eckhoff M and Behler J 2021 Insights into lithium manganese oxide–water interfaces using machine learning potentials *The Journal of Chemical Physics* **155** 244703

[9] Giannozzi P, Andreussi O, Brumme T, Bunau O, Buongiorno Nardelli M, Calandra M, Car R, Cavazzoni C, Ceresoli D, Cococcioni M, Colonna N, Carnimeo I, Dal Corso A, de Gironcoli S, Delugas P, DiStasio R A, Ferretti A, Floris A, Fratesi G, Fugallo G, Gebauer R, Gerstmann U, Giustino F, Gorni T, Jia J, Kawamura M, Ko H Y, Kokalj A, Küçükbenli E, Lazzeri M, Marsili M, Marzari N, Mauri F, Nguyen N L, Nguyen H V, Otero-de-la-Roza A, Paulatto L, Poncé S, Rocca D, Sabatini R, Santra B, Schlipf M, Seitsonen A P, Smogunov A, Timrov I, Thonhauser T, Umari P, Vast N, Wu X and Baroni S 2017 Advanced capabilities for materials modelling with Quantum ESPRESSO *Journal of Physics: Condensed Matter* **29** 465901

[10] Giannozzi P, Baroni S, Bonini N, Calandra M, Car R, Cavazzoni C, Ceresoli D, Chiarotti G L, Cococcioni M, Dabo I, Dal Corso A, De Gironcoli S, Fabris S, Fratesi G, Gebauer R, Gerstmann U, Gougoussis C, Kokalj A, Lazzeri M, Martin-Samos L, Marzari N, Mauri F, Mazzarello R, Paolini S, Pasquarello A, Paulatto L, Sbraccia C, Scandolo S, Sclauzero G, Seitsonen A P, Smogunov A, Umari P and Wentzcovitch R M 2009 QUANTUM ESPRESSO: a modular and open-source software project for quantum simulations of materials *Journal of Physics: Condensed Matter* **21** 395502-

[11] He C, Qi Z F and Zhang W X 2023 Design of transition metal carbonitrides (MCNs) as promising anchoring and high catalytic performance materials for lithium-sulfur batteries *Journal of Alloys and Compounds* **934** 167786

[12] Manopo J and Darma Y 2024 Regulating Spin-Valence States of Vanadium-Based Single-Atom Oxygen Reduction Catalysts through Substrate and Axial Ligand Engineering *Energy & Fuels* **38** 6260-8

[13] Manopo J, Kusumo P N, Sumboja A and Darma Y 2023 Coordination engineering of atomically dispersed zirconium on graphene for oxygen reduction reaction *Physical Chemistry Chemical Physics* **25** 25299-308

[14] Ong S P, Richards W D, Jain A, Hautier G, Kocher M, Cholia S, Gunter D, Chevrier V L, Persson K A and Ceder G 2013 Python Materials Genomics (pymatgen): A robust, open-source python library for materials analysis *Computational Materials Science* **68** 314-9

[15] Seh Z W, Sun Y, Zhang Q and Cui Y 2016 Designing high-energy lithium–sulfur batteries *Chemical Society Reviews* **45** 5605-34

[16] Shen Z, Jin X, Tian J, Li M, Yuan Y, Zhang S, Fang S, Fan X, Xu W, Lu H, Lu J and Zhang H 2022 Cation-doped ZnS catalysts for polysulfide conversion in lithium–sulfur batteries *Nature Catalysis* **5** 555-63

[17] Wang L, Wang Z-Y, Wu J-F, Li G-R, Liu S and Gao X-P 2020 To effectively drive the conversion of sulfur with electroactive niobium tungsten oxide microspheres for lithium–sulfur battery *Nano Energy* **77** 105173
[18] Wang Y, Cui X, Peng L, Li L, Qiao J, Huang H and Shi J 2021 Metal–Nitrogen–Carbon Catalysts of Specifically Coordinated Configurations toward Typical Electrochemical Redox Reactions *Advanced Materials* **33** 2100997
[19] Zhang Q, Wang Y, Seh Z W, Fu Z, Zhang R and Cui Y 2015 Understanding the Anchoring Effect of Two-Dimensional Layered Materials for Lithium–Sulfur Batteries *Nano Letters* **15** 3780-6
[20] Zhou G, Zhao S, Wang T, Yang S-Z, Johannessen B, Chen H, Liu C, Ye Y, Wu Y, Peng Y, Liu C, Jiang S P, Zhang Q and Cui Y 2020 Theoretical Calculation Guided Design of Single-Atom Catalysts toward Fast Kinetic and Long-Life Li–S Batteries *Nano Letters* **20** 1252-61

Journal of Physics: Conference Series **2980** (2025) 012037 doi:10.1088/1742-6596/2980/1/012037

Unlocking High-Entropy Alloys FeNiCrCoCu Design: Stacking Fault Energy Modeling with First-Principles DFT, Molecular Dynamics, Thermodynamics, and Machine Learning

Tria Laksana Achmad[1*], Ghanyyah Fahira Ritonga[1], Eri Immanuel Pane[1] and Leonardus Hans Sebastian[1]

[1] Department of Metallurgical Engineering, Institut Teknologi Bandung, Bandung, Indonesia

[*]E-mail: tria_laksana@itb.ac.id

Abstract. High-entropy alloys (HEAs) represent a groundbreaking development in alloy engineering. Among these, FeNiCrCoCu HEAs offer exceptional strength and toughness, making them promising candidates for nuclear applications due to their potential resistance to radiation damage. Understanding the Stacking Fault Energy (SFE) is vital for optimizing the mechanical properties of HEAs. SFE serves as a crucial indicator for alloy design, as it directly influences the deformation mechanism. A lower SFE not only strengthens HEAs but also enhances their ductility. Modeling SFE directly from first-principles density functional theory (DFT) and molecular dynamics (MD) can be computationally expensive. Then, we developed a machine learning (ML) model from the thermodynamics database that speeds up the SFE modeling. The ML model closely matches experimental data and offers a rapid screening method for HEAs. By integrating DFT and MD calculations with ML models, we can accurately predict SFEs in complex alloy systems, facilitating accelerated design. This groundbreaking research is focused on unlocking the full potential of FeNiCoCrCu HEAs by effectively modeling their SFE using a combination of theoretical calculations, simulations, and machine learning. Our research found that adjusting Cr, Co, and Cu will decrease its SFE while increasing Fe and Ni will increase its SFE. Specifically, the SFE ranges from 65.6 mJ/m^2 for the equiatomic alloy to 54.77 mJ/m^2 with the addition of 8.33 at.% Cu. MD simulations show that reducing SFE improves mechanical properties by forming stacking faults and twinning. DFT analysis reveals that the decline in SFE results from charge accumulation and reduced density of states, weakening interatomic bonding. Our machine learning model, particularly the Random Forest (RF) model, demonstrated high predictive accuracy with an accuracy rate of 98.8% and a root mean square error (RMSE) of 1.31. These metrics highlight the model's reliability in predicting SFE.

1. Introduction

Steam power plants have long been a reliable energy source, but their environmental impact, including greenhouse gas emissions and air pollution, cannot be ignored. Considering growing environmental concerns, nuclear energy has emerged as a promising alternative to meet our nation's electricity needs. Nuclear reactors, the heart of nuclear power plants (NPPs), harness the power of nuclear fission reactions to produce heat, which is then converted into steam to drive turbines and generate electricity. Operating in extreme conditions with high temperatures, pressures, and radiation exposure, nuclear reactors demand materials with exceptional resistance to irradiation damage and can withstand temperatures exceeding 1000 °C. These materials must also endure corrosive environments such as molten metal, molten salt, and high-pressure gases [1]. Additionally, they must resist radiation, creep, creep-fatigue, and corrosion while maintaining strength and ductility at operating temperatures. The Pressurized Water Reactor (PWR) is the most prevalent type of nuclear reactor used for electricity generation. Utilizing various conventional materials such as ferritic/martensitic steels, austenitic stainless steels, and zirconium alloys, the PWR's performance is directly impacted by material degradation and changes in mechanical properties caused by irradiation, including radiation hardening and creep. These challenges stem from the reactor's operation in extreme conditions, including intense radiation, high temperatures, and chemical solid corrosion [2].

To tackle these formidable challenges, extensive research has been dedicated to crafting new alloys capable of maintaining robust mechanical properties under extreme temperatures and pressures during operation. In recent decades, a groundbreaking alloy, High Entropy Alloys (HEA), has emerged, showcasing exceptional mechanical properties in the most demanding environmental conditions [3]. HEA is a unique alloy consisting of five or more main elements, with each element's concentration falling within the 5 - 35 at% range (typically near equiatomic) [4], and formed with a mixing entropy equal to or greater than 1.5 R, where R represents the gas constant [5]. HEA has displayed a wide array of promising properties for both structural and functional materials. This alloy exhibits markedly higher yield strength than nickel-based superalloys at elevated temperatures, toughness equivalent to the finest cryogenic steels (particularly certain austenitic stainless steels and steels with high Ni content), resistance to oxidation at high temperatures, and superplastic properties [6]. However, its most compelling attribute lies in its resistance to microstructural changes caused by radiation, rendering it a truly unique and promising material for nuclear applications. Some of the advantages of HEA regarding resistance to irradiation damage include heightened resistance to the formation of irradiation defects, reduced void formation in the crystal structure compared to conventional alloys, and more excellent microstructural stability compared to traditional alloys [7]. Therefore, the exceptional properties of HEA position it as an incredibly compelling candidate for deployment in nuclear reactor applications, presenting potential solutions to address national energy needs.

Using Ni-containing HEAs has shown high-temperature strength and excellent microstructural stability. Moreover, Ni-containing HEAs can reduce structural damage caused by radiation by carefully selecting the right combination and proportion of elements in the alloy. Various FeNiCrCo alloys offer significant advantages and better radiation resistance than other Ni alloys. The careful selection of these elements capitalizes on their remarkable properties, significantly enhancing the performance of the alloy for nuclear applications. Fe stands out as a crucial element in many structural materials due to its exceptional mechanical properties and wide availability. It provides a robust foundation for the alloy, ensuring strength and stability essential for demanding environments. Ni is renowned for its outstanding corrosion resistance

Journal of Physics: Conference Series **2980** (2025) 012037 doi:10.1088/1742-6596/2980/1/012037

and ability to maintain mechanical integrity at elevated temperatures. Its presence fortifies the alloy's resilience against radiation damage while improving thermal stability, making it indispensable in nuclear applications. Cr plays a vital role in significantly enhancing the alloy's corrosion resistance, a critical factor in materials intended for nuclear reactors. It forms a protective, stable oxide layer that guards against further degradation, ensuring long-lasting performance. Co elevates the high-temperature strength and stability of the alloy, enhancing wear resistance and overall toughness. This means the alloy can withstand extreme conditions, contributing to its durability. Cu is strategically added to bolster the alloy's resistance to radiation-induced void swelling and to improve thermal conductivity. This ensures the structural integrity of the alloy remains intact even under intense radiation exposure. The synergistic effects of these carefully selected elements result in an alloy that boasts superior mechanical properties, exceptional radiation resistance, and remarkable thermal stability. This makes it an outstanding choice for critical nuclear applications, ensuring safety and reliability in challenging environments. This study focused on developing a radiation-resistant FeNiCrCo HEA by introducing Cu elements to the FCC-structured HEA matrix. Adding Cu elements can enhance radiation resistance and prevent the formation of voids in the material and radiation-induced hardening. This suggests that FeNiCrCo HEAs, with the addition of Cu, have great potential for use in nuclear reactors as they are less susceptible to microstructural changes due to radiation [8].

Several researchers have studied the mechanical properties of FeNiCrCoCu HEA with variations in Cu element concentration and temperature [9,10]. These studies have provided valuable insights. For example, the work of Mukanov et al. [9] demonstrated that Cu doping in FeCrCoCuNi HEA can enhance wear resistance, but at low concentrations (5-10%), it may reduce corrosion resistance. However, the optimal copper concentration to achieve good mechanical properties and changes in microstructure has not been thoroughly studied. Understanding the Stacking Fault Energy (SFE) is vital for optimizing the mechanical properties of HEAs. SFE serves as a crucial indicator for alloy design, as it directly influences the deformation mechanism. A lower SFE not only strengthens HEAs but also enhances their ductility. Conducting this experimental study would be time-consuming and costly. Creating a mathematical model of the composition ratio and mechanical properties would also be challenging. Nevertheless, computational methods like molecular dynamics simulation offer a promising alternative. This method can provide detailed information on the microscopic mechanisms of atoms in HEA and predict material behavior, mechanical properties of alloys, microstructural changes, and alloy dislocations. Through this simulation, various variations in alloy composition can be tested, allowing for a comprehensive understanding of the effect of element variations on mechanical properties, phase stability, microstructure evolution, and dislocations without requiring direct experiments. Modeling SFE directly from first-principles density functional theory (DFT) and molecular dynamics can be computationally expensive. Then, we developed a machine learning (ML) model from the thermodynamics database that speeds up the SFE modeling. The ML model closely matches experimental data and offers a rapid screening method for HEAs. By integrating DFT and MD calculations with ML models, we can accurately predict SFEs in complex alloy systems, facilitating accelerated design. This groundbreaking research is focused on unlocking the full potential of FeNiCoCrCu HEAs by effectively modeling their SFE using a combination of theoretical calculations, simulations, and machine learning.

Journal of Physics: Conference Series **2980** (2025) 012037 doi:10.1088/1742-6596/2980/1/012037

2. Computational Methodology

We conducted a thorough series of calculations to delve into the effects of adding alloying elements and adjusting temperatures in the FeNiCoCrCu HEA on a range of crucial properties. These properties encompass lattice parameters, SFE, tensile strength, electronic structure, microstructure, and dislocation evolution. Our pioneering approach involved harnessing advanced molecular dynamics, first-principles, CALPHAD methods, and cutting-edge machine learning techniques, as vividly illustrated in Figure 1. Notably, the results of the SFE calculation from computational thermodynamics serve as pivotal inputs for machine learning models, empowering them to predict and categorize the SFE accurately.

2.1 Molecular Dynamics Methodology

The process of calculating the lattice parameters for the FeNiCrCoCu alloy involves creating an alloy structure using the Atomsk package. This package builds alloy structures by randomly distributing atoms [11]. In this simulation, the Face-Centered Cubic (FCC) crystal structure serves as the basis for the alloy atomic model. The concentration of each element in the alloy adheres to the provisions of a High Entropy Alloy (HEA), which specifies a range of 5-35 atomic percent for each element, and the composition is close to equiatomic. The atoms in FeNiCrCoCu are randomly positioned along the x, y, and z directions, respectively denoted as [100], [010], and [001]. The cell size is $20a \times 20a \times 20a$, containing a total of 32,000 atoms, as illustrated in Figure 2. Molecular dynamics simulations were carried out using the Large-scale Atomic/Molecular Massively Parallel Simulator (LAMMPS) software. LAMMPS optimizes the structure created by Atomsk to obtain the lattice parameter value with the lowest energy or in equilibrium, aiming to find the most stable configuration. This optimization involves adjusting the atom's position and the simulation cell's size (lattice constant) until the lowest energy state is reached. All atomic directions are relaxed to a temperature of 300 K at a pressure of 0 Pa. The atoms in the simulation move and adjust their positions until they reach the most stable atomic structure. Structure optimization simulations were conducted for MEA FeNiCrCo, Fe_{30}NiCrCoCu, and FeNiCrCoCu (with variations of Cu ranging from 0-30 atomic percent). The interatomic potential used is the Embedded Atom Method (EAM) potential with the FeNiCrCoCu-with-ZBL.eam.alloy file, developed by O. R. Deluigi et al. in 2021, which defines the interactions between atoms in the alloy [12].

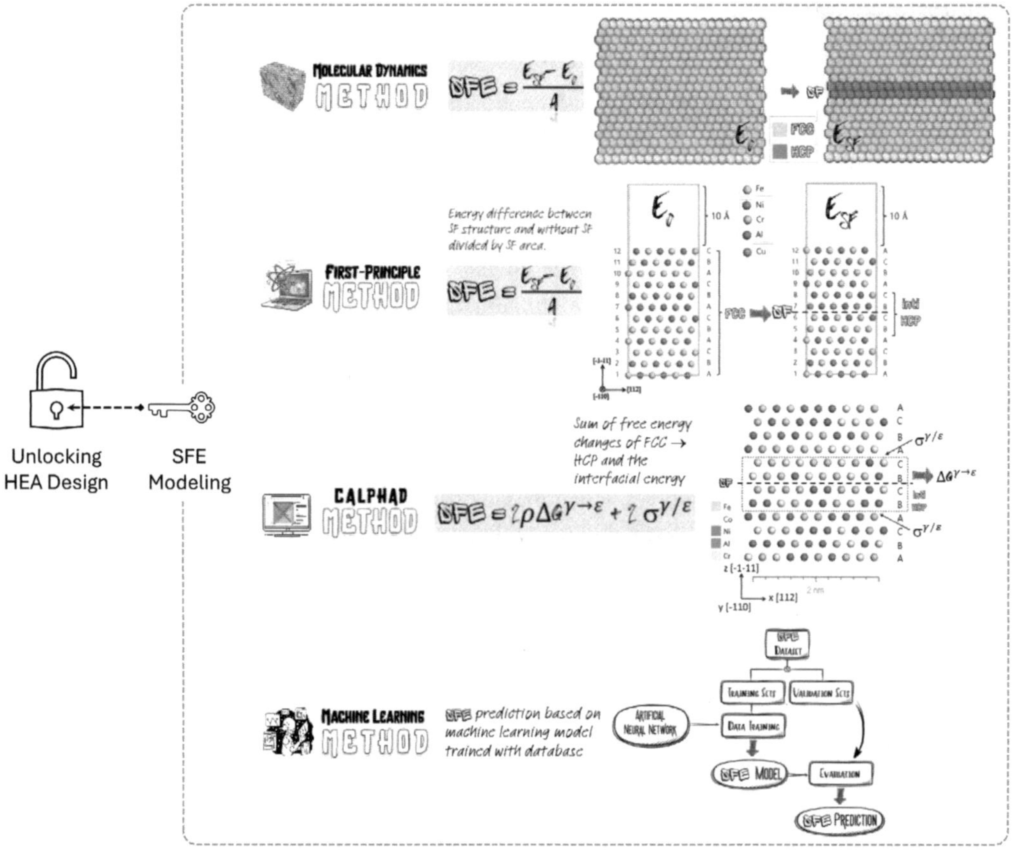

Figure 1. Illustration of SFE calculation methods in this study using molecular dynamics, first-principles, computational thermodynamics (CALPHAD), and machine learning.

The lattice parameters calculated using LAMMPS will be the basis for the uniaxial tensile test simulation. The simulation involves a cubic supercell with $20a \times 20a \times 12a$ dimensions, containing 28,800 FeNiCrCoCu atoms in an FCC structure. Under NPT ensemble conditions, we maintain constant pressure and temperature during deformation. The structure will be deformed in the x [112] direction with a strain rate of 0.01 1/ps. In our study, we deliberately induced a stacking fault in the alloy structure's [-1-11] plane and meticulously calculated the SFE based on the energy derived from both the initial and faulted structures. Furthermore, our approach involved utilizing the powerful OVITO visualization software [13] to analyze atomic trajectories using Common Neighbor Analysis (CNA)[14] and Dislocation Analysis (DXA)[15]. This comprehensive methodology has enabled us to gain invaluable insights that have the potential to revolutionize our understanding of material behavior.

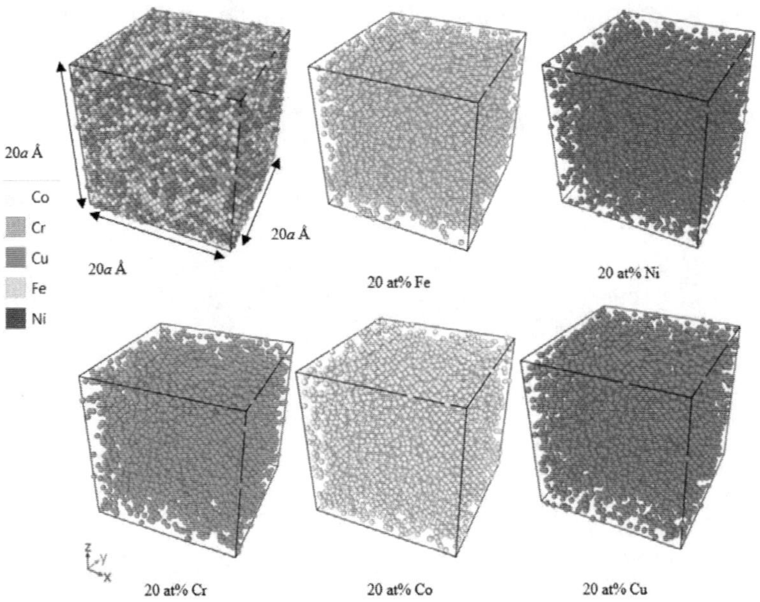

Figure 2. The element content distribution of equiatomic FeNiCrCoCu for latice parameter calculations.

2.2 First-principles DFT Methodology

The advanced first-principles method allows for a comprehensive calculation of the SFE by thoroughly calculating the total energy of a supercell structure with and without stacking faults using the equation in Figure 1. This approach employs FCC supercells in various dimensions and directions to ensure comprehensive SFE calculations. Additionally, SFE calculations were conducted using a 24-atom supercell, as illustrated in Figure 3. A deliberate 10 Å vacuum was introduced to prevent inter-periodic interactions. Furthermore, by intentionally shifting the supercell, we created the stacking fault structure, leading to the assertive formation of HCP nuclei within the FCC matrix. These precise calculations were carried out using the advanced CASTEP (Cambridge Sequential Total Energy Package) [16] software, which is based on Density Functional Theory (DFT). The electronic structure calculations utilized the highly accurate Perdew-Burke Ernzerhof (PBE) exchange-correlation function from the Generalized Gradient Approximation (GGA) [17]. Additionally, ultrasoft pseudopotentials from Vanderbilt were used to accurately describe the complex electron-nuclei interaction [18]. The criteria included a maximum energy difference of 4 x 10-6 eV/atom, an ionic Hellmann-Feynman force of 0.01 eV/Å, and an ionic displacement of 0.01 Å. Based on convergence tests, the Monkhorst-Pack [19] k-points grid sampling scheme is 6×6×1, and a plane wave energy cut-off value is 400 eV. Obtaining the lattice parameter is crucial for our research, and we use two methods to achieve this. We relax the atoms in the supercell through ionic minimization, while electronic minimization relaxes only the electrons. We use the Birch-Murnaghan EOS equation to generate the total energy versus volume curve by comparing the total energy obtained from the simulation with volume. Analyzing this curve allows us to determine the equilibrium lattice parameters, equilibrium volume, and bulk modulus [20]. Furthermore, the substitution of Co atoms by Cu occurs in the 4th and 7th layers, as illustrated in Figure 3(b). To gain valuable insights into the stability of complex HEA, SFE calculations were also performed on 5 random structures using Atomsk software. Variations in adding Cu were introduced by substituting the Co element in the supercell, which was limited to a maximum of 2 atoms.

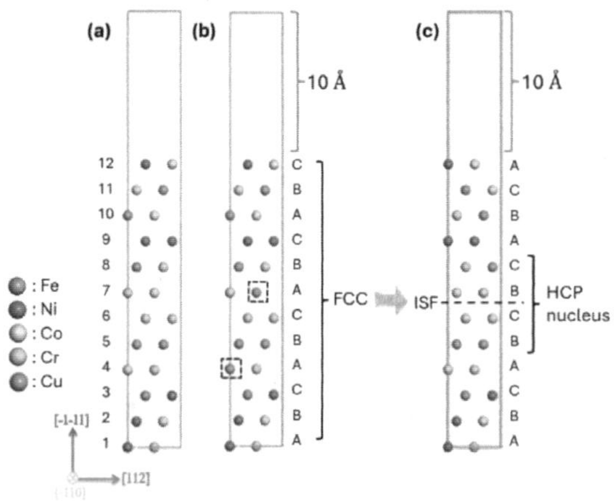

Figure 3. The geometry of the supercell used in SFE calculation, (a) fcc structure of FeNiCoCr, (b) position of Cu addition, (d) stacking fault structure of FeNiCoCr.

2.3 Computational Thermodynamic Methodology

Thermodynamic calculations are employed to determine the SFE using Olson and Cohen's equation [21], as illustrated in Figure 1. This equation is input into Matlab software. It is based on the phase transformation of a thin plate from FCC to HCP structures. The formulation considers SFE by taking into account the lattice parameters and the contribution of Gibbs free energy and assumes a constant interfacial energy per unit area of 8 mJ/m^2 for high-entropy alloys. The lattice parameter is assumed to be a constant value of 3.57 Å, determined using the embedded atom method (EAM) potential, with the assumption that lattice parameters do not significantly influence temperature and composition [22]. Our thermodynamic data are sourced from the Scientific Group Thermodata Europe (SGTE) as well as other publications related to phase diagram calculations, ensuring their thoroughness and reliability. Additionally, our previous studies, referenced in [24,25], provide the thermodynamic data utilized in these calculations.

2.4 Machine Learning Methodology

The results of our thermodynamic calculations form a robust database for machine learning modeling, totaling 4,860 data points. Our modeling utilizes key inputs, including temperature and atomic percentages of alloy compositions (Fe, Ni, Cr, Co, Cu), with the output focused on predicting the value and classification of SFE. We execute this modeling using Jupyter Notebook, leveraging powerful open-source libraries like Scikit-learn and TensorFlow. By systematically entering our thermodynamic results into a well-structured database, we split the data into training and test/validation sets at an 80:20 ratio. Training data is critical, allowing our model to uncover the underlying patterns and correlations between features and targets, enabling accurate predictions based on specific input values. The test/validation data ensures that the model's predictions are rigorously assessed against unseen data, bolstering our confidence in accuracy. We employ two highly effective models: random forest (RF) and decision tree regressor. These models are chosen for their proven ability to deliver excellent SFE predictions. Our objective is to identify the model that minimizes error and maximizes accuracy by evaluating predicted values against actual

Journal of Physics: Conference Series **2980** (2025) 012037 doi:10.1088/1742-6596/2980/1/012037

results. By carefully comparing the accuracy and root mean square error (RMSE), we identify the most effective model, equipping us to discover promising compositions for FeNiCrCoCu HEAs [25]. Our approach combines classification models for discrete labels with regression models for continuous SFE estimation, ensuring a comprehensive and accurate assessment that drives innovation in alloy design.

3. Results and Discussions

3.1 Molecular Dynamics Result

The results of the structure optimization reveal that the lattice parameter of FeNiCrCoCu at equiatomic composition is 3.552 Å. This value closely mirrors the findings of Zhang et al. [26], who also used the EAM potential and the FeNiCrCoCu-with-ZBL.eam.alloy file, reporting a lattice parameter of 3.56 Å. Furthermore, Hu et al. [27] studied the lattice parameters of FCC FeNiCrCoCu alloys at equiatomic compositions using XRD, which also yielded 3.56 Å. This strong agreement validates the accuracy of molecular dynamics simulations in predicting the lattice parameters of these alloys. Moreover, this study compiles multiple references for the lattice parameter values of FeNiCrCoCu under equiatomic conditions, calculated through both molecular dynamics simulations and experimental results, as detailed in Table 1. The impact of varying Cu concentrations from 0% to 30% on the lattice parameter of FeNiCrCoCu alloy is depicted in Figure 4(a). The figure shows a noticeable trend: the lattice parameter decreases with increasing Cu concentration, reaching a reduction of approximately 0.04 Å at 30% Cu concentration. This observation is consistent with Hu et al.'s research [27], which explored how Cu affects the microstructure and mechanical properties of the CrFeCoNiCux alloy through experimental approaches.

Table 1. Reference for lattice parameters of equiatomic HEA FeNiCrCoCu obtained from other studies.

Lattice Parameter (Å)	Methods	Reference
3.56	Molecular Dynamics (MD)	Xie dkk (2023)[28]
3.57	Molecular Dynamics (MD)	Su dkk (2023)[29]
3.56	Molecular Dynamics (MD)	Cao dkk (2022)[30]
3.57	Molecular Dynamics (MD)	Li dkk (2021)[31]
3.6	XRD Analysis	Mukanov (2023)[9]
3.58	XRD Analysis	An dkk (2015)[32]
3.55	Molecular Dynamics (MD)	This work

Differences in the atomic sizes of each constituent element can cause the behavior of decreasing lattice parameters in FeNiCrCoCu alloys. While the atomic radii of Fe, Ni, Cr, Co, and Cu are quite similar, the larger size of Cu atoms contributes to a notable reduction in the alloy's lattice parameters. This phenomenon cannot be solely attributed to Cu's size; it's essential to explore

Journal of Physics: Conference Series **2980** (2025) 012037
doi:10.1088/1742-6596/2980/1/012037

other factors that influence Cu's lattice behavior to gain a comprehensive understanding of the FeNiCrCoCu alloy system. The trend of decreasing lattice parameters with increased Cu content is further supported by the research of Mukanov et al. [9]. When Cu is added in high concentrations, it can lead to the formation of a dual-phase structure, featuring the primary FCC1 phase and a secondary copper-rich FCC2 phase. This is evidenced by TEM analysis, which reveals that an increase in secondary phases can significantly impact the lattice parameters of the CrFeCoNiCux alloy.

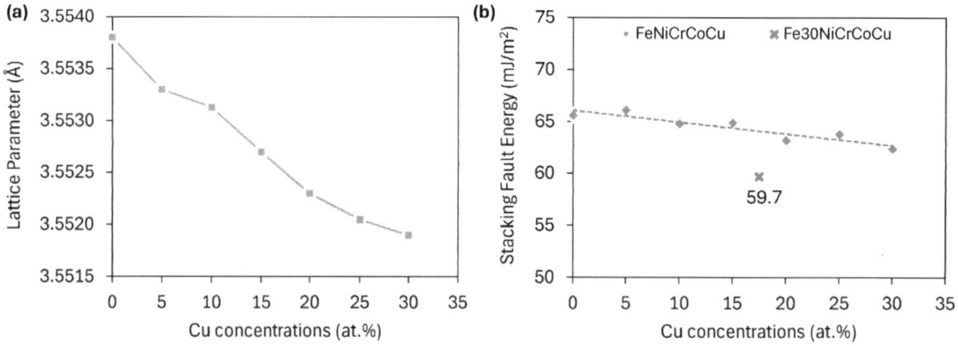

Figure 4. Effect of Cu composition on the (a) lattice parameter and (b) SFE of FeNiCrCoCu HEAs.

The findings regarding the SFE in relation to varying Cu concentrations are compelling and presented in Figure 4(b). It's clear that as Cu concentration rises, the SFE declines, though this change is not overly pronounced. For example, the equiatomic alloy MEA FeNiCrCo has an SFE of 65.6 mJ/m^2, which drops slightly to 63.2 mJ/m^2 upon adding 20 atomic percent (at.%) Cu. In contrast, increasing the Fe concentration to 30 at% results in a more substantial decrease to 59.7 mJ/m^2. Notably, the non-equiatomic HEA Fe_{30}NiCrCoCu exhibits the lowest SFE among the studied alloys. This decrease is significant, as a low SFE indicates a higher likelihood of forming the HCP structure, which correlates strongly with improved hardening and strength characteristics. Importantly, research by Mukanov et al. [9] reinforces these insights by illustrating the positive effects of Cu concentration on hardness, elastic modulus, and wear resistance. Their experiments have shown that hardness consistently increases with Cu additions from 0 to 10 at.%. Similarly, the elastic modulus rises in line with these findings, underscoring the value of optimizing Cu concentration for enhanced material properties.

Figure 5(a) presents a compelling comparison of the stress-strain curves for various alloys subjected to uniaxial tension. Notably, the stress-strain behavior during uniaxial compression is strikingly similar for the MEA FeNiCrCo, equiatomic HEA FeNiCrCoCu, and non-equiatomic HEA Fe_{30}NiCrCoCu. All exhibit a linear rise in stress against strain during the elastic deformation phase, followed by a sudden drop in stress after reaching maximum strength, ultimately stabilizing as strain increases. In Figure 5(b), we observe a clear trend: as the Cu concentration in the FeNiCrCoCu alloy increases, the ultimate tensile strength (UTS) decreases. UTS, the maximum stress a material can withstand before yielding, is significantly impacted by the addition of Cu, which forms more Cu-rich phases within the FCC matrix, thereby diminishing the UTS. These findings corroborate the research by Elgack et al. [34], who explored the mechanical properties of the FeNiCrCoCu alloy using an interatomic potential developed by Farkas and Caro [34]. Their comprehensive analysis focused on critical factors like grain size, crystal orientation, operating temperature, and strain rate, examining Cu concentrations ranging from 10 to 30 at% through advanced molecular dynamics and machine learning techniques. Moreover, the observed decrease in UTS with rising Cu levels is further validated by the molecular dynamics simulations reported by Yu et al. [35], reinforcing the notion that understanding these alloys' mechanical properties is essential for future innovations.

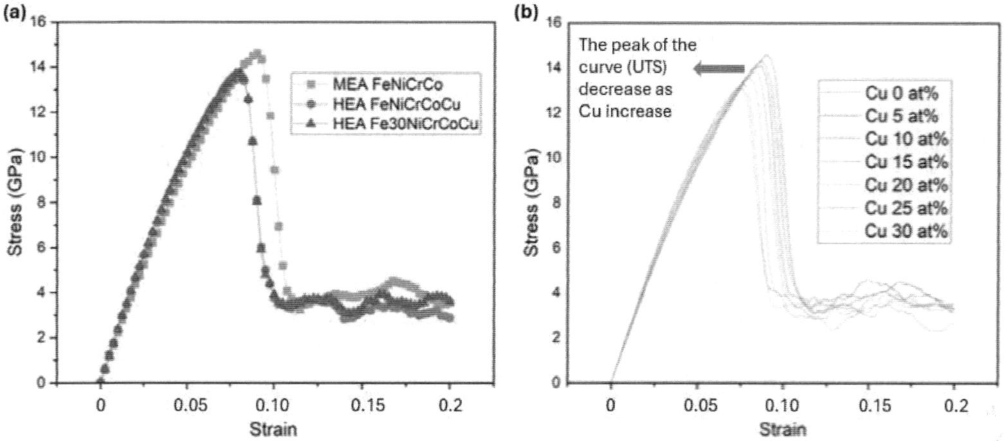

Figure 5. The calculated strain-stress curve during (a) uniaxial tension and (b) with various Cu compositions from the molecular dynamics at a strain rate of 10^{10}/s and 300 K.

To comprehensively explore the microstructural evolution of the FeNiCrCoCu HEA under uniaxial tensile stress, we utilized common neighbor analysis (CNA), as illustrated in Figure 6. CNA is a powerful technique that allows for a detailed examination of the microstructural composition and the distribution of various crystal structures in simulation results. In this analysis, atoms are color-coded for clarity: green represents face-centered cubic (FCC), red denotes hexagonal close-packed (HCP), blue signifies body-centered cubic (BCC), and gray indicates an amorphous structure. Initially, at a strain of approximately 0.05, the microstructure is purely FCC. However, as the strain increases to 0.0925, HCP structures begin to develop within the FCC matrix. At this stage, the FCC structure of the FeNiCrCoCu MEA demonstrates superior stability, offering greater strength compared to the equiatomic FeNiCrCoCu HEA, which makes it more resistant to dislocations and the subsequent formation of HCP structures. In contrast, the equiatomic alloy allows HCP structures to form with greater ease and speed. The HCP structure is characterized by twin boundaries (TB), where adjacent atomic layers adopt the HCP arrangement, indicating intrinsic stacking faults (ISF). Furthermore, combinations of two HCP layers and one FCC layer correspond to extrinsic stacking faults (ESF). The emergence of these twin boundaries and stacking faults results in a marked decrease in stress, and the slope of the stress-strain graph reveals their formation rate. Moreover, between strains of 0.125 and 0.2, the continual formation, growth, and disappearance of twin boundaries and stacking faults provoke significant stress instability within the alloy. This dynamic instability leads to stress fluctuations during plastic deformation, emphasizing the alloy's complex behavior under stress and its potential for advanced applications.

Dislocation Extraction Analysis (DXA) was employed to investigate the evolution of dislocations in OVITO, as illustrated in Figure 7. The color of the dislocation lines specifies the type of dislocation. A blue line indicates a perfect dislocation, a green line identifies a Shockley partial dislocation, a purple line represents a stair-rod dislocation, a yellow line signifies a Hirth dislocation and a sky blue line denotes a Frank dislocation. At a strain of 0.05, no dislocation activity was detected. However, Shockley partial dislocations begin to emerge as strain increases, with their quantity and length escalating significantly. Shockley partial dislocations are the main type formed, while other dislocation types, such as stair-rod, Hirth, and Frank, also appear in lesser numbers. Notably, at a strain of 0.0925, it becomes clear that the Shockley partial dislocations and other dislocations in the equiatomic HEA FeNiCrCoCu outnumber those in the MEA FeNiCrCo. This reveals that Shockley partial dislocations form more readily in the equiatomic HEA. Interestingly, the lengths of Shockley partial dislocations in the MEA FeNiCrCoCu surpass those in the equiatomic HEA FeNiCrCoCu. At a strain of 0.2, the length of these dislocations in MEA FeNiCrCoCu reaches 198 nm, while in equiatomic HEA FeNiCrCoCu, it measures 178 nm. This result emphasizes the different behaviors of microstructure and dislocations evolution during deformation in various alloys, further highlighting the significance of understanding these mechanisms.

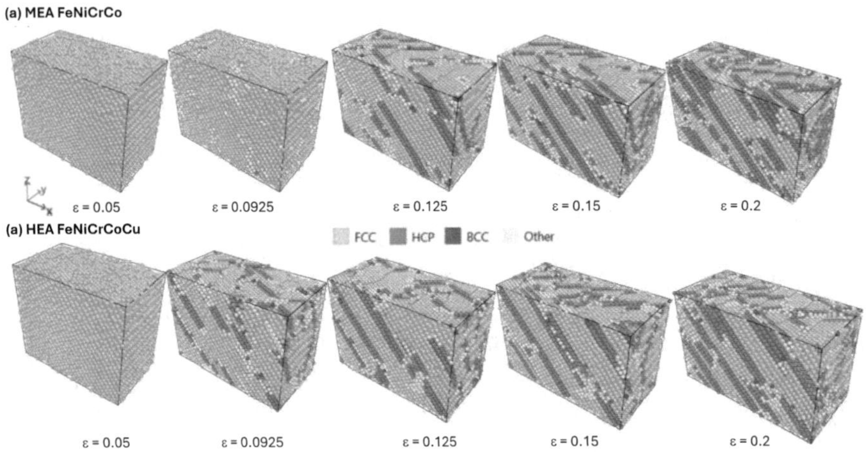

Figure 6. The microstructure of (a) MEA FeNiCrCo and (b) HEA FeNiCrCoCu during uniaxial tension at different strains of 0.05, 0.0925, 0.125, 0.15 and 0.2.

The reduction in SFE due to Cu addition significantly impacts dislocation dynamics and the hardening properties of the alloy. Lower SFE facilitates the formation of partial dislocations rather than perfect dislocations. These partial dislocations are more mobile and can easily nucleate stacking faults and twin boundaries, which are essential for deformation mechanisms such as twinning-induced plasticity (TWIP). The presence of partial dislocations and the subsequent formation of twin boundaries enhance the alloy's hardening properties. Twinning acts as a barrier to dislocation motion, increasing the material's strength and ductility. Our molecular dynamics simulations show that forming twin boundaries and stacking faults leads to significant stress fluctuations during plastic deformation, indicating enhanced hardening. The addition of Cu and the resulting lower SFE contribute to the stability of the FCC structure under stress. This stability is crucial for maintaining the alloy's mechanical properties under extreme conditions, such as nuclear applications. In summary, the addition of Cu to FeNiCrCoCu HEAs reduces SFE through lattice distortion, increased charge accumulation, and decreased DOS. This reduction in SFE promotes the formation of partial dislocations and twinning, leading to enhanced hardening properties and improved microstructural stability.

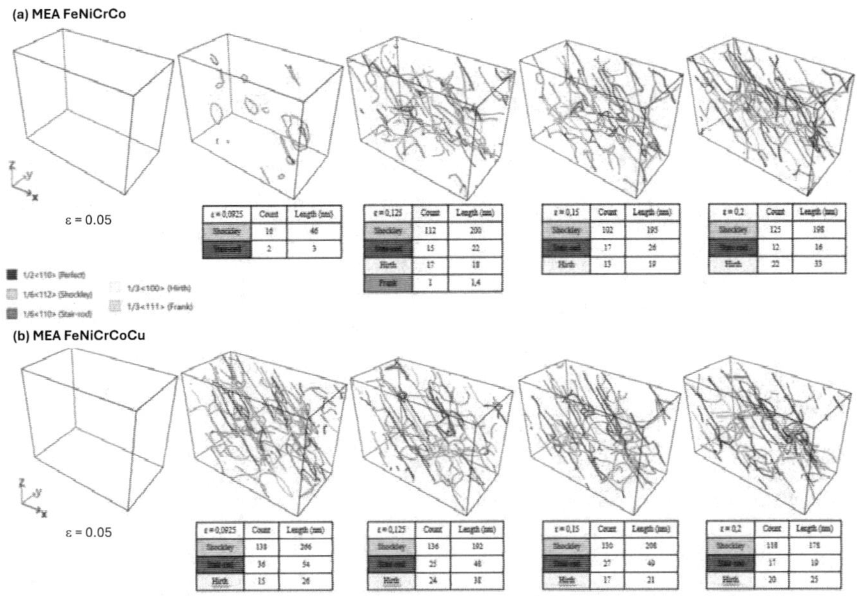

Figure 7. Dislocation evolution of (a) MEA FeNiCrCo and (b) HEA FeNiCrCoCu during uniaxial tension at different strains of 0.05, 0.0925, 0.125, 0.15 and 0.2.

3.2 First-principles DFT Result

Incorporating Cu atoms into the FeNiCoCr HEA by substituting Co with Cu offers significant advantages, particularly in increasing the alloy's lattice parameters. This enhancement is attributed to Cu's larger atomic size than Co, which elevates the average atomic size throughout the alloy. Moreover, replacing Co with larger Cu atoms induces lattice distortion within the crystal structure. This distortion compels the atomic framework to readjust for structural consistency, leading to a noteworthy lattice expansion. Ionic minimization reveals that the lattice parameter initially measured at 3.510 Å rises to 3.516 Å with 4.17 atomic percent (at.% Cu) and further escalates to 3.522 Å with 8.33 at.% Cu. The data demonstrate that the lattice parameters of the FeNiCoCr HEA significantly increase with Cu addition, affirming the substantial impact of this modification. These results are in agreement with prior studies, such as those conducted by Jiang et al. [36], which reported a similar lattice parameter of 3.59 Å. To delve deeper into the influence of Cu on local distortion, a calculation of the interlayer distances in the supercell structure was performed. This analysis is crucial for understanding the interactions between layers that contribute to lattice distortion. As illustrated in Figure 8, the distances between layers 4 and 5, as well as between layers 6 and 7, increase with the addition of Cu. Specifically, the distance between layers 4 and 5 grows from 1.99 Å to 2.00 Å upon adding one Cu atom (4.17 at.% Cu) and further to 2.03 Å when two Cu atoms (8.33 at.% Cu) are introduced. This compelling evidence reinforces the idea that Cu addition plays a pivotal role in enhancing the lattice parameters of the FeNiCoCr alloy, making it a valuable modification for future applications.

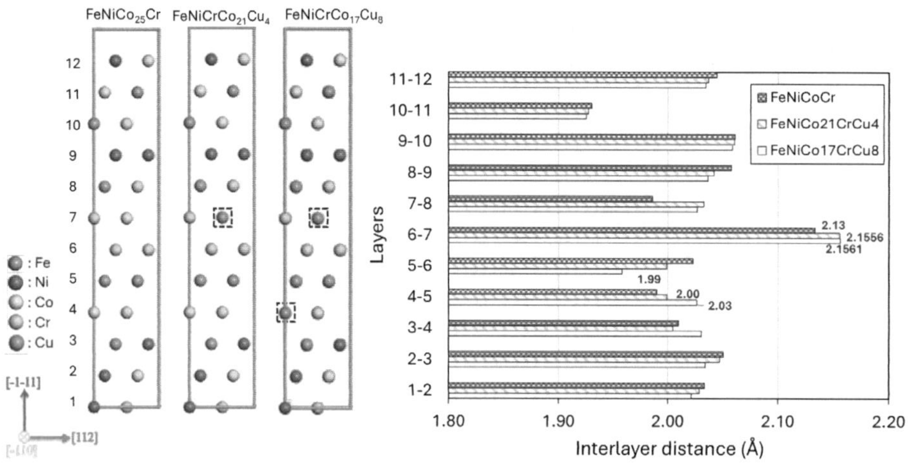

Figure 8. Comparison of the calculated interlayer distance with structure variations of HEA FeNiCoCrCu.

The addition of Cu to the FeNiCoCr alloy significantly enhances its mechanical properties, including compression, shear, and Young's modulus, as demonstrated in Figure 9(a). Notably, there is a marked improvement when 6.25 at.% Cu is added, while higher levels, specifically 12.5 at.%, resulting in a decline. This crucial finding highlights that the incorporation of 6.25 at.% Cu substantially boosts tensile strength, compressive strength, and stiffness. Figure 9(b) further underscores the positive effects of Cu on the elastic properties of the FeNiCoCr HEA. The graph reveals a rising Zener ratio with increased Cu content, indicating a progressive enhancement of anisotropy in the (100) and (110) planes. Such improvements can significantly benefit applications where strength and stability are paramount. Moreover, the Pugh and Poisson ratios illustrate intriguing trends: both decrease at 6.25 at.% Cu but then rise at 8.33 at.%. The Pugh ratio, for instance, dips from 1.88 to 1.83 but climbs back to 2.03. A B/G ratio above 1.75 suggests that all three alloy types remain ductile, a critical characteristic for various engineering applications. The Poisson ratio reflects a similar pattern. In conclusion, the data clearly indicate that increasing Cu content enhances anisotropy, ductility, and shear stability, particularly with the addition of 12.5 at.% Cu.

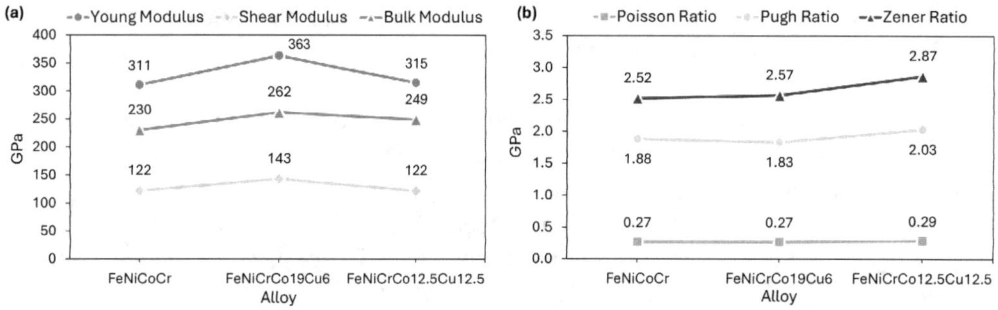

Figure 9. Effect of Cu compositions on the (a) bulk modulus (B), shear modulus (G), and Young modulus (E), (b) Poisson, Pugh, and Zener ratio from the first-principle calculation.

Figure 10(a) illustrates that as the amount of Cu added to the FeNiCoCr alloy increases, the SFE of the alloy decreases. Initially, the SFE is 79.32 mJ/m^2, which drops to 54.92 mJ/m^2 with a

4.17 at.% Cu addition, and further to 54.77 mJ/m^2 with an 8.33 at.% Cu addition. This decrease indicates that adding Cu facilitates the formation of partial dislocations in FeNiCoCr alloys. Moreover, since the SFE remains above 45 mJ/m^2, the primary deformation mechanism resulting from increased Cu addition is Shockley partial dislocations. This dislocation initiates the transformation of the FCC phase to the HCP structure, leading to the formation of ISF and HCP nuclei. As shown in Figure 10(b), this transformation enhances the hardness of the HEA. At relatively low Cu concentrations, the addition of copper strengthens the alloy through a solid solution mechanism, creating barriers to dislocation movement due to lattice distortion. However, at higher levels of Cu, the formation of unevenly distributed Cu-rich precipitates can occur, as reported in the research by Suprianto et al. [37]. This uneven distribution can ultimately reduce the material's hardness.

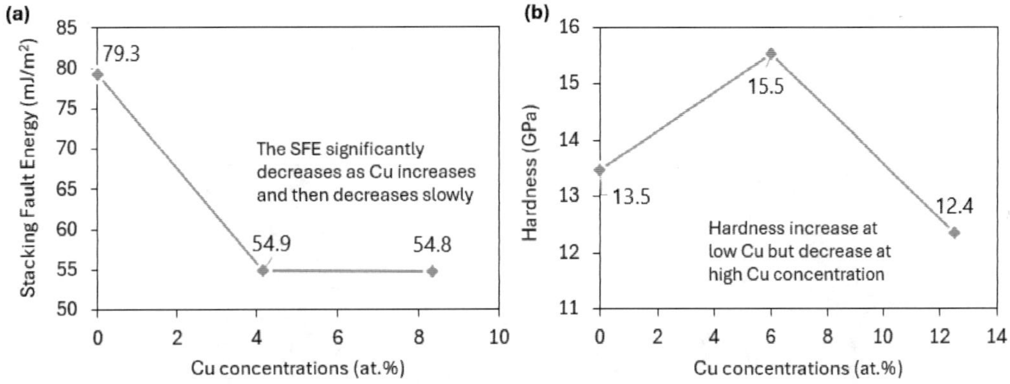

Figure 10. Effect of Cu concentrations on the (a) stacking fault energy, and (b) hardness from the first-principle calculation.

Figure 12(a) presents the total Density of States (DOS) for three alloys, highlighting the significant role of the d-shell in the Fermi level region (0 eV). It shows that the addition of Cu generally leads to a decrease in the DOS near the Fermi energy. This reduction suggests a decrease in the bond strength between atoms, which in turn affects the SFE by lowering it. This confirms that adding Cu to the FeNiCoCr alloy results in a decrease in SFE. Figure 11(b) illustrates an increase in charge accumulation. The initial red charge accumulation area of FeNiCoCr alloy is approximately 9.65% but with the addition of 4.17 at.% Cu and 8.33 at.% Cu, it rises to 22% and 23.2%, respectively. This increase in charge accumulation indicates that the bond energy between layers becomes weaker, consistent with the observed decrease in SFE. In summary, the first-principle calculations reveal that adding Cu to the FeNiCoCr alloy has a significant impact on lattice parameters, strength, hardness, and SFE. Most of these values increased, except for ductility and SFE, both of which decreased. However, it is important to note that the alloy remained in the ductile category, with the predominant mechanism of deformation being partial dislocations. The electronic analysis shows a notable shift in SFE, attributable to a decrease in the Fermi energy DOS and an increase in charge accumulation, ultimately leading to a decrease in SFE.

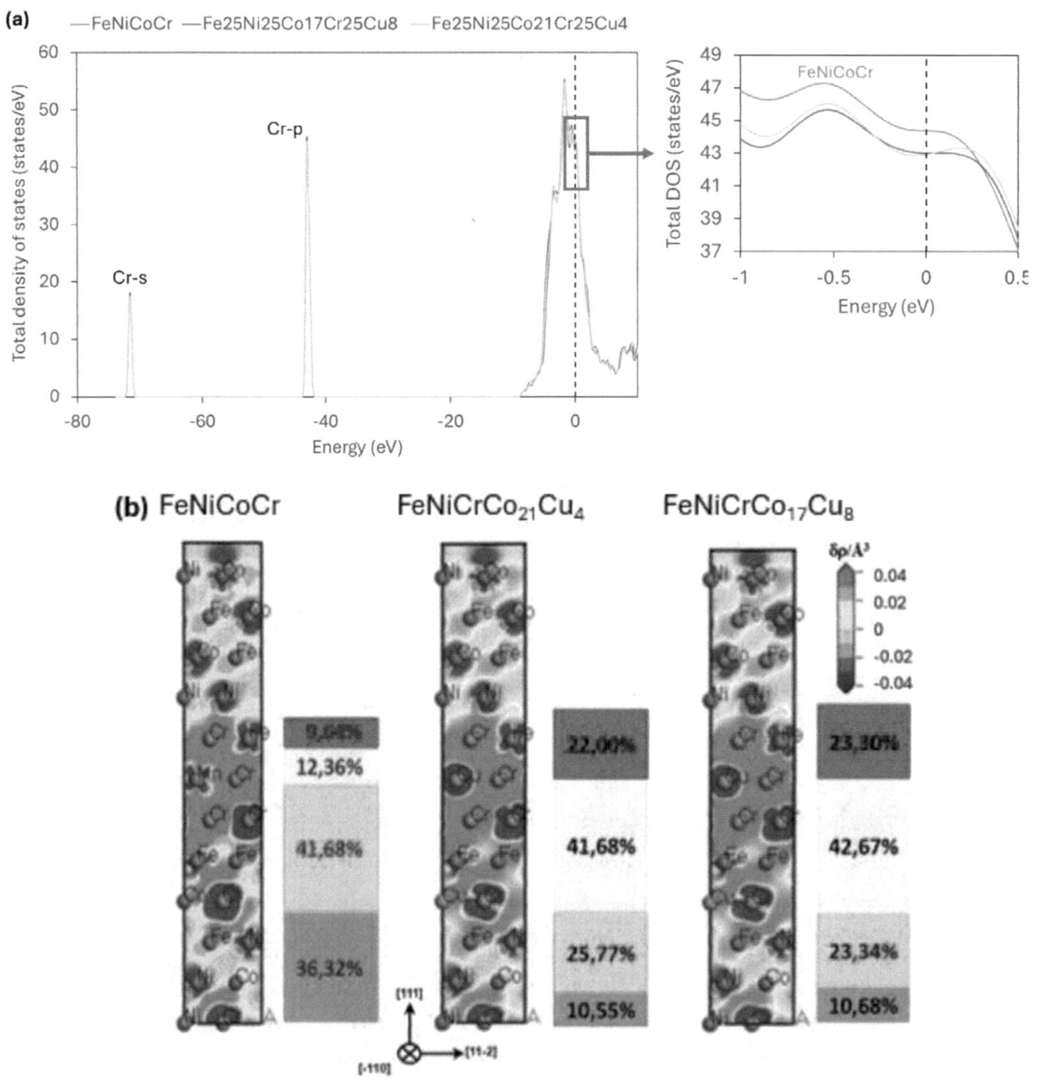

Figure 11. (a) Total density of states (DOS) curve and (b) charge density difference of FeNiCoCr, $FeNiCrCo_{21}Cu_4$, and $FeNiCrCo_{17}Cu_8$.

The addition of Cu to FeNiCrCoCu HEAs leads to a reduction in SFE due to several factors: (1) lattice distortion: Cu atoms (being larger than Co atoms) create substantial lattice distortion when they replace Co in the alloy. Our DFT calculations confirm that this distortion increases interlayer distances within the crystal structure. As the interlayer distances grow, interatomic bonds are weakened, directly leading to a notable reduction in SFE. (2) enhanced charge accumulation, our first-principles calculations indicate that adding Cu increases charge accumulation between atomic layers. This accumulation weakens the bond energy between layers, providing further evidence of the reduction in SFE and enhancing the alloy's structural integrity. (3) decreased DOS: Additionally, we observe a decrease in the total DOS near the Fermi level with Cu addition. This reduction indicates weakened bonding interactions, strongly correlated with a lower SFE. In summary, the combination of lattice distortion, enhanced charge

Journal of Physics: Conference Series **2980** (2025) 012037
doi:10.1088/1742-6596/2980/1/012037

accumulation, and reduced density of states clearly illustrates why Cu addition leads to a significant decrease in SFE of FeNiCrCoCu high-entropy alloys.

3.3 Computational Thermodynamic Result

Based on thermodynamic calculations, figure 12 illustrates a positive correlation between temperature increase and the SFE. As temperature rises, the SFE also increases. This rise in SFE is attributed to the increase in Gibbs free energy, a temperature function. The relationship indicates that as temperature increases, free energy also rises. In Figure 12(a), it is evident that higher temperatures decrease the number of low SFE classifications while the number of high SFE classifications increases. Similar findings have been reported in various studies. For instance, research by Molnár et al. demonstrated that rising temperatures directly contribute to greater SFE in austenitic stainless steel, which exhibits a positive thermodynamic relationship for SFE [38]. Additionally, Zhang et al. explained that the FCC phase is metastable at low temperatures, while the HCP phase is more stable [39]. This stability preference accounts for the lower or negative SFE at low temperatures. It is important to note that the Gibbs free energy function is consistently linear, whereas the SFE does not follow a linear pattern. This suggests that other factors, such as magnetic contributions, may influence the SFE. Figure 12(b) illustrates the effect of adding specific elements to an alloy and their relationship to the SFE. At 300 K, introducing Cr, Co, and Cu significantly lowers the SFE of the FeNi alloy. Notably, adding Cu to the FeNiCrCo alloy at this temperature also contributes to a decrease in SFE. The FeNiCr alloy exhibits the lowest SFE at room temperature compared to the FeNiCrCoCu alloy. As temperatures rise, especially to 750 K, the addition of Cu not only continues this trend but results in the lowest SFE among the alloys studied. Remarkably, higher Cu concentrations lead to a more pronounced decrease in SFE than in the equiatomic state. A detailed discussion of the effects of each element will follow in the next section.

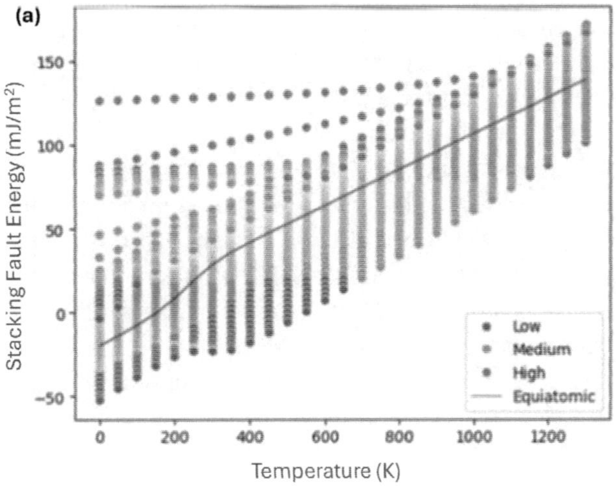

Journal of Physics: Conference Series **2980** (2025) 012037 doi:10.1088/1742-6596/2980/1/012037

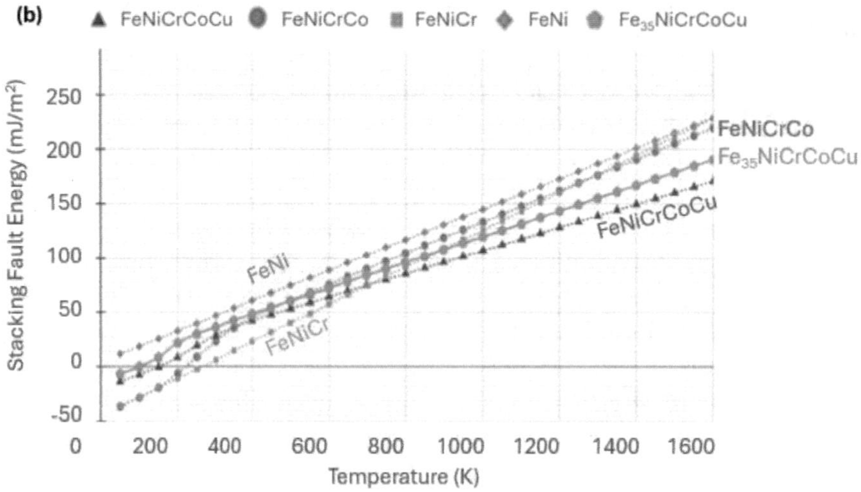

Figure 12. The calculated SFE as a function of temperature from computational thermodynamics of (a) equiatomic FeNiCrCoCu HEA and (b) comparison with other alloys.

In Figures 13(a-b), it is clear that higher concentrations of Cu lead to a significant decrease in the SFE. These findings are consistent with our molecular dynamics (MD) calculations shown in Figure 4(b) and the first principles presented in Figure 10(a). At temperatures below 300 K, higher Cu compositions may result in a low SFE classification, indicating a correlation between increased Cu concentrations and reduced SFE, particularly at lower temperatures. Figure 13(b) illustrates that at 800 K, the SFE decreases from 105 to 75 mJ/m^2. A branch line originating from the equiatomic compositions in Figure 13(b) highlights the influence of other elements as Cu concentration increases. The effect of each element on the SFE varies: a decrease in Cr raises the SFE, while reductions in Fe, Ni, and Co can lower it. The composition of the HEA FeNiCoCrCu significantly affects the SFE at 800 K, as shown in Figure 13(c). Increasing the percentages of Fe, Ni, and Co from 5% to 40% enhances the SFE, while higher concentrations of Cr and Cu reduce it.

Research on the relationship between Cu content and the SFE of FeNiCrCoCu alloys is still limited. However, studies related to the hardness of the alloy have been conducted. For example, Mukanov's research indicates that adding Cu to equiatomic CoCrNiFe alloys increases the hardness of the alloy, which is notable for HEA applications [9]. Additionally, research by Huang et al. found that increasing the Cu content in AlCrFeNiTiCux alloys enhances the intensity of the face-centered cubic (FCC) phase; this stability is attributed to a rise in SFE [40]. These findings present differing results compared to prior thermodynamic calculations. Furthermore, research by Cai et al. explains that the interactions among different elements in binary Cu alloys—such as Mg, Al, Si, and Ga—have varying impacts on the SFE of the alloy [41].

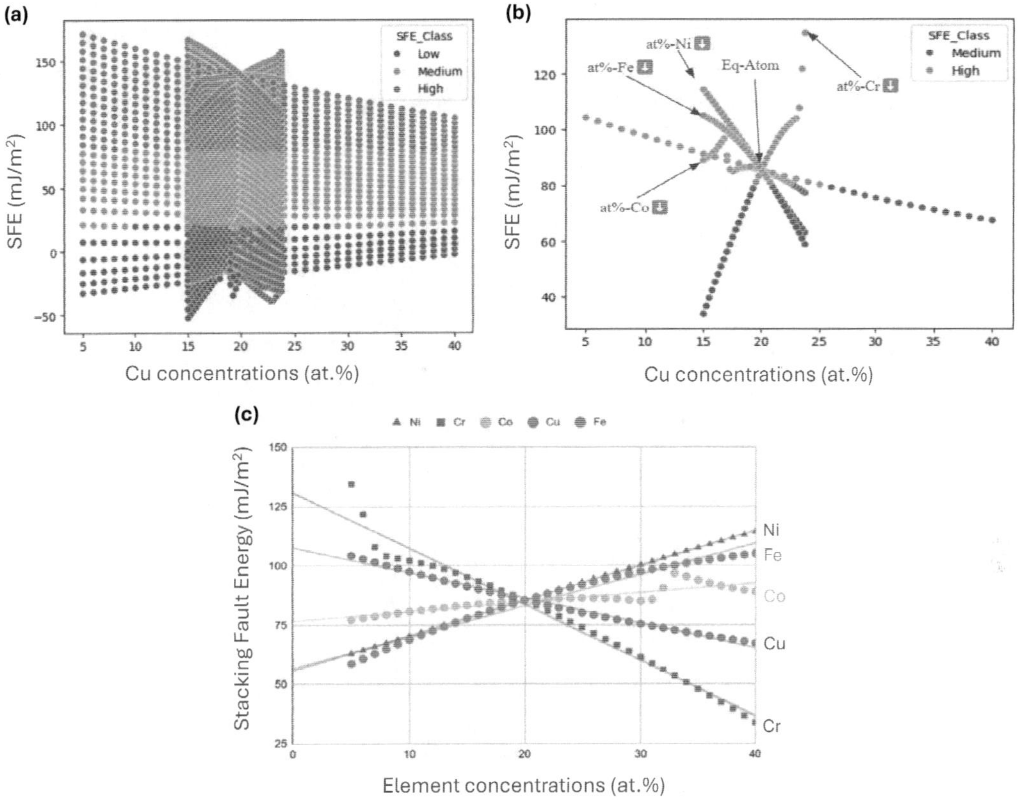

Figure 13. (a) Effect of Cu concentrations on the SFE at different temperatures from 1 to 1200 K and (b) at 800 K, (c) the calculated SFE at 800 K as a function of Ni, Fe, Co, Cu, and Cr.

3.4 Machine Learning Result

We utilized a confusion matrix to guide our analysis in our search for the most effective machine learning model. Accuracy is crucial for reliable predictions; higher accuracy improves classification capabilities. Similarly, a lower Root Mean Square Error (RMSE) signifies better regression performance by indicating the average error size. As shown in Figure 14, the Random Forest (RF) model stands out with the highest accuracy of 0.988 and the lowest RMSE at 1.31. These exceptional figures highlight its remarkable classification performance and ability to predict actual values more closely. Importantly, the RF model surpasses the Decision Tree in terms of both accuracy and RMSE, as well as in confusion matrix assessments. These compelling findings affirm the RF model's reliability and precision, positioning it as a critical tool for SFE predictions and showcasing its potential for significant advancements across various industries [42].

Feature importance refers to how significant a feature is in predicting an output based on a machine learning model without considering whether its influence is positive or negative. On the other hand, the Spearman rank correlation offers a robust, nonparametric method to assess the relationship between two variables, unaffected by the distribution or external variable interference [42]. As illustrated in Figure 15(a), Cr emerges as the most influential feature of the SFE, followed by Co, Cu, Ni, and Fe. Figure 15(b) reveals Cr's noteworthy negative impact on SFE at -0.21, while Cu follows closely at -0.08. In contrast, Fe and Ni positively contribute to SFE, registering at 0.09 and 0.06, respectively. Interestingly, even though Co demonstrates high importance based on these metrics, its influence on SFE is neutral, indicated by a value of 0.00. This underscores that while Co plays a pivotal role in the feature set due to its interactions, it does not establish a direct linear correlation with SFE, nor does it consistently drive changes in its value. This insight emphasizes the complexity of feature interactions in predictive modeling.

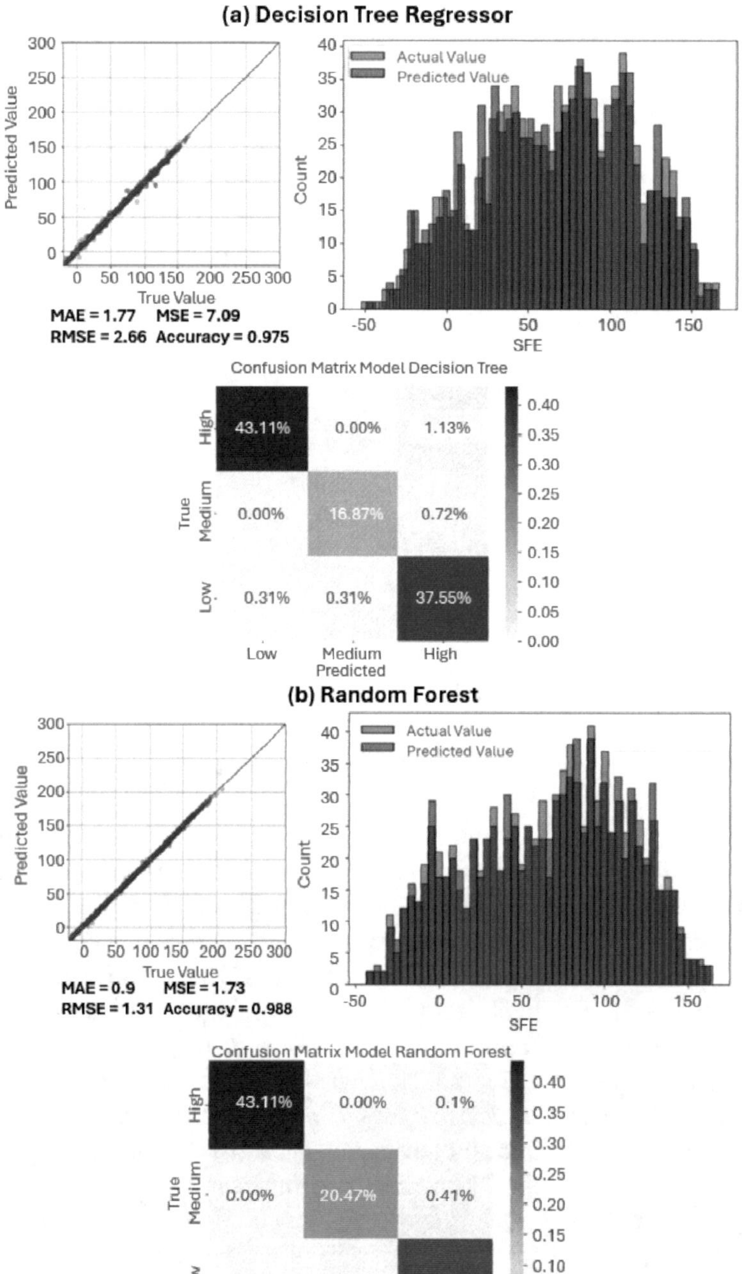

Figure 14. Machine learning model evaluation of (a) the Decision Trees, and (b) the Random Forest model.

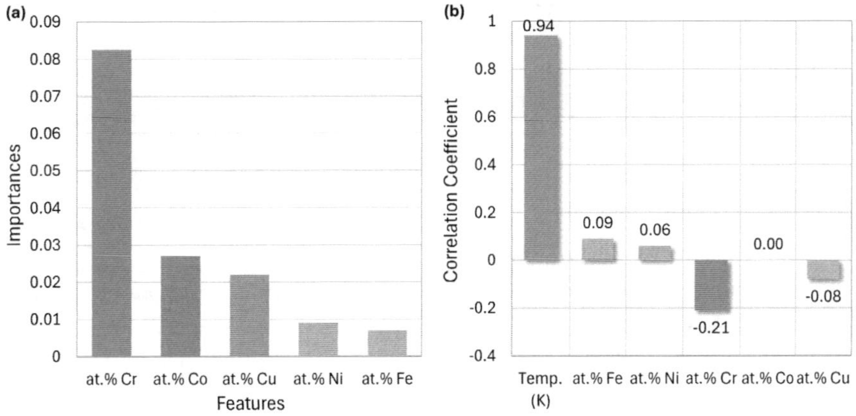

Figure 15. (a) Feature importance and (b) correlation coefficient of FeNiCrCoCu alloys from Random Forest model.

Figures 16(a-b) illustrate an example of a FeNiCrCoCu HEA design by varying the composition of two elements while maintaining the remaining elements in an equiatomic state. This analysis is conducted at a temperature of 800 K, which is relevant for the operating conditions of a nuclear reactor, allowing us to harness the strong mechanical properties of HEAs in irradiation environments. Figure 16(a) shows an alloy design that adjusts the atomic percentages of Cr and Cu, the two key factors that influence the SFE of the alloy. This guide aids in determining the optimal HEA composition. For example, in the equiatomic state, the SFE of the FeNiCrCoCu alloy is 85.44 mJ/m^2, indicating a tendency for unwanted deformation slip. To mitigate this, the alloy composition can be modified to $Fe_{20}Ni_{20}Cr_{30}Co_{10}Cu_{20}$, featuring a higher Cr content, which based on predictions from a random forest model, results in an SFE of 65.17 mJ/m^2, as indicated in the design guide in Figure 16(a). Similarly, Figure 16(b) presents an alloy design that varies the atomic percentages of Fe and Ni. For instance, in the $Fe_{20}Ni_{15}Cr_{21.67}Co_{21.67}Cu_{21.67}$ alloy—with a Ni content below the equiatomic state—the random forest model predicts an SFE of 75.45 mJ/m^2.

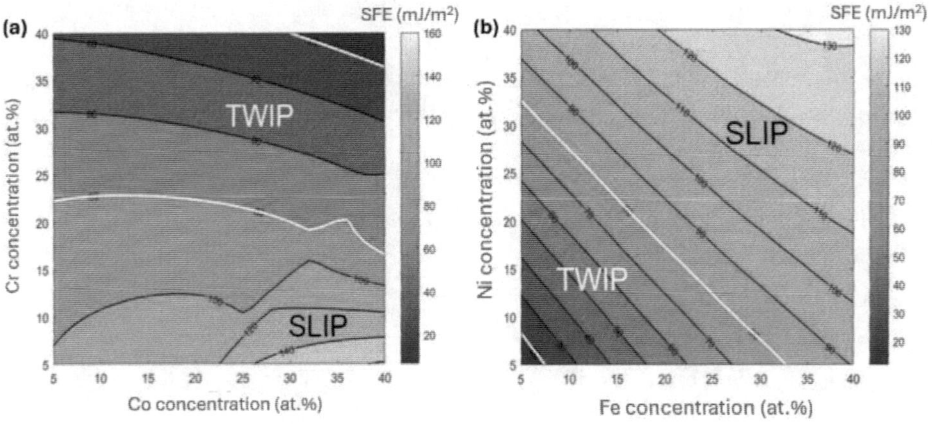

Figure 16. Deformation map and the calculated SFE as a function of (a) Cr and Co concentrations, (b) Ni and Fe concentrations for designing FeNiCrCoCu.

Additionally, Table 2 provides design guidelines by varying one alloy element while keeping the others in the equiatomic state. For example, the $Fe_{24}Ni_{24}Cr_{24}Co_4Cu_{24}$ alloy, which has a lower Co content, is predicted to have an SFE of 77.91 mJ/m^2. Furthermore, a specific composition range that allows for high Cr and Cu levels results in lower SFEs compared to the equiatomic state. For instance, the $Fe_{17.5}Ni_{17.5}Cr_{17.5}Co_{17.5}Cu_{30}$ alloy exhibits an SFE of 75.39 mJ/m^2. These examples present alternative alloy compositions that can achieve a lower SFE than the equiatomic state or meet specific SFE requirements.

Table 2. Optimum composition for designing FeNiCrCoCu HEAs based on the SFE calculations.

Element	Atomic percent (at%)
Fe	5 – 16
Ni	5 – 16
Cr	22 – 40
Co	5 – 9
Cu	26 - 40

4. Conclusions

The findings of this study are compelling and reveal important insights:

1. Molecular dynamics simulations clearly indicate that incorporating Cu leads to a reduction in lattice parameters and, consequently, the ultimate tensile strength (UTS). Notably, as Cu concentration increases, both the elastic modulus and yield strength significantly improve. The addition of Cu also notably decreases the alloy's SFE, and the FeNiCrCoCu HEA exhibits a rapid phase transition from FCC to HCP.
2. First-principles calculations further validate these results, demonstrating that introducing 4.17% and 8.33% at.% of Cu enhances the lattice parameters, raising the SFE from 79.3 to 54.7 mJ/m^2. This enhancement comes along with an increase in elastic modulus and hardness while maintaining ductility within acceptable limits.
3. The random forest model outperforms traditional decision tree models in predictive accuracy. Notably, the concentrations of Ni, Fe, and Co are shown to be less critical, suggesting that these elements can be minimized without significantly impacting the SFE—potentially even improving it. For instance, these elements can be capped at 5 at.%. The optimal composition for the FeNiCrCoCu alloy can effectively include Co (5-9 at.%), Cr (22-40 at.%), Fe (5-16 at.%), Cu (26-40 at.%), and Ni (5-16 at.%).

References

[1] M. Moschetti, P. Burr, E. Obbard, J. J. Kruzic, P. Hosemann, and B. Gludovatz, 2022, J. Nucl. Mater., vol. 567, p. 153814.
[2] S. J. Zinkle and G. S. Was, 2013, Acta Mater., vol. 61, no. 3, pp. 735–758.
[3] S. Praveen and H. S. Kim, 2018, Adv. Eng. Mater., vol. 20, no. 1.
[4] J. W. Yeh, 2004, Adv. Eng. Mater., vol. 6, no. 5, pp. 299–303.
[5] R. Carroll, 2015, Sci. Rep., vol. 5, pp. 1–12.

[6] A. V. Kuznetsov, D. G. Shaysultanov, N. D. Stepanov, G. A. Salishchev, and O. N. Senkov, 2012, Mater. Sci. Eng. A, vol. 533, pp. 107–118.
[7] E. J. Pickering, A. W. Carruthers, P. J. Barron, S. C. Middleburgh, D. E. J. Armstrong, and A. S. Gandy, 2021, Entropy, vol. 23, no. 1, pp. 1–28.
[8] T. Nagase, P. D. Rack, J. H. Noh, and T. Egami, 2015, Intermetallics, vol. 59, pp. 32–42.
[9] S. Mukanov, P. Loginov, A. Fedotov, M. Bychkova, M. Antonyuk, and E. Levashov, "The Effect of Copper on the Microstructure, Wear and Corrosion Resistance of CoCrCuFeNi High-Entropy Alloys Manufactured by Powder Metallurgy," Materials (Basel)., vol. 16, no. 3, 2023, doi: 10.3390/ma16031178.
[10] H. Zheng, 2020, J. Mater. Sci. Technol., vol. 38, pp. 19–27.
[11] P. Hirel, 2015, Comput. Phys. Commun., vol. 197, pp. 212–219.
[12] O. R. Deluigi, R. C. Pasianot, F. J. Valencia, A. Caro, D. Farkas, and E. M. Bringa, 2021, Acta Mater., vol. 213, p. 116951.
[13] Stukowski A and Albe K, 2010, Model. Simul. Mater. Sci. Eng., vol. 18, p. 085001.
[14] Honeycutt J D and Andersen H C, 1987, J. Phys. Chem. Vol. 91 pp. 4950-4963.
[15] Stukowski A and Albe K 2010, Model. Simul. Mater. Sci. Eng., vol. 18 p. 025016.
[16] Clark S, Segall M, Pickard C, Hasnip P, Probert M, Refson K and Payne M, 2005, Z. Kristallogr. 220.
[17] Perdew J P, Burke K and Ernzerhof M, 1996, Phys. Rev. Lett., vol. 77 (18), pp. 3865–3868.
[18] Vanderbilt D, 1990, Phys. Rev. B Condens. Matter, vol. 41 (11), pp. 7892–7895.
[19] Monkhorst H J and Pack J D, 1976, Phys. Rev. B Condens. Matter, vol. 13 (12), pp. 5188–5192.
[20] Nugraha F, Aulia M F and Achmad T L, 2024, E3S Web of Conferences, vol. 543, p. 03009.
[21] Olson G B and Cohen M, 1976, Metall. Trans. A, vol. 7, pp. 1897–1904
[22] Xie, H., Ma, Z., Zhao, H., and Ren, L., 2022, Tribol Int, vol. 174.
[23] Achmad T L, Fu W, Chen H, Zhang C and Yang Z G, 2018, Comput. Mater. Sci., vol. 143, pp. 112-117.
[24] Achmad T L, Fu W, Chen H, Zhang C and Yang Z G, 2018, J. Mater. Sci., vol. 53, pp. 10217 - 10230.
[25] Castañeda J A, Zambrano O A, Alcázar G A, Rodríguez S A and Coronado J J, 2021, Metals, vol. 11(11), pp. 1701.
[26] W. Zhang, 2024, Metals, vol. 14, no. 3.
[27] Q. Hu, H. Ling Wang, and X. Wang Liu, 2023, Trans. Nonferrous Met. Soc. China, vol. 33, no. 6, pp. 1803–1813.
[28] H. Xie, Z. Ma, W. Zhang, H. Zhao, and L. Ren, 2022, Int. J. Mech. Sci., vol. 238, p. 107855.
[29] Z. Su and Y. Zhang, 2023, J. Mater. Eng. Perform., vol. 32, no. 12, pp. 5653–5665.
[30] Y. Cao, J. Liu, S. Zhou, and Y. Xu, 2022, Solid State Commun., vol. 359, p. 115011.
[31] J. Li, 2021, Appl. Surf. Sci., vol. 570, p. 151236.
[32] Z. An 2015, Mater. Res. Lett., vol. 3, no. 4, pp. 203–209.
[33] O. Elgack, B. Almomani, J. Syarif, M. Elazab, M. Irshaid, and M. Al-Shabi, 2023, J. Mater. Res. Technol., vol. 25, pp. 5575–5585.
[34] D. Farkas and A. Caro, 2018, J. Mater. Res., vol. 33, no. 19, pp. 3218–3225.
[35] J. Yu, 2023, Nanomaterials, vol. 13, no. 6.
[36] Jiang, Z. F., Chen, W. P., Xia, Z. B., Xiong, W., & Fu, Z. Q., 2019, Intermetallics, vol. 108, pp. 45-56.
[37] Suprianto, and Chen, C.-L., 2022, Metals and Materials International, vol. 29, pp. 420–428.
[38] Molnár, D., Sun, X., Lu, S., Li, W., Engberg, G., and Vitos, L., 2019, Materials Science and Engineering: A, vol. 759, pp. 490–497.
[39] Zhang, Y. H., Zhuang, Y., Hu, A., Kai, J. J., and Liu, C. T., 2017, Scr Mater, vol. 130, pp. 96–99.
[40] Huang, L., Wang, X., Huang, B., Zhao, X., Chen, H., and Wang, C., 2022, Intermetallics, vol. 140, pp. 107397.
[41] Cai, T., Li, K. Q., Zhang, Z. J., Zhang, P., Liu, R., Yang, J. B., and Zhang, Z. F., 2020, J Mater Sci Technol, vol. 53, pp. 61–65.
[42] Achmad T L, Wibowo P A, Sukma F T, 2025, J. Alloys Compd., vol. 1010, 177514.

Journal of Physics: Conference Series **2980** (2025) 012038 doi:10.1088/1742-6596/2980/1/012038

Development of Quantum Key Distribution (QKD) with E91 Protocol for Future Secure Quantum Networks

Muhammad Fajri Zulfa and Khoirul Anwar

[1] The University Center of Excellence for Advanced Intelligent Communications (AICOMS), Telkom University, Bandung, Indonesia
[2] Research Collaboration Center for Quantum Technology 2.0 (PKR Kuantum 2.0), Labtek V, ITB, Bandung, Indonesia

E-mail: mfajrizulfa@student.telkomuniversity.ac.id, anwarkhoirul@telkomuniversity.ac.id

Abstract. This paper investigates practical performances of the Ekert 91 protocol (E91) for future secure quantum networks. To seek future possible improvement, we investigate the important different principles of E91 and Bennet-Brasard (BB84) protocols. We perform a series of computer simulations to evaluate the performances of them and investigate the practical parameters. Furthermore, we simulate the E91 protocol using real-world practical scenarios involving eavesdropping and quantum channel errors. We present two scenarios: (1) the performance with eavesdropper interception under a perfect channel and (2) the performances with eavesdropper interception combined with channel errors. We found that: (i) the E91 protocol produces fewer keys compared to the BB84 protocol but does not require verification between two parties, (ii) the E91 protocol enhances security due to the utilization of Clauser, Horne, Shimony, and Holt (CHSH) inequality, and (iii) both eavesdropper and channel errors can impact the less production of keys obtained during the keys production.

1. Introduction

Questions about cyber security and consumer privacy have become critical with the emergence of communications systems [1]. All of the communications systems require highly advanced security. Failure in the system could cause serious disruptions to essential service at any time. Classical cryptography is widely used in today's telecommunication systems, but it faces new threats from quantum computers, which are still rapidly developing. The RSA algorithm, currently used in practice, can be broken by quantum algorithms [2]. On the 5G and 6G future communications, quantum is one of the focusing study to give everything services with fast speed [3–5]. Quantum communications provides higher capacity, greater security, and lower latency, requiring quantum cryptography to support these needs [6]. Quantum cryptography is grounded in the principles of quantum mechanics and relies on the laws of physics to ensure secure communications. The main aim of quantum cryptography is to achieve information security and figure out any eavesdroppers in the communications channel.

Quantum Key Distribution (QKD) addresses security challenges for future applications in the sixth generation of telecommunications (6G) 2030 by leveraging entanglement and quantum teleportation. The E91 protocol, based on entanglement, is one such QKD method. Entanglement swapping, a fascinating

Journal of Physics: Conference Series **2980** (2025) 012038 doi:10.1088/1742-6596/2980/1/012038

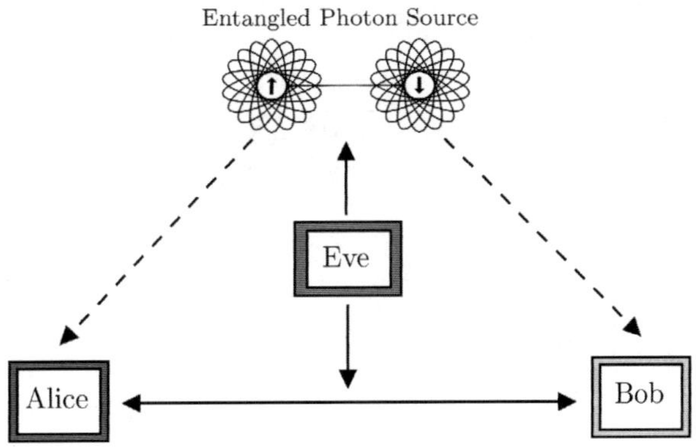

Fig. 1. QKD based on entanglement in quantum communications systems.

phenomenon in quantum mechanics, involves manipulating entangled particles like Einstein-Podolsky-Rosen (EPR) pairs [7], which can generate sequences of random bits in two remote locations. QKD uses this concept for key exchange, ensuring secure communications between distant nodes. The E91 protocol exemplifies that how quantum entanglement and measurement can enhance security in communications. Fig. 1 shows that how QKD serves as a key for secure communications through quantum mechanics. The E91 protocol, devised by Artur Ekert in 1991, is a notable quantum key distribution method that utilizes quantum entanglement and measurements to ensure the secure exchange of cryptographic keys. This protocol is particularly robust against attacks, guaranteeing a high level of security.

Several developments of E91 protocols have been conducted and reported. Authors in [8] developed E91 protocol for long distances communications. Authors in [9] explore a generalization of Ekert quantum cryptographic protocol, where qubits are substituted with qudits, which are systems of N- or d-dimensions. Authors in [10] developed E91 protocol with generated on demand by a quantum dot. Authors in [11] reviews security of E91 protocol with several attacking strategies. Authors in [12] proposed two new protocol inspired from E91 protocol for improving efficiency key exchange. Authors in [13] investigated of strength of E91 QKD protocol. In securities issue era, several developments for developing E91 QKD protocol have been developed to solve classical cryptography problem and security for entanglement communications. In the era of securities issues, several advancements have been made to develop the E91 QKD protocol to address the challenges of classical cryptography and enhance security for entanglement-based communications. In addition to these efforts, Authors in [14] proposed the utilization of quantum cryptography with two non-orthogonal states, while Authors in [15] introduced the concepts of anonymous-key quantum cryptography and unconditionally secure quantum bit commitment.

This paper investigates the impact of practical performance factors on the keys production in the E91 protocol. Our analysis evaluates more real-world scenario, providing insights into how practical challenges. On the other hand, this paper also clarifies the final key of keys productions process in the E91 protocol and BB84 in [16], and elaborates on the differences between them. We also develop E91 protocol to detailed investigation of the E91 protocol performance in the presence of eavesdropper interception and erroneus channels. Therefore, in this paper, we focus on the performance evaluation to the real-world scenario of the E91 protocol by doing a series computer sumulation. We also analyze the impact of eavesdropper interception and erroneus channels, which is widely regarded as the most adverse case in QKD or quantum communications.

2. Method and Scenario

Quantum cryptography in the E91 protocol is based on Bell theorem and entanglement. Introduced by Artur Ekert in 1991, this protocol employs the concept of entangled quantum particles to detect eavesdropping attempts. This approach makes E91 a highly secure QKD method, with the strength of its security rooted in the non-locality properties of quantum mechanics.

2.1. Quantum Gates

Quantum gates are the basic components of quantum circuits. Quantum bits or qubits, can be in a superposition of states, unlike classical bits. In this study, We utilize several quantum gates expressed

Journal of Physics: Conference Series **2980** (2025) 012038 doi:10.1088/1742-6596/2980/1/012038

Table 1. Bell States in E91 protocol.

In	Out
$\|00\rangle$	$(\|00\rangle + \|11\rangle)/\sqrt{2} \equiv \|\beta_{00}\rangle$
$\|01\rangle$	$(\|01\rangle + \|10\rangle)/\sqrt{2} \equiv \|\beta_{01}\rangle$
$\|00\rangle$	$(\|00\rangle - \|11\rangle)/\sqrt{2} \equiv \|\beta_{10}\rangle$
$\|00\rangle$	$(\|01\rangle - \|10\rangle)/\sqrt{2} \equiv \|\beta_{11}\rangle$

mathematically as

$$\mathbf{H} = \frac{1}{\sqrt{2}}\begin{pmatrix} 1 & 1 \\ 1 & -1 \end{pmatrix}, \quad \mathbf{CNOT} = \begin{bmatrix} 1 & 0 & 0 & 0 \\ 0 & 1 & 0 & 0 \\ 0 & 0 & 0 & 1 \\ 0 & 0 & 1 & 0 \end{bmatrix}. \tag{1}$$

The Hadamard gate, represented by $\mathbf{H}$ in (1), is an essential single-qubit gate in quantum computing. It converts the basis states $|0\rangle$ and $|1\rangle$ into equal superpositions. The CNOT gate in (1) performs a NOT operation on the target qubit (qubit 2) only when the control qubit (qubit 1) is in state $|1\rangle$. It is pivotal for creating entanglement and implementing conditional operations in quantum circuits. Other utilize gates in this paper are Pauli matrices, which are expressed as

$$\mathbf{Z} = \begin{bmatrix} 1 & 0 \\ 0 & -1 \end{bmatrix}, \quad \mathbf{X} = \begin{bmatrix} 0 & 1 \\ 1 & 0 \end{bmatrix}, \quad \mathbf{Y} = \begin{bmatrix} 0 & -i \\ i & 0 \end{bmatrix}. \tag{2}$$

The Gate $\mathbf{Z}$, also called the Pauli-Z gate in (2), applies a phase flip by altering the sign of the second qubit state. Specifically, it leaves the state $|0\rangle$ intact and changes the sign of $|1\rangle$, converting it to $-|1\rangle$. The Gate $\mathbf{X}$ or the Pauli-X gate performs a bit flip by swapping the states $|0\rangle$ and $|1\rangle$. It acts like a classical NOT gate flipping the qubit state.

The Gate $\mathbf{Y}$ or the Pauli-Y gate combines a flip in the bit and a flip in the phase. The Pauli-Y adds a complex phase factor. It maps $|0\rangle$ to $i|1\rangle$ and $|1\rangle$ to $-i|0\rangle$, where i is the imaginary unit. Alice and Bob measurement bases are linear combinations of the Pauli matrices $\mathbf{X}$ and $\mathbf{Z}$ normalized by $\frac{1}{\sqrt{2}}$ expressed mathematically as

$$\mathbf{W} = \frac{\mathbf{X} + \mathbf{Z}}{\sqrt{2}}, \quad \mathbf{V} = \frac{-\mathbf{X} + \mathbf{Z}}{\sqrt{2}}. \tag{3}$$

In the E91 protocol, these gates are used to define the measurement bases and simulate the erroneus channels for real-world E91 protocol scenarios.

2.2. Bell States or EPR Pair

Entanglement is a situation where two objects influence each other no matter how far the distance between the two objects. In mathematical, the states are entangled when if there is no factorization as expressed as

$$\alpha|10\rangle + \beta|11\rangle = |1\rangle(\alpha|0\rangle + \beta|1\rangle) \text{ (not entangled)}; \tag{4}$$

$$\alpha|00\rangle + \beta|11\rangle \text{ (entangled)}. \tag{5}$$

These Bell States can be generated using quantum circuits and are represented as shown in Table 1.

2.3. CHSH Inequality

John Stewart Bell, in his groundbreaking work [17], introduced a mathematical framework to differentiate the forecasts of quantum mechanics from local hidden variable theories, encapsulated in Bell inequality. The measurement results for two particles are given by $A(\vec{a}, \lambda) = \pm 1$ and $B(\vec{b}, \lambda) = \pm 1$, where A for particle 1 is independent of the setting $\vec{b}$ for particle 2, and the reverse. The expectation value of the product of the two measurements is

$$P(\vec{a}, \vec{b}) = \int d\lambda\, \rho(\lambda) A(\vec{a}, \lambda) B(\vec{b}, \lambda), \tag{6}$$

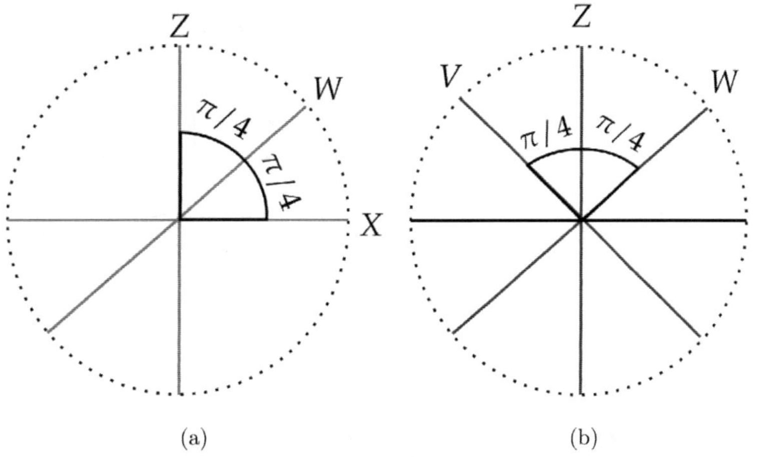

Fig. 2. The use of (a) Alice measurement angle $\phi_a^i = \{0, \frac{1}{4}\pi, \frac{1}{2}\pi\}$ and (b) Bob measurement angle $\phi_b^j = \{\frac{1}{4}\pi, \frac{1}{2}\pi, \frac{3}{4}\pi\}$.

where $\rho(\lambda)$ defines the probability distribution of the hidden variable λ.

The CHSH inequality is a specific formulation of bell inequality, developed by John Clauser, Michael Horne, Abner Shimony, and Richard Holt (CHSH). It provides a way to test the predictions of quantum mechanics in the context of entangled particles. The CHSH inequality is expressed as

$$\langle S\rangle = |E(A_1B_1) + E(A_1B_2) + E(A_2B_1) - E(A_2B_2)| \leq 2, \quad (7)$$

where A_i and B_j represent measurement settings chosen by two observers (Alice and Bob) and $E(A_iB_j)$ represents the expectation value of the product of measurements taken along settings A_i and B_j. Specifically, $E(A_iB_j)$ describes the average value of the product of outcomes from Alice measurement at setting A_i and Bob's measurement at setting B_j. We use E because the outcomes of the measurements are discrete rather than continuous as assumed in (6). For any classical system governed by local hidden variables, this inequality sets an upper bound of 2 for the combined correlation. However, quantum mechanics, due to the phenomenon of entanglement, allows for the maximal violation of this inequality. Specifically, for the quantum mechanical scenario, the $\langle S\rangle$ value is calculated as

$$\langle S\rangle = \langle \mathbf{Z}\otimes\mathbf{W}\rangle + \langle \mathbf{Z}\otimes\mathbf{V}\rangle + \langle \mathbf{X}\otimes\mathbf{W}\rangle - \langle \mathbf{X}\otimes\mathbf{V}\rangle = 2\sqrt{2}. \quad (8)$$

Thus, we obtain $\langle S\rangle = 2\sqrt{2}$, which clearly violates the CHSH version of bell inequality in (7).

Violation in (8) benefit for security of the E91 protocol to detect an eavesdropper. The security of the E91 protocol is deeply rooted in CHSH inequality. If an adversary like Eve attempts to gain knowledge of the keys by performing measurements on the entangled particles, her intervention would inevitably disturb the quantum system, leading to a drop in the CHSH value to below the quantum bound of $2\sqrt{2}$. Thus, the CHSH inequality acts as an additional safeguard, ensuring the security of keys production.

2.4. The E91 Protocol

In this section, we provide a summary on how the E91 protocol works. Bases in this protocol refers to quantum gates for generate quantum state. The gates are uses is **Z**, **X**, **W** and **V**. These gates are used for bases measurement. Steps for constructing a simple E91 protocol are as follows:

- The E91 protocol uses entangled states, where in this paper, we use singlet state of

$$|\Phi^+\rangle = \frac{|\uparrow\uparrow\rangle + |\downarrow\downarrow\rangle}{\sqrt{2}}. \quad (9)$$

- A trustful source generates pairs of $\frac{1}{2}$-spin particles in a singlet or an entangled state and sends them to Alice and Bob.

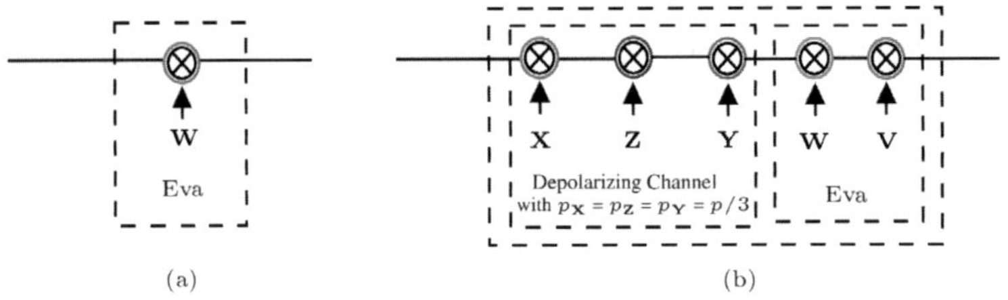

Fig. 3. Channel simulations scenario for (a) the performance with eavesdropper interception under a perfect channel and (b) the performances with eavesdropper interception under the erroneus channels.

- Of each particle, Alice and Bob measure an observable, chosen randomly and independently from each other from a set of complementary attributes.
- Alice measures with angles $\phi_a^i = \{0, \frac{1}{4}\pi, \frac{1}{2}\pi\}$ or similar to $\{\mathbf{X}, \mathbf{W}, \mathbf{Z}\}$ as shown in Fig. 2(a), while Bob measures with $\phi_b^j = \{\frac{1}{4}\pi, \frac{1}{2}\pi, \frac{3}{4}\pi\}$ or similar to $\{\mathbf{W}, \mathbf{Z}, \mathbf{V}\}$ as shown in Fig. 2(b).
- Each random measurements can result in a spin state of +1 (spin up) or -1 (spin down), whereby they show an correlation coefficient E of a certain observable to have a certain value calculated by

$$E(a_i, b_j) = P_{++}(a_i, b_j) + P_{--}(a_i, b_j) - P_{+-}(a_i, b_j) - P_{-+}(a_i, b_j). \tag{10}$$

- After that, they exchange the bases of their measurement over the public channel and divide the results in two groups:
  - one with according bases
  - one with disaccording bases
- The ones with according bases determine the key as they are perfectly correlated (which means they have the exact value) due to their entangled state expressed as

$$E(a_2, b_1) = E(a_3, b_2) = 1. \tag{11}$$

- The disaccording particles can still be used to test either Eve listens to the qubits on the quantum channel or not.
- The Bell inequality is able to test the presence of an eavesdropper.
- The protocol uses a variant in [18] to test the presence of an eavesdropper.
- Alice and Bob can now exchange the disaccording particles openly (over the public channel), since they do not participate in the composition of the key.
- These values determine whether or not S is fulfilled.
  - if so, an entangled state is present and the key derived from the group of corresponding result is legitimate
  - if not, Eve may have altered some or all of the particles and the key is not secure.

2.5. Scenario of E91 Protocol with Eavesdropper under Erroneous Channels

In this subsection, we explore two scenarios of practical parameters in potential vulnerabilities in the E91 protocol as shown in Figs. 3(a) for perfect channels and 3(b) for erroneous channels. These scenarios examine the impacts of eavesdropping and channel errors on secure communications. The E91 protocol relies on entangled quantum states to facilitate secure key exchange between Alice and Bob. However, these entangled states are susceptible to interference from eavesdroppers and simultaneously can be disrupted by errors in the communication channel. Here, we discuss how these vulnerabilities affect the reliability of keys production and compromise the security of the exchange process.

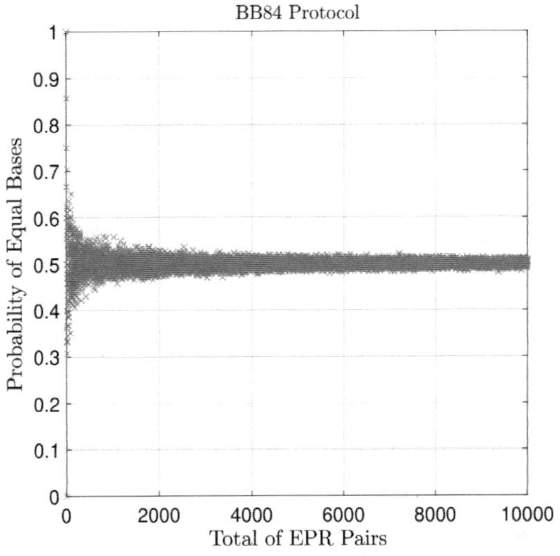

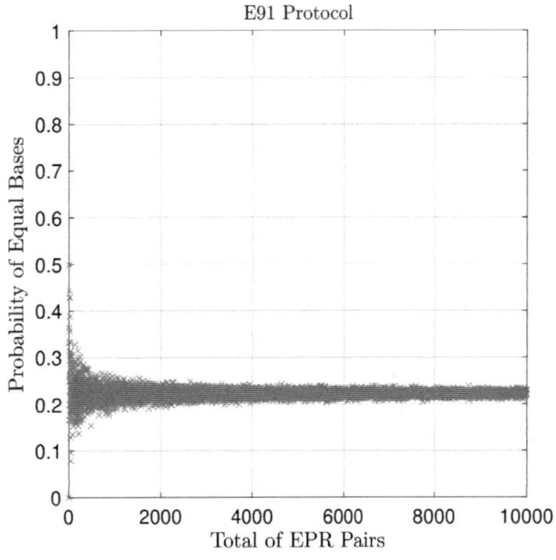

Fig. 4. Probability of resulted keys from: (a) BB84 protocol and (b) E91 protocol.

2.5.1. Scenario with Eavesdropper Interception under Perfect Channels

In this scenario, we consider the E91 protocol in the involvement of an eavesdropper. The E91 protocol is simulated with a perfect channel, where no transmission errors occur. A trusted source generates 10,000 blocks of entangled qubits. Each blocks contains 100 pairs represented as L. The total number of keys production per set for E91 protocol is expected to be $3L$ assuming that 1 EPR pair represent 1 classical bit.[1] Eve is intercepting the communications channel between Alice and Bob. The initial steps for constructing the protocol remain the same as described in the previous subsection. A trustfull source generates entangled states and one qubit from each entangled pair is sends to Alice while the other is sends to Bob. However, in this scenario, Eve intercepts the qubits being sent to Alice as shown in Fig. 3(a). After intercepting, Eve measures the qubits and then sends her own qubits to Alice, attempting to remain undetected. Despite Eve interception, Alice and Bob proceed with the protocol as usual, they measure their qubits using randomly chosen measurement bases and subsequently share their measurement results over a public channel. To detect the presence of Eve, Alice and Bob compare a subset of their bases measurement results. in this results, we analyze a deviations from the key production due to the eve interception.

2.5.2. Scenario with Eavesdropper Interception Combined under the Erroneus Channels

In this scenario, we extend the E91 protocol to account for both the involvement of an eavesdropper and the emergence of depolarizing errors within the communications channels between Alice and Bob, as shown in Fig. 3(b). The same entangled state parameters are used: 10,000 blocks, 100 pairs per block, and a total of 300 keys per set. The initial steps for constructing the protocol remain consistent with the previous subsections. A trusfull source generates entangled states. From each pair, one qubit is directed to Alice and the other to Bob. During the transmission, the channel is considered the worst-case scenario because $p_X = p_Z = p_Y = \frac{p}{3}$, representing equal error probabilities for all axes. Simultaneously, Eve intercepts the qubits on both channels, measures them, and resends them in an attempt to remain undetected. Intercept and error can happen on either or both. Despite these challenges, Alice and Bob proceed with the protocol as usual. They measure their qubits using randomly chosen measurement bases

[1]The standard ASCII system uses 7 bits per character. However, in practice, ASCII typically employs 8 bits (1 byte) for simplicity and compatibility. For example, a password consisting of 8 characters would require 8 bytes. Using 8 bits per character, the total size for such a password would be 64 bits. In this simulation, each EPR pair represents 1 classical bit. Given that the E91 protocol achieves a success rate of 22 %, a block of 100 pairs yields approximately 22 bits. Consequently, to generate a password requiring 64 bits, at least 3 blocks are needed, calculated as $64/22 = 2.9$, which is rounded up to 3 blocks.

Table 2. The differences of E91 and BB84 protocols.

BB84	E91
Polarization state of a single photon	Polarization state of two entangled photons
Need to verify between two parties	No need to verify between two parties
Probability of same basis of resulted key much more than E91	Probability of same basis of resulted key is very small
Unable to detect unauthorized interception	Capable to detect unauthorized interception
Lower than E91 for high security and much more time consumption than E91	Maintaining security and minimizing time consumption.
BB84 use single particle but vulnerable of a few attack (like Photons Number Splitting)	E91 leverages the CHSH inequality to offer extra layer security.
The implementation of BB84 protocol more viable	E91 protocol faces technological difficulties of computing resources.

and then publicly share a subset of their measurement results. By analyzing these results, we can detect deviations from the key production due to both the channel errors and Eve interception.

3. Simulation Results and Analysis

In this section, we demonstrate and analyze the findings from the simulation. We begin by demonstrating a computer simulations of the keys production process in the E91 protocol. Then we delve into a comparative analysis of the E91 and BB84 protocols. We also highlight their respective strengths and weaknesses as well as the resulted keys from each. Finally, we explore the effects of eavesdropping and channel errors on the E91 protocol. Beside that, we also provide insights into how these challenging communications conditions influence key distribution and the overall robustness of the system.

3.1. Resulted Key

A notable difference is the final key rate for the E91 protocol as depicted in Fig. 4(b). This result indicates a probability of around $\frac{2}{9}$ for choosing equal measurement bases. This reflects thorough verification steps that enhance security by detecting eavesdropping through entangled states, albeit at the cost of a lower resultant key. Conversely, the BB84 protocol shown in Fig. 4(a) stabilizes at a probability of $\frac{1}{2}$ for equal measurement bases. This sign give a more straightforward and efficient resultant key. However, this simplicity makes BB84 slightly less robust against eavesdropper interception compared to E91. These parameters underscore the importance of selecting the appropriate protocol based on specific communications needs and balancing the requirements for security, speed, and final key efficiency.

3.2. Comparison of The BB84 and E91 Protocols

the E91 and BB84 protocols exhibit notable differences in their approach to security and efficiency. One of the key distinctions is the type of quantum states employed. The E91 protocol leverages the entangled states of particles, which inherently provide stronger security features due to their sensitivity to external disturbances. In contrast, BB84 uses single-particle polarization states, which, while easier to implement, offer less robustness in revealing the presence of an eavesdropper.

A significant advantage of the E91 protocol lies in its ability to detect eavesdropping through quantum entanglement correlations. The protocol applies the CHSH inequality to test for deviations from the expected quantum correlations between the entangled particles. Any violation of these correlations serves as a direct indicator of tampering, making E91 highly effective in identifying potential threats. This contrasts with BB84, which requires statistical post-processing to infer the likelihood of an eavesdropper's presence, leading to potential delays and higher risks of undetected intrusion during transmission.

An important aspect to highlight is that the E91 protocol does not require verification between two parties, contributing to reduced time for practical implementations. However, this efficiency comes at a cost. Unlike E91, the BB84 protocol cannot inherently detect the presence of an eavesdropper and relies on statistical post-processing to infer whether communications have been compromised. BB84 involves the exchange of verification information, adding to the overhead of the key exchange process. However, this efficiency in E91 comes with the tradeoff of a lower probability of basis alignment for keys production

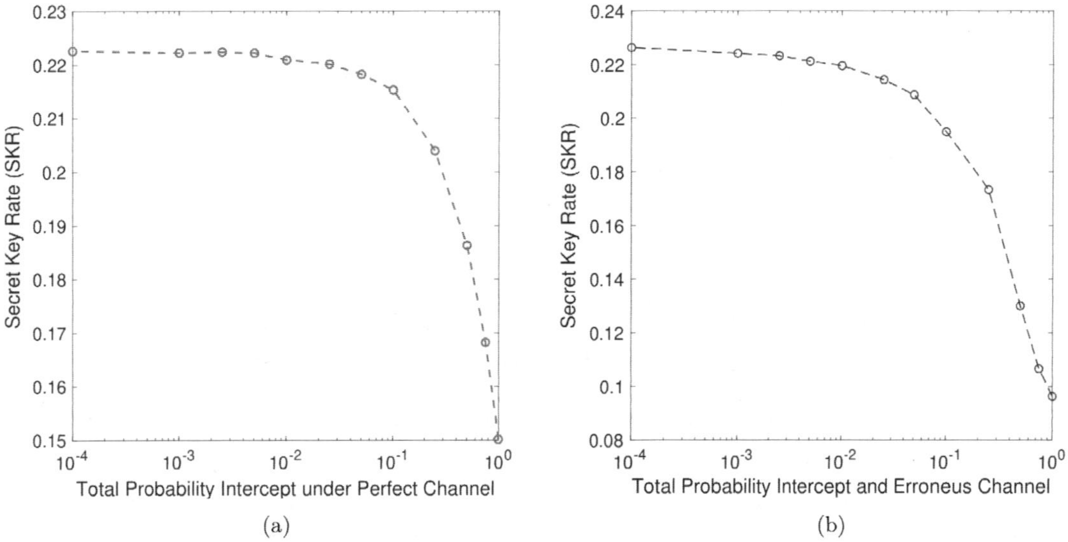

Fig. 5. The performances of secret key rate to total probability of (a) intercept under perfect channel and (b) both interception and erroneus channels.

due to the randomness of entanglement measurements, which could lead to delays and a higher risk if an eavesdropper is manipulating the transmission.

The E91 protocol has a lower probability of measurement basis alignment in keys production due to the use of the CHSH inequality. Its reliance on entanglement and the CHSH inequality provides stronger protection against eavesdropping compared to the BB84 protocol, where bases are chosen independently, resulting in higher alignment and efficiency. Additionally, the BB84 protocol is particularly susceptible to Photon Number Splitting (PNS) attacks. This vulnerability arises because, in BB84, an eavesdropper can leverage multi-photon signals to gather partial knowledge about the keys without revealing their presence. In contrast, the quantum correlations in the E91 protocol make such attacks more difficult to perform without detection, enhancing the protocol's overall security.

However, practical implementation of the E91 protocol in real-world scenarios presents certain challenges. One of the primary constraints is the hardware requirement. Quantum entanglement-based systems, like those needed for E91, require more advanced and sensitive equipment, such as entanglement sources and detectors with high precision and low noise. These systems are more expensive and complex than those used for BB84, which can rely on relatively simpler single-photon detectors. Furthermore, E91's reliance on entanglement makes it more susceptible to environmental noise, such as photon loss and decoherence over long distances, which can significantly degrade the quality of the entangled states and compromise the security of the transmission. To address these issues, potential practical applications include the use of integrated photonic circuits to simplify hardware [19], superconducting nanowire detectors to enhance detection efficiency [20], polarization-maintaining fibers to mitigate environmental interference [21], and error correction algorithms [22]. Furthermore, investing in scalable sources of entangled photons, leveraging shared research infrastructure, and developing hybrid protocols that integrate quantum and classical systems can optimize practicality and scalability [23].

In summary, the E91 protocol offers enhanced security due to its reliance on quantum entanglement and the CHSH inequality, providing more robust detection of eavesdropping compared to BB84. While BB84 may be more efficient in terms of basis alignment and practical for a broader range of applications, it lacks the inherent security advantages of E91, especially in real-time eavesdropping detection. The choice between E91 and BB84 ultimately depends on the specific priorities of the application, whether the focus is on maximizing security or optimizing time and computational resources. Additionally, hardware requirements and environmental noise tolerance are critical factors in determining the practical viability of the E91 protocol in real-world networks.

3.3. The Performances of Secret Key Rate to Total Probability of Intercept under Perfect Channel

The secret key rate (SKR) performance is strongly influenced by the probability of interception, which reflects the likelihood of a successful eavesdropping attempt on the quantum communication channel. In Fig. 5(a), the x-axis represents the total probability of interception under ideal (perfect) channel conditions, while the y-axis shows the resulted SKR. This figure reveals that at low intercept probabilities (approximately 10^{-4} to 10^{-2}), the SKR remains consistently high at around 0.22, indicating that the protocol is resilient to minor interceptions. However, once the probability of interception rises above 10^{-2}, the SKR starts to decline noticeably. This sharp decline highlights the balance between maintaining a high SKR and tolerating higher interception probabilities. At intercept probabilities nearing 1, the SKR decreases significantly to around 0.15, underscoring how high intercept levels can severely compromise the protocol's efficiency in generating secure keys. This relationship emphasizes the need to minimize interception probability to preserve a robust SKR and ensure protocol effectiveness.

3.4. The Performances of Secret Key Rate to Total Probability of Eavesropper Interception and Erroneus Channels

The combined impact of interception probability and channel errors is crucial in assessing the overall performance of the QKD protocol. Minimizing both factors is essential to maintain the integrity and security of keys production in practical applications. In Fig. 5(b), the x-axis represents the combined probability of interception and channel errors, while the y-axis displays the corresponding SKR. At low combined probabilities (approximately 10^{-4} to 10^{-2}), the SKR is stable and high, around 0.22, suggesting that the protocol can withstand minor errors and interceptions with minimal performance loss. However, as the probability of combined interception and error surpasses 10^{-2}, the SKR declines more steeply than in the perfect channel scenario, indicating that errors further degrade protocol performance. As this probability approaches 1, the SKR drops to approximately 0.09, demonstrating that both interception and channel errors critically reduce the protocol's effectiveness in generating secure keys. This finding underlines the importance of controlling both interception and error rates to maintain a high SKR, essential for effective protocol performance in real-world conditions.

4. Conclusion

In this paper, we have investigated and compared the E91 and BB84 protocols to evaluate their practical performances for future secure quantum networks. We have demonstrated a series of computer simulations of: (1) probability of resulted keys from E91 and BB84 protocol and (2) the performance of secret key rate to total probability of: (a) intercept under perfect channel and (b) both interception and erroneus channels. The key findings from our study are: (i) the E91 protocol produced fewer keys with small probability compared to the BB84 protocol, which had a keys production probability of higher. However, the E91 protocol does not require verification between the communicating parties, (ii) the E91 protocol enhances security due to its utilization of the CHSH inequality, which provides a stronger basis for detecting eavesdropping, and (iii) both eavesdropper interception and quantum channel errors significantly impact keys generation when the error probability exceeds certain value. The combination of eavesdropper interception and channel errors exacerbates the reduction in keys production compared to a perfect channel, emphasized the need for robust error correction mechanisms to improve the performance of the SKR to enhance the security of future quantum networks. Our simulations offer insights into keys production under various real-world scenarios, underscoring the importance of selecting between E91 and BB84 based on specific security needs and technological capabilities. Advancements in the E91 protocol, particularly its enhanced resilience to eavesdropping, have significant implications for future networks like 6G. As the next generation communication systems increasingly rely on quantum technologies for enhanced security, the integration of E91 could provide stronger defenses against interception and improve overall network security, even in the presence of errors. These findings highlight the protocol's potential to play a key role in the development of quantum-secure 6G networks and beyond.

5. Acknowledgment

This research is partially supported by The B5.5G Laboratory.

References

[1] Mehic M, *et al.* Quantum cryptography in 5g networks: A comprehensive overview. *IEEE Commun Surveys Tut.* 2023:302-46.

[2] Geddada VJ, Lakshmi PV. Distance Based Security using Quantum Entanglement:a survey. *Int Conf Comput Commun Netw Technol (ICCCNT)*. 2022.

[3] Zhang Z, *et al.* 6G Wireless Networks: Vision, Requirements, Architecture, and Key Technologies. *IEEE Veh Technol Mag.* 2019;**14**:28-41.

[4] Ahadiansyah D, Anwar K, Budiman G. Investigation on Shor Codes as Degenerate Codes but Correct All Single Quantum Errors. *IEEE Symp Future Telecommun Technol (SOFTT).* 2022.

[5] Anwar K, Ramadhan M. The Smallest Perfect Quantum Accumulate Codes. *IEEE Asia-Pac Conf Commun (APCC).* 2021.

[6] Basudewa MI, Anwar K, Meylani L. Study on the Design of Simple Quantum Communications Based on Orbital Angular Momentum. *2022 IEEE Symp Future Telecommun Technol (SOFTT).* 2022:8-14.

[7] Einstein A, Podolsky B, Rosen N. Can Quantum-Mechanical Description of Physical Reality Be Considered Complete? *Phys Rev.* 1935;**47**:777-80.

[8] Geddada VJ, Lakshmi PV. Distance Based Security using Quantum Entanglement:a survey. *Int Conf Comput Commun Netw Technol (ICCCNT).* 2022.

[9] Durt T, Kaszlikowski D, Chen JL, Kwek LC. Security of quantum key distributions with entangled qudits. *Phys Rev.* 2004;**69**.

[10] Basset B, *et al* . Quantum key distribution with entangled photons generated on demand by a quantum dot. *Sci Adv.* 2021;**7**.

[11] Madaan A, Raj G. Analysis of Quantum Key Distribution using Key Distribution and Attacking Strategies over Security Protocol. *Int Conf Cloud Comput, Data Sci Eng (Confluence).* 2018.

[12] Parakh A, Verma P. Improving the Efficiency of Entanglement Based Quantum Key Exchange. *Int Conf Comput Commun Netw (ICCCN).* 2014.

[13] Begimbayeva Y, Zhaxalykov T, Ussatova O. Investigation of Strength of E91 Quantum Key Distribution Protocol. *Int Asian Sch-Seminar Optim Complex Syst (OPCS).* 2023.

[14] Bennett CH. Quantum cryptography using any two nonorthogonal states. *Phys Rev Lett.* 1992 May;**68**:3121-4.

[15] Yuen HP. Anonymous-Key Quantum Cryptography and Unconditionally Secure Quantum Bit Commitment. In: Quantum Communication, Computing, and Measurement 3. Springer; 2002. p. 285-93.

[16] Bennett H, Brassard G. Quantum Cryptography: Public Key Distribution and Coin Tossing. *Int Conf Comput, Syst Signal Process.* 1984;**1**.

[17] Bell JS. On the Einstein Podolsky Rosen Paradox. *Phys Physiq.* 1964;**1**(3):195-200.

[18] Clauser JF, Horne MA, Shimony A, Holt RA. Proposed Experiment to Test Local Hidden-Variable Theories. *Phys Rev Lett.* 1969 Oct;**23**:880-4.

[19] Ning S, Zhu H, Feng C, Gu J, Jiang Z, Ying Z, et al. Photonic-Electronic Integrated Circuits for High-Performance Computing and AI Accelerators. Journal of Lightwave Technology. 2024;42(22):7834-59.

[20] Wang Q, Renema JJ, Engel A, de Dood MJA. Design of NbN Superconducting Nanowire Single-Photon Detectors with Enhanced Infrared Detection Efficiency. Phys Rev Appl. 2017 Sep;8:034004.

[21] Yu Z, Yang J, Lin C, Zhang X, Dang F, Yuan Y, et al. Distributed Polarization Measurement for Fiber Sensing Coils: A Review. Journal of Lightwave Technology. 2021;39(12):3699-710.

[22] Ginting MB, Anwar K, Maryopi D. Constructing Quantum Surface Codes for Arbitrary Surface Forms. In: 2021 IEEE Symposium On Future Telecommunication Technologies (SOFTT); 2021. p. 75-80.

[23] Hevia JL, Peterssen G, Piattini M. Dynamic integration for hybrid quantum/Classical software systems. Journal of Systems and Software. 2024;214:112061.

AUTHOR INDEX